STRING THEORY IN CURVED SPACE TIMES

STRING THEORY IN CURVED SPACE TIMES

A COLLABORATIVE RESEARCH REPORT

Paris ⋆ Geneva ⋆ Madrid ⋆ Milano ⋆ Torino ⋆ Thessaloniki

Editor

N. SANCHEZ
Observatoire de Paris, Demirm
61, Avenue de l'Observatoire
75014 Paris, FRANCE

World Scientific
Singapore • New Jersey • London • Hong Kong

Published by

World Scientific Publishing Co. Pte. Ltd.

P O Box 128, Farrer Road, Singapore 912805

USA office: Suite 1B, 1060 Main Street, River Edge, NJ 07661

UK office: 57 Shelton Street, Covent Garden, London WC2H 9HE

Library of Congress Cataloging-in-Publication Data
String theory in curved space times : a collaborative research report
/ editor N. Sánchez.
p. cm.
ISBN 9810234392
1. String models. 2. Special relativity (Physics) 3. Space and
time. I. Sanchez, N. (Norma), 1952–
QC794.6.S85S75 1998
530.11--dc21 98-2754
CIP

British Library Cataloguing-in-Publication Data
A catalogue record for this book is available from the British Library.

Photos of the cover: K. Blumenfeld

This book is printed on acid-free paper.

Printed in Singapore by Uto-Print

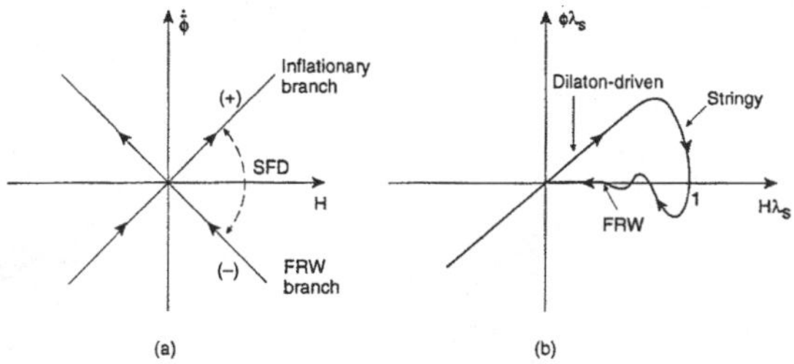

Figure 2. Phase diagrams for the perturbative (a) regime and a conjectured non-perturbative solution (b) to the branch-change problem.

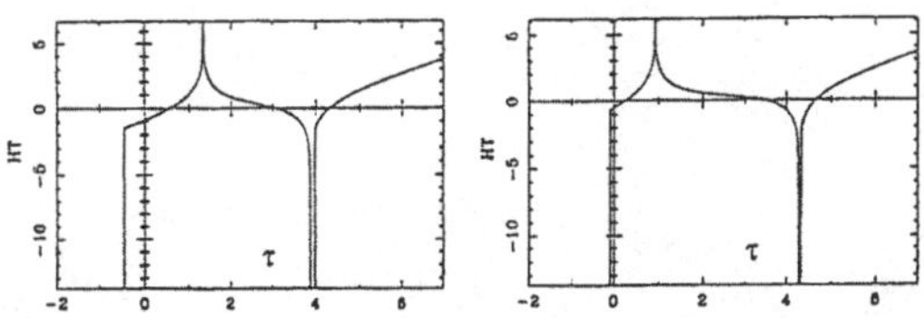

Figure 2: Same as fig.1, for $n = 4, |x^0 >= (1, -1, i, 1)$. Because of a degeneracy, there are now only three strings.

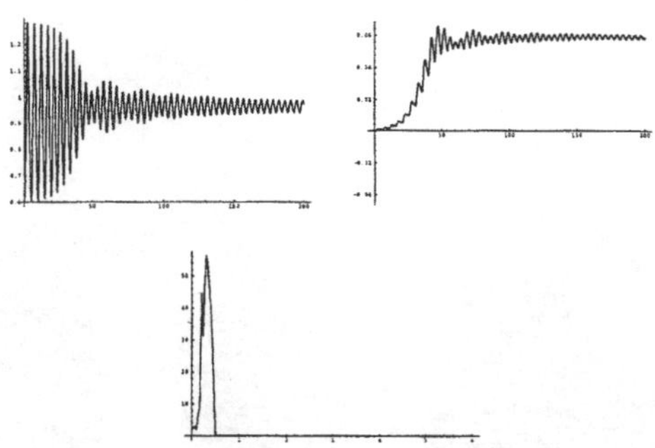

Figs.2 : Scalars: broken symmetry for the values of the parameters ,
$y = 0$; $\lambda/8\pi^2 = 0.2$; $g/\lambda = 0.05$; $m_\sigma = 0.2\,|m_\phi|$; $\eta(0) = 0.6$; $\dot{\eta}(0) = 0$. *Upper Left:* $\eta(\tau)$ *vs* τ . *Upper Right:* $\mathcal{N}_\sigma(\tau)$ *vs.* τ. *Lower:* $\mathcal{N}_{q,\sigma}(\tau = 120)$ *vs.* q

Photographs:
K. Blumenfeld

PREFACE

The main goal and impact of modern string theory is to provide a consistent quantum theory of gravity. This book provides an updated collection of original new developments and fundamental research in string theory in connection with gravity and physics at the Planck energy scale.

Topics covered in this book by pioneering researchers in the field include: classical and quantum string dynamics in strong gravitational fields, space-time singularities, black holes and cosmological backgrounds; particle and string scattering at the Planck energy scale; string cosmology and its observational consequences; the new features of multistrings and of quantum particle transmutation for strings in curved spacetimes.

The book deals with (i) the several new methods developed to solve the highly nonlinear string dynamics in curved spactimes: approximative perturbative methods, asymptotic expansions, exact local expansions and exact global (over the whole world sheet) string solitonic solutions. (ii) The string energy momentum tensor and the equation of state for the string matter, the stretching of the string size in spacetimes with event horizons and near spacetime singularities. (iii) The canonical and semiclassical quantization of strings in curved spacetimes and the physical effects found for: the mass spectrum, structure of levels, scattering amplitudes, number operator and particle transmutation.

Perhaps the main challenge of theoretical physics today is the quantum unification of all interactions including gravity. At present, string theories appear as the most promising candidates to achieve such a quantum unification. However, many conceptual and technical problems remain and a full quantum theory of gravity is still non-existent. Continuous effort has demonstrated the many difficulties encountered in constructing such a theory and has also indicated some of the particular properties which an eventual complete theory would have to possess.

Besides the non-renormalizability problem, probably solvable in the string context, the quantization of gravity involves fundamental conceptual problems such as those arising in black holes and primordial cosmology.

Whatever the final theory of the world would be, if it is to be a theory of everything, one would like to know what new understanding it would give us about the singularities of classical general relaivity, and how boundary conditions, e.g. those consistent with the observational data associated with the Big Bang must be chosen.

If string theory could provide us with a theory of quantum gravity, it should give us a proper theory for describing the ultimate state of quantum black holes and the initial (very early) state of the universe. That is, a theory describing the physics (and the relevant framework) at the Planck energy and length.

Since the most relevant new physics provided by strings concerns the quantization of gravity, we must, at least, understand string quantization in curved spacetimes to start. Actually, one would like to extract the spacetime and the particle spectrum from the solution of string theory, but we are still very far from doing that explicitly (or implicitly).

Curved spacetimes, besides their evident relevance in classical gravitation, are also important at energies of the order of the Planck scale. At the Planck energy scale, gravitational interactions are at least as important as the rest and cannot anymore be neglected as is usually the case in particle physics.

This book accounts for the main joint collaborative work achieved within the Research Network Collaboration "String Gravity and Physics at the Planck Energy Scale", supported by the European Commission, Contract ERBCHRXCT940488, with the participation of the following Teams:

> Observatoire de Paris-DEMIRM (N. Sanchez), University of Paris VI-LPTHE (H. J. de Vega), CERN-Theory Division (G. Veneziano), Universidad Complutense de Madrid (M. Ramon Medrano), University of Milano-Dip. di Fisica (C. Destri), University of Torino-INFN (M. Gasperini), University of Thessaloniki-Theoretical Physics (A. Nicolaidis).

This book also accounts for the main contributions presented at the String Gravity Meeting within this Network, which took place in June 1996 at the beautiful Parisian campus of Observatoire de Paris (built on orders from Colbert and to plans by Claude Perrault from 1667 to 1672). Sessions took place at the "Salle du Conseil" in the historic Perrault building under the portraits of Laplace, Le Verrier, Lalande, Arago, Delambre and Louis XIV.

We express our grateful thanks to the Human Capital & Mobility and TMR Programme of the Commission of European Communities for its support, as well as to the Observatoire de Paris, to the Network Team Institutions, and to all who participated, so harmoniously and so efficiently, in this exciting collaboration.

Norma Sanchez

CONTENTS

STRING THEORY IN COSMOLOGICAL SPACETIMES

H.J. de Vega[(a)] **and N. Sánchez**[(b)]

(a) Laboratoire de Physique Théorique et Hautes Energies, Université Pierre et Marie Curie (Paris VI) et Université Denis Diderot (Paris VII), Tour 16, 1er. étage, 4, Place Jussieu 75252 Paris, cedex 05, France. Laboratoire Associé au CNRS URA280.

(b) Observatoire de Paris, DEMIRM, 61, Avenue de l'Observatoire, 75014 Paris, France. Laboratoire Associé au CNRS URA336, Observatoire de Paris et École Normale Supérieure.

Abstract

Recent progress on string theory in curved spacetimes is reviewed. The string dynamics in cosmological and black hole spacetimes is investigated. The different methods available to solve the string equations of motion and constraints in curved spacetimes are described. That is, the string perturbation approach, the null string approach, the τ-expansion, and the construction of global solutions (for instance by inverse scattering methods).

The classical behaviour of strings in FRW and inflationary spacetimes is now understood in a large extent from the various types of explicit string solutions. Three different types of behaviour appear: **unstable, dual** to unstable and **stable.** For the unstable strings, the energy and size grow for large scale factors $R \to \infty$, proportional to R. For the dual to unstable strings, the energy and size blow up for $R \to 0$ as $1/R$. For stable strings, the energy and proper size are bounded. (In Minkowski spacetime, all string solutions are of the stable type).

Recent progress on self-consistent solutions to the Einstein equations for string dominated universes is reviewed. The energy-momentum tensor for a gas of strings is then considered as source of the spacetime geometry and from the above string behaviours the string equation of state is **derived**. The self-consistent string solution exhibits the realistic matter dominated behaviour $R \simeq T^{2/3}$ for large times and the radiation dominated behaviour $R \simeq T^{1/2}$ for early times (T being the cosmic time).

We report on the **exact integrability** of the string equations plus the constraints in de Sitter spacetime that allows to systematically find **exact** string solutions by soliton methods and the multistring solutions. **Multistring solutions** are a new feature in curved spacetimes. That is, a single world-sheet simultaneously describes many different and independent strings. This phenomenon has no analogue in flat spacetime and follows to the coupling of the strings with the geometry.

CONTENTS

I. INTRODUCTION

The construction of a sensible quantum theory of gravitation is probably the greatest challenge in theoretical physics for the end of this century and most probably for the next century too.

Another problem (the most often discussed in connection with gravity quantization) is the one of the renormalizability of the Einstein theory (or its various generalizations) when quantized as a local quantum field theory. Actually, even deeper conceptual problems arise when one tries to combine quantum concepts with General Relativity. For example, statistical phenomena like Hawking's radiation arise when free fields are quantized in black-hole backgrounds. This points out a lack of quantum coherence even keeping the gravitational field classical.

It may be very well that a quantum theory of gravitation needs new concepts and ideas. Of course, this future theory must have the today's General Relativity and Quantum Mechanics (and QFT) as limiting cases. In some sense, what everybody is doing in this domain (including string theories approach) may be something analogous to the developpment of the old quantum theory in the 10's of this century. Namely, people at that time imposed quantization conditions (the Bohr-Sommerfeld conditions) to hamiltonian mechanics but keeping the concepts of classical mechanics.

The main drawback to develop a quantum theory of gravitation is clearly the **total lack of experimental guides** for the theoretical developpment. Just from dimensional reasons, physical effects combining gravitation and quantum mechanics are relevant only at energies of the order of $M_{Planck} = \sqrt{\hbar c/G} = 1.22\,10^{16}$Tev. Such energies were available in the Universe at times $t < t_{Planck} = 5.4\,10^{-44}$sec. Anyway, as a question of principle, the construction of a quantum theory of gravitation is a problem of fundamental relevance for theoretical physics . In addition, one cannot rule out completely the possibility of some "low energy" ($E \ll M_{Planck}$) physical effect that could be experimentally tested. One may speculate about effects analogous to the presence of magnetic monopoles in some grand unified theories. [Monopoles can be detected by low energy experiments in spite of their large mass].

Let us now see what are the consequences of Heisenberg's principle in quantum mechanics combined with the notion of gravitational (Schwarzschild) radius in General Relativity. Assume we make two measurements at a very small distance Δx . Then,

$$\Delta p \sim \Delta E \sim 1/\Delta x$$

where we set $\hbar = c = 1$. For sufficiently large ΔE, particles with masses $m \sim 1/\Delta x$ will be produced. The gravitational radius of such particles are of the order

$$R_G \sim Gm \sim \frac{(l_{Planck})^2}{\Delta x}$$

where $l_{Planck} \sim 10^{-33}$ cm. Now, General Relativity allows measures at a distance Δx , provided

$$\Delta x > R_G \quad \to \quad \Delta x > \frac{(l_{Planck})^2}{\Delta x} \quad .$$

That is,

$$\Delta x > l_{Planck} \quad \text{and} \quad m < M_{Planck} \tag{1.1}$$

This means that no measurements can be made at distances smaller than the Planck length and that no particle can be heavier than M_{Planck} . This is a simple consequence of relativistic quantum mechanics combined with General Relativity. In addition, the notion of locality and hence of spacetime becomes meaningless at the Planck scale. Notice that the equality in eq.(1.1) means that the Compton length equals the Schwarzschild radius of a particle. Since M_{Planck} is the heaviest possible particle scale, a theory valid there (necessarily involving quantum gravitation) will also be valid at any lower energy scale. One may ignore higher energy phenomena in a low energy theory, but not the opposite. In other words, a theory of quantum gravity will be a 'theory of everything'. We think that this is the **key point** on the quantization of gravity. A theory that holds till the Planck scale must describe **all** what happens at lower energies including all known particle physics as well as what we do not know yet (that is, beyond the standard model) [1]. Notice that this conclusion is totally independent of the use or not of string models. A direct important consequence of this conclusion, is that it may not make physical sense to quantize **pure gravity**. A physically sensible quantum theory cannot contain only gravitons. To give an example, a theoretical prediction for graviton-graviton scattering at energies of the order of M_{Planck} must include all particles produced in a real experiment. That is, in practice, **all** existing particles in nature, since gravity couples to all matter.

In conclusion : a consistent quantum theory of gravitation must be a finite theory [1] and must include all other interactions. That is, it must be a theory of everything (TOE). This is a very ambitious project. In particular it needs the understanding of the present desert between 1 and 10^{16} TeV. There is an additional dimensional argument about the inference Quantum Theory of Gravitation $\rightarrow$ TOE. There are only three dimensional physical magnitudes in nature: length, energy and time and correspondingly only three dimensional constants in nature: $c, \hbar$ and G. All other physical constants like $\alpha = 1/137,04...$, $M_{proton}/m_{electron}$, $\theta_{WS}, \ldots$ etc. are pure numbers and they must be calculable in a TOE. This is a formidable, but extremely appealing problem. From the theoretical side, the **only serious candidate** for a TOE is at present string theory. This is why we think that strings desserve a special attention in order to quantize gravity.

String theory is therefore an appropriate arena to work out the quantization of gravity consistently. It provides an unified theory of all interactions overcoming at the same time the nonrenormalizable character of quantum fields theories of gravity.

As a first step in the understanding of quantum gravitational phenomena in a string framework, we started in 1987 a programme of string quantization on curved spacetimes [2,9]. The investigation of strings in curved spacetimes is currently the best framework to study the physics of gravitation in the context of string theory, in spite of its limitations. First, the use of a continuous Riemanian manifold to describe the spacetime cannot be valid at scales of the order of l_{Planck}. More important, gravitational backgrounds effectively provide classical or semiclasical descriptions even if the matter backreaction to the geometry is included through semiclassical Einstein equations (or stringy corrected Einstein equations) by inserting the expectation value of the string energy-momentum tensor in the r.h.s. One would want a full quantum treatment for matter and geometry. However, to find a formulation of string theory going beyond the use of classical backgrounds is a very difficult (but fundamental) problem. One would like to derive the spacetime geometry as a classical and low energy ($\ll M_{Planck}$) limit from the solution of (quantum) string theory.

After a short introduction on strings in Minkowski and curved spacetimes, we focus on strings in cosmological spacetimes.

Substantial results were achieved in this field since 1992. The classical behaviour of strings in FRW and inflationary spacetimes is now understood in a large extent [3]. This understanding followed the finding of various types of exact, asymptotic and numerical string solutions in FRW and inflationary spacetimes [5] - [12]. For inflationary spacetimes, the exact integrability of the string propagation equations plus the string constraints in de Sitter spacetime [4] is indeed an important help. This allowed to systematically find **exact** string solutions by soliton methods using the linear system associated to the problem (the so-called dressing method in soliton theory) and the multistring solutions [5] - [8].

In summary, three different types of behaviour are exhibited by the string solutions in cosmological spacetimes: **unstable, dual** to unstable and **stable.** For the unstable strings, the energy and size grow for large scale factors $R \to \infty$, proportional to R. For the dual to unstable strings, the energy and size blow up for $R \to 0$ as $1/R$. For stable strings, the energy and proper size are bounded. (In Minkowski spacetime, all string solutions are of the stable type). The equation of state for these string behaviours take the form

- (i) **unstable** for $R \to \infty$ $p_u = -E_u/(D-1) < 0$
- (ii) **dual to unstable** for $R \to 0$, $p_d = E_d/(D-1) > 0$.
- (iii) **stable** for $R \to \infty$, $p_s = 0$.

Here E_u and E_d stand for the corresponding string energies and $D-1$ for the number of spatial dimensions where the string solutions lives. For example, $d-1 = 1$ for a straight string, $d-1 = 2$ for a ring string, etc.

As we see above, the dual to unstable string behavior leads to the same equation of state than radiation (massless particles or hot matter). The stable string behavior leads to the equation of state of massive particles (dust or cold matter). The unstable string behavior is a purely 'stringy' phenomenon. The fact that in entails a negative pressure is however physically acceptable. For a gas of strings, the unstable string behaviour dominates in inflationary universes when $R \to \infty$ and the dual to unstable string behavior dominates for $R \to 0$.

The unstable strings correspond to the critical case of the so called *coasting universe* [19,32]. That is, classical strings provide a *concrete* realization of such cosmological models. The 'unstable' behaviour is called 'string stretching' in the cosmic string literature [20,21].

It must be stressed that while time evolves, a **given** string solution may exhibit two and even three of the above regimes one after the other (see sec. III). Intermediate behaviours are also observed in ring solutions [7,12]. That is,

$$P = (\gamma - 1)\, E \quad \text{with} \ -\frac{1}{D-1} < \gamma - 1 < +\frac{1}{D-1} \, .$$

We also report here on the exact integrability of the string equations plus the constraints in de Sitter spacetime which allows to systematically find **exact** string solutions by soliton methods and the multistring solutions. **Multistring solutions** are a new feature in curved spacetimes. That is, a single world-sheet simultaneously describes many different and independent strings. This phenomenon has no analogue in flat spacetime and appears as a consequence of the coupling of the strings with the spacetime geometry.

The world-sheet time τ turns out to be an multi-valued function of the target string time X^0 (which can be the cosmic time T, the conformal time η or for de Sitter universes it can be the hyperboloid time q^0). Each branch of τ as a function of X^0 corresponds to a different string. In flat spacetime, multiple string solutions are necessarily described by multiple world-sheets. Here, a single world-sheet describes one string, several strings or even an infinite number of different and independent strings as a consequence of the coupling with the spacetime geometry. These strings do not interact among themselves; all the interaction is with the curved spacetime. One can decide to study separately each of them (they are all different) or consider all the infinite strings together.

Of course, from our multistring solution, one *could* just choose only one interval in τ (or a subset of intervals in τ) and describe just one string (or several). This will be just a **truncation** of the solution.

The really remarkably fact is that all these infinitely many strings come **naturally together** when solving the string equations in de Sitter spacetime as we did in refs. [5] - [7].

Here, interaction among the strings (like splitting and merging) is neglected, the only interaction is with the curved background.

The multistring property appears associated to the presence of a cosmological constant (whatever be its sign) [14]. Multistring solutions have not been found in black-hole backgrounds (without cosmological constant). More recently, new classes of dynamical and stationary multistring solutions in curved spacetimes have been found and classified and their physical properties analyzed [14]. Multistrings has been found for all inflationary spacetimes [15] but not in FRW universes.

The study of string propagation in curved spacetimes provide essential clues about the physics in this context but is clearly not the end of the story. The next step beyond the investigation of **test** strings, consist in finding **self-consistently** the geometry from the strings as matter sources for the Einstein equations or better the string effective equations (beta functions). This goal is achieved in ref. [3] for cosmological spacetimes at the classical level. Namely, we used the energy-momentum tensor for a gas of strings as source for the Einstein equations and we solved them self-consistently.

To write the string equation of state we used the behaviour of the string solutions in cosmological spacetimes. Strings continuously evolve from one type of behaviour to another, as is explicitly shown by our solutions [4] - [7]. For intermediate values of R, the equation of state for gas of free strings is clearly complicated but a formula of the type:

$$\rho = \left(u_R \, R + \frac{d}{R} + s \right) \frac{1}{R^{D-1}} \tag{1.2}$$

where

$$\lim_{R\to\infty} u_R = \begin{cases} 0 & \text{FRW} \\ u_\infty \neq 0 & \text{Inflationary} \end{cases} \tag{1.3}$$

This equation of state is qualitatively correct for all R and becomes exact for $R \to 0$ and $R \to \infty$. The parameters u_R, d and s are positive constants and the u_R varies smoothly with R.

The pressure associated to the energy density (1.2) takes then the form

$$p = \frac{1}{D-1}\left(\frac{d}{R} - u_R\, R\right)\frac{1}{R^{D-1}} \tag{1.4}$$

Inserting this source into the Einstein-Friedman equations leads to a self-consistent solution for string dominated universes (see sec. VI) [3]. This solution exhibits the realistic matter dominated behaviour $R \simeq T^{2/(D-1)}$ for large times and the radiation dominated behaviour $R \simeq T^{2/D}$ for early times.

For the sake of completeness we analyze in sec. IV the effective string equations [3]. These equations have been extensively treated in the literature [28] and they are not our central aim.

It must be noticed that there is no satisfactory derivation of inflation in the context of the effective string equations. This does not mean that string theory is not compatible with inflation, but that the effective string action approach *is not enough* to describe inflation. The effective string equations are a low energy field theory approximation to string theory containing only the *massless* string modes. The vacuum energy scales to start inflation are typically of the order of the Planck mass where the effective string action approximation breaks down. One must also consider the *massive* string modes (which are absent from the effective string action) in order to properly get the cosmological condensate yielding inflation. De Sitter inflation does not emerge as a solution of the the effective string equations.

In conclusion, the effective string action (whatever be the dilaton, its potential and the central charge term) is not the appropriate framework in which to address the question of string driven inflation.

Early cosmology (at the Planck time) is probably the best place to test string theory. In one hand the quantum treatment of gravity is unavoidable at such scales and in the other hand, observable cosmological consequences are derivable from the inflationary stage. The natural gravitational background is an inflationary universe as, for instance, de Sitter spacetime. Such geometries are not string vacua. This means that conformal and Weyl symmetries are broken at the quantum level. In order to quantize consistently strings in such case, one must enlarge the physical phase space including, in particular, the factor $\exp\phi(\sigma,\tau)$ in the world-sheet metric [see eq.(2.6)]. This is a very interesting and open problem. Physically, the origin of such difficulties in quantum string cosmology comes from the fact that one is not dealing with an **empty** universe since a cosmological spacetime necessarily contains matter. In the other hand, conformal field theory techniques are till now only adapted to backgrounds for which the beta functions are identically zero, i. e. sourceless geometries. A (quantum) string theory treatment of early cosmology necessarily implies **excited** states, not just string vacua. This problem is completely open today.

The outline of these lectures is as follows. Section II presents an introduction to strings in curved spacetimes including basic notions on classical and quantum strings in Minkowski spacetime and introducing the main physical string magnitudes: energy-momentum and invariant string size.

Section III deals with the several methods of resolution of the string propagation in curved spacetimes. (In sections III.A and III.B we treat the perturbative approaches, the τ-expansion and the global solutions.

Section V deals with strings in cosmological spacetimes, the τ-expansion at work there and we present the perfect gas of strings as a model for string matter.

In section V we treat self-consistent string cosmology including the string equations of state. (Section V.A deals with general relativity, V.B with the string thermodynamics).

Section VI discuss the effective (beta functions) string equations in the cosmological perspective and the search of inflationary solutions.

In sec. VII, we briefly review the systematic construction of string solutions in de Sitter universe *via* soliton methods and the new feature of multistring solutions.

II. STRINGS IN CURVED AND MINKOWSKI SPACETIMES.

Let us consider bosonic strings (open or closed) propagating in a curved D-dimensional spacetime defined by a metric $G_{AB}(X), 0 \leq A, B \leq D-1$. The action can be written as

$$S = \frac{1}{2\pi\alpha'} \int d\sigma d\tau \sqrt{g}\, g_{\alpha\beta}(\sigma,\tau)\, G_{AB}(X)\, \partial^{\alpha} X^A(\sigma,\tau)\, \partial^{\beta} X^B(\sigma,\tau) \tag{2.1}$$

Here $g_{\alpha\beta}(\sigma,\tau)$ ($0 \leq \alpha, \beta \leq 1$) is the metric in the worldsheet, α' stands for the string tension. As in flat spacetime, $\alpha' \simeq (M_{Planck})^{-2} \simeq (l_{Planck})^2$ fixes the scale in the theory. There are no other free parameters like coupling constants in string theory.

We will start considering given gravitational backgrounds $G_{AB}(X)$. That is, we start to investigate *test* strings propagating on a given spacetime. In section IV, the back reaction problem will be studied. That is, how the strings may act as source of the geometry.

String propagation in massless backgrounds other than gravitational (dilaton, antisymmetric tensor) can be investigated analogously.

The string action (2.1) classically enjoys Weyl invariance on the world sheet

$$g_{\alpha\beta}(\sigma,\tau) \to \lambda(\sigma,\tau)\, g_{\alpha\beta}(\sigma,\tau) \tag{2.2}$$

plus the reparametrization invariance

$$\sigma \to \sigma' = f(\sigma,\tau) \quad , \quad \tau \to \tau' = g(\sigma,\tau) \tag{2.3}$$

Here $\lambda(\sigma,\tau), f(\sigma,\tau)$ and $g(\sigma,\tau)$ are arbitrary functions.

The dynamical variables being here the string coordinates $X_A(\sigma,\tau)$, $(0 \leq A \leq D-1)$ and the world-sheet metric $g_{\alpha\beta}(\sigma,\tau)$.

Extremizing S with respect to them yields the classical equations of motion:

$$\begin{aligned}\partial^{\alpha}[\sqrt{g}\, G_{AB}(X)\, \partial_{\alpha} X^B(\sigma,\tau)] &= \frac{1}{2}\sqrt{g}\, \partial_A G_{CD}(X)\, \partial_{\alpha} X^C(\sigma,\tau)\, \partial^{\alpha} X^D(\sigma,\tau) \\ & 0 \leq A \leq D-1\end{aligned} \tag{2.4}$$

$$\begin{aligned}T_{\alpha\beta} \equiv\ & G_{AB}(X)[\,\partial_{\alpha} X^A(\sigma,\tau)\, \partial_{\beta} X^B(\sigma,\tau) \\ & - \frac{1}{2} g_{\alpha\beta}(\sigma,\tau)\, \partial_{\gamma} X^A(\sigma,\tau)\, \partial^{\gamma} X^B(\sigma,\tau)] = 0 \ , \quad 0 \leq \alpha, \beta \leq 1.\end{aligned} \tag{2.5}$$

Eqs. (2.5) contain only first derivatives and are therefore a set of constraints. Classically, we can always use the reparametrization freedom (2.3) to recast the world-sheet metric on diagonal form

$$g_{\alpha\beta}(\sigma,\tau) = \exp[\phi(\sigma,\tau)]\ \mathrm{diag}(-1,+1) \tag{2.6}$$

In this conformal gauge, eqs. (2.4) - (2.5) take the simpler form:

$$\partial_{-+}X^A(\sigma,\tau) + \Gamma^A_{BC}(X)\,\partial_+X^B(\sigma,\tau)\,\partial_-X^C(\sigma,\tau) = 0\ , \quad 0 \leq A \leq D-1, \tag{2.7}$$

$$T_{\pm\pm} \equiv G_{AB}(X)\,\partial_\pm X^A(\sigma,\tau)\,\partial_\pm X^B(\sigma,\tau) \equiv 0\,, \quad T_{+-} \equiv T_{-+} \equiv 0 \tag{2.8}$$

where we introduce light-cone variables $x_\pm \equiv \sigma \pm \tau$ on the world-sheet and where $\Gamma^A_{BC}(X)$ stand for the connections (Christoffel symbols) associated to the metric $G_{AB}(X)$.

Notice that these equations in the conformal gauge are still invariant under the conformal reparametrizations:

$$\sigma + \tau \to \sigma' + \tau' = f(\sigma+\tau) \quad , \quad \sigma - \tau \to \sigma' - \tau' = g(\sigma-\tau) \tag{2.9}$$

Here $f(x)$ and $g(x)$ are arbitrary functions.

The string boundary conditions in curved spacetimes are identical to those in Minkowski spacetime. That is,

$$\begin{aligned} X^A(\sigma+2\pi,\tau) &= X^A(\sigma,\tau) \quad \text{closed strings} \\ \partial_\sigma X^A(0,\tau) &= \partial_\sigma X^A(\pi,\tau) = 0 \quad \text{open strings.} \end{aligned} \tag{2.10}$$

A. A brief review on strings in Minkowski spacetime

In flat spacetime eqs.(2.7) become linear

$$\partial_{-+}X^A(\sigma,\tau) = 0\ , \quad 0 \leq A \leq D-1, \tag{2.11}$$

and one can solve them explicitly as well as the quadratic constraint (2.8) [see below]:

$$\left[\partial_\pm X^0(\sigma,\tau)\right]^2 - \sum_{j=1}^{D-1}\left[\partial_\pm X^j(\sigma,\tau)\right]^2 = 0 \tag{2.12}$$

The solution of eqs.(2.11) is usually written for closed strings as

$$X^A(\sigma,\tau) = q^A + 2p^A\alpha'\tau + i\sqrt{\alpha'}\sum_{n\neq 0}\frac{1}{n}\{\alpha^A_n \exp[in(\sigma-\tau)] + \tilde{\alpha}^A_n \exp[-in(\sigma+\tau)]\} \tag{2.13}$$

where q^A and p^A stand for the string center of mass position and momentum and α^A_n and $\tilde{\alpha}^A_n$ describe the right and left oscillator modes of the string, respectively. Since the string coordinates are real,

$$\bar{\alpha}^A_n = \alpha^A_{-n} \quad , \quad \bar{\tilde{\alpha}}^A_n = \tilde{\alpha}^A_{-n}$$

This resolution is no more possible in general for curved spacetime where the equations of motion (2.7) are non-linear. In that case, right and left movers interact with themselves and with each other.

In Minkowski spacetime we can also write the solution of the string equations of motion (2.11) in the form

$$X^A(\sigma,\tau) = l^A(\sigma+\tau) + r^A(\sigma-\tau) \tag{2.14}$$

where $l^A(x)$ and $r^A(x)$ are arbitrary functions. Now, making an appropriate conformal transformation (2.9) we can turn any of the string coordinates $X^A(\sigma,\tau)$ (but only one of them) into a constant times τ. The most convenient choice is the light-cone gauge where

$$U \equiv X^0 - X^1 = 2\,p^U\alpha'\tau. \tag{2.15}$$

That is, there are no string oscillations along the U direction in the light-cone gauge. We have still to impose the constraints (2.12). In this gauge they take the form

$$\pm 2\alpha' p^U\,\partial_\pm V(\sigma,\tau) = \sum_{j=2}^{D-1}\left[\partial_\pm X^j(\sigma,\tau)\right]^2 \tag{2.16}$$

where $V \equiv X^0 + X^1$. This shows that V is not an independent dynamical variable since it expresses in terms of the transverse coordinates $X^2,\ldots,X^{D-1}$. Only q^V is an independent quantity.

The physical picture of a string propagating in Minkowski spacetime clearly emerges in the light-cone gauge. The gauge condition (2.15) tells us that the string 'time' τ is just proportional to the physical null time U. Eqs.(2.13) shows that the string moves as a whole with constant speed while it oscillates around its center of mass. The oscillation frequencies are all integers multiples of the basic one. The string thus possess an infinite number of normal modes; $\alpha_n^A, \tilde\alpha_n^A$ classically describe their oscillation amplitudes. Only the modes in the direction of the transverse coordinates $X^2,\ldots,X^{D-1}$ are physical. This is intuitively right, since a longitudinal or a temporal oscillation of a string is meaningless. In summary, the string in Minkowski spacetime behaves as an extended and composite relativistic object formed by a $2(D-2)$-infinite set of harmonic oscillators.

Integrating eq.(2.16) on σ from 0 to 2π and inserting eq. (2.13) yields the classical string mass formula:

$$m^2 \equiv p^U p^V - \sum_{j=2}^{D-1}(p^j)^2 = \frac{1}{\alpha'}\sum_{j=2}^{D-1}\sum_{n=1}^{\infty}\left[\alpha_n^j\,\alpha_{-n}^j + \tilde\alpha_n^j\,\tilde\alpha_{-n}^j\right] \tag{2.17}$$

We explicitly see how the mass of a string depends on its excitation state. The classical string spectrum is continuous as we read from eq.(2.17). It starts at $m^2 = 0$ for an unexcited string: $\alpha_n^j = \tilde\alpha_n^j = 0$ for all n and j.

The independent string variables are:
the transverse amplitudes $\{\alpha_n^j, \tilde\alpha_n^j, n\varepsilon\mathcal{Z}, n\neq 0,\ 2\leq j\leq D-1\}$,
the transverse center of mass variables $\{q^j, p^j,\ 2\leq j\leq D-1\}$,
q^V and p^U.

Up to now we have considered a classical string.
At the quantum level one imposes the canonical commutation relations (CCR)

$$[\alpha^i_n, \alpha^j_m] = n\,\delta_{n,-m}\,\delta^{i,j} \quad , \quad [\tilde{\alpha}^i_n, \alpha^j_m] = n\,\delta_{n,-m}\,\delta^{i,j} \;,$$

$$[\tilde{\alpha}^i_n, \alpha^j_m] = 0 \;,$$

$$[q^i, p^j] = i\,\delta^{i,j} \quad , \quad [q^V, p^U] = i \tag{2.18}$$

All other commutators being zero. An order prescription is needed to unambiguously express the different physical operators in terms of those obeying the CCR. The symmetric ordering is the simplest and more convenient.

The space of string physical states is the the tensor product of the Hilbert space of the $D-1$ center of mass variables $q_V, p_U, \{q^j, p^j, 2 \leq j \leq D-1\}$, times the Fock space of the harmonic transverse modes. The string wave function is then the product of a center of mass part times a harmonic oscillator part. The center of mass can be taken, for example, as a plane wave. The harmonic oscillator part can be written as the creation operators $\alpha^{j\,\dagger}_n, \tilde{\alpha}^{j\,\dagger}_n$, $n \geq 1,\ 2 \leq j \leq D-1$ acting on the oscillator ground state $|0>$. This state is defined as usual by

$$\alpha^j_n\,|0> = \tilde{\alpha}^j_n\,|0> = 0, \quad \text{for all } n \geq 1,\ 2 \leq j \leq D-1$$

Notice that a string describes **one particle**. The kind of particle described depends on the oscillator wave function. The mass and spin can take an infinite number of different values. That is, there is an infinite number of different possibilities for the particle described by a string.

Let us consider the quantum mass spectrum. Upon symmetric ordering the mass operator becomes,

$$m^2 = \frac{1}{2\alpha'} \sum_{j=2}^{D-1} \sum_{n=1}^{\infty} \left[\alpha^j_n\,\alpha^j_{-n} + \alpha^j_{-n}\,\alpha^j_n + \tilde{\alpha}^j_n\,\tilde{\alpha}^j_{-n} + \tilde{\alpha}^j_{-n}\,\tilde{\alpha}^j_n \right] \,. \tag{2.19}$$

Using the commutation rules (2.18) yields

$$m^2 = \frac{D-2}{\alpha'} \sum_{n=1}^{\infty} n + \frac{1}{2\alpha'} \sum_{j=2}^{D-1} \sum_{n=1}^{\infty} \left[\alpha^{j\,\dagger}_n\,\alpha^j_n + \tilde{\alpha}^{j\,\dagger}_n\,\tilde{\alpha}^j_n \right] \tag{2.20}$$

The divergent sum in the first term can be defined through analytic continuation of the zeta function

$$\zeta(z) \equiv \sum_{n=1}^{\infty} \frac{1}{n^z} \tag{2.21}$$

One finds $\zeta(-1) = -1/12$ [30]. Thus,

$$m^2 = -\frac{D-2}{12\alpha'} + \frac{1}{2\alpha'} \sum_{j=2}^{D-1} \sum_{n=1}^{\infty} \left[\alpha^{j\,\dagger}_n\,\alpha^j_n + \tilde{\alpha}^{j\,\dagger}_n\,\tilde{\alpha}^j_n \right] \tag{2.22}$$

Hence, the string ground state $|0>$ has a negative mass squared

$$m_0^2 = -\frac{D-2}{12\alpha'} \tag{2.23}$$

Such particles are called tachyons and exhibit unphysical behaviours. When fermionic degrees of freedom are associated to the string the ground state becomes massless (superstrings) [29].

Notice that the appearance of a negative mass square yields a dispersion relation $E^2 = p^2 - |m_0^2|$ similar to classical waves when gravity (even newtonian) is taken into account (Jeans unstabilities) [31].

Let us consider now excited states.

The constraints (2.12) integrated on σ from 0 to 2π impose

$$\sum_{j=2}^{D-1} \sum_{n=1}^{\infty} (\alpha_n^j)^\dagger \alpha_n^j = \sum_{j=2}^{D-1} \sum_{n=1}^{\infty} (\tilde{\alpha_n^j})^\dagger \tilde{\alpha}_n^j \tag{2.24}$$

on the physical states. This means that the number of left and right modes coincide in all physical states.

The first excited state is then described by

$$|i,j>= (\tilde{\alpha_1^i})^\dagger (\alpha_1^j)^\dagger \, |0> \tag{2.25}$$

times the center of mass wave function. We see that this wavefunction is a symmetric tensor in the space indices i, j. It describes therefore a spin two particle plus a spin zero particle (the trace part).

From eqs.(2.22-2.25) follows that

$$m^2 \, |i,j>= -\frac{D-26}{12\alpha'} |i,j> \tag{2.26}$$

This state is then a massless particle only for $D = 26$. In such critical dimension we have then a graviton (massless spin 2 particle) and a dilaton (massless spin 0 particle) as string modes of excitation. For superstrings the critical dimension turns to be $D = 10$ [29].

We shall consider, as usual, that only four space-time dimensions are uncompactified. That is, we shall consider the strings as living on the tensor product of a curved four dimensional space-time with lorentzian signature and a compact space which is there to cancel the anomalies. From now on strings will propagate in the curved (physical) four dimensional space-time. However, we will find instructive to study the case where this curved space-time has dimensionality D, where D may be 2, 3, 4 or arbitrary.

B. The string energy-momentum tensor and the string invariant size

The spacetime string energy-momentum tensor follows (as usual) by taking the functional derivative of the action (2.1) with respect to the metric G_{AB} at the spacetime point X. This yields,

$$\sqrt{-G}\, T^{AB}(X) = \frac{1}{2\pi\alpha'} \int d\sigma d\tau \left(\dot{X}^A \dot{X}^B - X'^A X'^B\right) \delta^{(D)}(X - X(\sigma,\tau)) \tag{2.27}$$

where dot and prime stands for $\partial/\partial\tau$ and $\partial/\partial\sigma$, respectively.

Notice that X in eq.(2.27) is just a spacetime point whereas $X(\sigma, \tau)$ stands for the string dynamical variables. One sees from the Dirac delta in eq.(2.27) that $T^{AB}(X)$ vanishes unless

X is exactly on the string world-sheet. We shall not be interested in the detailed structure of the classical strings. It is the more useful to integrate the energy-momentum tensor (2.27) on a volume that completely encloses the string. It takes then the form [17]

$$\Theta^{AB}(X^0) = \frac{1}{2\pi\alpha'}\int d\sigma d\tau \left(\dot{X}^A\dot{X}^B - X'^A X'^B\right)\delta(X^0 - X^0(\tau,\sigma)). \tag{2.28}$$

When X^0 depends only on τ, we can easily integrate over τ with the result,

$$\Theta^{AB}(X^0) = \frac{1}{2\pi\alpha'|\dot{X}^0(\tau)|}\int_0^{2\pi} d\sigma \left[\dot{X}^A\dot{X}^B - X'^A X'^B\right]_{\tau=\tau(X^0)} \tag{2.29}$$

Another relevant physical magnitude for strings is the invariant size. We define the invariant string size ds^2 using the metric induced on the string world-sheet:

$$ds^2 = G_{AB}(X)\, dX^A\, dX^B \tag{2.30}$$

Inserting $dX^A = \partial_+ X^A\, dx^+ + \partial_- X^A\, dx^-$, into eq.(2.30) and taking into account the constraints (2.8) yields

$$ds^2 = 2\, G_{AB}(X)\, \partial_+ X^A\, \partial_- X^B \left(d\tau^2 - d\sigma^2\right) = G_{AB}(X)\, \dot{X}^A\, \dot{X}^B \left(d\tau^2 - d\sigma^2\right). \tag{2.31}$$

Thus, we define the string size as the integral of

$$S(\sigma,\tau) \equiv \sqrt{G_{AB}(X)\, \dot{X}^A\, \dot{X}^B} \tag{2.32}$$

over σ at fixed τ. For a causal choice of the string time τ we must have

$$G_{AB}(X)\, \dot{X}^A\, \dot{X}^B \geq 0 \tag{2.33}$$

The equality sign here corresponds to a string behaving as radiation. Such type of solutions always exist. For example any σ-independent solution of eqs.(2.7-2.8). Such solutions describe massless geodesics.

Notice that the trace of the energy-momentum tensor eq.(2.27) is just the integral of $S(\sigma,\tau)^2$,

$$\begin{aligned}\sqrt{-G}\; T^A_A(X) &= \frac{1}{\pi\alpha'}\int d\sigma d\tau\; G_{AB}(X)\dot{X}^A\dot{X}^B\; \delta^{(D)}(X - X(\sigma,\tau)) \\ &= \frac{1}{\pi\alpha'}\int d\sigma d\tau\; S(\sigma,\tau)^2\, \delta^{(D)}(X - X(\sigma,\tau))\end{aligned} \tag{2.34}$$

For strings behaving as radiation both $T^A_A(X)$ and S vanish.

C. Simple String Solutions in Minkowski Spacetime

Let us now consider a circular string as a simple example of a string solution in Minkowski spacetime.

$$X^0(\sigma,\tau) = \alpha' E\tau \quad , \quad X^3(\sigma,\tau) = \alpha' p\tau$$

$$X^1(\sigma,\tau) = \alpha' m \cos\tau \cos\sigma = \frac{\alpha' m}{2}\left[\cos(\tau+\sigma) + \cos(\tau-\sigma)\right] \tag{2.35}$$

$$X^2(\sigma,\tau) = \alpha' m \cos\tau \sin\sigma = \frac{\alpha' m}{2}\left[\sin(\tau+\sigma) - \sin(\tau-\sigma)\right]$$

This is obviously a solution of eqs.(2.11) where only the modes $n = \pm 1, j = 1, 2$ are excited. The constraints (2.12) yields

$$E^2 = p^2 + m^2$$

Eqs.(2.35) describe a circular string in the X^1, X^2 plane, centered in the origin and with an oscillating radius $\rho(\tau) = \alpha' m \cos\tau$. In addition the string moves uniformly in the z-direction with speed p/E. (That is, p is its momentum in the z-direction). The oscillation amplitude m can be identified with the string mass and E with the string energy. Notice that the string time τ is here proportional to the physical time X^0 [this solution is not in the light-cone gauge (2.15)].

It is instructive to compute the integrated energy-momentum tensor (2.29) for this string solution. We find in the rest frame ($p = 0$) that it takes the fluid form

$$\Theta_A^B = \begin{pmatrix} \rho & 0 & 0 & 0 \\ 0 & -p & 0 & 0 \\ 0 & 0 & -p & 0 \\ 0 & 0 & 0 & 0 \end{pmatrix} \tag{2.36}$$

where

$$\rho = E = m \quad , \quad p = -\frac{m}{2}\cos(2\tau) \tag{2.37}$$

We see that the total energy coincides with m as one could expect and that the (space averaged) pressure oscillates around zero. That is, the string pressure goes through positive and negative values. The time average of p on a period vanishes. The string behaves then as cold matter (massive particles).

The upper value of p equals $E/2$. This is precisely the relation between E and p for radiation (massless particles). (Notice that this circular strings lives on a two-dimensional plane). The lower value of p correspond to the limiting value allowed by the strong energy condition in General Relativity [16]. We shall see below that these two extreme values of p appear for strings in general cosmological spacetimes.

The invariant size of the string solution (2.35) follows by inserting eq.(2.35) into eq.(2.31). We find

$$ds^2 = (\alpha' m)^2 \left(d\tau^2 - d\sigma^2\right) \tag{2.38}$$

Therefore, this string solution has a constant size $2\pi\alpha' m$.

Another simple but instructive solution in Minkowski spacetime is a rotating straight string (a rotating rod) given by

$$X^0(\sigma,\tau) = \alpha' m \tau \quad ,$$

$$X^1(\sigma,\tau) = \alpha' m \cos\tau \cos\sigma \ , \tag{2.39}$$

$$X^2(\sigma,\tau) = \alpha' m \sin\tau \cos\sigma \ ,$$

or in polar coordinates

$$\rho = \alpha' m \ |\cos\sigma| \quad , \quad \phi = \tau \ .$$

That is, a straight string on the $X^1 - X^2$ plane rotating around the origin with an angular speed $\frac{1}{\alpha' m}$. Eqs. (2.39) identically fulfil the string equations and constraints (2.11-2.12).

The energy momentum tensor for this rotating string takes the form:

$$\Theta_A^B = \begin{pmatrix} m & 0 & 0 & 0 \\ 0 & \frac{m}{2}\cos(2\tau) & \frac{m}{2}\sin(2\tau) & 0 \\ 0 & \frac{m}{2}\sin(2\tau) & -\frac{m}{2}\cos(2\tau) & 0 \\ 0 & 0 & 0 & 0 \end{pmatrix} . \tag{2.40}$$

This means an energy density $\rho = m$. The spatial part of Θ_A^B becomes diagonal in a rotating frame and has $\pm\frac{m}{2}$ as eigenvalues. That is, we again find $\pm\frac{\rho}{2}$ as extreme values for the pressure in two space dimensional string solutions. The time average of the stress tensor Θ_i^j here vanishes indicating zero pressure (cold matter behaviour).

One obtains the invariant size for a generic solution in Minkowski spacetime inserting the general solution (2.13) into $\partial_+ X^A \partial_- X_A$. This gives constant plus oscillatory terms. In any case the invariant string size is always **bounded** in Minkowski spacetimes. We shall see how differently behave strings in curved spacetimes.

III. HOW TO SOLVE THE STRING EQUATIONS OF MOTION IN CURVED SPACETIMES?

There is no general method to solve the string equations of motion and constraints for arbitrary curved spacetime.

The so-called τ-expansion method provides **exact local** solutions for any background. The basic idea goes as follows. Suppose one is interested on the string behaviour near a given point of the curved spacetime. Then, one chooses a conformal gauge such that $\tau = 0$ corresponds to such a point. For example, to study strings in cosmological spacetimes near the initial singularity (cosmic time $T = 0$), one chooses $T(\sigma,\tau)$ such that $T(\sigma,0) = 0$. It is shown below that this is indeed possible for generic string solutions. This expansion was developped first in ref. [10,11] for inflationary universes.

Similarly, in order to study strings near the black hole singularity $r = 0$, one chooses $r(\sigma,\tau)$ such that $r(\sigma,0) = 0$.

Once this gauge choice is done, the string equations of motion and constraints can be solved in powers of τ. These powers may not be integer powers. For example, one finds powers of $\tau^{-\frac{2}{\alpha+1}}$ in FRW universes with scale factor $R(T) = c\, T^\alpha$. For Schwarzschild black holes powers of $\tau^{\frac{2}{5}}$ appear.

An approximate but general method is the expansion around center of mass solutions [2,9]. In this method one starts from an exact solution of the geodesic equations

$$\ddot{q}^A(\tau) + \Gamma^A_{BC}(q)\,\dot{q}^B(\tau)\,\dot{q}^C(\tau) = 0 \tag{3.1}$$

The world-sheet time variable is here identified with the proper time of the center of mass trajectory. [Notice that eqs.(3.1) just follow from the string equations (2.7) by dropping the σ-dependence].

Then one develops in perturbations around it. That is, one sets

$$X^A(\sigma,\tau) = q^A(\tau) + \eta^A(\sigma,\tau) + \xi^A(\sigma,\tau) + \ldots \tag{3.2}$$

Here $\eta^A(\sigma,\tau)$ obeys the linearized perturbation around $q^A(\sigma,\tau)$ and $\xi^A(\sigma,\tau)$ the second order perturbation equations [2]. These fluctuations obey coupled ordinary differential equations that can be written systematically inserting eq.(3.2) into eqs.(2.7-2.8). [See ref. [9] for more details].

Another general approximation method is the null string approach [22]. In such approach the string equations of motion and constraints are systematically expanded in powers of c (the speed of light in the world-sheet). This corresponds to a small string tension expansion. At zeroth order, the string is effectively equivalent to a continuous beam of massless particles labelled by the parameter σ. The points on the string do not interact between them but they interact with the gravitational background.

For several spacetimes one can construct explicit string solutions using specific properties of the background. This is the case of singular plane waves, shock-waves, conical spacetimes and the de Sitter universe. The string equations of motion and constraints in the de Sitter spacetime are integrable in the inverse scattering sense as shown in ref. [4]. [The de Sitter universe is a symmetric space and hence the string equations there correspond to a two dimensional integrable sigma model].

A. The τ-expansion

Let us consider the intersection of the world-sheet with a singular or non-singular point (or surface) of the spacetime like $T(\sigma,\tau) = 0$ or $T(\sigma,\tau) = T_o$ in a cosmological spacetime or $r(\sigma,\tau) = 0$ or $r(\sigma,\tau) = r_o$ in a Schwarzschild black-hole.

We can write the curve describing such intersection with the world-sheet as

$$x_+ = \chi(x_-), \tag{3.3}$$

whenever this intersection is nondegenerate. (Here $x_\pm \equiv \tau \pm \sigma$).

Upon a conformal transformation,

$$x_+ \to x'_+ = f(x_+) \quad , \quad x_- \to x'_- = g(x_-), \tag{3.4}$$

we can map the curve (3.3) into $\tau' = 0$ by an appropriate choice of f and g [18]. For example, we can choose

$$f(x_+) = x_+ \quad , \quad g(x_-) = -\chi(x_-).$$

This defines our choice of gauge. From now on, we rename τ' and σ' by τ and σ, respectively. Notice that this choice does not completely fix the gauge. We can still perform transformations that leave the line $\tau = 0$ unchanged. This is the case for the following class of conformal mappings

$$x_+ \to x'_+ = \varphi(x_+) \quad , \quad x_- \to x'_- = -\varphi(-x_-), \tag{3.5}$$

where $\varphi(x)$ is an arbitrary function. Eq. (3.5) can be written as,

$$\tau' = \frac{1}{2}\left[\varphi(\tau+\sigma) - \varphi(\sigma-\tau)\right] = \tau\,\varphi'(\sigma) + \frac{1}{6}\tau^3\,\varphi'''(\sigma) + O(\tau^4)$$
$$\sigma' = \frac{1}{2}\left[\varphi(\tau+\sigma) + \varphi(\sigma-\tau)\right] = \varphi(\sigma) + \frac{1}{2}\tau^2\,\varphi''(\sigma) + O(\tau^4) \tag{3.6}$$

The transformations (3.5) represent a diagonal subgroup of the set of left-right conformal transformations (3.4).

In summary, any (non-degenerate) intersection of the world-sheet with a spacetime submanifold $T(\sigma,\tau) = T_o$ can be mapped into $\tau = 0$. This mapping is not unique, it is invariant under the diagonal conformal transformations.

Once this gauge has been imposed one studies the string equations of motion and constraints (2.7-2.8) in powers of τ. The equations themselves determine the precise values of the powers [10,11,18].

B. Global Solutions

There is no general method to find solutions valid in the whole world-sheet. However, many global solutions have been found in physically relevant spacetimes.

First, there are spacetimes where the **general** solution of the string equations and constraints has been found. That is, shock-waves [44], singular plane waves [46] and conical spacetimes [47].

Second, by making specific ansatz according to the symmetry of the background, the string equations of motion can be reduced to ordinary differential equations. Then, these ordinary differential equations can be solved globally by numerical methods. In this way, solutions valid in the whole worldsheet has been found in cosmological spacetimes and black holes [7,12] - [14].

Third, the de Sitter spacetime can be treated by inverse scattering methods. In this way exact string solutions has been constructed systematically (see sec. VII and refs. [5] - [8]).

In all cases where global solutions can be found, the τ-expansion results are confirmed.

IV. STRING PROPAGATION IN COSMOLOGICAL SPACETIMES

We obtain in this section physical string properties from the string solutions in cosmological spacetimes.

We consider strings in spatially homogeneous and isotropic universes with metric

$$ds^2 = (dT)^2 - R(T)^2 \sum_{i=1}^{D-1} (dX^i)^2 \,, \tag{4.1}$$

where T is the cosmic time and the function $R(T)$ is called the scale factor. In terms of the conformal time

$$\eta = \int^T \frac{dT}{R(T)} , \tag{4.2}$$

the metric (4.1) takes the form

$$ds^2 = R(\eta)^2 \left[(d\eta)^2 - \sum_{i=1}^{D-1} (dX^i)^2 \right] . \tag{4.3}$$

The classical string equations of motion can be written here as

$$\begin{aligned} \partial^2 T + \ R(T) \frac{\mathrm{d}R}{\mathrm{d}T} \sum_{i=1}^{D-1} (\partial_\mu X^i)^2 = 0 \, , \\ \partial_\mu \left[R^2 \partial^\mu X^i \right] = 0 \, , \qquad 1 \le i \le D-1 \, , \end{aligned} \tag{4.4}$$

and the constraints are

$$T_{\pm\pm} = (\partial_\pm T)^2 - R(T)^2 \, (\partial_\pm X^i)^2 = 0 \, . \tag{4.5}$$

The most relevants universes correspond to power type scale factors. That is,

$$R(T) = a \, T^\alpha = A \, \eta^{k/2} \, , \tag{4.6}$$

where $\alpha = \frac{k}{k+2}$ and $k = \frac{2\alpha}{1-\alpha}$.

For different values of the exponents we have either FRW or inflationary universes.

$$\mathrm{FRW} : 0 < k \le \infty, \ 0 < \alpha \le 1 = \begin{cases} \alpha = 1/2, \ k = 2, & \text{radiation dominated,} \\ \alpha = 2/3, \ k = 4, & \text{matter dominated,} \\ \alpha = 1, \ k = \infty, & \text{`stringy'.} \end{cases}$$

$$\text{Inflationary} : -\infty < k < 0, \ \alpha < 0 \text{ and } \alpha > 1 = \begin{cases} \alpha = \infty, \ k = -2, & R(T) = e^{HT}, \text{ de Sitter,} \\ \alpha > 1, \ k < -2, & \text{power inflation,} \\ \alpha < 0, \ -2 < k < 0, & \text{superinflationary.} \end{cases}$$

The denomination 'stringy' comes from the fact that such backgrounds follow as solution of the string effective equations [24]. Inflationary universes are those with accelerated expansion. That is,

$$\frac{d^2 R(T)}{dT^2} > 0 \, .$$

The string equations of motion and constraints take then the following form using the conformal time η:

$$\begin{aligned} \ddot\eta - \eta'' + \frac{k}{2\eta} \left\{ \dot\eta^2 - \eta'^2 + \sum_{i=1}^{D-1} \left[(\dot X^i)^2 - (X'^i)^2 \right] \right\} = 0 \, , \\ \ddot X^i - X''^i + \frac{k}{\eta} \left[\dot\eta \dot X^i - \eta' X'^i \right] = 0 \, , \quad 1 \le i \le D-1 \, , \\ (\dot\eta \pm \eta')^2 - \sum_{i=1}^{D-1} (\dot X^i \pm X'^i)^2 = 0 \, , \end{aligned} \tag{4.7}$$

where prime and dot stand for ∂_σ and ∂_τ, respectively.

As we will see below, once appropriately averaged $\Theta^{AB}(X)$ takes the fluid form for string solutions in cosmological spacetimes, allowing us to define the string pressure p and energy density ρ :

$$< \Theta_A^B >= \begin{pmatrix} \rho & 0 \cdots & 0 \\ 0 & -p \cdots & 0 \\ \cdots & \cdots \cdots & 0 \\ 0 & 0 \cdots & -p \end{pmatrix} . \tag{4.8}$$

Notice that the continuity equation

$$D^A \, \Theta_A^B = 0 \, ,$$

takes here the form

$$\frac{d\rho}{dT} + (D-1)\, H\, (p+\rho) = 0 \tag{4.9}$$

where $H \equiv \frac{1}{R}\frac{dR}{dT}$.

For an equation of state of the type of a perfect fluid, that is

$$p = (\gamma - 1)\, \rho \qquad , \qquad \gamma = \text{ constant}, \tag{4.10}$$

eqs.(4.9) and (4.10) can be easily integrated with the result

$$\rho = \rho_0 \; R^{\gamma(1-D)} \quad . \tag{4.11}$$

For $\gamma = 1$ this corresponds to cold matter ($p = 0$) and for $\gamma = \frac{D}{D-1}$ this describes radiation with $p = \frac{\rho}{D-1}$.

The speed of sound in a fluid is given by

$$v_s = \sqrt{\left(\frac{\partial p}{\partial \rho}\right) |_s} = \sqrt{\gamma - 1}$$

This relation makes sense for $1 \geq \gamma - 1 \geq 0$. For $\gamma - 1 < 0$, there are no sound waves since the perturbations in p and ρ obey elliptic evolution equations. For $\gamma - 1 > 1$ causality would be violated. Actually, for strings we always find $-\frac{1}{D-1} \leq \gamma - 1 \leq \frac{1}{D-1}$.

A. Strings in cosmological universes: the τ-expansion at work

Let us consider strings in inflationary universes with scale factor (4.6) and $k < 0$. In order to apply the τ-expansion we fix the gauge such that

$$\eta(\tau = 0, \sigma) = 0 \, . \tag{4.12}$$

As explained above, this is always possible for generic string solutions and it leaves still the freedom of the transformations (3.5). Notice that $\eta \to 0$ corresponds in the inflationary case to large scale factors $R \to \infty$.

The behaviour of $\eta(\tau,\sigma)$ and $X^i(\tau,\sigma)$ for $\tau\to 0$ follows from eq.(4.7) where we use eq.(4.12) and assume that $X^i(\tau,\sigma)$ is regular at $\tau=0$. One finds [11]

$$\begin{aligned}\eta(\tau,\sigma) &\overset{\tau\to 0}{=} \eta_o(\sigma)\,\tau\,\left[1+O(\tau^2)\right]+\lambda_o(\sigma)\,\tau^{2-k}\left[1+O(\tau^2)\right]\\ &+\ \zeta_o(\sigma)\,\tau^{1-2k}\left[1+O(\tau^2)\right]+O(\tau^{1-3k})\,,\\ X^i(\tau,\sigma) &\overset{\tau\to 0}{=} A^i(\sigma)\left[1+O(\tau^2)\right]+\tau^{1-k}\,B^i(\sigma)\left[1+O(\tau^2,\tau^{1-k})\right],\\ &1\le i\le D-1\,.\end{aligned}\tag{4.13}$$

The solutions appear as a series in powers of τ^2 and τ^{1-k}. In the special case where k is rational, say $k=-\frac{l}{n}$, l,n =integers, logarithmic terms in τ appear in addition. This happens, for example in de Sitter spacetime ($k=-2$) and in Minkowski spacetime ($k=0$).

The coefficients in eq.(4.15) result related as follows

$$\begin{aligned}&\sum_{i=1}^{D-1} B^i(\sigma)\,A'^i(\sigma)=0\,,\ \eta_o(\sigma)=\sqrt{\sum_{i=1}^{D-1}[A'^i(\sigma)]^2}\\ &\lambda_o(\sigma)=\frac{2k}{(2-k)(1-k)}\,\frac{\sum_{i=1}^{D-1}B'^i(\sigma)\,A'^i(\sigma)}{\eta_o(\sigma)}\,.\end{aligned}\tag{4.14}$$

Moreover, one can use the residual conformal invariance (3.5) to set $\eta_o(\sigma)\equiv 1$. One finally obtains for inflationary universes [11],

$$\begin{aligned}\eta(\tau,\sigma) &\overset{\tau\to 0}{=} \tau\,\left[1+O(\tau^2)\right]+\lambda_o(\sigma)\,\tau^{2-k}\left[1+O(\tau^2)\right]\\ &+\ \zeta_o(\sigma)\,\tau^{1-2k}\left[1+O(\tau^2)\right]+O(\tau^{1-3k})\,,\\ X^i(\tau,\sigma) &\overset{\tau\to 0}{=} A^i(\sigma)\left[1+O(\tau^2)\right]+\tau^{1-k}\,B^i(\sigma)\left[1+O(\tau^2,\tau^{1-k})\right],\\ &1\le i\le D-1\,.\end{aligned}\tag{4.15}$$

where

$$\begin{aligned}&\sum_{i=1}^{D-1} B^i(\sigma)\,A'^i(\sigma)=0\quad,\quad\sum_{i=1}^{D-1}\left[A'^i(\sigma)\right]^2=1\\ &\lambda_o(\sigma)=\frac{2k}{(2-k)(1-k)}\sum_{i=1}^{D-1}B'^i(\sigma)\,A'^i(\sigma)\quad,\quad\zeta_o(\sigma)=\frac{(1-k)^2}{2(1-2k)}\sum_{i=1}^{D-1}\left[B'^i(\sigma)\right]^2\,.\end{aligned}\tag{4.16}$$

Here $A^i(\sigma)$, $B^i(\sigma)$, $1\le i\le D-1$ are the initial ($\tau=0$) string coordinates and momenta.

We see that the solution depends on $2(D-2)$ independent functions among the $A^i(\sigma)$ and $B^i(\sigma)$, $1\le i\le D-1$. All coefficients (including the higher orders not written in eq.(4.15)) express in terms of the $A^i(\sigma)$ and $B^i(\sigma)$. Therefore, the counting of degrees of freedom turns to be the same as in Minkowski spacetime: only the $2(D-2)$ **transverse** coordinates are physical.

It must be noticed that

$$\dot X^i(\tau,\sigma)\overset{\tau\to 0}{=}(1-k)\,\tau^{-k}\,B^i(\sigma)\left[1+O(\tau^2,\tau^{1-k})\right]\to 0$$

$$X'^i(\tau,\sigma)\overset{\tau\to 0}{=}A'^i(\sigma)\neq 0$$

That is, $\partial_\sigma X^i$ is **larger** than $\partial_\tau X^i$ for $R \to \infty$. This is the opposite to a point particle behaviour. For a point particle, $\partial_\sigma X^i \equiv 0$ and $\partial_\tau X^i = p^i \neq 0$.

Let us now apply the τ-expansion to strings in FRW universes. That is $k > 0$ in the scale factor (4.6).

We fix again the gauge according to eq.(4.12). It must be noticed that now $\eta \to 0$ corresponds to $R \to 0$ since $k > 0$. That is the τ-expansion applies near the singularity (big bang) of the spacetime.

After calculations analogous to the inflationary case, one finds from eqs.(4.7) for FRW universes ($k > 0$) [11]

$$\begin{aligned} &\eta(\tau,\sigma) \overset{\tau\to 0}{=} \tau^{\frac{1}{k+1}} \left[1 + O(\tau^2)\right] + \eta_1(\sigma)\, \tau^{2-\frac{1}{k+1}} \left[1 + O(\tau^2)\right] \\ &X^i(\tau,\sigma) \overset{\tau\to 0}{=} A^i(\sigma) \left[1 + O(\tau^2)\right] + \tau^{\frac{1}{k+1}}\, B^i(\sigma) \left[1 + O(\tau^2)\right] + \tau^{2-\frac{1}{k+1}}\, C^i(\sigma) \left[1 + O(\tau^2)\right] , \\ &\qquad 1 \le i \le D-1 . \end{aligned} \tag{4.17}$$

where the residual conformal invariance (3.5) has been used.

The string equations of motion and constraints impose the following relations on the coefficients:

$$\begin{aligned} &\sum_{i=1}^{D-1} \left[B^i(\sigma)\right]^2 = 1 \quad , \quad \sum_{i=1}^{D-1} B^i(\sigma)\, A'^i(\sigma) = 0 , \\ &C^i(\sigma) = -\eta_1(\sigma)\, B^i(\sigma) \quad , \quad \eta_1(\sigma) = \frac{(k+1)^2}{4(2k+1)} \sum_{i=1}^{D-1} \left[A'^i(\sigma)\right]^2 . \end{aligned} \tag{4.18}$$

$A^i(\sigma)$, $B^i(\sigma)$, $1 \le i \le D-1$ turn to be the initial ($\tau = 0$) string coordinates and momenta.

We again find that the string solution is determined by $2(D-2)$ independent functions indicating that only the transverse coordinates are physical. As we see, the τ-expansion produces for FRW universes the string solution as a series in powers of τ^2 and $\tau^{\frac{2k}{1+k}}$.

For large R the spacetime curvature tends to zero as $T^{-2} \simeq R^{-2/\alpha}$ in FRW spacetimes. That is, for $T \to \infty$. In order to analyze the string behaviour in such regime it is convenient to choose the gauge

$$\eta = \eta(\tau) \to \infty \quad , \quad \text{for } \tau \to \infty \tag{4.19}$$

[This is a slight generalization of the previous gauge choices at finite τ].

We then find from eq.(4.7)

$$\begin{aligned} &\eta(\tau,\sigma) \overset{\tau\to\infty}{=} \tau^{\frac{2}{k+2}} \to \infty , \\ &X^i(\tau,\sigma) \overset{\tau\to\infty}{=} \frac{1}{k+2}\, \tau^{-\frac{k}{k+2}} \left[f_i^+(\sigma+\tau) + f_i^-(\sigma-\tau)\right] , \\ &\qquad 1 \le i \le D-1 . \end{aligned} \tag{4.20}$$

where the $f_i^\pm(x)$ are arbitrary periodic functions of x

$$f_i^\pm(x + 2\pi) = f_i^\pm(x) ,$$

obeying the pair of constraints:

$$\sum_{i=1}^{D-1} \left[f_i'^{\pm}(x)\right]^2 = 1 \,. \tag{4.21}$$

We obtain from eq.(4.20)

$$T(\tau) \overset{\tau\to\infty}{=} \frac{2\sqrt{A}}{k+2}\,\tau \to \infty \quad \text{and} \quad R \overset{\tau\to\infty}{=} \tau^{\frac{k}{k+2}} \to \infty \,. \tag{4.22}$$

In short, the string solutions in this regime are asymptotically Minkowski solutions (2.13) *scaled* by a factor R^{-1}. This is not unexpected since the spacetime curvature vanishes for this regime. The counting of degrees of freedom is again as in Minkowski spacetime.

We have determined the string behaviour for $R \to \infty$ in inflationary and FRW universes and for $R \to 0$ in FRW universes. Let us now compute for such regimes the string physical properties, energy-momentum and size.

The calculation of $\Theta^{AB}(T)$ in the different limiting regimes follows from eq.(2.29) since $\eta = \eta(\tau)$ asymptotically (both for $\tau \to 0$ and $\tau \to \infty$).

Let us start by considering the inflationary universes for $R \to \infty$. We find for the (integrated) energy-momentum tensor from eqs.(2.29) and (4.15)

$$\rho(T) \;=\; \Theta^{00}(T) \overset{\tau\to 0}{=} \frac{R}{\alpha'} \to +\infty \,, \tag{4.23}$$

$$\Theta^i_j(T) \overset{\tau\to 0}{=} \frac{R}{\alpha'} \int_0^{2\pi} \frac{d\sigma}{2\pi}\, A'^i(\sigma)\, A'^j(\sigma) \;\to \infty \,,$$

$$\Theta^0_i(T) \overset{\tau\to 0}{=} -\frac{1}{\alpha'} \int_0^{2\pi} \frac{d\sigma}{2\pi}\, B^i(\sigma) \tag{4.24}$$

where we also used the spacetime metric $G_{00} = 1,\; G_{ij} = -R^2\,\delta_{ij}$.

Recall that $\sum_{i=1}^{D-1} [A'^i(\sigma)]^2 = 1$.

We see that the energy density $\rho(T)$ diverges for $R \to \infty$. The stress tensor $[\Theta^i_j(T)]$ is not diagonal but it is given by a **positive definite** matrix. Such matrix has then positive eigenvalues and therefore tells us that the pressure is **negative** [compare with eq.(4.8)]. This is the **unstable** string behaviour. That is, the energy tends to $+\infty$ and the pressure to $-\infty$. At the same time the energy flux density $\Theta^0_i(T)$ stands bounded.

The string size S in cosmological spacetimes (4.1) takes the form

$$S^2 = G_{AB}(X)\,\dot{X}^A\,\dot{X}^B = \dot{T}^2 - R^2 \sum_{i=1}^{D-1} (\dot{X}^i)^2 \tag{4.25}$$

For $\tau \to 0$, $R \to \infty$ in inflationary spacetimes, we find using eq.(4.15) that the first term dominates in eq.(4.25)

$$S \stackrel{\tau\to 0}{=} c\,\tau^{k/2} \simeq R \to \infty\,, \tag{4.26}$$

where c is a constant. We see that the string grows infinitely big when the universe inflates. The string size being proportional to the scale factor and also to the string energy [ρ in eq.(4.23)]. These explosive growings characterize the string unstable behaviour.

Let us now consider FRW universes ($k > 0$) for $\tau \to 0,\ R \to 0$. The (integrated) energy-momentum tensor in such regime takes the form

$$\rho(T) = \Theta^{00}(T) \stackrel{\tau\to 0}{=} \frac{1}{\alpha'(k+1)\,R} \to +\infty\,, \tag{4.27}$$

$$\Theta^{i}_{j}(T) \stackrel{\tau\to 0}{=} -\frac{1}{\alpha'(k+1)\,R}\int_0^{2\pi}\frac{d\sigma}{2\pi}\,B^i(\sigma)\,B^j(\sigma) \to \infty\,,$$

$$\Theta^{0}_{i}(T) \stackrel{\tau\to 0}{=} \frac{1}{\alpha'(k+1)\,R}\int_0^{2\pi}\frac{d\sigma}{2\pi}\,B^i(\sigma) \tag{4.28}$$

where we used eqs.(2.29) and (4.17). [Recall that $\sum_{i=1}^{D-1}[B^i(\sigma)]^2 = 1$].

We see in eq.(4.27) that the stress tensor $(\Theta^i_j(T))$ is not diagonal but it is given by a **negative definite** matrix. Such matrix has then negative eigenvalues and therefore tells us that the pressure is **positive**. This string behaviour is dual to the previous unstable behaviour.

We find from eq.(4.25) for the string size S in FRW universes for $R \to 0$,

$$S \stackrel{\tau\to 0}{=} \sqrt{\sum_{i=1}^{D-1}[A'^i(\sigma)]^2}\;\tau^{\frac{k}{2(k+1)}} \simeq \sqrt{\sum_{i=1}^{D-1}[A'^i(\sigma)]^2}\;R \to 0 \tag{4.29}$$

In this dual to unstable behaviour, the string size **vanishes**. That is, the string starts at the big bang with zero size.

In summary, the energy tends to $+\infty$, the pressure also to $+\infty$ and the size tends to zero in the **dual to unstable** behaviour.

In this regime strings behave as radiation (massless particles). Recall that the string size is proportional to the trace of the energy momentum tensor [see eq.(2.34)].

Let us finally consider FRW universes ($k > 0$) for $\tau \to \infty,\ R \to \infty$. There, the (integrated) energy-momentum tensor takes the form

$$\rho(T) = \Theta^{00}(T) \stackrel{\tau\to\infty}{=} \frac{2\sqrt{A}}{\alpha'(k+2)}\,, \tag{4.30}$$

$$\Theta^{i}_{j}(T) \stackrel{\tau\to\infty}{=} \frac{R}{\alpha'}\int_0^{2\pi}\frac{d\sigma}{2\pi}\left[f'^{+}_i\,f'^{-}_j + f'^{-}_i\,f'^{+}_j\right],$$

$$\Theta^{0}_{i}(T) \stackrel{\tau\to\infty}{=} 0\,. \tag{4.31}$$

where we used eqs.(2.29) and (4.20). The string energy here tends to a bounded constant. Since the $f'^{\pm}_i(\sigma \pm \tau)$ are periodic functions, their average on a period of time vanishes:

$$\int_0^{2\pi} d\tau d\sigma\; f'^{+}_i(\sigma+\tau)\,f'^{-}_i(\sigma-\tau) = \frac{1}{2}\int_{-2\pi}^{+2\pi} dx_-\int_{|x_-|}^{4\pi-|x_-|} dx_+\; f'^{+}_i(x_+)\,f'^{-}_i(x_-) = 0\,.$$

Hence, the pressure vanishes when averaged over a string oscillation. This is the **stable** string behaviour. Here strings behave as dust (cold matter) with $p = 0$ as equation of state.

The string size S follows eq.(4.25) and eq.(4.20),

$$S^2 \overset{\tau\to\infty}{=} \frac{2}{(k+2)^2}\left[1 + \sum_{i=1}^{D-1} f_i'^{+} f_i'^{-}\right] \tag{4.32}$$

The string size is thus bounded for $R \to \infty$. Moreover, averaging over a period of time, we find

$$\bar{S} = \frac{\sqrt{2}}{k+2} \, .$$

B. The perfect gas of strings

Our aim is to provide a string description of matter appropriate to the early universe.

Let us consider classical strings interacting with the cosmological spacetime background and neglect their mutual interactions. That is, we consider a perfect gas of strings under the cosmic gravitational field. The energy-momentum of such gas is just the sum over individual string solutions. For each string the results of section IV.A apply.

We assume arbitrary initial data for the strings. Therefore, summing over solutions is equivalent to *average* over the initial data $A^i(\sigma)$, $B^i(\sigma)$, $1 \leq i \leq D-1$.

For inflationary spacetimes the relevant quantity to average is the integral

$$\int_0^{2\pi} \frac{d\sigma}{2\pi}\, A'^i(\sigma)\, A'^j(\sigma)$$

that appears in the energy-momentum tensor eq.(4.23).

The average will vanish for $i \neq j$ and will be i-independent for $i = j$ because we treat independently the different $A^i(\sigma)$. Taking into account the constraint (4.16)

$$\sum_{i=1}^{D-1} \left[A'^i(\sigma)\right]^2 = 1 \, ,$$

finally yields,

$$< \int_0^{2\pi} \frac{d\sigma}{2\pi}\, A'^i(\sigma)\, A'^j(\sigma) >= \frac{\delta_{ij}}{D-1} \, . \tag{4.33}$$

Therefore the stress tensor (4.23) takes for unstable strings the fluid form for $R \to \infty$,

$$< \Theta^i_j(T) > \overset{R\to\infty}{=} -\frac{p}{D-1}\,\delta_{ij} \, ,$$

with

$$p \equiv -\frac{R}{\alpha'(D-1)} = -\frac{\rho}{D-1} \to -\infty \, , \tag{4.34}$$

where we used the expression for ρ in eq.(4.23). Recall that the string size also grows as R for $R \to \infty$ [eq.(4.26)].

This equation of state exactly saturates the strong energy condition in general relativity.

The unstable string behaviour corresponds to the critical case of the so-called coasting universe [19,32]. In other words, the perfect gas of strings provide a *concrete* matter realization of such cosmological model. Till now, no form of matter was known to describe coasting universes [19].

The quantity to average in FRW spacetimes for small R is

$$\int_0^{2\pi} \frac{d\sigma}{2\pi}\, B^i(\sigma)\, B^j(\sigma)$$

which appears in the energy-momentum tensor eq.(4.27).

Following the same argument as above for the average over $A^i(\sigma)$ and recalling that (4.18)

$$\sum_{i=1}^{D-1} \left[B^i(\sigma)\right]^2 = 1 ,$$

we find

$$< \int_0^{2\pi} \frac{d\sigma}{2\pi}\, B^i(\sigma)\, B^j(\sigma) >= \frac{\delta_{ij}}{D-1} \tag{4.35}$$

Hence the string energy-momentum tensor (4.27) for dual to unstable strings takes the fluid form for $R \to 0$,

$$< \Theta^i_j(T) > \overset{R\to 0}{=} -\frac{p}{D-1}\,\delta_{ij}$$

with

$$p \equiv \frac{1}{\alpha'(D-1)(k+1)\,R} \overset{R\to 0}{=} +\frac{\rho}{D-1} \to +\infty$$

where we used the expression for ρ in eq.(4.27). Recall that the string size vanishes as R, as R vanishes [eq.(4.29)].

Therefore in the dual to unstable case, strings behave as radiation (massless particles).

For large R in FRW spacetimes we must average independently over the functions f_i^+ and f_j^-, $1 \leq i,j \leq D-1$. We find then from eqs.(4.30)

$$< \Theta^i_j(T) > \overset{R\to\infty}{=} 0 .$$

That is, the equation of state

$$p = 0 .$$

The string energy and size are bounded in this regime. The strings behave for the stable regime as dust (cold matter). That is, they behave as massive particles.

In conclusion, an ideal gas of classical strings in cosmological universes exhibit three different thermodynamical behaviours, all of perfect fluid type:

- (1) For inflationary universes and $R \to \infty$ unstable strings: negative pressure gas with $p = -\frac{\rho}{D-1}$.
- (2) Dual behaviour in FRW universes and $R \to 0$: positive pressure gas similar to radiation, $p = +\frac{\rho}{D-1}$.
- (3) Stable strings in FRW universes and $R \to \infty$: positive pressure gas similar to cold matter, $p = 0$.

Tables 1 and 2 summarize the main string properties for any scale factor $R(T)$.

TABLE 1. String energy and pressure as obtained from exact string solutions for various expansion factors $R(T)$.

STRING PROPERTIES FOR ARBITRARY $R(T)$

D-Dimensional spacetimes: three asymptotic behaviours (u, d, s)	Energy	Pressure	Equation of State:
(i) unstable for $R \to \infty$	$E_u \overset{R\to\infty}{=} u\ R \to \infty$	$P_u = -\frac{E_u}{D-1} \to -\infty$	'stringy'
(ii) dual to (i) for $R \to 0$	$E_d \overset{R\to 0}{=} d/R \to \infty$	$P_d = +\frac{E_d}{D-1} \to \infty$	radiation
(iii) stable for $R \to \infty$	$E_s =$ constant	$P_s = 0$	dust (cold matter)

TABLE 2. The string energy density and pressure for a gas of strings can be summarized by the formulas below which become exact for $R \to 0$ and for $R \to \infty$.

STRING ENERGY DENSITY AND PRESSURE FOR ARBITRARY $R(T)$

	Energy density: $\rho \equiv E/R^{D-1}$	Pressure
Qualitatively correct formulas for all R and D	$\rho = \left(u\ R + \frac{d}{R} + s\right)\frac{1}{R^{D-1}}$	$p = \frac{1}{D-1}\left(\frac{d}{R} - u\ R\right)\frac{1}{R^{D-1}}$

Finally, notice that strings continuously evolve from one type of behaviour to the other two. This is explicitly seen from the string solutions in refs. [5] - [7] . For example the string described by $q_{-}(\sigma,\tau)$ for $\tau > 0$ shows unstable behaviour for $\tau \to 0$, dual behaviour for $\tau \to \tau_0 = 1.246450...$ and stable behaviour for $\tau \to \infty$.

TABLE 3. The **self-consistent** cosmological solution of the Einstein equations in General Relativity with the string gas as source.

STRING COSMOLOGY IN GENERAL RELATIVITY

Einstein equations (no dilaton field)	Expansion factor $R(T)$	Temperature $T(R)$
$T \to 0$	$\frac{D}{2}\left[\frac{2d}{(D-1)(D-2)}\right]^{\frac{1}{D}} T^{\frac{2}{D}}$	$\frac{dD}{S(D-1)}\, 1/R$
$T \to \infty$	$\left[\frac{(D-1)s}{2(D-2)}\right]^{\frac{1}{D-1}} T^{\frac{2}{D-1}}$	usual matter dominated behaviour

V. SELF-CONSISTENT STRING COSMOLOGY

In the previous section we investigated the propagation of test strings in cosmological space-times. Let us now investigate how the Einstein equations in General Relativity and the effective equations of string theory (beta functions) can be verified **self-consistently** with our string solutions as sources.

We shall assume a gas of classical strings neglecting interactions as string splitting and coalescing. We will look for cosmological solutions described by metrics of the type (4.1). It is natural to assume that the background will have the same symmetry as the sources. That is, we assume that the string gas is homogeneous, described by a density energy $\rho = \rho(T)$ and a pressure $p = p(T)$. In the effective equations of string theory we consider a space independent dilaton field. Antisymmetric tensor fields wil be ignored.

A. String Dominated Universes in General Relativity (no dilaton field)

The Einstein equations for the geometry (4.1) take the form

$$\frac{1}{2}\,(D-1)(D-2)\; H^2 = \rho \quad ,$$
$$(D-2)\dot{H} + p + \rho = 0 \quad . \tag{5.1}$$

where $H \equiv \frac{dR}{dT}/R$. We know p and ρ as functions of R in asymptotic cases. For large R, the unstable strings dominate [eq.(4.34)] and we have for inflationary spacetimes

$$\rho = u\, R^{2-D} \quad , \quad p = -\frac{\rho}{D-1} \quad \text{for } R \to \infty \tag{5.2}$$

For small R, the dual regime dominates with

$$\rho = d\, R^{-D} \quad , \quad p = +\frac{\rho}{D-1} \quad \text{for } R \to 0 \tag{5.3}$$

We also know that stable solutions may be present with a contribution $\simeq R^{1-D}$ to ρ and with zero pressure. For intermediate values of R the form of ρ is clearly more complicated but a formula of the type

$$\rho = \left(u_R\, R + \frac{d}{R} + s\right) \frac{1}{R^{D-1}} \tag{5.4}$$

where

$$\lim_{R\to\infty} u_R = \begin{cases} 0 & \text{FRW} \\ u_\infty \neq 0 & \text{Inflationary} \end{cases} \tag{5.5}$$

This equation of state is qualitatively correct for all R and becomes exact for $R \to 0$ and $R \to \infty$. The parameters u_R, d and s are positive constants and the u_R varies smoothly with R.

The pressure associated to the energy density (5.4) takes then the form

$$p = \frac{1}{D-1}\left(\frac{d}{R} - u_R\, R\right) \frac{1}{R^{D-1}} \tag{5.6}$$

Inserting eq.(5.4) into the Einstein-Friedmann equations [eq.(5.1)] we find

$$\frac{1}{2}\,(D-1)(D-2)\,\left(\frac{dR}{dT}\right)^2 = \left(u_R\, R + \frac{d}{R} + s\right) \frac{1}{R^{D-3}} \tag{5.7}$$

We see that R is a monotonic function of the cosmic time T. Eq.(5.7) yields

$$T = \sqrt{\frac{(D-1)(D-2)}{2}} \int_0^R dR\, \frac{R^{D/2-1}}{\sqrt{u_R\, R^2 + d + s\, R}} \tag{5.8}$$

where we set $R(0) = 0$.

It is easy to derive the behavior of R for $T \to 0$ and for $T \to \infty$.

For $T \to 0$, $R \to 0$, the term d/R dominates in eq.(5.7) and

$$R(T) \overset{T\to 0}{\simeq} \frac{D}{2}\left[\frac{2d}{(D-1)(D-2)}\right]^{\frac{1}{D}} T^{\frac{2}{D}} \tag{5.9}$$

For $T \to \infty$, $R \to \infty$ and the term $u_R\, R$ dominates in eq.(5.7). Hence,

$$R(T) \stackrel{T\to\infty}{\simeq} \left[\frac{(D-2)u_\infty}{2(D-1)}\right]^{\frac{1}{D-2}} T^{\frac{2}{D-2}} \tag{5.10}$$

[u_R tends to a constant u_∞ for $R \to \infty$]. This expansion is faster than (cold) matter dominated universes where $R \simeq T^{\frac{2}{D-1}}$. For example, for $D = 4$, R grows linearly with T whereas for matter dominated universes $R \simeq T^{2/3}$. However, eq.(5.10) **is not** a self-consistent solution. Assuming that the term $u_R R$ dominates for large R we find a scale factor $R(T) \simeq T^{\frac{2}{D-2}} \simeq \eta^{\frac{2}{D-4}}$ for $D \neq 4$ and $R(T) \simeq T \simeq e^\eta$ at $D = 4$. This **is not an inflationary universe** but a FRW universe. The term $u_R R$ is absent for large R in FRW universes as explained before. Therefore, we must instead use for large R

$$\rho = \left(\frac{d}{R} + s\right) \frac{1}{R^{D-1}} \tag{5.11}$$

Now, for $T \to \infty, R \to \infty$ and we find a matter dominated regime:

$$R(T) \stackrel{T\to\infty}{\simeq} \left[\frac{(D-1)s}{2(D-2)}\right]^{\frac{1}{D-1}} T^{\frac{2}{D-1}} \tag{5.12}$$

For intermediate values of T , $R(T)$ is a continuous and monotonically increasing function of T.

In summary, the universe starts at $T = 0$ with a singularity of the type dominated by radiation. (The string behaviour for $R \to 0$ is like usual radiation). Then, the universe expands monotonically, growing for large T as $R \simeq T^{\frac{2}{D-1}}$. In particular, this gives $R \simeq T^{2/3}$ for $D = 4$.

It must be noticed that the qualitative form of the solution $R(T)$ does not depend on the particular positive values of u_R, d and s.

We want to stress that we achieve a **self-consistent** solution of the Einstein equations with string sources since the behaviour of the string pressure and density given by eqs.(5.4)-(5.6) precisely holds in universes with power like $R(T)$.

In ref. [3] similar results were derived using arguments based on the splitting of long strings.

B. Thermodynamics of strings in cosmological spacetimes

Let us consider a comoving volume R^{D-1} filled by a gas of strings. The entropy change for this system is given by:

$$\mathcal{T} dS = d(\rho\, R^{D-1}) \;+\; p\, d(R^{D-1}) \tag{5.13}$$

The continuity equation (4.9) and (5.13) implies that dS/dT vanishes. That is, the entropy per comoving volume stays constant in time. Using now the thermodynamic relation [25]

$$\frac{dp}{d\mathcal{T}} = \frac{p+\rho}{\mathcal{T}} \tag{5.14}$$

it follows [26] that

$$S = \frac{R^{D-1}}{\mathcal{T}}(p+\rho) + \text{constant} \tag{5.15}$$

Eq.(5.15) together with eqs.(5.4) and (5.6) yields the temperature as a function of the expansion factor R. That is,

$$\mathcal{T} = \frac{1}{S}\left\{s + \frac{1}{D-1}\left[\frac{D\,d}{R} + (D-2)\,u_R\,R\right]\right\} \tag{5.16}$$

where S stands for the (constant) value of the entropy.

Eq.(5.16) shows that for small R, $\mathcal{T}$ scales as $1/R$ whereas for large R it scales as R. The small R behaviour of $\mathcal{T}$ is the usual exhibited by radiation.

For large R, in FRW universes $u_R \to 0$ and the constant term in s dominates. We just find a cold matter behaviour for large R.

For large R in inflationary universes, $u_R \to u_\infty$ and eq.(5.16) would indicate a temperature that **grows** proportionally to R. However, as stressed in ref. [3], the decay of long strings (through splitting) makes u_R exponentially decreasing with R.

VI. EFFECTIVE STRING EQUATIONS WITH THE STRING SOURCES INCLUDED

Let us consider now the cosmological equations obtained from the low energy string effective action including the string matter as a classical source. In D spacetime dimensions, this action can be written as

$$\begin{aligned} S &= S_1 + S_2 \\ S_1 &= \frac{1}{2}\int d^Dx\,\sqrt{-G}\,e^{-\Phi}\,\left[\,R + G_{AB}\,\partial^A\Phi\,\partial^B\Phi\, + 2\,U(G,\Phi) - c\,\right] \\ S_2 &= -\frac{1}{4\pi\alpha'}\sum_{strings}\int d\sigma d\tau\; G_{AB}(X)\,\partial_\mu X^A\,\partial^\mu X^B \qquad , \end{aligned} \tag{6.1}$$

Here $A, B = 0, \ldots, D-1$. This action is written in the so called 'Brans-Dicke frame' (BD) or 'string frame', in which matter couples to the metric tensor in the standard way. The BD frame metric coincides with the sigma model metric to which test strings are coupled.

Eq.(6.1) includes the dilaton field (Φ) with a potential $U(G,\Phi)$ depending on the dilaton and graviton backgrounds; c stands for the central charge deficit or cosmological constant term. The antisymmetric tensor field was not included, in fact it is irrelevant for the results obtained here. Extremizing the action (6.1) with respect to G_{AB} and Φ yields the equations of motion

$$\begin{aligned} R_{AB} + \nabla_{AB}\Phi + 2\,\frac{\partial U}{\partial G_{AB}} - \frac{G_{AB}}{2}\left[R + 2\,\nabla^2\Phi - (\nabla\Phi)^2 - c + 2\,U\right] &= e^{\phi}\,T_{AB} \\ R + 2\,\nabla^2\Phi - (\nabla\Phi)^2 - c + 2U - \frac{\partial U}{\partial\Phi} &= 0\ , \end{aligned} \tag{6.2}$$

which can be more simply combined as

$$R_{AB} + \nabla_{AB}\Phi + 2\,\frac{\partial U}{\partial G_{AB}} - G_{AB}\,\frac{\partial U}{\partial\Phi} = e^{\Phi}\,T_{AB}$$

$$R + 2\,\nabla^2\Phi - (\nabla\Phi)^2 - c + 2\,U - \frac{\partial U}{\partial \Phi} = 0 \tag{6.3}$$

Here T_{AB} stands for the energy momentum tensor of the strings as defined by eq.(2.27). It is also convenient to write these equations as

$$R_{AB} - \frac{G_{AB}}{2}\,R = T_{AB} + \tau_{AB} \tag{6.4}$$

where τ_{AB} is the dilaton energy momentum tensor :

$$\tau_{AB} = -\nabla_{AB}\Phi + \frac{G_{AB}}{2}\left[2\,\frac{\partial U}{\partial \Phi} - R\right]$$

The Bianchi identity

$$\nabla^A\left(R_{AB} - \frac{G_{AB}}{2}\,R\right) = 0$$

yields, as it must be, the conservation equation,

$$\nabla^A\,(T_{AB} + \tau_{AB}) = 0 \tag{6.5}$$

It must be noticed that eqs.(6.3) do not reduce to the Einstein equations of General Relativity even when $\Phi = U = 0$. Eqs. (6.3) yields in that case the Einstein equations *plus* the condition $R = 0$.

A. Effective String Equations in Cosmological Universes

For the homogeneous isotropic spacetime geometries described by eq.(4.1) we have

$$\begin{aligned} R_0^0 &= -(D-1)(\dot H + H^2) \\ R_i^k &= -\delta_i^k\,[\dot H + (D-1)\,H^2] \\ R &= -(D-1)(2\,\dot H + D\,H^2). \end{aligned} \tag{6.6}$$

where $H \equiv \frac{1}{R}\frac{dR}{dT}$.

The equations of motion (6.3) read

$$\begin{aligned} \ddot\Phi - (D-1)(\dot H + H^2) - \frac{\partial U}{\partial \Phi} &= e^{\Phi}\,\rho \\ \dot H + (D-1)\,H^2 - H\,\dot\Phi + \frac{\partial U}{\partial \Phi} + \frac{R}{D-1}\frac{\partial U}{\partial R} &= e^{\Phi}\,p \\ 2\,\ddot\Phi + 2(D-1)\,H\,\dot\Phi - \dot\Phi^2 - (D-1)(2\,\dot H + D\,H^2) - 2\,\frac{\partial U}{\partial \Phi} - c + 2\,U &= 0 \end{aligned} \tag{6.7}$$

where dot $\cdot$ stands for $\frac{d}{dT}$, and

$$\rho = T_0^0 \quad , \quad -\delta_i^k\,p = T_i^k\,. \tag{6.8}$$

The conservation equation takes the form of eq.(4.9)

$$\dot{\rho} + (D-1)\,H\,(p+\rho) = 0 \quad . \tag{6.9}$$

By defining,

$$\begin{array}{l} \Psi \equiv \Phi - \log\sqrt{-G} = \Phi - (D-1)\,\log R \\ \bar{\rho} = e^{\Phi}\,\rho \quad , \quad \bar{p} = e^{\Phi}\,p \,, \end{array} \tag{6.10}$$

equations (6.7) can be expressed in a more compact form as

$$\begin{array}{c} \ddot{\Psi} - (D-1)\,H^2 - \left.\dfrac{\partial U}{\partial \Psi}\right|_R = \bar{\rho} \\ \dot{H} - H\,\dot{\Psi} + \dfrac{R}{D-1}\left.\dfrac{\partial U}{\partial R}\right|_{\Psi} = \bar{p} \\ \dot{\Psi}^2 - (D-1)\,H^2 - 2\,\bar{\rho} - 2U + c = 0 \,, \end{array} \tag{6.11}$$

The conservation equation reads

$$\dot{\bar{\rho}} - \dot{\Psi}\,\bar{\rho} + (D-1)\,H\,\bar{p} = 0 \tag{6.12}$$

As is known, under the duality transformation $R \longrightarrow R^{-1}$, the dilaton transforms as $\Phi \longrightarrow \Phi + (D-1)\,\log R$. The shifted dilaton Ψ defined by eq.(6.10) is invariant under duality.
The transformation

$$R' \equiv R^{-1} \quad , \tag{6.13}$$

implies

$$\Psi' = \Psi \quad , \; H' = -H \quad , \; \bar{p}' = -p \quad , \; \bar{\rho}' = \bar{\rho} \tag{6.14}$$

provided $u = d$, that is, a duality invariant string source. This is the duality invariance transformation of eqs.(6.11).

Solutions to the effective string equations have been extensively treated in the literature [28] and they are not our main purpose. For the sake of completeness, we briefly analyze the limiting behaviour of these equations for $R \to \infty$ and $R \to 0$.

It is difficult to make a complete analysis of the effective string equations (6.11) since the knowledge about the potential U is rather incomplete. For weak coupling (e^{Φ} small) the supersymmetry breaking produces an effective potential that decreases very fast (as the exponential of an exponential of Φ) for $\Phi \to -\infty$.

Let us analyze the asymptotic behavior of eqs.(6.11) for $R \to \infty$ and $R \to 0$ assuming that the potential U can be ignored. It is easy to see that a power behaviour Ansatz both for R and for e^{Ψ} as functions of T is consistent with these equations. It turns out that the string sources do not contribute to the leading behaviour here, and we find for $R \to 0$

$$R_{\mp} = C_1\,T^{\pm 1/\sqrt{D-1}} \to 0 \quad ,$$

$$e^{\Psi_\mp} = \quad C_2\, T^{-1} \;\to \begin{cases} \infty \\ 0 \end{cases} \tag{6.15}$$

Where C_1 and C_2 are constants. Here the branches $(-)$ and $(+)$ correspond to $T \to 0$ and to $T \to \infty$ respectively. In both regimes $R_\mp \to 0$ and $e^{\Phi_\mp} \to 0$.

The potential $U(\Phi)$ is hence negligible in these regimes. In terms of the conformal time η , the behaviours (6.15) result

$$\begin{aligned} R_\mp = &\quad C_1'\, \eta^{\pm\frac{1}{\sqrt{D-1}\mp 1}} \to 0 \\ e^{\Psi_\mp} &= C_2'\, \eta^{-\frac{\sqrt{D-1}}{\sqrt{D-1}\mp 1}} \to \begin{cases} \infty \\ 0 \end{cases} \end{aligned} \tag{6.16}$$

Where C_1' and C_2' are constants. The branch $(-)$ would describe an expanding non-inflationary behaviour near the initial singularity $T = 0$, while the branch $(+)$ describes a 'big crunch' situation and is rather unphysical.

Similarly, for $R \to \infty$ and $e^\Phi \to \infty$, we find

$$\begin{aligned} R_\mp &= D_1\, T^{\mp 1/\sqrt{D-1}} \;\to \infty \quad , \\ e^{\Psi_\mp} &= \quad D_2\, T^{-1} \;\to \begin{cases} \infty \\ 0 \end{cases} \end{aligned} \tag{6.17}$$

Where D_1 and D_2 are constants. Here again, the branches $(-)$ and $(+)$ correspond to $T \to 0$ and to $T \to \infty$ respectively, but now in both regimes $R_\mp \to \infty$ and $e^{\Phi_\mp} \to \infty$. (In this limit, one is not guaranteed that U can be consistently neglected). In terms of the conformal time, eqs.(6.17) read

$$\begin{aligned} R_\mp = &\quad D_1'\, \eta^{\mp\frac{1}{\sqrt{D-1}\pm 1}} \to \infty \\ e^{\Psi_\mp} &= D_2'\, \eta^{-\frac{\sqrt{D-1}}{\sqrt{D-1}\pm 1}} \to \begin{cases} \infty \\ 0 \end{cases} \end{aligned} \tag{6.18}$$

The branch $(+)$ describes a noninflationary expanding behaviour for $T \to \infty$ faster than the standard matter dominated expansion, while the branch $(-)$ describes a super-inflationary behaviour $\eta^{-\alpha}$, since $0 < \alpha < 1$, for all D.

The behaviours (6.15) for $R_\mp \to 0$ and (6.17) for $R_\mp \to \infty$ are related by duality $R \leftrightarrow 1/R$.

B. String driven inflation?

Let us consider now the question of whether de Sitter spacetime may be a self-consistent solution of the effective string equations (6.7) with the string sources included. The strings in cosmological universes like de Sitter spacetime have the equation of state (5.4)-(5.6). Since $e^\Psi = e^\Phi\, R^{1-D}$:

$$\bar{\rho} = e^\Psi \left(u\, R + \frac{d}{R} + s \right) \tag{6.19}$$

$$\bar{p} = \frac{e^\Psi}{D-1} \left(\frac{d}{R} - u\, R \right) \tag{6.20}$$

In the absence of dilaton potential and cosmological constant term, the string sources do not generate de Sitter spacetime as discussed in sec. V.A. We see that for $U = c = 0$, and $R = e^{HT}$, eqs.(6.11) yields to a contradiction (unless $D = 0$) for the value of Ψ , required to be $-HT +$ constant.

A self-consistent solution describing asymptotically de Sitter spacetime self-sustained by the string equation of state (6.19)-(6.20) is given by

$$\begin{aligned} R &= e^{HT} \quad , \quad H = \text{constant} > 0 \quad , \\ 2U - c &= D\, H^2 = \text{constant} \\ \Psi_\pm &= \mp HT \pm i\pi + \log \frac{(D-1)\, H^2}{\rho_\pm} \\ \rho_+ &\equiv u \quad , \quad \rho_- \equiv d \end{aligned} \tag{6.21}$$

The branch Ψ_+ describes the solution for $R \to \infty$ ($T \to +\infty$), while the branch Ψ_- corresponds to $R \to 0$ ($T \to -\infty$). De Sitter spacetime with lorentzian signature self-sustained by the strings necessarily requires a constant imaginary piece $\pm i\pi$ in the dilaton field. This makes $e^\Psi < 0$ telling us that the gravitational constant $G \sim e^\Psi < 0$ here describes antigravity.

Is interesting to notice that in the euclidean signature case, i. e. $(+++\ldots++)$, the Ansatz $\dot{H} = 0$, $2U - c =$constant, yields a constant curvature geometry with a real dilaton, but which is of Anti-de Sitter type. This solution is obtained from eqs.(6.20)-(6.21) through the transformation

$$\hat{X}^0 = iT \quad , \quad \hat{H} = -iH \quad , \quad X^i = X^i \quad , \quad \Psi = \Psi \tag{6.22}$$

which maps the Lorentzian de Sitter metric into the positive definite one

$$d\hat{s}^2 = (d\hat{X}^0)^2 + e^{\hat{H}\hat{X}^0}\, (d\vec{X})^2. \tag{6.23}$$

The equations of motion (6.11) within the constant curvature Ansatz ($\dot{\hat{H}} = \ddot{\Psi} = 0$) are mapped onto the equations

$$\begin{aligned} (D-1)\, \hat{H}^2 - \left.\frac{\partial U}{\partial \Psi}\right|_R &= \bar{\rho} \\ \hat{H}\, \frac{d\Psi}{d\hat{X}^0} + \left.\frac{R}{D-1}\frac{\partial U}{\partial R}\right|_\Psi &= \bar{p} \\ -(\frac{d\Psi}{d\hat{X}^0})^2 + (D-1)\, \hat{H}^2 - 2\, \bar{\rho} - 2U + c &= 0 \, , \end{aligned} \tag{6.24}$$

with the solution

$$\begin{aligned} R &= e^{\hat{H}\hat{X}^0} \quad , \quad \hat{H} = \text{constant} > 0 \quad , \\ c - 2U &= D\, \hat{H}^2 = \text{constant} \\ \Psi_\pm &= \mp \hat{H}\hat{X}^0 + \log \frac{(D-1)\, \hat{H}^2}{\rho_\pm} \\ \rho_+ &\equiv u \quad , \quad \rho_- \equiv d \end{aligned} \tag{6.25}$$

Both solutions (6.25) and (6.21) are mapped one into another through the transformation (6.22).

It could be recalled that in the context of (point particle) field theory, de Sitter spacetime (as well as anti-de Sitter) emerges as an exact selfconsistent solution of the semiclassical Einstein equations with the back reaction included [34] - [35]. (Semiclassical in this context, means that matter fields including the graviton are quantized to the one-loop level and coupled to the (c-number) gravity background through the expectation value of the energy-momentum tensor T_A^B . This expectation value is given by the trace anomaly: $< T_A^A >= \bar{\gamma}\ R^2$). On the other hand, the α' expansion of the effective string action admits anti-de Sitter spacetime (but not de Sitter) as a solution when the quadratic curvature corrections (in terms of the Gauss-Bonnet term) to the Einstein action are included [36]. It appears that the corrections to the anti-de Sitter constant curvature are qualitatively similar in the both cases, with α' playing the rôle of the trace anomaly parameter $\bar{\gamma}$ [35].

The fact that de Sitter inflation with true gravity $G \sim e^{\Psi} > 0$ does not emerge as a solution of the effective string equations does not mean that string theory excludes inflation. What means is that the effective string equations are not enough to get inflation. The effective string action is a low energy field theory approximation to string theory containing only the *massless* string modes (*massless* background fields).

The vacuum energy scales to start inflation (physical or true vacuum) are typically of the order of the Planck mass [26] - [27] where the effective string action approximation breaks down. One must consider the massive string modes (which are absent from the effective string action) in order to properly get the cosmological condensate yielding de Sitter inflation. We do not have at present the solution of such problem.

TABLE 4. Asymptotic solution of the string effective equations (including the dilaton).

EFFECTIVE STRING EQUATIONS SOLUTIONS IN COSMOLOGY

Effective String equations	$R(T) \to 0$ behaviour	$R(T) \to \infty$ behaviour
$T \to 0$	$\simeq T^{+1/\sqrt{D-1}}$	$\simeq T^{-1/\sqrt{D-1}}$
$T \to \infty$	$\simeq T^{-1/\sqrt{D-1}}$	$\simeq T^{+1/\sqrt{D-1}}$

VII. MULTI-STRINGS AND SOLITON METHODS IN DE SITTER UNIVERSE

Among the cosmological backgrounds, de Sitter spacetime occupies a special place. This is, in one hand relevant for inflation and on the other hand string propagation turns to be specially interesting there [2] - [8]. String unstability, in the sense that the string proper length grows indefinitely is particularly present in de Sitter. The string dynamics in de Sitter universe is described by a generalized sinh-Gordon model with a potential unbounded from below [4]. The sinh-Gordon function $\alpha(\sigma,\tau)$ having a clear physical meaning : $H^{-1}e^{\alpha(\sigma,\tau)/2}$ determines the string proper length. Moreover the classical string equations of motion (plus the string constraints) turn to be integrable in de Sitter universe [4,5]. More precisely, they are equivalent to a non-linear sigma model on the grassmannian $SO(D,1)/O(D)$ with periodic boundary conditions (for closed strings). This sigma model has an associated linear system [37] and using it, one can show the presence of an infinite number of conserved quantities [38]. In addition, the string constraints imply a zero energy-momentum tensor and these constraints are compatible with the integrability.

The so-called dressing method [37] in soliton theory allows to construct solutions of non-linear classically integrable models using the associated linear system. In ref. [6] we systematically construct string solutions in three dimensional de Sitter spacetime. We start from a given exactly known solution of the string equations of motion and constraints in de Sitter [5] and then we "dress" it. The string solutions reported there indeed apply to cosmic strings in de Sitter spacetime as well.

The invariant interval in D-dimensional de Sitter space-time is given by

$$ds^2 = dT^2 - \exp[2HT] \sum_{i=1}^{D-1}(dX^i)^2. \tag{7.1}$$

Here T is the so called cosmic time. In terms of the conformal time η,

$$\eta \equiv -\frac{\exp[-HT]}{H} \quad , -\infty < \eta \leq 0 ,$$

the line element becomes

$$ds^2 = \frac{1}{H^2\eta^2}[d\eta^2 - \sum_{i=1}^{D-1}(dX^i)^2] .$$

The de Sitter spacetime can be considered as a D-dimensional hyperboloid embedded in a D+1 dimensional flat Minkowski spacetime with coordinates $(q^0, ..., q^D)$:

$$ds^2 = \frac{1}{H^2}[-(dq^0)^2 + \sum_{i=1}^{D}(dq^i)^2] \tag{7.2}$$

where

$$\begin{aligned} q^0 &= \sinh HT + \frac{H^2}{2}\exp[HT] \sum_{i=1}^{D-1}(X^i)^2 , \\ q^1 &= \cosh HT - \frac{H^2}{2}\exp[HT] \sum_{i=1}^{D-1}(X^i)^2 , \\ q^{i+1} &= H\exp[HT]\, X^i \quad , \quad 1 \leq i \leq D-1,\ -\infty < T,\ X^i < +\infty. \end{aligned} \tag{7.3}$$

The complete de Sitter manifold is the hyperboloid

$$-(q^0)^2 + \sum_{i=1}^{D}(q^i)^2 = 1.$$

The coordinates (T, X^i) and (η, X^i) cover only the half of the de Sitter manifold $q^0 + q^1 > 0$.

We will consider a string propagating in this D-dimensional space-time.The string equations of motion (2.7) in the metric (7.2) take the form:

$$\partial_{+-} q + (\partial_+ q . \partial_- q)\, q = 0 \quad \text{with} \quad q.q = 1, \tag{7.4}$$

where . stands for the Lorentzian scalar product $a.b \equiv -a_0 b_0 + \sum_{i=1}^{D} a_i b_i$, $x_\pm \equiv \frac{1}{2}(\tau \pm \sigma)$ and $\partial_\pm q = \frac{\partial q}{\partial x_\pm}$. The string constraints (2.8) become for de Sitter universe

$$T_{\pm\pm} = \frac{\partial q}{\partial x_\pm} . \frac{\partial q}{\partial x_\pm} = 0. \tag{7.5}$$

Eqs.(7.4) describe a non compact O(D,1) non-linear sigma model in two dimensions. In addition, the (two dimensional) energy-momentum tensor is required to vanish by the constraints eqs.(7.5) . This system of non-linear partial differential equations can be reduced by choosing an appropriate basis for the string coordinates in the $(D+1)$-dimensional Minkowski space time $(q^0, ..., q^D)$ to a noncompact Toda model [4].

These equations can be rewritten in the form of a chiral field model on the Grassmanian $G_D = SO(D,1)/O(D)$. Indeed, any element $\mathbf{g} \in G_D$ can be parametrized with a real vector $q\rangle$ of the unit pseudolength

$$\mathbf{g} = 1 - 2|q\rangle\langle q|J, \quad \langle q|J|q\rangle = 1. \tag{7.6}$$

In terms of $\mathbf{g}$, the string equations (7.4) have the following form

$$2\,\mathbf{g}_{\xi\eta} - \mathbf{g}_\xi\, \mathbf{g}\, \mathbf{g}_\eta - \mathbf{g}_\eta\, \mathbf{g}\, \mathbf{g}_\xi = 0\ , \tag{7.7}$$

and the conformal constraints (7.5) become

$$\mathrm{tr}\, \mathbf{g}_\xi^2 = 0, \quad \mathrm{tr}\, \mathbf{g}_\eta^2 = 0\ . \tag{7.8}$$

The fact that $\mathbf{g} \in G_D$ implies that $\mathbf{g}$ is a real matrix with the following properties:

$$\mathbf{g} = J\mathbf{g}^t J, \quad \mathbf{g}^2 = I, \quad tr\, \mathbf{g} = D-1\ , \quad \mathbf{g} \in SL(D+1, R). \tag{7.9}$$

These conditions are equivalent to the existence of the representation (7.6). Equation (7.7) is the compatibility condition for the following overdetermined linear system:

$$\Psi_\xi = \frac{U}{1-\lambda}\Psi, \quad \Psi_\eta = \frac{V}{1+\lambda}\Psi, \tag{7.10}$$

where

$$U = \mathbf{g}_\xi\, \mathbf{g}, \quad V = \mathbf{g}_\eta\, \mathbf{g}\ \ . \tag{7.11}$$

Or, in terms of the vector $q\rangle$

$$U = 2\, q_\xi\rangle\langle q\, J - 2\, q\rangle\langle q_\xi\, J,$$

$$V = 2\, q_\eta\rangle\langle q\, J - 2\, q\rangle\langle q_\eta\, J.$$

Eq. (7.6) can be easily inverted yielding q in terms of the matrix g:

$$q_0 = \sqrt{\frac{g_{00}-1}{2}} \quad , \quad q_i = \sqrt{\frac{1-g_{ii}}{2}} \quad 1 \leq i \leq D \text{ (no sum over } i) \tag{7.12}$$

The use of overdetermined linear systems to solve non-linear partial differential equations associated to them goes back to refs. [39]. (See refs. [40] - [41] for further references).

In order to fix the freedom in the definition of Ψ we shall identify

$$\Psi(\lambda = 0) = \mathbf{g}. \tag{7.13}$$

This condition is compatible with the above equations since the matrix function Ψ at the point $\lambda = 0$ satisfies the same equations as $\mathbf{g}$. Thus the problem of constructing exact solutions of the string equations is reduced to finding compatible solutions of the linear equations (7.10) such that $\mathbf{g} = \Psi(\lambda = 0)$ satisfies the constraints eqs.(7.8) and (7.9).

We concentrate below on the linear system (7.10) since this is the main tool to derive new string solutions in de Sitter spacetime.

In ref. [6] the dressing method was applied as follows. We started from the exact ring-shaped string solution $q_{(0)}$ [5] and we find the explicit solution $\Psi^{(0)}(\lambda)$ of the associated linear system, where λ stands for the spectral parameter. Then, we propose a new solution $\Psi(\lambda)$ that differs from $\Psi^{(0)}(\lambda)$ by a matrix rational in λ. Notice that $\Psi(\lambda = 0)$ provides in general a new string solution.

We then show that this rational matrix must have at least **four** poles, $\lambda_0, 1/\lambda_0, \lambda_0^*, 1/\lambda_0^*$, as a consequence of the symmetries of the problem. The residues of these poles are shown to be one-dimensional projectors. We then prove that these projectors are formed by vectors which can all be expressed in terms of an arbitrary complex constant vector $|x_0\rangle$ and the complex parameter λ_0. This result holds for arbitrary starting solutions $q_{(0)}$.

Since we consider closed strings, we impose a 2π-periodicity on the string variable σ . This restricts λ_0 to take discrete values that we succeed to express in terms of Pythagorean numbers. In summary, our solutions depend on two arbitrary complex numbers contained in $|x_0\rangle$ and two integers n and m . The counting of degrees of freedom is analogous to 2+1 Minkowski spacetime except that left and right modes are here mixed up in a non-linear and precise way.

The vector $|x_0\rangle$ somehow indicates the polarization of the string. The integers (n, m) determine the string winding. They fix the way in which the string winds around the origin in the spatial dimensions (here S^2). Our starting solution $q_{(0)}(\sigma, \tau)$ is a stable string winded $n^2 + m^2$ times around the origin in de Sitter space.

The matrix multiplications involved in the computation of the final solution were done with the help of the computer program of symbolic calculation "Mathematica". The resulting solution $q(\sigma, \tau) = (q^0, q^1, q^2, q^3)$ is a complicated combination of trigonometric functions of σ and hyperbolic functions of τ. That is, these string solitonic solutions do not oscillate in time. This is a typical feature of string unstability [5] - [11] - [9]. The new feature here is that strings (even stable solutions) do not oscillate neither for $\tau \to 0$, nor for $\tau \to \pm\infty$.

We plot in figs. 1-7 the solutions for significative values of $|x_0\rangle$ and (m, n) in terms of the comoving coordinates (T, X^1, X^2)

$$T = \frac{1}{H}\log(q^0 + q^1) \quad , \quad X^1 = \frac{1}{H}\frac{q^2}{q^0 + q^1} \quad , \quad X^2 = \frac{1}{H}\frac{q^3}{q^0 + q^1} \tag{7.14}$$

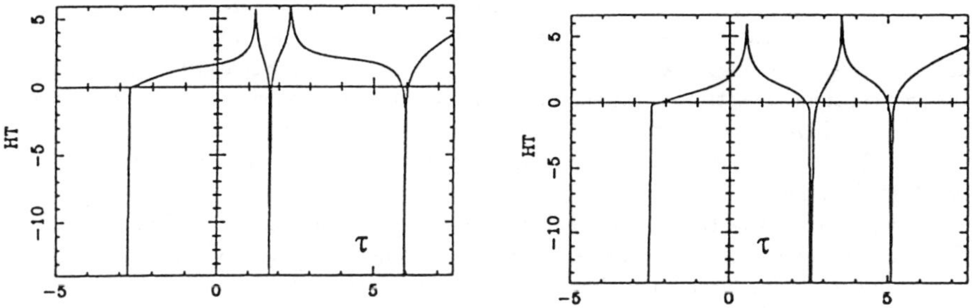

Figure 1: Plot of the function $HT(\tau)$, for two values of σ, for $n = 4, |x^0>= (1+i, .6+.4i, .3+.5i, .77+.79i)$. The function $\tau(T)$ is multivalued, revealing the presence of five strings.

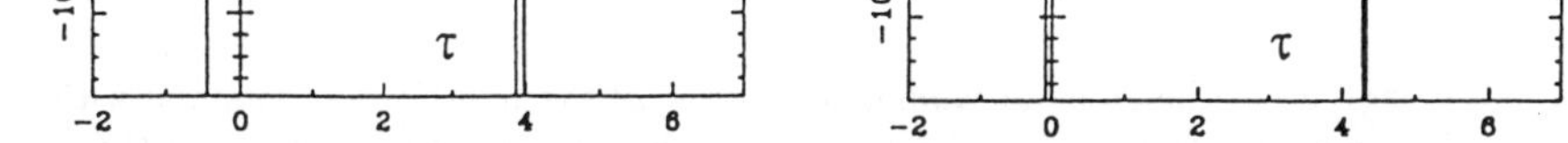

Figure 2: Same as fig.1, for $n = 4, |x^0>= (1, -1, i, 1)$. Because of a degeneracy, there are now only three strings.

The first feature to point out is that our solitonic solutions describe **multiple** (here five or three) strings, as it can be seen from the fact that for a given time T we find several different values for τ. That is, τ is a **multivalued** function of T for any fixed σ (fig.1-2). Each branch of τ as a function of T corresponds to a different string. This is a entirely new feature for strings in curved spacetime, with no analogue in flat spacetime where the time coordinate can always be chosen proportional to τ. In flat spacetime, multiple string solutions are described by multiple world-sheets. Here, we have a **single** world-sheet describing several independent and simultaneous strings as a consequence of the coupling with the spacetime geometry. Notice that we consider *free* strings. (Interactions among the strings as splitting or merging are not considered). Five is the generic number of strings in our dressed solutions. The value five can be related to the fact that we are dressing a one-string solution ($q_{(0)}$) with *four* poles. Each pole adds here an unstable string.

In order to describe the real physical evolution, we eliminated numerically $\tau = \tau(\sigma, T)$ from the solution and expressed the spatial comoving coordinates X^1 and X^2 in terms of T and σ.

We plot $\tau(\sigma, T)$ as a function of σ for different fixed values of T in fig.3-4. It is a sinusoidal-type function. Besides the customary closed string period 2π, another period appears which varies on τ. For small τ , $\tau = \tau(\sigma, T)$ has a convoluted shape while for larger τ (here $\tau \leq 5$), it becomes a regular sinusoid. These behaviours reflect very clearly in the evolution of the spatial coordinates and shape of the string.

The evolution of the five (and three) strings simultaneously described by our solution as a function of T, for positive T is shown in figs. 5-7. One string is stable (the 5th one). The other four are unstable. For the stable string, (X^1, X^2) contracts in time precisely as e^{-HT}, thus keeping the proper amplitude $(e^{HT}X^1, e^{HT}X^2)$ and proper size constant. For this stable string $(X^1, X^2) \leq \frac{1}{H}$. ($1/H$ = the horizon radius). For the other (unstable) strings, (X^1, X^2) become very fast constant in time, the proper size expanding as the universe itself like e^{HT} . For these strings $(X^1, X^2) \geq \frac{1}{H}$. These exact solutions display remarkably the asymptotic string behaviour found in refs. [4,11].

In terms of the sinh-Gordon description, this means that for the strings outside the horizon the sinh-Gordon function $\alpha(\sigma, \tau)$ is the same as the cosmic time T up to a function of σ. More precisely,

$$\alpha(\sigma,\tau) \overset{T>>\frac{1}{H}}{=} 2H\,T(\sigma,\tau) + \log\left\{2H^2\left[(A^1(\sigma)')^2 + (A^2(\sigma)')^2\right]\right\} + O(e^{-2HT}). \tag{7.15}$$

Here $A^1(\sigma)$ and $A^2(\sigma)$ are the X^1 and X^2 coordinates outside the horizon. For $T \to +\infty$ these strings are at the absolute *minimum* $\alpha = +\infty$ of the sinh-Gordon potential with infinite size. The string inside the horizon (stable string) corresponds to the *maximum* of the potential, $\alpha = 0$. $\alpha = 0$ is the only value in which the string can stay without being pushed down by the potential to $\alpha = \pm\infty$ and this also explains why only one stable string appears (is not possible to put more than one string at the maximum of the potential without falling down). These features are *generically* exhibited by our one-soliton multistring solutions, independently of the particular initial state of the string (fixed by $|x^0>$ and (n, m)). For particular values of $|x^0>$, the solution describes three strings, with symmetric shapes from $T = 0$, for instance like a rosette or a circle with festoons (fig. 5-7).

The string solutions presented here trivially embedd on D-dimensional de Sitter spacetime ($D \geq 3$). It must be noticed that they exhibit the essential physics of strings in D-dimensional

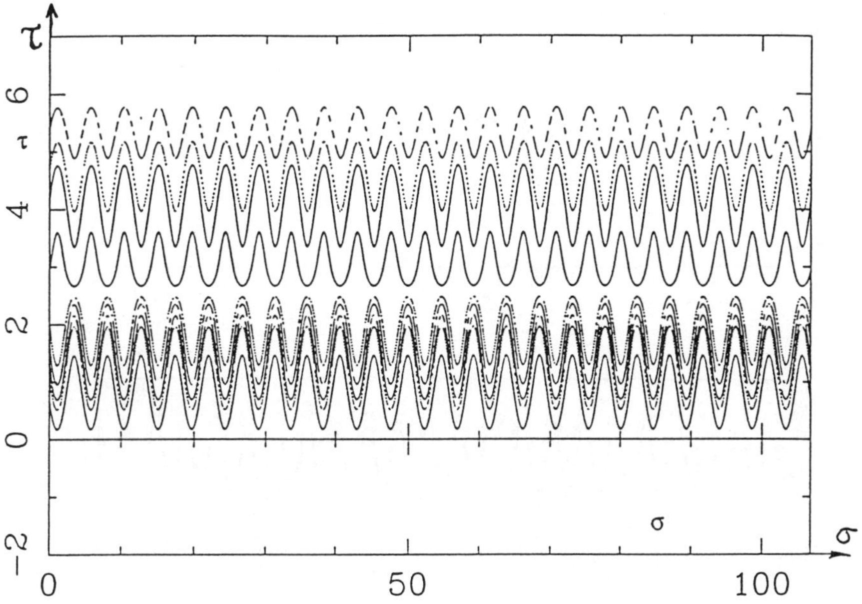

Figure 3: $\tau = \tau(\sigma, T)$ for fixed T for $n = 4, |x^0 >= (1, -1, i, 1)$. Three values of HT are displayed, corresponding to HT=0 (full line), 1 (dots), and 2 (dashed line). For each HT, three curves are plotted, which correspond to the three strings. They are ordered with τ increasing.

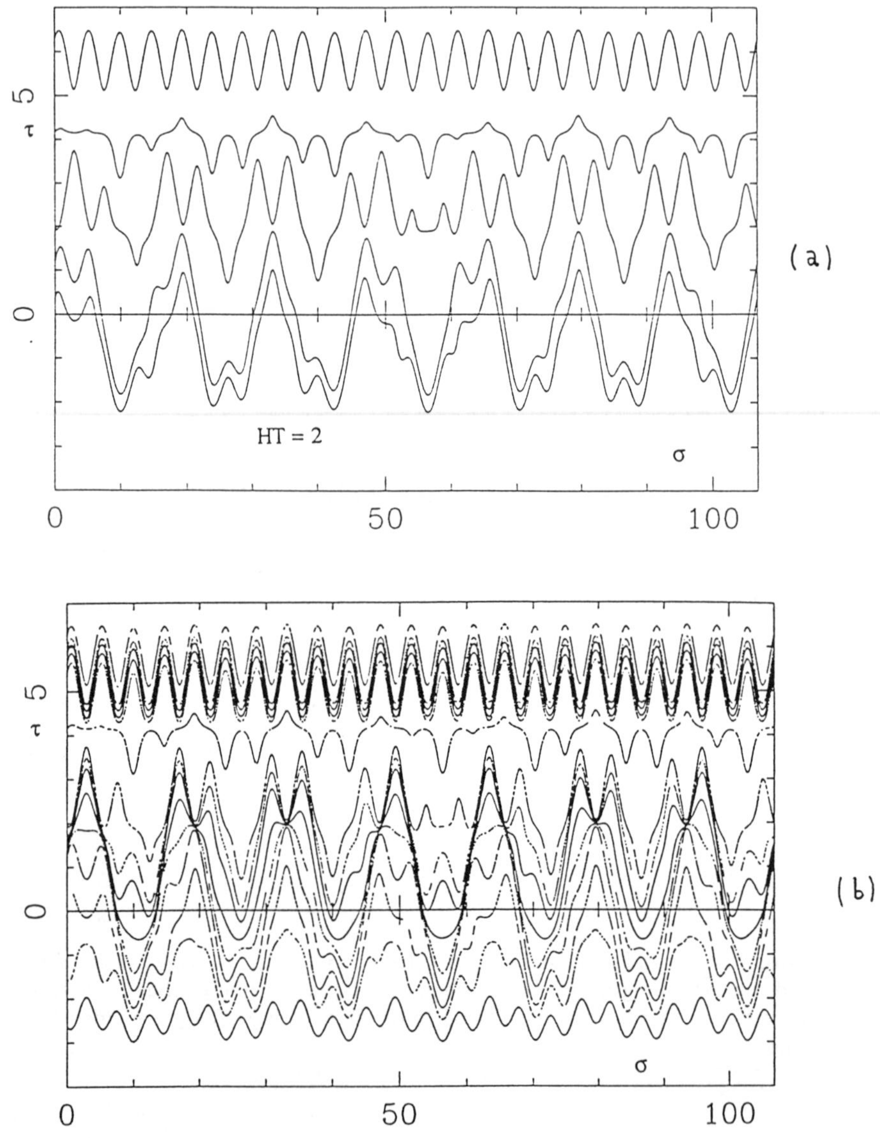

Figure 4: Same as fig. 3 for $n = 4, x^0 >= (1 + i, .6 + .4i, .3 + .5i, .77 + .79i)$. a) The five curves corresponding to the five strings at HT=2. b) The five curves for three values of HT: HT=0 (full line), 1 (dots), and 2 (dashed line).

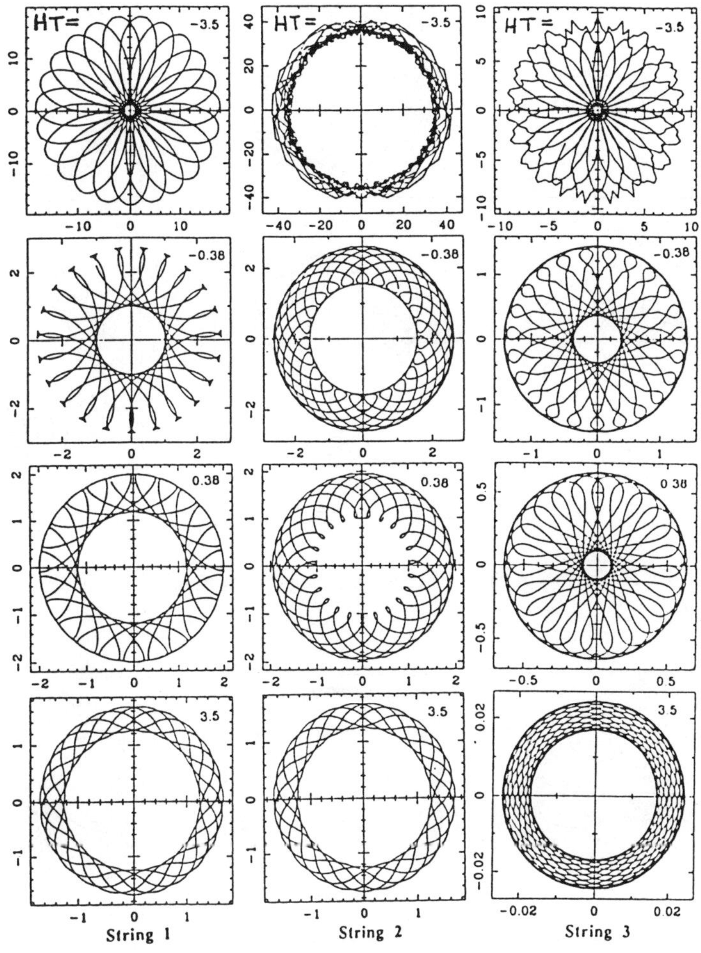

Figure 5: Evolution of the three strings, for $n = 4, |x^0 >= (1, -1, i, 1)$. The comoving size of string (1) stays constant for $HT < -3$, then decreases around $HT = 0$, and stays constant again after $HT = 1$. The invariant size of string (2) is constant for negative HT, then grows as the expansion factor for $HT > 1$, and becomes identical to string (1). The string (3) has a constant comoving size for $HT < -3$, then collapses as e^{-HT} for positive HT.

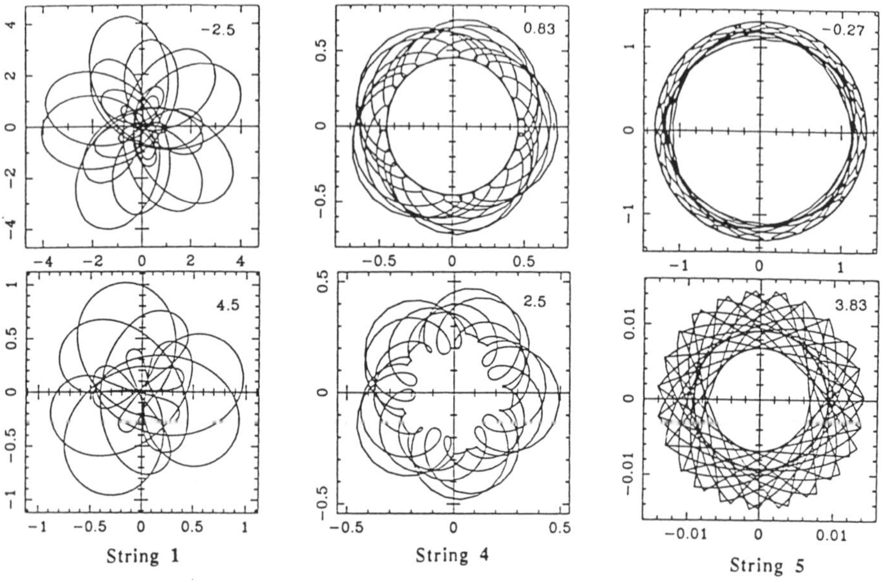

Figure 6: Evolution of three of the five strings for $n = 4, |x^0 >= (1 + i, .6 + .4i, .3 + .5i, .77 + .79i)$.

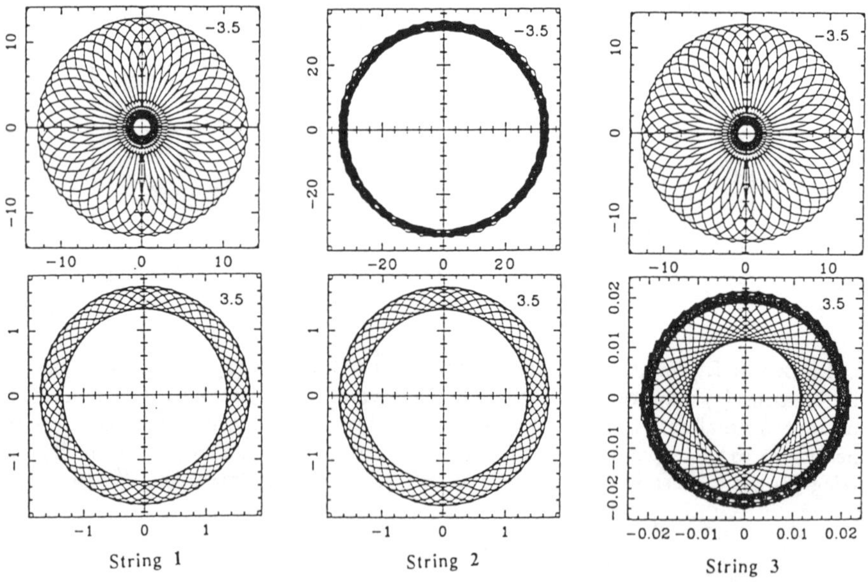

Figure 7: Evolution of the three strings for the degenerate case $n = 6, |x^0 >= (1, -1, i, 1)$.

de Sitter universe. Moreover, the construction method used here works in any number of dimensions.

New classes of multistring solutions in curved spacetime has been recently found in [14].

REFERENCES

[1] Lectures by H. J. de Vega and N. Sánchez in 'String Quantum Gravity and the Physics at the Planck Scale', Proceedings of the Erice Workshop held in June 1992. Edited by N. Sánchez, World Scientific, 1993. Pages 73-185, and references given therein.
Lectures by H. J. de Vega and N. Sánchez in 'Current Topics in Astrofundamental Physics: The Early Universe', Proceedings of the Nato ASI Third D. Chalonge School, 4-16 September 1994, p. 99-128, edited by N. Sánchez and A. Zichichi, Kluwer, 1995.

[2] H. J. de Vega and N. Sánchez, Phys. Lett. **B 197**, 320 (1987).

[3] H. J. de Vega and N. Sánchez, Phys. Rev. **D50**, 7202 (1994).

[4] H. J. de Vega and N. Sánchez, Phys. Rev. **D47**, 3394 (1993).

[5] H. J. de Vega, A. V. Mikhailov and N. Sánchez, Teor. Mat. Fiz. **94** (1993) 232.

[6] F. Combes, H. J. de Vega, A. V. Mikhailov and N. Sánchez, Phys. Rev. **D50**, 2754 (1994).

[7] H. J. de Vega, A. L. Larsen and N. Sánchez, Nucl. Phys. **B 427**, 643 (1994).

[8] I. Krichever, Funct. Anal. and Appl. **28**, 21 (1994), [Funkts. Anal. Prilozhen. **28**, 26 (1994)].

[9] H. J. de Vega and N. Sánchez, Nucl. Phys. **B309**, 552 and 577 (1988).

[10] N. Sánchez and G. Veneziano, Nucl. Phys. **B333**, 253 (1990).

[11] M. Gasperini, N. Sánchez and G. Veneziano, Int. J. Mod. Phys. **A 6**, 3853 (1991) and Nucl. Phys. **B364**, 365 (1991).

[12] H. J. de Vega and I. L. Egusquiza, Phys. Rev. **D49**, 763 (1994).

[13] A. L. Larsen and N. Sánchez, Phys. Rev. **D50**, 7493 (1994).

[14] A. L. Larsen and N. Sánchez, Phys. Rev. **D51**, 6929 (1995).

[15] H. J. de Vega and I. L. Egusquiza, Class. and Quantum Gravity **13**, 1041 (1996).

[16] S. Hawking and G. F. R. Ellis, 'The large scale structure of the spacetime', Cambridge Univ. Press, 1973.

[17] H. J. de Vega and N. Sánchez, Int. J. Mod. Phys. **A 7**, 3043 (1992).

[18] H. J. de Vega and I. L. Egusquiza, 'Strings next and inside black holes', hep-th/9506214, to appear in Phys. Rev. D.

[19] G. F. R. Ellis, Banff Lectures 1990, *in* Gravitation, eds. R. Mann and P. Wesson , World Scientific 1991.

[20] See for a review, T. W. B. Kibble, Erice Lectures at the Chalonge School in Astrofundamental Physics, N. Sánchez editor, World Scientific, 1992.

[21] A. Vilenkin, Phys. Rev. **D 24**, 2082 (1981) and Phys. Rep. **121**, 263 (1985).
N. Turok and P. Bhattacharjee, Phys. Rev. **D 29**, 1557 (1984).

[22] H. J. de Vega and A. Nicolaidis, Phys. Lett. **B 295**, 214 (1992).
H. J. de Vega, I. Giannakis and A. Nicolaidis, Mod. Phys. Lett. **A 10**, 2432 (1995).

[23] C. Loustó and N. Sánchez, 'String dynamics in cosmological and black hole backgrounds: the null string approach', in preparation.

[24] R. Myers, Phys. Lett. **B199**, 371 (1987).
M. Mueller, Nucl. Phys. **B337**, 37 (1990).
See for a review: A.A. Tseytlin in the Proceedings of the Erice School "String Quantum Gravity and Physics at the Planck Energy Scale", 21-28 June 1992, Edited by N. Sánchez, World Scientific, 1993.

[25] S. Weinberg, 'Gravitation and Cosmology', J. Wiley, 1972.
[26] E. W. Kolb and M. S. Turner, 'The Early Universe', Addison-Wesley, 1990.
[27] A. D. Linde, 'Particle Physics and Inflationary Cosmology', Harwood (1990).
[28] See for example,
I. Antoniadis, C. Bachas, J. Ellis and D. V. Nanopoulos,
Nucl. Phys. **B 328**, 117 (1989) and Phys. Lett. **B 257**, 278 (1991).
B. A. Campbell, A. Linde and K. A. Olive, Nucl. Phys. **B 355**, 146 (1991).
B. A. Campbell, N. Kaloper and K. A. Olive, Phys. Lett. **B 277**, 265 (1992).
A.A. Tseytlin, Mod. Phys. Lett **A 6**, 1721 (1991).
A.A. Tseytlin and C. Vafa, Nucl. Phys. **B 372**, 443 (1992).
R. Brustein and P. J. Steinhardt, Phys. Lett. **B 302**, 196 (1993).
M. Gasperini and G. Veneziano, Mod. Phys. Lett. **A 8**, 370 (1993),
Phys. Lett. **B 277**, 256 (1992) and Astroparticle Physics **1**, 317 (1993).
E. Raiten, Nucl. Phys. **B 416**, 881 (1994).
R. Brustein and G. Veneziano, Phys. Lett. **B 329**, 429 (1994).
V. A. Kostelecký and M. J. Perry, Nucl. Phys. **B 414**, 174 (1994).
See in addition ref. [24].
[29] See for example:
M. Green, J. Schwarz, E. Witten, 'Superstring Theory'.
Cambridge University Press. 1987.
[30] I.S. Gradshteyn and I.M. Ryzhik, Table of Integrals Series and Products,
(Academic Press, New York, fourth edition, 1965).
[31] See for example,
S. Weinberg, 'Gravitation and Cosmology', J. Wiley, 1972.
[32] F. Müller-Holstein, Class. Quant. Grav. **3**, 665 (1986).
K. G. Akdeniz et al. Mod. Phys. Lett. **A 6**, 1543 (1991) and
Phys. Lett. **B 321**, 329 (1994).
[33] M. V. Fischetti, J. B. Hartle and B. L. Hu, Phys. Rev. **D 20**, 1757 (1979).
[34] S. Wada and T. Azuma, Phys. Lett. **B 132**, 313 (1983).
V. Sahni and L. A. Kofman, Phys. Lett. **A 117**, 275 (1986).
[35] M. A. Castagnino, J. P. Paz and N. Sánchez, Phys. Lett. **B 193**, 13 (1987).
[36] D. G. Boulware and S. Deser, Phys. Rev. Lett. **55**, 2656 (1985).
[37] V. E. Zakharov and A. V. Mikhailov, JETP, **75**, 1953 (1978).
[38] H. J. de Vega, Phys. Lett. **B 87**, 233 (1979).
[39] C. S. Gardner, J. M. Greene, M. D. Kruskal and R. M. Miura,
Phys. Rev. Lett. **19**, 1095 (1967).
P. D. Lax, Comm. Pure and Appl. Math. **21** 467 (1968).
[40] M. J. Ablowitz and H. Segur, "Solitons and the Inverse scattering transformation",
SIAM Philadelphia 1981.
V.E. Zakharov, S.V. Manakov, S.P. Novikov and L.P. Pitaevsky,
"Soliton Theory; The Inverse Method", Nauka, Moscow, 1980.
[41] A. C. Scott, F. Y. F. Chu and D. W. MacLaughlin, Proc. IEEE, **61**, 1443 (1973).
G. L. Lamb, Elements of Soliton Theory, J. Wiley, NY (1980).
[42] C. Loustó and N. Sánchez, Phys. Rev. **D47**, 4498 (1993).
[43] See for example,

C. W. Misner, K. S. Thorne and J. A. Wheeler, 'Gravitation', Freeman, 1973.
[44] H. J. de Vega and N. Sánchez, Nucl. Phys. **B 317**, 706 (1989) .
D. Amati and K. Klimĉik, Phys. Lett. **B 210** , 92 (1988) ,
see also: ref. [45].
[45] M. Costa and H. J. de Vega, Ann. Phys. **211**, 223 and 235 (1991).
[46] H. J. de Vega and N. Sánchez, Phys. Rev. D 45 , 2783 (1992).
H. J. de Vega, M. Ramón Medrano and N. Sánchez,
Class. and Quantum Grav. **10**, 2007 (1993).
G. Horowitz and A.R. Steif, Phys. Rev. Lett. **64**, 260 (1990) and
Phys. Rev. **D 42** , 1950 (1990).
[47] H. J. de Vega and N. Sánchez, Phys. Rev. **D 42**, 3969 (1990) and
H. J. de Vega, M. Ramón Medrano and N. Sánchez, Nucl. Phys. **B 374**, 405 (1992).
[48] H.J. de Vega, M. Ramón Medrano and N. Sánchez, Nucl. Phys. **B351**, 277 (1991).

EVOLUTION OF A STRING NETWORK IN BACKGROUNDS WITH ROLLING HORIZONS

M. Gasperini, M. Giovannini
Dipartimento di Fisica Teorica, Università di Torino,
Via P.Giuria 1, 10125 Turin, Italy,
K. A. Meissner
Institute of Theoretical Physics, ul. Hoza 69, 00-681 Warsaw, Poland
and
G. Veneziano
TH. Division, CERN, CH-1211 Geneva 23, Switzerland

ABSTRACT

We discuss the temporal variation of the equation of state of a classical string network, evolving in a background in which the Hubble radius H^{-1} shrinks to a minimum and then re-expands to infinity. We also present a method to look for self-consistent non-vacuum string backgrounds, corresponding to the simultaneous solution of the gravi-dilaton background equations and of the string equations of motion.

1. Introduction.

Extensive studies of string propagation in cosmological backgrounds [1-4], have shown that the qualitative behaviour of classical strings depends dramatically on the presence of event horizons in the given background geometry. In this case it is possible to define two physically distinct regimes for the classical solutions, according to whether the proper size L of a string, $L \sim \alpha' E$, is smaller or larger than the size of the horizon L_h. It turns out that, for $L << L_h$, the solutions are oscillatory with constant proper size, while the regime $L >> L_h$ is characterized by non-oscillatory solutions, in which the string proper size evolves in time like the scale factor, $a(t)$, of the background geometry (see also [5]).

This latter regime, first discussed in the context of a de Sitter-type manifold [1,2,5], is of crucial importance for all those metric backgrounds in which the size of the event horizon shrinks to zero. Classically, in such geometries, strings of any size asymptotically become larger than the horizon. The classical description, however, is expected to break down at some point. Thus, while the string proper size is expected to be always larger than $\lambda_s = \sqrt{\alpha' \hbar}$ as a result of the uncertainty principle [6], classical geometries with horizon size smaller than λ_s are also expected to suffer large quantum corrections. Since we want to keep our discussion here entirely classical we shall use λ_s as a short distance cut-off both for the string size distribution and for the horizon of the background geometry.

The shrinking of the event horizon is typical of inflationary scenarios (also called "pre-big-bang" scenarios) which follow from the low energy string effective action [7-8], the accelerated evolution of the metric being driven by the kinetic energy of the dilaton field. A question which arises naturally in that context is whether the back reaction of string matter, evolving in the presence of shrinking horizons, could lead to a string-plus-dilaton-driven inflationary scenario which solves simultaneously, and self-consistently, both the background field equations and the string equations of motion (very recent work on similar issues can be found in [9]).

In order to avoid confusion, we recall that the shrinking of the event horizon can occur in two types of backgrounds [8], characterized, respectively, by a superinflationary accelerated expansion,

$$\dot{a} > 0, \quad \ddot{a} > 0, \quad \dot{H} > 0 \tag{1.1}$$

or by an accelerated contraction

$$\dot{a} < 0, \quad \ddot{a} < 0, \quad \dot{H} < 0 \tag{1.2}$$

of the scale factor $a(t)$. Here $H = \dot{a}/a$, and a dot denotes differentiation with respect to cosmic time t.

Consider, in fact, a spatially flat homogeneous and isotropic metric,

$$g_{\mu\nu} = diag(1, -a^2 \delta_{ij}), \qquad i, j = 1, ..., d \geq 3 \tag{1.3}$$

parameterized by

$$a(t) \sim (-t)^{\alpha} \tag{1.4}$$

for $t \to 0_-$. The proper size of the event horizon

$$L_h(t) = a(t) \int_t^0 dt' a^{-1}(t') \tag{1.5}$$

is finite, and shrinks linearly in cosmic time, $L_h(t) \simeq (-t)$, for all $\alpha < 1$. The negative part of this range ($\alpha < 0$) corresponds to superinflation, eq.(1.1), while the positive part ($0 < \alpha < 1$) corresponds to accelerated contraction, eq.(1.2).

Both superinflation and accelerated contraction provide a representation of the pre-big-bang scenario [7,8], the former in the so-called string frame, (in which weakly coupled test strings move along geodesic surfaces of the background metric), the latter in the Einstein frame (in which the dilaton is minimally coupled to the metric and the corresponding effective action is diagonalized in the standard canonical form).

Obviously, the abovementioned phenomena, being of purely geometrical nature, refer just to the string-frame metric (the dilaton controls instead the strength of the mutual interaction among strings, assumed to be negligible in the following). For all these reasons we will limit our attention to a background which describes (in the string frame) an initial superinflationary era followed by the time-reversed version of an accelerated contraction: the latter phase describes nothing more than the standard (non-inflationary) decelerated expansion ($\dot{a} > 0$, $\ddot{a} < 0$) of the present epoch. It was shown in [10] that a transition between the two epochs cannot occur in the low curvature regime: an intermediate genuinly stringy epoch is needed as well.

Consider now the string equations of motion which, in the gauge in which the world-sheet metric is conformally flat, can be written explicitly as [1-3]

$$\frac{d^2x^\mu}{d\tau^2} - \frac{d^2x^\mu}{d\sigma^2} + \Gamma^\mu_{\alpha\beta}(\frac{dx^\alpha}{d\tau} + \frac{dx^\alpha}{d\sigma})(\frac{dx^\beta}{d\tau} - \frac{dx^\beta}{d\sigma}) = 0$$

$$g_{\mu\nu}(\frac{dx^\mu}{d\tau}\frac{dx^\nu}{d\tau} + \frac{dx^\mu}{d\sigma}\frac{dx^\nu}{d\sigma}) = 0, \quad g_{\mu\nu}\frac{dx^\mu}{d\tau}\frac{dx^\nu}{d\sigma} = 0 \tag{1.6}$$

where $x^\mu(\sigma,\tau)$ are the string coordinates, τ and σ the usual world-sheet time and space variables, and $\Gamma^\mu_{\alpha\beta}$ is the Christoffel connection for the background metric (1.3). When the string proper size is negligible with respect to the horizon, the exact solution of these equations can be expressed as an expansion around the point-particle motion of the string center of mass [1]. If the horizon is shrinking, however, such an expansion breaks down, and the explicit solution of eqs.(1.6) shows that the string comoving size becomes frozen, asymptotically, while the proper size evolves in time like the scale factor, $L(t) \sim a(t)$ for $L(t) >> L_h(t) \sim (-t)$.

This occurs both for superinflation and for accelerated contraction [3]. The crucial difference between the two types of background, however, is that, in an expanding geometry, the asymptotic solutions of eqs.(1.6) are characterized by

$$\left|\frac{\partial x^0}{\partial\tau}\right| >> \left|\frac{\partial x^0}{\partial\sigma}\right|, \qquad \left|\frac{\partial x^i}{\partial\tau}\right| << \left|\frac{\partial x^i}{\partial\sigma}\right| \tag{1.7}$$

while, in the contracting case, they satisfy

$$\left|\frac{\partial x^0}{\partial \tau}\right| >> \left|\frac{\partial x^0}{\partial \sigma}\right|, \qquad \left|\frac{\partial x^i}{\partial \tau}\right| >> \left|\frac{\partial x^i}{\partial \sigma}\right| \tag{1.8}$$

As a consequence, the string energy-momentum tensor

$$T^{\mu\nu}(x) = \frac{1}{\pi\alpha'\sqrt{|g|}}\int d\sigma d\tau\left(\frac{dx^\mu}{d\tau}\frac{dx^\nu}{d\tau} - \frac{dx^\mu}{d\sigma}\frac{dx^\nu}{d\sigma}\right)\delta^{d+1}(x - x(\sigma,\tau)) \tag{1.9}$$

($(\alpha')^{-1}$ is the string tension) satisfies respectively the asymptotic conditions

$$T_0{}^0 \simeq \sum_i T_i{}^i, \quad \alpha < 0, \qquad T_0{}^0 \simeq -\sum_i T_i{}^i, \quad 0 < \alpha < 1, \tag{1.10}$$

In the perfect fluid approximation, $T_0{}^0 = \rho$, $T_i{}^j = -p\delta_i^j$, one thus obtain for a diluted gas of classical strings the effective equations of state [3]

$$\frac{p}{\rho} \simeq -\frac{1}{d}, \quad \alpha < 0, \qquad \frac{p}{\rho} \simeq \frac{1}{d}, \quad 0 < \alpha < 1, \tag{1.11}$$

The relationship between the two regimes can be easily understood in terms of the so-called scale-factor duality property of string motions in cosmological backgrounds [7].

We note, for comparison, that in the opposite regime $L(t) << L_h(t)$ the string solutions are instead characterized by the condition

$$\left|\frac{\partial x^0}{\partial \tau}\right| >> \left|\frac{\partial x^0}{\partial \sigma}\right|, \qquad \left|\frac{\partial x^i}{\partial \tau}\right| \simeq \left|\frac{\partial x^i}{\partial \sigma}\right| \tag{1.12}$$

which implies

$$T_0{}^0 >> \sum_i T_i{}^i, \qquad \frac{p}{\rho} \simeq 0 \tag{1.13}$$

As it is clear from the way they are obtained, the effective equations of state (1.11) and (1.13), characterized by a constant ratio $\gamma = p/\rho$, are only valid asymptotically when strings are, respectively, far outside or well inside the horizon. The purpose of this paper is to discuss instead the time-variation of the ratio p/ρ (for a gas of classical strings) versus the time-variation of the event horizon, considering in particular a background in which the Hubble radius H^{-1} shrinks to a minimum and then re-expands to infinity. To this aim we shall propose in Section 2 an effective evolution equation for the density of strings of arbitrary proper size, and we shall present in Section 3 some examples of explicit solution of such equation.

A possible approach to the problem of obtaining simultaneous and self-consistent solutions to the system of background field equations and string equations of motion will be discussed in Section 4. Our main results will be finally summarized in Section 5.

2. Evolution equation for the string distribution.

Let us consider a cosmological metric background of the type (1.3) which, starting from an asymptotically flat initial state at $t = -\infty$, evolves towards a high curvature regime [7-8]. The curvature reaches at $t = 0$ a maximum scale of order $\lambda_s^{-1} \lesssim \ell_p^{-1}$ (where ℓ_p is the Planck length), and then decreases approaching zero as $t \to +\infty$. In such a background, the Hubble length $D(t) \equiv H^{-1}$ (which we shall assume here to be always positive) shrinks to a minimum $D(0) \simeq \lambda_s$, and then re-expands to infinity.

Consider, in this background, a string of initial proper size $L_{-\infty}$. According to the results reviewed in the previous Section [1-5], the string size stays constant inside the horizon, and evolves like $a(t)$ outside the horizon. The time-dependence of the string proper size $L(t)$ can thus be fixed by the equation

$$L(t) = L_{-\infty}\theta(-t)\theta(D(t) - L(t)) + L_{-\infty}\theta(L(t) - D(t))$$

a(t)$\frac{}{a(t_{out})+} + L_{+\infty}\theta(D(t) - L(t))\theta(t)(2.1) where \theta$ is the Heaviside step function,

$$L_{+\infty} = L_{-\infty}\frac{a(t_{in})}{a(t_{out})} \tag{2.2}$$

and t_{out} and t_{in} are, respectively , the times at which the string leaves and re-enters the horizon, determined by the conditions $D(t_{out}) = L_{-\infty}$, $D(t_{in}) = L_{+\infty}$.

Consider next, in the same background, a network of N such non-interacting strings, occupying initially a certain comoving volume. Let us denote by $L^i(t)$ $(i = 1, 2, ..., N)$ the individual proper sizes at time t. The number distribution $n(L, t)$ of strings of proper size L, at time t, such that $N = \int n dL$, is obviously given by

$$n(L,t) = \sum_{i=1}^{N} \delta(L - L^i(t)) \tag{2.3}$$

We now differentiate this distribution with respect to L and t and take into account that, from eq.(2.1),

$$\dot{L}^{(i)} = H(t)\theta(L^{(i)}(t) - D(t))L^{(i)} \tag{2.4}$$

We thus find that the number distribution of a diluted gas of classical strings, in the considered background, must satisfy the approximate evolution equation

$$\frac{\partial n(L,t)}{\partial t} + H(t)\frac{\partial}{\partial L}[n(L,t)L\theta(L-D)] = 0 \tag{2.5}$$

Equivalently, by using a instead of t as evolution parameter,

$$a\frac{\partial n(L,a)}{\partial a} + \frac{\partial}{\partial L}[n(L,a)L\theta(L-D)] = 0 \tag{2.6}$$

The general solution of this equation can be conveniently expressed in terms of the initial string distribution, $n_{-\infty}(L)$, as

$$n(L,a) = n_{-\infty}(L)\theta(D-L) + \frac{1}{a}h(\frac{L}{a})\theta(L-D) \tag{2.7}$$

where

$$h(\xi) = a(\xi)n_{-\infty}(D(\xi))\left[1 + \frac{\partial \ln a(\xi)}{\partial \ln H(\xi)}\right]^{-1} \tag{2.8}$$

($\xi = \frac{D}{a}$). Once the background $a(t)$ and the initial string distribution $n_{-\infty}$ are given one can thus compute all physical observables, and in particular the energy density of strings which are smaller and larger than the horizon (which we shall call for simplicity stable and unstable, respectively).

By recalling that the string energy is proportional to the proper size L, the energy of stable and unstable strings in our ensemble can be estimated, respectively, as

$$\begin{aligned} E_s &= \int d^3x\sqrt{|g|}\rho_s = \frac{1}{\pi\alpha'}\int Ln(L,a)\theta(D-L)dL \\ E_u &= \int d^3x\sqrt{|g|}\rho_u = \frac{1}{\pi\alpha'}\int Ln(L,a)\theta(L-D)dL \end{aligned} \tag{2.9}$$

(the integral extends of course over the whole range of L, from the minimal allowed proper size λ_s up to infinity). The corresponding pressure, on the other hand, is given asymptotically by the effective equations of state (1.13) and (1.11), respectively. The time evolution of the ration $\gamma = p/\rho$ for the full string distribution, therefore, can be approximated as

$$\gamma(t) = \pm\frac{E_u}{dE} = \pm\frac{1}{d}\frac{\int_{\lambda_s}^{\infty} Ln\theta(L-D)dL}{\int_{\lambda_s}^{\infty} LndL} \tag{2.10}$$

(here $E = E_s + E_u$, and the sign of γ depends on the type of background, according to eq.(1.11)).

The time-dependence of γ determined in this way relies on the evolution equation (2.6), obtained in the sudden approximation in which the transition from the stable to unstable regime is very roughly parameterized by the step function $\theta(L-D)$. Such a transition, however, could be be approximated in a continuous way by introducing an appropriate function $\theta(L,a)$ which goes to 0 for $LH << 1$ and to 1 for $LH >> 1$, and which interpolates smoothly between strings smaller and larger than the horizon. The evolution equation for n, written in terms of such a smooth function,

$$a\frac{\partial n(L,a)}{\partial a} + \frac{\partial}{\partial L}[n(L,a)L\theta(L,a)] = 0 \tag{2.11}$$

has then the general solution

$$n(L,a) = -\sigma(z)\frac{\partial z}{\partial L} \tag{2.12}$$

where σ is an arbitrary function of z, and $z(L,a)$ is a particular solution of the differential equation

$$a\frac{\partial z}{\partial a} + \theta(L,a)L\frac{\partial z}{\partial L} = 0 \tag{2.13}$$

In the following Section we shall provide examples of solution of the evolution equation for the string number distribution, considering both the sudden approximation, eq.(2.6), and the continuous one, eq.(2.11).

3. Examples of exact solutions.

We shall consider first an accelerated background corresponding to a phase of pre-big-bang evolution, [7-8], parameterized by the scale factor (1.4) with $\alpha < 1$, and $-\infty \le t \le -t_1$, where $t_1 \simeq \lambda_s$. We shall assume, moreover, a power-law behavior for the initial string distribution,

$$n_{-\infty}(L) = \Lambda^{\beta-1}L^{-\beta} \tag{3.1}$$

Here $\beta > 2$ for the convergence of the energy integral $\int^\infty LndL$, and Λ is an appropriate length parameter, related to the total string number N by

$$N = \int_{\lambda_s}^\infty n_{-\infty}(L)dL = \frac{1}{\beta-1}\left(\frac{\Lambda}{\lambda_s}\right)^{\beta-1} \tag{3.2}$$

In this background $(\partial \ln a/\partial \ln H) = -\alpha$, and from the general solution (2.7) we find that the energy of stable and unstable strings is given, respectively, by

$$\pi\alpha' E_s = \int_{\lambda_s}^{D} LndL = \frac{\Lambda^{\beta-1}}{\beta-2}\left(\lambda_s^{2-\beta} - D^{2-\beta}\right)$$

$$\pi\alpha' E_u = \int_{D}^{\infty} LndL = \frac{\Lambda^{\beta-1}}{\beta+\alpha-2} D^{2-\beta} \tag{3.3}$$

(we have supposed $\beta + \alpha > 2$). Note that $\dot{E} = HE_u$, in agreement with the covariant conservation of the stress tensor $(\dot{\rho} + dH(\rho + p) = 0)$, and with the fact that only unstable strings contribute to the total pressure, with $p = -\rho/d$. As a consequence of eqs.(3.2), (3.3), the ratio $\gamma(t)$ evolves in time (according to eq.(2.10)) as

$$\gamma(t) = \pm\frac{1}{d}\frac{\beta-2}{(\beta+\alpha-2)|H\lambda_s|^{2-\beta} - \alpha}, \quad \beta > 2, \quad \beta+\alpha > 2 \tag{3.4}$$

Asymptotically,

$$\gamma = \pm\frac{1}{d}\frac{\beta-2}{\beta+\alpha-2}|H\lambda_s|^{\beta-2}, \qquad |H\lambda_s| \to 0 \tag{3.4a}$$

$$\gamma = \pm\frac{1}{d}, \qquad |H\lambda_s| \to 1 \tag{3.4b}$$

Here $sign\{\gamma\} = sign\{\alpha\}$, so that in case of superinflationary expansion γ approaches zero from negative values as $t \to -\infty$, while in case of accelerated contraction γ approaches zero from positive values. If we consider a background which evolves from a phase of pre-big-bang accelerated expansion,

$$a(t) \sim (-t)^{\alpha}, \qquad \alpha < 0, \quad t < 0 \tag{3.5}$$

to a phase of standard decelerated expansion,

$$\tilde{a}(t) \sim t^{\delta}, \qquad 0 < \delta < 1, \quad t > 0 \tag{3.6}$$

we must expect, therefore, a change of trend in the equation of state, corresponding to the evolution of γ which starts from 0 at $t = -\infty$, reaches a negative minimum of order unity at $t = 0$ (when $H_1 \simeq \lambda_s^{-1}$), and then starts to increase towards positive values, approaching zero as $t \to +\infty$.

In order to describe the time evolution of γ in the decelerated regime following the phase of pre-big-bang inflationary expansion, we write first the general solution for n in the background (3.6) and then demand that it joins smoothly to the final distribution originating from the preceding inflationary background (3.5). One easily gets, at $t > 0$

$$n(L, \tilde{a}) = n_{+\infty}(L)\theta(D - L) + \frac{1}{\tilde{a}}h(\frac{L}{\tilde{a}})\theta(L - D) \tag{3.7}$$

where $n_{-\infty}$ has been replaced by the final asymptotic string distribution for $t \to +\infty$,

$$n_{+\infty}(\xi) = (1 - \delta)\frac{h(\xi)}{\tilde{a}(\xi)} \tag{3.8}$$

(according to eq.(2.8)), and

$$h(\xi) = A\xi^{(\alpha-\beta)/(1-\alpha)} \tag{3.9}$$

Here the ξ-dependence of h is obtained from eq.(2.8) written in the background (3.5), and A is a constant coefficient which we shall fix by normalizing the total number of strings to that of the initial distribution, $N = \int n_{-\infty} dL$.

For the decelerated background (3.6) $\xi \sim D^{1-\delta}$. From eq.(3.8) we get, therefore,

$$n_{+\infty}(L) = c_{+\infty}L^{[\delta(\beta-1)+\alpha-\beta]/(1-\alpha)} \tag{3.10}$$

where the number $c_{+\infty}$ depends on A, α, β and δ. On the other hand, if we start at $t = -\infty$ with the initial distribution (3.1), we have

$$n_{+\infty}(L)dL = \frac{c_{+\infty}}{\Lambda^{\beta-1}}\left(\frac{1-\alpha}{1-\delta}\right)n_{-\infty}(L^{\frac{1-\delta}{1-\alpha}})d(L^{\frac{1-\delta}{1-\alpha}}) \tag{3.11}$$

so that we can always choose A in such a way that the total number of strings is conserved in the transition from a to $\tilde{a}$. This in agreement with a classical description of string propagation in curved backgrounds, where any quantum string creation or string decay process is completely neglected (see [9] for a discussion which includes the effects of quantum decay through string splitting).

We note, finally, that for the general solution (3.7) the energy of stable and unstable strings is given, respectively, by

$$\pi\alpha' E_s = \frac{c_{+\infty}}{k-2}\left(\lambda_s^{2-k} - D^{2-k}\right)$$

$$\pi\alpha' E_u = \frac{c_{+\infty}}{k+\delta-2}D^{2-k}, \qquad k = -\frac{\delta(\beta-1)+\alpha-\beta}{1-\alpha} \tag{3.12}$$

($k > 2$ for a finite total energy). The corresponding ratio $\gamma(t)$ increases towards zero, for $t \to +\infty$, as

$$\gamma(t) = -\frac{1}{d}\frac{k-2}{k+\delta-2}(H\lambda_s)^{k-2} \tag{3.13}$$

The transition from a phase of shrinking horizons to a phase of expanding horizons is associated, therefore, to a switching of $\dot{\gamma}$ from negative to positive values, and to a subsequent decreasing of $|\gamma|$ as more and more strings re-enter the horizon.

This behavior of γ, obtained in the context of the sudden approximation, can be qualitatively confirmed by using a different approach based on the smoothed evolution equation (2.11). Consider in fact the continuous scale factor

$$a = \left(\frac{t+\sqrt{t^2+t_1^2}}{t_1}\right)^{1/2}, \quad D \equiv H^{-1} = t_1(a^2+a^{-2}), \quad -\infty \le t \le +\infty \tag{3.14}$$

which is self-dual [7,8], i.e. satisfies $a(t) = a^{-1}(-t)$. It connects smoothly the standard radiation-dominated expansion, $a \sim t^{1/2}$ for $t \to +\infty$, to the duality-related inflationary regime, $a \sim (-t)^{-1/2}$ for $t \to -\infty$. By choosing

$$\theta(L,a) = \frac{L}{L+H^{-1}} \tag{3.15}$$

as interpolating function between the stable ($LH >> 1$) and unstable ($LH << 1$) regime, one can easily check that

$$z(L,a) = \frac{2t_1}{L} + \arctan\left(a^2 + \frac{L}{2t_1}\right) - \frac{\pi}{2} \tag{3.16}$$

is a particular solution of eq.(2.13). We choose, moreover,

$$\sigma(z) = \frac{N\pi z}{\left(z+\frac{\pi}{2}\right)^3} \tag{3.17}$$

as simple example of distribution which leads to a finite total number of strings and to a finite total energy for $L \to 0$ and $L \to \infty$.

From the solution (2.12) of the smoothed evolution equation we obtain, in particular,

$$N = \int_0^\infty dL\; n(L,a) = \int_0^\infty \sigma(z)dz$$

$$\pi\alpha' E = \int_0^\infty dL\; Ln(L,a) \to c_1 + c_2 a^2, \qquad a \to 0$$

$$\pi\alpha' E = \int_0^\infty dL\; Ln(L,a) \to \frac{4Nt_1}{\pi} - \frac{16Nt_1}{\pi^2 a^2}\ln\left(\frac{\pi a^2}{2}\right), \qquad a \to \infty$$

where c_1 and c_2 are positive constants. On the other hand, in d isotropic dimensions, the pressure to energy density ratio for a perfect gas contained inside a proper volume V can be expressed as

$$\gamma = \frac{p}{\rho} = -\frac{V}{E}\frac{\partial E}{\partial V} = -\frac{a}{dE}\frac{\partial E}{\partial a} \tag{3.19}$$

The behavior of γ over the whole range of t can then be found by integrating numerically the given string energy distribution. For what concerns the asymptotic behavior of γ, however, we can easily obtain from eqs.(3.18) that

$$t \to -\infty, \quad a \to 0 \quad \Longrightarrow \quad \gamma \sim -a^2 \sim -\frac{1}{(-t)} \sim -H_\infty$$

$$t \to +\infty, \quad a \to \infty \quad \Longrightarrow \quad \gamma \sim -\frac{1}{a^2}\ln a^2 \sim -H_\infty \ln H_\infty \tag{3.20}$$

where $H_\infty \equiv H(t = \pm\infty) = 1/(2|t|)$ (see eq.(3.14). Therefore, γ starts decreasing from zero at $t \to -\infty$, and then eventually increases to approach zero from negative values as $t \to +\infty$, in full qualitative agreement with our previous conclusions. It should be noted, moreover, that γ is always negative, in agreement with eq.(3.19) and with the fact that the total string energy is growing,

$$\dot{E} = -H\int_0^\infty LdL\frac{\partial}{\partial L}(nL\theta) = H\int_0^\infty dLnL\theta \equiv HE_u > 0 \tag{3.21}$$

over the whole time range $-\infty \leq t \leq +\infty$.

The initial distributions considered in this section consist,

essentially, of short tests strings. The mean length of the strings is $O(\lambda_s)$ in

the

case of the distribution (3.2), and $O(t_1)$ in the case of the distribution (3.17). It's interesting to consider instead a distribution of strings

with arbitrary mean length L_0. In this case, if L_0 is

sufficiently large compared to λ_s, or t_1

the ratio $\frac{p}{\rho}$ approaches the minimum value

$-\frac{1}{d}$ as soon as D becomes much smaller than L_0.

Consider for instance an accelerated

expanding background described by the scale factor (3.5), and assume the following initial distribution :

$$n_{-\infty}(L) = \frac{N}{L_0}e^{-\frac{L}{L_0}} \tag{3.22}$$

In this case the mean string length is:

$$< L >= \frac{\int_{\lambda_s}^{\infty} L n_{-\infty}(L) dL}{}$$

$\int_{\lambda_s}^{\infty} n_{-\infty}(L)dL = L_0 + \lambda_s$ (3.23). Then, from eq. (2.7)

$$\frac{1}{N}\pi\alpha' E_s = \int_{\lambda_s}^{D} dLLn(L,a) =$$

$(\lambda_s + L_0)e^{-\frac{\lambda_s}{L_0}}$
- $(D+L_0)e^{\frac{-D}{L_0}}$ (3.24) $\frac{1}{N}\pi\alpha' E_u = \int_D^{+\infty} dLLn(L,a) = (\frac{D}{L_0})^{\alpha} L_0$
$\Gamma \Big(2 - \alpha, \frac{1}{L_0}(-\alpha)^{\frac{-\alpha}{\alpha-1}}$
$(D\frac{}{a)^{\frac{1}{1-\alpha}} (3.25)}$ (Γ is the incomplete Euler Gamma function).
In order to give a particularly simple example we will choose now an inflationary background with $\alpha = -1$,
so that E_u reduces to:

$$\frac{1}{N}\pi\alpha' E_u =$$

$L_0(\frac{L_0}{D})[(\frac{D}{L_0})^2 + 2(\frac{D}{L_0}) + 2]e^{\frac{-D}{L_0}}$ and then $\gamma(t)$ becomes, according to equation (2.10),

$$\gamma(t) = -\frac{1}{d}\frac{1}{(1 + \frac{(\frac{\lambda_s}{L_0}+1)e^{-\frac{\lambda_s}{L_0}} - (\frac{D}{L_0}+1)e^{-\frac{D}{L_0}}}{(\frac{D}{L_0}+2\frac{L_0}{D}+2)e^{\frac{-D}{L_0}}})} \tag{3.27}$$

From this equation we can easily check that, if $\lambda_s \simeq L_0$,
$\gamma \to 0$ for $D >> \lambda_s$ and $\gamma \to -\frac{1}{d}$ for $D \to \lambda_s$, in agreement with our previous examples. If, on the contrary, we assume an initial distribution with $L_0 >> \lambda_s$ then we have $\gamma \simeq -\frac{1}{d}$
as soon as $D << L_0$. This means that, for a gas of sufficiently long strings,
the minimal negative pressure of the unstable regime can be reached before attaining the quantum string limit.

4. Self-consistent string sources for the background equations.

The field equations obtained from the low energy string effective action for a spatially flat, homogeneous and isotropic metric and dilaton background, with matter sources but vanishing dilaton potential, can be written (in d spatial dimensions) as [7,8,11]

$$\dot{\overline{\phi}}^2 - 2\ddot{\overline{\phi}} + dH^2 = 0$$
$$\dot{\overline{\phi}}^2 - dH^2 = \overline{\rho} e^{\overline{\phi}}$$
$$2(\dot{H} - H\dot{\overline{\phi}}) = \overline{p} e^{\overline{\phi}} \tag{4.1}$$

Here $\overline{\phi}$ is the duality-invariant shifted dilaton field,

$$\overline{\phi} = \phi - \ln\sqrt{|g|} = \phi - d\ln a \tag{4.2}$$

and the rescaled energy density and pressure

$$\overline{\rho} = \rho\sqrt{|g|}, \qquad \overline{p} = p\sqrt{|g|} \tag{4.3}$$

satisfy the conservation equation

$$\dot{\overline{\rho}} + dH\overline{p} = 0\ . \tag{4.4}$$

As already discussed in [8,11], these equations can be integrated a first time to give

$$\frac{1}{a}\frac{da}{dx} = \frac{2\Gamma}{(x+x_0)^2 - d\Gamma^2} \tag{4.5}$$

$$\frac{d\overline{\phi}}{dx} = -\frac{2(x+x_0)}{(x+x_0)^2 - d\Gamma^2} \tag{4.6}$$

$$\overline{\rho} = \frac{e^{\overline{\phi}}}{4\ell^2}[(x+x_0)^2 - d\Gamma^2] \tag{4.7}$$

where ℓ, x_0 are integration constants, x a new (dimensionless) time parameter related to cosmic time by

$$\overline{\rho} = \frac{1}{\ell}\frac{dx}{dt} \tag{4.8}$$

and

$$\Gamma(x) = \int^x \gamma(x')dx' \tag{4.9}$$

($\gamma = p/\rho = \overline{p}/\overline{\rho}$ as before). For any given time-dependent equation of state $\gamma(x)$ one can thus obtain, from eqs.(4.5)-(4.7), the general exact solution for $a(x), \phi(x)$ and $\rho(x)$, which can be eventually expressed in cosmic time through eq.(4.8).

If we are looking, however, for a simultaneous solution of the background equations and of the string equations of motion, the functional form of γ to be inserted into eq.(4.9) must be consistent with the time-evolution of the given initial string distribution. In particular, in the context of the fluid model discussed in the previous Sections, γ is to be obtained by solving the evolution equation (2.6) or (2.11), and it will be expressed in general as a function of the background scale factor, $\gamma = \gamma(a)$. It turns out, therefore, that a more convenient variable to look for solutions of the background equations (4.5)-(4.7), with a string gas as a consistent matter sources, is $y = \ln a$ instead of x.

By defining a new function $\omega(y)$ such that

$$(x + x_0) = -2\Gamma \frac{\omega'}{\omega} \tag{4.10}$$

(a prime denotes differentiation with respect to y), eq.(4.5) implies then

$$\omega'' + \frac{\Gamma'}{\Gamma}\omega' - \frac{d}{4}\omega = 0 \tag{4.11}$$

For any given $\Gamma(y)$ this equation uniquely determines $\omega(y)$, which inserted into eq.(4.10) provides $x = x(y)$. By inverting such relation we can eventually obtain $a = e^y = a(x)$. Moreover, the dilaton equation (4.6) can be rewritten

$$\overline{\phi}' = -\frac{2(x+x_0)}{(x+x_0)^2 - d\Gamma^2}\left(\frac{dx}{dy}\right) = -\frac{x+x_0}{\Gamma} = 2\frac{\omega'}{\omega} \tag{4.12}$$

so that $\overline{\phi}$ is also completely determined in terms of y as

$$\overline{\phi}(y) = \phi_0 + 2\ln\omega(y) \tag{4.13}$$

(ϕ_0 is an integration constant). The corresponding energy density is then, from eqs.(4.7), (4.10),

$$\overline{\rho}(y) = \left(\frac{\Gamma}{\ell}\right)^2 e^{\phi_0}\left(\omega'^2 - \frac{d}{4}\omega^2\right) \tag{4.14}$$

We note, finally, that for any given $\Gamma(y)$ the corresponding time-evolution of the ratio $\gamma = p/\rho$ is determined by eqs.(4.9) and (4.5) as

$$\gamma(y) = \frac{d\Gamma(y)}{dx} = \frac{\Gamma'}{2\Gamma}\left(\frac{\omega'^2}{\omega^2} - \frac{d}{4}\right)^{-1} \tag{4.15}$$

A simple, but important application of this method to obtain self-consistent string-driven backgrounds, is the linear case $\Gamma(y) \sim y$. This case corresponds, according to eqs.(4.8), (4.9), to a phase in which $dx/dt = const$, and in which

$\gamma \sim dy/dt = d(\ln a)/dt = H$, as suggested by the results of the previous Section (see eqs.(3.20) and (3.4a) with $\beta = 3$). In this case the solution of eq.(4.11) can be expressed in terms of modified Bessel functions, and the full system of equations can be easily integrated.

On the other hand, a background with $dx/dt \sim \overline{\rho} = const$ corresponds to an exact solution of the cosmological equations (4.1) obtained by imposing, as initial conditions at $t = -\infty$, the string perturbative vacuum (flat spacetime, $\phi = -\infty$), with a small but finite initial density of dust-like sources, $\rho > 0$, $p = 0$ [11]. Such a solution can be written explicitly, in the isotropic case, as

$$a(t) = a_0 \left|\frac{t - 2T}{t}\right|^{\pm 1/\sqrt{d}}, \qquad e^{\overline{\phi}} = \frac{16\ell^2 e^{-\phi_0}}{|t(t - 2T)|},$$

$$\overline{\rho} = \frac{1}{\ell}\frac{dx}{dt} = \frac{e^{\phi_0}}{4\ell^2} = const, \qquad p = 0, \quad t < 0 \tag{4.16}$$

where a_0, ϕ_0 and T are positive integration constants (the plus and minus sign in the exponent corresponds, for $t \to 0_-$, to a metric describing accelerated expansion and contraction, respectively).

This background is certainly consistent with with a string-driven evolution in the limit $t \to -\infty$, and in particular for $|t| >> T$. Indeed, in this regime the metric becomes flat,

$$a = const, \qquad \phi \sim -2\ln(-t),, \qquad \rho = const \tag{4.17}$$

so that there are no horizons, all strings are stable and are thus represented by a pressureless stress tensor. When $|t| \sim T$, however, the curvature scale begins to increase, and all strings progressively enter the non-oscillating unstable regime. The time variation of the ratio p/ρ has then to be taken into account, in general, for an exact string-driven solution which may describe the background evolution away from the string perturbative vacuum, consistently with the string equations of motion. What is remarkable, however, is that the simple background (4.16) may remain a good zeroth-order approximation to the exact string-driven solution, if the duration of the accelerated regime (from T to λ_s) is long enough to satisfy the phenomenological constraints on inflation [11].

By adopting an iterative procedure, let us assume indeed the solution (4.16) to be a zeroth-order approximation, and let us compute the first-order corrections by inserting that solution into the string evolution equation (2.6), in order to obtain the corresponding value of $\gamma(t)$. We shall take for simplicity an initial string distribution with $\beta = 3$, but our final conclusion turns out to be independent of the particular choice of β. Following the procedure illustrated in Section 3 we find, for $|t| >> T$, that $\xi \sim t^2$, so that $\pi\alpha' E_u = \Lambda^2 D^{-1}$. As a consequence we have, from eq.(2.10),

$$\gamma(t) = -\frac{1}{d} H \lambda_s \tag{4.18}$$

Note that γ goes to zero like $-H$ as $t \to -\infty$, in agreement with our previous results.

We now insert this expression of $\gamma(t)$ into the right-hand side of the background equations (4.5)-(4.7), by recalling that for the pressureless background (4.16) one has, to zeroth-order,

$$\Gamma^{(0)} = \pm \frac{T}{4\ell\sqrt{d}} e^{\phi_0} \tag{4.19}$$

Therefore, to next order,

$$\Gamma = \Gamma^{(0)} + \int_{-\infty}^{x} \gamma(x')dx' = \Gamma^{(0)} - \frac{\lambda_s e^{\phi_0}}{d4\ell} \int_{-\infty}^{t} H(t')dt' =$$

$$= \Gamma^{(0)} \left(1 - \frac{\lambda_s}{dT} \ln \left| \frac{t - 2T}{t} \right| \right) \tag{4.20}$$

According to our iterative approach, the integration of eqs. (4.5)-(4.7) with the new expression (4.20) for γ provides a first-order approximation to the background fields $a(t)$, $\phi(t)$. The first-order corrections to (4.16) due to a non-vanishing string effective pressure are certainly negligible in the regime $|t| >> T$. However, as clearly shown by eq. (4.20), the corrections may remain small over the whole time range if $T >> \lambda_s$.

In connection with this last point, it is important to note that the ratio T/λ_s measures the duration of the inflationary phase associated to the solution (4.16). For $|t| < T$ the metric background describes indeed a dilaton-dominated, accelerated evolution, which can be approximated by

$$a(t) \sim (-t)^{\mp 1/\sqrt{d}} \tag{4.21}$$

During such a phase the event horizon (1.5) shrinks linearly, and the ratio r of the proper size of a causally connected region to the proper size of the horizon grows in time like $r(t) \sim (-t)^{-1 \mp 1/\sqrt{d}}$, for $t \to 0_-$. On the other hand, the horizon problem of the standard cosmological model is solved if the growth of $r(t)$, for $|t|$ ranging from T to λ_s, is large enough to compensate the decrease of $r(t)$ in the subsequent phase of expanding horizons, down to the present time t_0. This requires [11], by assuming that the end of inflation at $t = -\lambda_s$ is followed by the standard radiation-dominated and matter-dominated evolution,

$$\left(\frac{\lambda_s}{T}\right)^{-1\mp 1/\sqrt{d}} \gtrsim 10^{30} \left(\frac{\ell_p}{\lambda_s}\right)^{1/2} \tag{4.22}$$

We can thus conclude that if the integration constant T is chosen in such a way that the solution (4.16) may describe a phenomenologically interesting inflationary background, then $T >> \lambda_s$. In particular, for $\lambda_s \sim \ell_p$,

$$T \gtrsim 10^{30\sqrt{d}/(\sqrt{d}\pm 1)} \lambda_s \tag{4.23}$$

In that case, according to eq.(4.20), the background (4.16) also represents a good zeroth-order approximation to an exact simultaneous solution of the background equations and of the string equations of motion, having the string perturbative vacuum as initial condition.

5. Conclusion.

In this paper we have studied the number distribution of a classical string network, evolving in a background in which the causal correlation length H^{-1} shrinks to a minimum and then re-expands. Such kind of background emerges naturally from the low energy string effective action when looking for cosmological models based on the string perturbative vacuum as initial condition.

In such a background, the string proper size L is constant inside the horizon, while it varies like the scale factor outside the horizon. As a consequence, the number distribution $n(L)$ varies in time, leading to a variation of the effective equation of state associated to the string network. We have presented two possible approaches to compute the ratio $\gamma = p/\rho$ in terms of the initial distribution and of the background scale factor. In both cases we have checked that, for a test string distribution, γ goes to zero with a power-like behaviour as $|H|^{-1}$ increases; moreover, γ ranges over negative values when the metric describes accelerated expansion, and over positive values when the metric describes accelerated contraction. In the first case, the transition from shrinking to expanding horizons is associated with a transition from decreasing to increasing pressure. The pressure, however, stays always negative, unless quantum effects (such as string decay into radiation) are included, which would eventually bring the equation of state to that of relativistic matter, $\gamma = 1/d$.

We have discussed, finally, a possible method to look for simultaneous solutions of the background equations and of the string propagation equation. We have shown that the simple dust-like solution can approximate, self-consistently, the exact string driven background describing the evolution from the initial string perturbative vacuum to a final highly curved, strong coupling regime. The corrections to such a zeroth-order approximation keep small over the whole temporal range, provided the time extension T of the inflationary regime is much larger than one when measured in string units, $T/\lambda_s >> 1$.

References.

1. H. J. De Vega and N. Sánchez, Phys. Lett. B197 (1987) 320.
2. N. Sánchez and G. Veneziano, Nucl. Phys. B333 (1990) 253.
3. M. Gasperini, N. Sánchez and G. Veneziano, Int. J. Mod. Phys. A6 (1991) 3853; Nucl. Phys. B364 (1991) 365;
   Nguyen Suan Han and G. Veneziano, Mod. Phys. Lett A6 (1991) 1993.
4. M. Gasperini, Phys. Lett. B258 (1991) 70.
5. N. Turok and P. Bhattacharjiee, Phys. Rev. D29 (1984) 1557;
   N. Turok, Phys. Rev. Lett. 60 (1988) 543.
6. G. Veneziano, Europhys. Lett. 2 (1986) 133.
7. G. Veneziano, Phys. Lett. B265 (1991) 287.
8. M. Gasperini and G. Veneziano, Astropart. Phys. 1 (1993) 317; Mod. Phys. Lett. A8 (1993) 3701.
9. H. J. De Vega and N. Sánchez, Phys. Rev. D50 (1994) 7202.
10. R. Brustein and G. Veneziano, Phys. Lett. B329 (1994) 429.
11. M. Gasperini and G. Veneziano, Phys. Rev. D50 (1994) 2519.

PARTICLE TRANSMUTATION AND FERMION NUMBER VIOLATION FROM THE SCATTERING OF STRINGS AND SUPERSTRINGS IN CURVED SPACETIMES

H.J. de Vega

M. Ramón-Medrano

N. Sánchez

ABSTRACT

We study bosonic and fermionic first quantized strings that propagate in a background. New phenomena appear called particle transmutations. Fermion number is violated in supergravity backgrounds.

CONTENTS

I. INTRODUCTION

The study of bosonic strings in (purely "bosonic") curved spacetimes provides new fundamental insights with respect to the study of bosonic strings in flat space-time [1-9]. However, it turns out that for the study of fermionic strings (superstrings), in curved spacetimes, a purely "bosonic" background allows only a partial description [10], as spinor string variables do not couple to bosonic string variables through the background. It is then physically richer to consider backgrounds containing both "fermionic" and "bosonic" degrees of freedom, i.e. supergravity backgrounds [11]. In this way, the spinor superstring coordinates are able to play a much more significant role through their coupling with the given background.

One of the most interesting physical phenomenon, that appears in the study of the scattering of a string by a curved space-time, is **particle transmutation** [12-13]. Let us consider a first quantized (super) string that propagates in a (classical) non-trivial background which admits flat space-time regions, at least asymptotically. In these regions, the string satisfies free equations and we can define ingoing $\left(X^o \rightarrow -\infty\right)$ and outgoing $\left(X^o \rightarrow \infty\right)$ particle states, which are generated by the in and out creation string mode operators acting on their respective vacua $\left(|0\rangle_{in}, |0\rangle_{out}\right)$. We remark that we have a **single** test (super) string, that is, initial and final states are one-particle but different states. The question here is that, due to the interaction of the string with the given geometry, the infinite set of oscillator string modes becomes excited, and **mixing** of particle and antiparticle modes takes place. Then, outgoing operators will be related to ingoing creation and annihilation operators in a non trivial (and, in general, non linear) way, which depends on the given space-time geometry.

As a consequence of this interaction, we will find: (a) **pair creation** of modes, coming from the in to out vacuum transition; (b) **elastic** processes in which initial and final states describe the **same** particle, although momentum and spin polarization may change; (c) inelastic processes in which we have a non-vanishing transition amplitude for a particle, of mass m and spin s, going to a **different** particle of mass m' and spin s'. This last, physically relevant, case is what we have called **Particle Transmutation**.

To reveal the above phenomena, we have analyzed bosonic and fermionic (Neveu-Schwarz-Ramond and Green-Schwarz) strings which propagate not only in curved space-times with "bosonic" degrees of freedom, but also in backgrounds with "fermionic" degrees of freedom. The different nature of the background turns out to have a clear physical relevance. If the background is "purely bosonic", particle transmutation occurs only from **boson** to **boson** and from **fermion** to **fermion** [12], as spinor string variables do **not** couple to bosonic string variables through the background. On the contrary, if the background has "fermionic" degrees of freedom as well, i.e. supergravity backgrounds, spinor string variables **do** couple non trivially to the bosonic string variables trough the "fermionic" background, and then **boson** to **fermion** and **fermion** to **boson** transitions are allowed [13]. In other words, the presence of a "bosonic-fermionic" background provides in string theory a natural mechanism **to violate fermion number**, wich is an entirely new feature of particle transmutation. This kind of background "injects" half- integer spin into the particle states. Notice that this is a pure stringy effect since local free fields in such a background do not exhibit these phenomena.

The carrying of this program has practical difficulties, as the string equations of motion can be solved exactly only in some particular backgrounds: among them, we consider here space-times generated by gravitational shockwaves (Aichelburg-Sexl geometry) [3, 6, 7, 8] and by cosmic strings (conical space-times) [9, 14]. For a general background, string equations have to be solved approximately (e.g for black holes space-times). Our perturbative scheme is as follows [1]: we treat the space-time **exactly**, and take, as perturbations, the string oscillations around its centre of mass solutions; this corresponds to low excitations of the string compared with the energy scales of the geometry. The perturbation parameter will be the dimensionless constant $g = \sqrt{\alpha'} / R_c$, where α' is the string constant and R_C characterizes the curvature radius of the given space-time. At first order, outgoing (creation or annihilation) operators will be expressed as **linear** combinations of the ingoing (creation and annihilation) operators; the coefficients of this linear, or Bogoliubov, transformation will give us the scattering amplitudes for pair creation, elastic and inelastic processes.

Finally, according to what we have said before, we organize our subject as follows:

In section II, we study (open and closed) bosonic and fermionic strings whch propagate in a "bosonic" bakground. This section is divided in two main subsections, II.1 and II.2.

In subsect II.1 (and in the framework of the perturbative scheme), Bogoliubov transformations are given for a **general** curved space-time. Scattering amplitudes are calculated for (open and closed) bosonic strings and superstrings. As examples, we present the more relevant inelastic amplitudes for black holes and shock waves space-time metrics (e.g dilaton to graviton transmutation amplitude for a closed string) [12].

In subsect II.2, particle transmutation and creation are analyzed for the Neveu-Schwarz-Ramond (NRS) and Green-Schwarz (GS) superstring models in the presence of a straight **cosmic string**, of zero thickness and infinite lenght, **with** or **without spin** [14]. Previously, we have solved these models **exactly** in this conical space-time. We remind here that this geometry is locally flat, but has a cone-like singularity at the cosmic string with a deficit angle Δ equals to $8\pi G\mu$ (μ is the cosmic string tension, and G the Newton constant). We see that inelastic scattering happens only when the string **touches** the cosmic string; if there is no collision the string suffers only a deflection ($\pm \Delta/2$ for NSR; $\pm \Delta/4$ for GS) and the oscillator modes have only a polarization change.

In section III, we consider a superstring scattering by a background which has "bosonic" **and** "fermionic" degrees of freedom. This section is also divided in two parts.

The first step is the construction of the "fermionic-bosonic" background. In subsect III.3, we propose an N=1 supergravity version of an Aichelburg-Sexl type metric in ten dimensions [11]. We should point out here that for this purely "bosonic" shockwave background it has been shown that the string fermionic coordinates **decouple** and obey **free** equations [10].

Due to the complexity of the supergravity equations, we solve a linearized D-dimensional Rarita-Schwinger or gravitino field equation, but we keep the **exact** (full non-linear) bosonic shockwave background; also, trivial solutions for the rest of the fields of the supergravity multiplet are chosen. This set of solutions would correspond physically to a weak matter source background. The gravitino background provides a classical description for a coherent state of gravitinos. This classical approximation is valid when the number of gravitinos is very large, in the same may as the bosonic shockwave background describes a superposition of a very large number of gravitons. Moreover, conformal invariance would hold in this background, since the N=1 supergravity equations coincide with the vanishing of the one-loop supertstring βeta function.

In subsect III.4, we investigate a heterotic Green-Schwarz superstring coupled to the above background [11]. We take a heterotic G-S superstring as one cannot accomodate both the right and left movers of the G-S superstring. The solution of the equations of motion for the string variables shows that there is a **mixing** between fermionic **and** bosonic oscillator modes, due to the presence of the "fermionic" background (Rarita-Schwinger or gravitino field). As a consequence, we find non-vanishing transition amplitudes from an ingoing bosonic state to an outgoing fermionic state, and viceversa [13]. In particular, ground state to ground state transition means that a massless **fermion** initial state can **transmute** into a massless **boson** final state, and viceversa. Therefore, we have **fermion number violation**.

II. BOSONIC AND FERMIONIC STRINGS IN A PURELY "BOSONIC" BACKGROUND

II.1 GENERAL FORMALISM

II.1.1 Bosonic strings in curved space-time: "In" and "out" formulation

For a bosonic string in a curved space time, the action is given by

$$S = \frac{1}{4\pi\alpha'} \int d\sigma\, d\tau \sqrt{g}\; g^{AB} G_{\mu\nu}(X)\, \partial_A X^\mu \partial_B X^\nu \tag{1.1}$$

$$0 \leq A, B \leq 1 \qquad , \quad 0 \leq \mu, \nu \leq D-1$$

and the equations of motion are

$$\partial^2 X^\mu + \Gamma^\mu_{\nu\lambda}(X)\, \partial_+ X^\nu\, \partial_- X^\lambda = 0 \tag{1.2}$$

Here $\Gamma^\mu_{\nu\lambda}$ (X) are the Christoffel connections of the space-time metric $G_{\mu\nu}$ (X)

$$\zeta_\pm = \frac{1}{2}(\sigma \pm \tau) \quad , \quad \partial_\pm = \partial_\sigma \pm \partial_\tau \quad , \quad \partial^2 = \partial_+ \partial_- \tag{1.3}$$

and the conformal gauge on the world sheet metric g_{AB} has been taken; $(2\pi\alpha')^{-1}$ is the string tension. For strings propagating in a curved space-time, the equations of motion are non-linear, so right and left movers interact with each other and also with themselves. For a given $G_{\mu\nu}$, the general solution of eqs.(1.2) is not exactly known. The string equations of motion can be solved approximately in a perturbative scheme [1] in which eqs. (1.2) are linearized around an exact particular solution (usually taken as the center of mass of the string). In this scheme, the space of solutions is represented as

$$X^\mu(\sigma,\tau) = q^\mu(\tau) + \eta^\mu(\sigma,\tau) + \xi(\sigma,\tau) + \cdots \tag{1.4}$$

where $q^\mu(\tau)$ is the exact point-particle solution

$$\ddot{q}^\mu(\tau) + \Gamma^\mu_{\nu\lambda}\, \dot{q}^\nu \dot{q}^\lambda = 0 \tag{1.5}$$

with τ describing the proper time of the center of mass trajectory and $\eta^\mu(\sigma,\tau)$ and $\xi^\mu(\sigma,\tau)$ the first and second order fluctuations respectively, obtained by linearization around q^μ. Higher order fluctuations can also be considered, but for our purposes here it is enough to limit ourselves to fluctuations η^μ. They satisfy

$$D^\mu_\nu(\tau)\ \eta^\nu(\sigma,\tau) = 0 \tag{1.6}$$

$$D^\mu_\nu(\tau) \equiv \delta^\mu_\nu\,\partial^2 - 2\,\Gamma^\mu_{\nu\lambda}(q)\,\dot q^\lambda\,\partial_\tau - \dot q^\lambda\,\dot q^\sigma\,\partial_\nu\,\Gamma^\mu_{\lambda\sigma}(q)$$

Since q^μ only depends on τ, η^μ admits the Fourier expansion

$$\eta^\mu(\sigma,\tau) = \sum_{n\in Z} e^{in\sigma}\, f^\mu_n(\tau) \tag{1.7}$$

where f^μ_n satisfy *decoupled* ordinary differential equations

$$\ddot f^\mu_n + n^2 f^\mu_n + 2\,\Gamma^\mu_{\nu\lambda}(q)\,\dot q^\nu\,\dot f^\lambda_n + \dot q^\nu\,\dot q^\lambda\,\partial_\lambda\Gamma^\mu_{\nu\rho}(q)\,f^\rho_n = 0 \tag{1.8}$$

These first order fluctuations describe the interaction between the string modes and the space-time geometry. Interactions between the string modes themselves start to appear from second order ξ^μ.

In this perturbative scheme we treat the space-time metric *exactly* and take the string oscillations around its centre of mass solution as perturbations. This corresponds to low excitations of the string compared with the energy scales of the geometry; in other words, this is a strong gravitational field expansion. The perturbation parameter is the dimensionless constant:

$$g = \sqrt{\alpha'}\,/\,R_c \tag{1.9}$$

where R_c characterizes the curvature radius of the space-time under consideration.

If the metric $G_{\mu\nu}$ admits flat space time regions, say (in) and (out) (i.e. $G_{\mu\nu} \to \eta_{\mu\nu}$), as it is the case for the shock wave space-times or for asymptotically flat geometries, then eqs. (1.2) become in these regions

$$\partial^2 X^\mu_{in,\,out} = 0 \tag{1.10}$$

In particular $\partial^2 X^+_{\substack{in\\out}} = 0$, and the light cone gauge identification can be chosen

$$X^+_{\substack{in\\out}} = \frac{1}{\sqrt{2}}(X^0 + X^1)_{\substack{in\\out}} = 2\alpha' p^+_{\substack{in\\out}} \tau \tag{1.11}$$

This enables us to define scattering ingoing (<) and outgoing (>) solutions for $\tau \to -\infty$ and $\tau \to +\infty$ respectively, namely $(0 \lessgtr \sigma \lessgtr \pi)$

$$X^\mu_{\lessgtr} = q^\mu_{\lessgtr} + 2\alpha' p^\mu_{\lessgtr}\tau + i\sqrt{\frac{\alpha'}{2}} \sum_{n\neq 0} \frac{1}{n}\left(\alpha^\mu_{n\lessgtr} e^{-2in(\tau-\sigma)} + \tilde{\alpha}^\mu_{n\lessgtr} e^{-2in(\tau+\sigma)}\right)$$

(closed string) (1.12a)

$$X^\mu_{\lessgtr} = q^\mu_{\lessgtr} + 2\alpha' p^\mu_{\lessgtr}\tau + i\sqrt{2\alpha'} \sum_{n\neq 0} \frac{1}{n} \alpha^\mu_{n\lessgtr} e^{-in\tau} \cos n\sigma$$

(open string) (1.12b)

where

$$[\alpha^\mu_n, \alpha^\nu_m] = n\,\delta_{n+m,0}\,\eta^{\mu\nu}, \quad \alpha^{\mu +}_{n\lessgtr} = \alpha^\mu_{-n\lessgtr} \qquad (\alpha \to \tilde{\alpha})$$

The connection between the out-operators $\alpha^\mu_{n>}$ and the in-operators $\alpha^\mu_{n<}$ depends on the interaction in the non-zero curvature regions. In general, the exact relation between $\alpha^\nu_{n>}$ and $\alpha^\nu_{n<}$ will be highly non-trivial involving all the in -oscillators $\alpha^\mu_{k<}$ and $\alpha^\mu_{-k<}$ $(1 \leq k \leq \infty)$ through a non-linear transformation

$$\alpha^\mu_{n>} = F^\mu_n(\ldots\ldots \alpha^\nu_{k<}, \alpha^\nu_{-k<} \ldots) \tag{1.13}$$

This transformation contains in principle, all the information about the scattering (elastic and inelastic) of the string by the geometry. However, it is very difficult to derive the functions F_n except for the exactly solvable cases [3,9,10,14]. It is possible to perform detailed analysis in the context of the linearization scheme described above (thus the transformation eq.(1.13) will become a linear, i.e. Bogoliubov, transformation). In this scheme we will have

$$\partial^2 \eta^\mu_{\lessgtr} = 0 \tag{1.14}$$

Let us consider first *closed* strings. Then we can define a basis of solutions $f^{\mu\nu}_{n,\pm}(\tau)$ of eq.(1.8) by selecting the asymptotic behaviour:

$$f^{\mu\nu}_{n,\pm}(\tau) \underset{\tau\to\pm\infty}{=} e^{-2in\tau}\,\eta^{\mu\nu} \tag{1.15}$$

The choice of a positive frequency factor $e^{-2in\tau}$ corresponds to in or out particle states for $\tau \to -\infty$ or $\tau \to +\infty$ respectively. For each value of μ,ν we have two basis (-) and (+). Since we have free oscillators in both asymptotic limits $\tau \to \pm\infty$,

$$f^{\mu\nu}_{n,\pm}(\tau) \underset{\tau\to\mp\infty}{=} e^{-2in\tau} A^{\mu\nu}_{n,\pm} + e^{2in\tau} B^{\mu\nu}_{n,\pm} \tag{1.16}$$

where the constant coefficients A , B depend on the detailed scattering for all τ.

We can expand the string fluctuations η^μ in the basis $f^{\mu\nu}_{n,\pm}$ as:

$$\eta^\mu(\sigma,\tau) = i\sqrt{\frac{\alpha'}{2}}\sum_{n\neq 0} f^{\mu\nu}_{n,\pm}(\tau)\frac{1}{n}\left[\alpha_{n\lessgtr,\nu}\, e^{2in\sigma} + \tilde{\alpha}_{n\lessgtr,\nu}\, e^{-2in\sigma}\right] \tag{1.17}$$

Then, from eqs.(1.15) - (1.17) and (1.14) we have

$$\eta^\mu_<(\sigma,\tau) = i\sqrt{\frac{\alpha'}{2}}\sum_{n\neq 0}\frac{1}{n}\left[\alpha^\mu_{n<}\, e^{-2in(\tau-\sigma)} + \tilde{\alpha}^\mu_{n<}\, e^{-2in(\tau+\sigma)}\right] \tag{1.18a}$$

$$\eta^\mu_>(\sigma,\tau) = i\sqrt{\frac{\alpha'}{2}}\sum_{n\neq 0}\frac{1}{n}\left[\alpha^\mu_{n>}\, e^{-2in(\tau-\sigma)} + \tilde{\alpha}^\mu_{n>}\, e^{-2in(\tau+\sigma)}\right] \tag{1.18b}$$

where

$$\begin{aligned}
\alpha^\mu_{n>} &= A^\mu_{n\nu}\,\alpha^\nu_{n<} - B^\mu_{-n\nu}\,\tilde{\alpha}^\nu_{-n<} \\
\tilde{\alpha}^\mu_{n>} &= A^\mu_{n\nu}\,\tilde{\alpha}^\nu_{n<} - B^\mu_{-n\nu}\,\tilde{\alpha}^\nu_{-n<} \\
A^{\mu\nu}_n &\equiv A^{\mu\nu}_{n,-} \quad , \quad B^{\mu\nu}_n \equiv B^{\mu\nu}_{n,-}
\end{aligned} \tag{1.19}$$

This is a Bogoliubov transformation, giving the out operators $\alpha^{\mu}_{n>}$ as a *linear* superposition of the creation and annhilation ingoing operators. We see that, for fixed n, transitions take place between the internal oscillatory modes of the string. The scattering of the string by the curved geometry involves two main effects:

(i) Polarization changes in the modes (without changing their right or left character).

(ii) A mixing of particle and antiparticle modes, changing at the same time their right or left character.

In order words: if the string in the ingoing state has a right (or left) excited mode with a given polarization ν, there will be in the outgoing state:

(i) an amplitude $\mathcal{A}^{\mu\nu}_n$ for a right (or left) mode polarized in the μ-direction and

(ii) an amplitude $\mathcal{B}^{\mu\nu}_n$ for a left (or right) antiparticle mode polarized in the μ-direction $(0 \leq \mu \leq D-1)$.

The Bogoliubov transformation eq.(1.19) can be witten as

$$\alpha_{n>} = e^{G} \, \alpha_{n<} \, e^{-G} \tag{1.20}$$

where, since the coefficients of the transformation eq.(1.19) are not necessarily real, G is given by

$$G = \sum_{n=1}^{\infty} \frac{1}{n} \Big\{ \left(D^{\mu\nu}_n \alpha_{n\mu} \tilde{\alpha}_{n\nu} - D^{*\mu\nu}_n \alpha_{-n\mu} \tilde{\alpha}_{-n\nu} \right) + \left(C^{\mu\nu}_n - C^{*\nu\mu}_n \right) \left(\alpha_{n\mu} \alpha_{-n\nu} + \tilde{\alpha}_{n\mu} \tilde{\alpha}_{-n\nu} \right) \Big\} \tag{1.21}$$

with $(D^{*\mu\nu}_n = D^{*\nu\mu}_n)$, which satisfies $G = -G^{+}$ $([\alpha^{\mu}_n, \tilde{\alpha}^{\nu}_n] = 0)$. In eq.(1.20), it is understood that G refers to the operators $\alpha_{n<}$.

At the first order -in which we are working- we just have

$$\alpha^{\mu}_{n>} = \left[\eta^{\mu\nu} - \left(C^{\nu\mu}_{n} - C^{*\mu\nu}_{n}\right)\right] \alpha_{n<,\nu} + D^{*\mu\nu}_{n}\, \tilde{\alpha}_{-n<,\nu} \tag{1.22}$$

$$(\alpha_n \rightarrow \tilde{\alpha}_n)$$

since $C^{\mu\nu}$ and $D^{\mu\nu}$ are of order g.

From eqs.(1.19) and (1.22) we see that

$$\begin{aligned} A^{\mu\nu}_{n} &= \eta^{\mu\nu} + \left(C^{*\mu\nu}_{n} - C^{\nu\mu}_{n}\right) \\ B^{\mu\nu}_{-n} &= -\, D^{*\mu\nu}_{n} \end{aligned} \tag{1.23}$$

Similarly, the in (<) and out (>) vacua

$$\alpha^{\mu}_{n<} |0_{in}\rangle = 0 \quad , \quad \alpha^{\mu}_{n>} |0_{out}\rangle = 0 \quad , \quad n > 0 \tag{1.24}$$

are related by

$$|0_{out}\rangle = e^{G} |0_{in}\rangle \tag{1.25}$$

Therefore, from eqs.(1.21) - (1.25), the in-out vacuum transition amplitude is given at first order by

$$\begin{aligned} \langle 0_{out} | 0_{in} \rangle &= \langle 0_{in} | 1 - G | 0_{in} \rangle \\ &= 1 - 4i \sum_{n=1}^{\infty} \mathrm{Im}\, C^{\mu}_{n,\mu} \end{aligned} \tag{1.26}$$

We see that the out-vacuum follows from the in-vacuum by filling it with ingoing mode pairs (here a pair mode is formed by a right and left mode). Therefore, pair creation of modes out of the in-vacuum takes place for $\tau \rightarrow +\infty$ as a consequence of the scattering by the curved space time geometry. This is also a manifestation of the composite character of strings. The infinite set of oscillator modes, constituting the string, becomes excited during the scattering by the influence of the gravitational field. In fact, any localized external field (gravitational or not) would lead to qualitatively similar effects.

A similar treatment holds for the *open* string. We now have

$$\eta^{\mu}(\sigma,\tau) = i\sqrt{2\alpha'} \sum_{n\neq 0} f^{\mu\nu}_{n,\pm}(\tau)\,\frac{\alpha_{n,\nu}}{n}\cos n\sigma \tag{1.27}$$

The basis $f^{\mu\nu}_{n,\pm}$ is defined as in eqs.(1.15) and (1.16) but without the factor 2 in the τ-exponentials. The Bogoliubov transformation is now

$$\begin{aligned} \alpha^{\mu}_{n>} &= e^{G}\,\alpha^{\mu}_{n<}\,e^{-G} \\ &\simeq A^{\mu}_{n,\nu}\,\alpha^{\nu}_{n<} - B^{\mu}_{-n,\nu}\,\alpha^{\nu}_{-n<} \end{aligned} \tag{1.28}$$

with G given by

$$\begin{aligned} G_{\rm open} = \sum_{n=1}^{\infty} \Big\{ & \frac{1}{2n}\left(D_{n,\mu\nu}\,\alpha^{\mu}_{n<}\,\alpha^{\nu}_{n<} - D^{*}_{n,\mu\nu}\,\alpha^{\mu}_{-n<}\,\alpha^{\nu}_{-n<}\right) \\ & + \frac{1}{n}\left(C_{n,\mu\nu} - C^{*}_{n,\nu\mu}\right)\alpha^{\mu}_{n<}\,\alpha^{\nu}_{-n<} \Big\} \end{aligned} \tag{1.29}$$

II.1.2 Superstrings in curved space-time: In and out formulation

Let us consider now open superstrings. In the light-cone gauge, the Green-Schwarz action in flat space-time reads

$$S = \frac{1}{2\pi}\int d\sigma\, d\tau \left[-\frac{1}{2\alpha'}\,\partial_A X^j\,\partial^A X^j + i\,\bar{S}^a \rho^A \partial_A S^a\right] \tag{1.30}$$

where $S^a = \begin{pmatrix} S^{1a} \\ S^{2a} \end{pmatrix}$ is a two component Majorana world-sheet spinor and a ($1 \le a \le 8$) is a spinor index in a ten dimensional space time. S^a is a Weyl and Majorana spinor transforming under the representation 8_s [15]; ρ^A represents two-dimensional Dirac matrices (A = 0,1)

$$\rho^o = \sigma_2 = \begin{pmatrix} 0 & -i \\ i & 0 \end{pmatrix} \quad , \quad \rho^1 = i\sigma_1 = \begin{pmatrix} 0 & i \\ i & 0 \end{pmatrix} \tag{1.31}$$

$$\{ \rho^A , \rho^B \} = -2\, \eta^{AB} \quad , \quad \eta^{AB} = \mathrm{diag}\,(-1,1) \tag{1.32}$$

The spinor fields obey the field equations

$$(\partial_\sigma + \partial_\tau)\, S^{1a} = 0 \qquad , \quad (\partial_\sigma - \partial_\tau)\, S^{2a} = 0 \tag{1.33}$$

and the open type I superstring (N=1 supersymmetry) satisfies the boundary conditions

$$S^{1a}(0,\tau) = S^{2a}(0,\tau) \qquad , \quad S^{1a}(\pi,\tau) = S^{2a}(\pi,\tau) \tag{1.34}$$

Therefore, the solution may be written in general as

$$S^{1a}(\sigma,\tau) = \frac{1}{\sqrt{2}} \sum_{n\in Z} S_n^a \, e^{-in(\tau-\sigma)}$$

$$S^{2a}(\sigma,\tau) = \frac{1}{\sqrt{2}} \sum_{n\in Z} S_n^a \, e^{-in(\tau+\sigma)} \tag{1.35}$$

where

$$S_{-n}^a = S_n^{a+}$$

and

$$\{ S_n^a , S_m^b \} = \delta^{ab}\, \delta_{n+m,0} \tag{1.36}$$

It is well known that the Fock space of states for the open superstring consists of 16 massless ground states and an infinite number of excited states obtained by applying the S_{-n}^{a} $(n \geq 1)$, on them. These are eight massless Bose particles $|i\rangle$, transforming under the 8_v representation of the Spin (8) Group, and eight Fermi particles $|\dot{a}\rangle$ transforming under 8_c.

$$|\Phi_0\rangle = \{|i\rangle, |\dot{a}\rangle\}$$

They are normalized as

$$\begin{aligned} \langle i | j \rangle &= \delta_{ij} \quad , \quad 1 \leq i, j \leq 8 \\ \langle \dot{a} | \dot{b} \rangle &= \delta_{\dot{a}\dot{b}} \quad , \quad 1 \leq \dot{a}, \dot{b} \leq 8 \end{aligned} \tag{1.37}$$

and satisfy

$$\alpha_n^j \, , \, S_n^a \, |\Phi_0\rangle = 0 \quad , \quad n > 0$$

The bosonic and fermionic particle states $|i\rangle$ and $|a\rangle$ transform into each other under the action of the supercharges.. i.e. the ground states form a supersymmetric multiplet.

In curved space-time, the superstring action is much more involved since there are several space time and world-sheet invariants that can be built using the metric tensor, Christoffel connections and the zehn-beine. On the other hand, it is not possible to impose the light-cone gauge $X^0 = 2\alpha' p\tau$ in the whole space-time. Nevertheless, since we assume our space-time to be flat in the ingoing and outgoing regions, we can use there the usual flat space treatment (eqs. (1.35) - (1.37)). That is, $S^a(\sigma,\tau)$ will obey eqs. (1.33) and (1.34) only for $\tau \to -\infty$ and $\tau \to +\infty$. By analogy with the bosonic case [eq.(1.14)], $S^a(\sigma,\tau)$ obeys - at first order in $g = \sqrt{\alpha'}/R_c$ - a partial differential equation with σ-independent coefficients, since the curved space-time enters here through the center of mass motion. A basis of positive energy solutions can be written as

$$\left\{ f_{n-}^{ab}(\tau)\, e^{in\sigma} \, , \, g_{n-}^{ab}(\tau)\, e^{in\sigma} \, ; n \in Z \, , \, 1 \leq a, b \leq 8 \right\} \tag{1.38}$$

where

$$f_{n-}^{ab}(\tau) \underset{\tau \to -\infty}{=} \delta^{ab} \begin{pmatrix} e^{-in\tau} \\ e^{in\tau} \end{pmatrix} \quad , \quad g_{n-}^{ab}(\tau) \underset{\tau \to -\infty}{=} i\,\delta^{ab} \begin{pmatrix} e^{-in\tau} \\ -e^{in\tau} \end{pmatrix} \tag{1.39}$$

The precise form of the functions $f_n^{ab}(\tau)$ and $g_n^{ab}(\tau)$ being determined by the fermionic analogue of eq. (1.8). We shall not consider them here. It will be enough for us to write their asymptotic behaviour for $\tau \to +\infty$. Since eqs. (1.33) and (1.34) hold asymptotically in the outgoing limit, we have

$$f_{n-}^{ab}(\tau) \underset{\tau\to\infty}{=} \begin{pmatrix} \mathcal{A}_n^{ab} e^{-in\tau} \\ \mathcal{A}_n^{*ab} e^{in\tau} \end{pmatrix} , \quad g_{n-}^{ab}(\tau) \underset{\tau\to\infty}{=} \begin{pmatrix} \mathcal{B}_n^{ab} e^{-in\tau} \\ \mathcal{B}_n^{*ab} e^{in\tau} \end{pmatrix} \tag{1.40}$$

The coefficients $\mathcal{A}_n^{ab}$ and $\mathcal{B}_n^{ab}$ are in general complex and of order $\sqrt{\alpha'}/R_c$. In addition, we used the fact that the exact differential equation for $S^a(\sigma,\tau)$ has real coefficients.

Alternatively, we could define an outgoing basis

$$\left\{ f_{n+}^{ab}(\tau)\, e^{in\sigma} ,\ g_{n+}^{ab}(\tau)\, e^{in\sigma} ;\ n \in Z ,\ 1 \le a,b \le 8 \right\} \tag{1.41}$$

by conditions similar to eqs (1.40).

The ingoing and outgoing fields are defined as usual as

$$S_<^a(\sigma,\tau) \underset{\tau\to-\infty}{=} S^a(\sigma,\tau) , \quad S_>^a(\sigma,\tau) \underset{\tau\to+\infty}{=} S^a(\sigma,\tau) \tag{1.42}$$

They can be expanded as

$$\begin{aligned} S_<^a(\sigma,\tau) &= \frac{1}{\sqrt{2}} \sum_{n\in Z} S_{n<}^a \begin{pmatrix} e^{-in(\tau-\sigma)} \\ e^{-in(\tau+\sigma)} \end{pmatrix} \\ &= \frac{1}{\sqrt{2}} \sum_{n\in Z} e^{in\sigma} \left\{ \begin{pmatrix} e^{-in\tau} \\ e^{in\tau} \end{pmatrix} S_{n<}^{(R)a} + i \begin{pmatrix} e^{-in\tau} \\ -e^{in\tau} \end{pmatrix} S_{n<}^{(I)a} \right\} \end{aligned} \tag{1.43}$$

where $S_{n<}^{(R)a} = \dfrac{S_{n<} + S_{-n<}}{2}$ and $S_{n<}^{(I)a} = \dfrac{S_{n<} - S_{-n<}}{2i}$

Therefore we have

$$S^a(\sigma,\tau) = \frac{1}{\sqrt{2}} \sum_{n\in Z} e^{in\sigma} \sum_{b=1}^{8} \left\{ f_{n-}^{ab}(\tau)\, S_{n<}^{(R)b} + g_{n-}^{ab}\, S_{n<}^{(L)b} \right\} \tag{1.44}$$

Taking the $\tau \to +\infty$ limit of eq. (1.44) and with the help of eq.(1.40), $S^a_>$ can be written in the form

$$S^a_>(\sigma,\tau) = \frac{1}{\sqrt{2}} \sum_{n\in Z} e^{in\sigma} \sum_{b=1}^{8} \begin{pmatrix} (A_n^{ab} S_{n<}^b + B_n^{ab} S_{-n<}^b) e^{-in\tau} \\ (B_n^{*ab} S_{n<}^b + A_n^{*ab} S_{-n<}^b) e^{in\tau} \end{pmatrix} \tag{1.45}$$

which can be recast in the usual form

$$S^a_>(\sigma,\tau) = \frac{1}{\sqrt{2}} \sum_{n\in Z} S^a_{n>} \begin{pmatrix} e^{-in(\tau-\sigma)} \\ e^{-in(\tau+\sigma)} \end{pmatrix} \tag{1.46}$$

where

$$S^a_{n>} = \sum_{b=1}^{8} \left(C_n^{ab} S_{n<}^b + D_n^{ab} S_{-n<}^b \right)$$

and

$$C_n^{ab} = A_n^{ab} - i B_n^{ab} \quad , \quad D_n^{ab} = A_n^{ab} + i B_n^{ab}$$

Therefore, the resultant effect of the space-time geomety on the fermionic string is a Bogoliubov transformation mixing also the different spinor components. Notice that at this order, in $\sqrt{\alpha'}/R_c$, there is no mixing of modes of different n. In particular, for n = 0 we find

$$S^a_{0>} = 2 \sum_{b=1}^{8} A_0^{ab} S_{0<}^b$$

Namely, the outgoing zero modes only suffer a rotation.

Let us now analyze the physical consequences that are independent of the detailed form of the $\mathcal{C}_n^{ab}$ and $\mathcal{D}_n^{ab}$ - coefficients, but which depend only on their tensor structure and the fact that they are of order $\sqrt{\alpha'}/R_c$.

In order to preserve the canonical anticommutation rules, eq. (1.36), the coefficients $\mathcal{C}_n{}^{ab}$ and $\mathcal{D}_n{}^{ab}$ must satisfy

$$\sum_{b=1}^{8} \left(\mathcal{C}_n^{ab} \mathcal{C}_n^{bc} + \mathcal{D}_n^{ab} \mathcal{D}_n^{bc} \right) = \delta^{ac} \tag{1.47}$$

It is interesting to write down the operator that generates the transformation eq.(1.46) in the Fock space. Then we express eq. (1.46) as

$$S_{n>}^{a} = e^{G} S_{n<}^{a} e^{-G} \tag{1.48}$$

where

$$G = \frac{1}{2} \sum_{n=1}^{\infty} \sum_{a,b=1}^{\infty} \left[(\delta^{ab} - \mathcal{C}_n^{ab}) S_{n<}^{a} S_{-n<}^{b} - \mathcal{D}_n^{*ab} S_{n<}^{a} S_{n<}^{b} - \mathcal{D}_n^{ab} S_{-n<}^{a} S_{-n<}^{b} \right] \tag{1.49}$$

II.1.3 Particle transmutations for bosonic strings

Let us analyze now the particle proceses induced by the interaction of the string with the geometry. As in previous sections we study this at first order in our perturbation scheme i.e. at first order in $g = \sqrt{\alpha'}/R_c$.

For bosonic strings the initial particle states $|\Psi_{in}(m,s)\rangle$ are described by appropriated linear combinations of products of in-oscillator operators $\alpha_{-n.<}^{\mu}$ $(n \geq 1)$ acting on $|0_{in}\rangle$. These vectors must be eigenstates of the mass operators (m^2) and of the spins (s). The final states $|\Psi_{out}(m',s')\rangle$ are constructed similarly from the out-oscillators. Since the ingoing and outgoing operators are inequivalent, ingoing and outgoing observers see different particle states. The scalar product

$$\langle \Psi_{out}(m',s') \mid \Psi_{in}(m,s) \rangle$$

describes the probability amplitude for an outgoing observer to find a particle (m',s') when the initial state was a particle (m,s). When (m,s) = (m',s') this corresponds to a elastic process where we have the same particle in the initial and final states, although the momentum and spin polarizations change. Otherwise, we have an inelastic process where the initial particle (m,s) *transmutes* into a different final particle (m',s') = (m,s).

Let us consider now the more relevant transitions:

(i) *Ground state-ground state transition (for both closed and open strings)*

This is the scalar product $< 0_{in} | 0_{out} >$ computed in section II.1.1.

$$< 0_{in} | 0_{out} > = 1 - 4i \sum_{n=1}^{\infty} \sum_{i=2}^{D-1} \mathrm{Im}(C_n^{ii}) \qquad \text{(closed)} \qquad (1.50)$$

$$< 0_{in} | 0_{out} > = 1 - 2i \sum_{n=1}^{\infty} \sum_{i=2}^{D-1} \mathrm{Im}(C_n^{ii}) \qquad \text{(open)} \qquad (1.51)$$

(ii) *Ground state-vector transition (open strings)*

This is the scalar product

$$\begin{aligned} \langle \psi^i_> | 0_{in} \rangle &= \langle 0_{out} | \alpha^i_{n>} | 0_{in} \rangle \\ &= \langle 0_{in} | \alpha^i_{n<} e^{-G} | 0_{in} \rangle = 0 \end{aligned} \qquad (1.52)$$

where we have used eqs. (1.20), (1.25) and (1.21). Therefore, the transition from the ground state to an arbitrary vector particle state vanishes identically at first order in g. One should notice that this amplitude will be non zero at next order (g^2), due to terms higher than linear in the operator transformation (1.13).

(iii) *Ground state-high massive state transitions (open strings)*

More generally, the transition from the ground state to a state with an *odd* number of creation operators

$$| \psi^i_{2K+1\,out} \rangle = \prod_{s=1}^{2K+1} \alpha^{i_s}_{-n_s>} | 0_{out} \rangle \qquad (1.53)$$

vanishes identically to this order.

On the contrary, the transition to a state with an *even* number of creation operators

$$|\psi^{i}_{2K\,out}\rangle = \prod_{s=1}^{2K} \alpha^{i_s}_{-n_s>}\,|0_{out}\rangle \tag{1.54}$$

is in general non-zero at first order. Such transition will be proportional to $(\beta^{ij}_{n})^{k}$
These final states describe highly massive states ($m \gtrsim m_{Planck}$)

(iv) *Ground state-graviton and scalar transitions (closed strings)*

Let us consider

$$|\psi^{ij}_{out}\rangle = \alpha^{i}_{-n>}\,\tilde{\alpha}^{j}_{-n>}\,|0_{out}\rangle$$

From eqs. (1.25), (1.21) and (1.23) we find

$$\langle\psi^{ij}_{out}|\,0_{in}\rangle = -\,n\,\mathcal{B}^{ij}_{-n} \tag{1.55}$$

The symmetric and traceless part of this scalar product describes the transition from the ground state to a final graviton state with ij polarization, namely

$$<\text{graviton, ij}_{out}\,|\,0_{in}> = -\frac{1}{2}\left(\mathcal{B}^{ij}_{-1} + \mathcal{B}^{ji}_{-1}\right) + \frac{\delta^{ij}}{D-2}\sum_{K=2}^{D-1}\mathcal{B}^{KK}_{-1} \tag{1.56a}$$

The trace part and the antisymmetric parts of eq.(1.55) describe respectively the transitions from the ground state to a massless scalar (dilaton) and to a massles antisymmetric second rank tensor, that is

$$<\text{dilaton}_{out}\,|\,0_{in}> = -\sum_{i=2}^{D-1}\mathcal{B}^{ii}_{-1} \tag{1.56b}$$

$$<\text{antisymm tensor ij}_{out}\,|\,0_{in}> = -\frac{1}{2}\left(\mathcal{B}^{ij}_{-1} - \mathcal{B}^{ji}_{-1}\right) \tag{1.56c}$$

(v) *Ground state-high massive state transitions (closed strings)*

The transition from the ground state to a highly massive state

$$|\psi^{ij}_{k\,out}\rangle = \prod_{s=1}^{k} \alpha^{i_s}_{n>} \tilde{\alpha}^{j_s}_{m>} |0_{out}\rangle$$

is in general non-zero at first order. This amplitude turns out to be proportional to $(\beta^{ij}_n)^k$.

(vi) *Vector-vector state transitions (open strings)*

The transition amplitude from the initial state

$$|\psi^i_m\rangle = \alpha^i_{-m<} |0_{in}\rangle$$

to the final state

$$|\psi^j_{out}\rangle = \alpha^j_{-n>} |0_{out}\rangle$$

is given by

$$\langle \psi^i_{in} | \psi^j_{out} \rangle = \delta_{nm} A^{*ij}_n \tag{1.57}$$

(vii) *Dilaton-graviton transmutation amplitude (closed strings)*

The transmutation of a dilaton in the initial state into an outgoing graviton is given by the matrix element

$$\langle \psi_{dil,in} | \psi^{ij}_{grav,out} \rangle = -4i\, \mathrm{Im}\,(C^{ij}_1 + C^{ji}_1) + \frac{8i}{D-2} \delta^{ij} \sum_{k=2}^{D-1} \mathrm{Im}(C^{kk}_1) \tag{1.58}$$

where

$$|\psi_{dil,in}\rangle = \sum_{i=2}^{D-1} \alpha^i_{-1<} \tilde{\alpha}^i_{-1<} |0_{in}\rangle$$

$$|\psi_{grav,out}\rangle = \left[\alpha^i_{-1>} \tilde{\alpha}^j_{-1>} + \alpha^j_{-1>} \tilde{\alpha}^i_{-1>} - \frac{2}{D-2} \delta^{ij} \sum_{k=2}^{D-1} \alpha^k_{-1>} \tilde{\alpha}^k_{-1>} \right] |0_{out}\rangle$$

II.1.4 Particle transmutations for superstrings

a/ OPEN SUPERSTRING:

Let us discuss first the particle transmutation proceses for open superstrings, in a similar way as we did for bosonic strings. The more relevant processes are:

(i) *Ground state-ground state transitions*

These are the scalar products

$$< j_{out} \mid i_{in} > \ , \qquad < \dot{a}_{out} \mid \dot{b}_{in} > \ , \qquad < j_{out} \mid b_{in} > \ ,$$

where $|j>$ and $|\dot{a}>$ stand respectively for the boson and fermion massless particle states, transforming under the representations 8_v and 8_c respectively.

From eqs. (1.37), (1.49) we find

$$\langle j_{out} | i_{in} \rangle = \delta_{ij} \left[1 - 2i \, \mathrm{Im} \sum_{n=1}^{\infty} \sum_{i'=1}^{8} F_n^{i'i'} \right] \tag{1.59a}$$

$$\langle \dot{a}_{out} | \dot{b}_{in} \rangle = \delta_{ab} \left[1 - 2i \sum_{n=1}^{\infty} \sum_{c=1}^{8} F_n^{cc} \right] \tag{1.59b}$$

$$\langle j_{out} | \dot{a}_{in} \rangle = 0 \tag{1.59c}$$

where $$2i \, \mathrm{Im} \, F_n^{aa} + 1 = \mathcal{C}_n^{aa}$$

That is, we find non-zero amplitudes only for elastic processes between the vacua. We notice that the vanishing of the inelastic amplitudes in an exact property to all orders in the constant g of our perturbation scheme. This follows from the fact that the exact (non-linear) transformation between the $S_{n<}^{a}$ and $S_{n>}^{b}$ operator preserves the boson-fermion parity. Therefore, the operator U generating such transformation is *even* in the S_n^a ($n \in Z$). As a consequence

$$< i_{in} \mid j_{out} > = < i_{in} \mid U \mid j_{out} > = \delta_{ij} \, \mathcal{F} \tag{1.60a}$$

$$< a_{in} \mid b_{out} > = < a_{in} \mid U \mid b_{in} > = \delta_{ab} \, \mathcal{G} \tag{1.60b}$$

$$< a_{in} \mid j_{out} > = 0 \tag{1.60c}$$

where the amplitudes $\tilde{\mathcal{F}}$ and $\mathcal{G}$ at first order in $\sqrt{\alpha'}/R_c$ are given by eqs. (1.59a,b). Therefore, the *massless* particles *cannot* transmute between themselves. *Only elastic* transitions are allowed.

(ii) *Ground state-first excited state transitions*

These are given by the transition amplitudes from the ground states $\mid \dot{a}_{in} >$, $\mid i_{in} >$ to the excited states

$$|\psi^{ij}_{out}> = \alpha^i_{-1>} |j_{out}> \quad , \quad S^a_{-1>} |\dot{b}_{out}> \qquad \text{bosonic states}$$

$$|\psi^{ia}_{out}> = \alpha^i_{-1>} |\dot{a}_{out}> \quad , \quad S^a_{-1>} |i_{out}> \qquad \text{fermionic states}$$

By using eqs. (1.48) - (1.49) we find that all these transition amplitudes are zero. Once more, this property is exact to all orders in perturbations, as it follows from analogous arguments to the ones used in (i).

(iii) *Ground state-high massive states*

At the second excited level ($\alpha' m^2 = 2$), the spectrum contains

$$\alpha^i_{-2} |\Phi_0> \quad , \quad S^a_{-2} |\Phi_0> \quad , \quad \alpha^i_{-1} S^a_{-1} |\Phi_0>$$

and

$$S^a_{-1} S^b_{-1} |\Phi_0> \quad , \quad \alpha^i_{-1} \alpha^j_{-1} |\Phi_0>$$

where $\mid \Phi_0 >$ stands for any of the boson or fermion massless ground states. The transition amplitudes from a massless particle state to the above first three types of states are identically zero, that is

$$
\begin{aligned}
&< \Phi_{0\,out} | \alpha^i_{2>} | \Phi_{0\,in} > = 0 \\
&< \Phi_{0\,out} | S^a_{2>} | \Phi_{0\,in} > = 0 \\
&< \Phi_{0\,out} | \alpha^i_{1>} S^a_{1>} | \Phi_{0\,in} > = 0
\end{aligned} \tag{1.61}
$$

On the contrary, we find non-zero amplitudes for the last two types of states

$$\langle \Phi_{o\,out} | S^a_{1>} S^b_{1>} | \Phi_{o\,in} \rangle = - \mathcal{D}^{ba}_1 \tag{1.62a}$$

$$\langle \Phi_{o\,out} | \alpha^i_{1>} \alpha^j_{1>} | \Phi_{o\,in} \rangle = - \mathcal{B}^{*ij}_1 \tag{1.62b}$$

More generally, the transition from the ground state to a state with an *odd* number of fermion (S^a_{-n}) or boson (α^i_{-m}) creation operators vanishes. The transition from the ground state to a state with an *even* number of S^a_{-n} and an even number of α^i_{-m}, is non-zero.

b/ CLOSED SUPERSTRINGS

A similar analysis may be performed for closed strings. The ground states are described by a direct product $|\Phi_o\rangle \otimes |\tilde{\Phi}_o\rangle$. As in the open superstring case, only *elastic* transitions between in and out ground state particles take place. The excited (massive) states for closed superstrings are given by an equal number of right moving and left moving creation operators ($S_{-n}{}^a$, $\tilde{S}_{-n}{}^a$ and/or α^j_{-m} , $\tilde{\alpha}_{-m}{}^j$) acting on the ground states. The transition from a ground state to such a massive state is in general non-zero at first order.More precisely, we find that the transition from a ground state to a state

$$|\psi^{ij\,ab}_{KL\,out}\rangle = \prod_{r=1}^{K} \prod_{s=1}^{L} \alpha^i_{-n_r>} \tilde{\alpha}^j_{-m_r>} S^a_{-n'_s} \tilde{S}^b_{-m'_s} \left(|\Phi_{o\,out}\rangle \otimes |\tilde{\Phi}_{o\,out}\rangle \right)$$

is proportional to $(\mathcal{B}_n{}^{*ij})^K \times (\mathcal{D}^{ab*}_{n'})^L$.

II.1.5 Examples

(a) BLACK-HOLES

The small fluctuation equations of a bosonic string in the Schwarzschild geometry were studied in ref.[2]. We choose the X^1, X^2 coordinates to lie in the scattering plane. In a light-cone type gauge

$$X^+ \underset{\tau \to -\infty}{=} 2\alpha' p^+ \tau$$

the physical (transverse) modes turn out to be associated to the X^2-coordinate and to those coordinates X^j, $3 \leq j \leq D - 1$, orthogonal to the scattering plane. The Bogoliubov coefficients $\mathcal{B}_n^{ij}$ have the following structure

$$\mathcal{B}_n^{ij} = \begin{cases} 0 & \text{if} \quad i = j \\ \mathcal{B}_n^{II} & \text{if} \quad i = j = 2 \\ \mathcal{B}_n^{\perp} & \text{if} \quad 3 \leq i = j \leq D - 1 \end{cases} \tag{1.63}$$

Here II and $\perp$ mean paralell and orthogonal to the scattering plane (X^1, X^2). A similar structure is exhibited by the $\mathcal{A}_n^{ij}$ coefficients. The coefficients $\mathcal{A}_n^{ij}$ and $\mathcal{B}_n^{aij}$ are functions of the string impact parameter b, of the string energy E and of the black hole radius R_s. Their explicit expressions have been derived in ref. [2] for $b >> R_s$. From eq. (1.56), we find here that the more relevant inelastic transmutation amplitudes, whose forms depend on the coefficients $\mathcal{B}_n^{ij}$, are equal to

$$< \text{graviton}_{\text{out}}^{ij} \mid 0_{\text{in}} > = \frac{\delta^{ij}}{D-2} \left(\mathcal{B}_1^{\perp} - \mathcal{B}_1^{\parallel} \right) \times \begin{cases} 1 & , \ 3 \leq i \leq D-1 \\ 3-D & , \ i = 2 \end{cases} \tag{1.64a}$$

$$< \text{dilaton}_{\text{out}} \mid 0_{\text{in}} > = (D-3)\, \mathcal{B}_1^{\perp} + \mathcal{B}_1^{\parallel} \tag{1.64b}$$

$$< \text{antisymm.tensor}_{\text{out}}^{ij} \mid 0_{\text{in}} > = 0 \tag{1.64c}$$

$$< \text{dilaton}_{\text{out}} \mid \text{graviton}_{\text{in}}^{ij} > = \frac{\delta^{ij}}{D-2}\, 8i \left(C_1^{\parallel} - C_1^{\perp} \right) \times \begin{cases} 1 & , \ 3 \leq i \leq D-1 \\ 3-D & , \ i = 2 \end{cases} \tag{1.65}$$

(b) AICHELBURG-SEXL GEOMETRY

In this geometry the string equations of motion are exactly solvable [3]. In the linear approximation, the Bogoliubov coefficients are given by

$$A^i_{nj} = \delta^i_j + B^i_{nj} \quad , \quad B^i_{nj} = \frac{b}{n} \Delta^i_j \qquad 1 \le i, j \le D-2 \tag{1.66}$$

where

$$b = \frac{1}{\rho_o^{D-2}} \, 2i\alpha' p_v G \not{p} \pi^{1-D/2} \Gamma\left(\frac{D}{2} - 1\right)$$

$$\Delta^i_j = \delta^i_j - (D-2) \frac{q^i q^j}{\rho_o^2} \tag{1.67}$$

$\Delta_{j,}{}^{i}$ is traceless

$$\sum_{i=1}^{D-2} \Delta^i_i = 0$$

$\rho_o = \left(\sum_{l=1}^{D-2} q_l^2 \right)^{1/2}$ is the impact parameter of the string, $\not{p}$ is the energy of the plane wave source of the A-S geometry and p_v is the light-cone component of the string momentum. Here $1 \le i, j \le D-2$ stand for the physical string components. From eqs. (1.56 - 1.66) and (1.67) we see that the relevant transmutation amplitudes are

$$< \text{graviton}_{\text{out}},{}^{ij} \mid 0_{in} > = b \ \Delta^i_j \tag{1.68a}$$

$$< \text{dilaton}_{\text{out}} \mid 0_{in} > = 0 \tag{1.68b}$$

$$< \text{antisymm.tensor}_{\text{out}},{}^{ij} \mid 0_{in} > = 0 \tag{1.68c}$$

$$< \text{dilaton}_{\text{out}} \mid \text{graviton}_{\text{in}},{}^{ij} > = \frac{4b}{n} \ \Delta^i_j \tag{1.69}$$

II.2 EXACT SOLUTIONS IN CONICAL SPACE-TIMES

II.2.1 Neveu-Schwarz-Ramond (NSR) model: Elastic and inelastic scattering

Conical space-time in D dimensions is defined by the metric

$$(dS)^2 = -(dX^o)^2 + (dR)^2 + R^2(d\phi)^2 + (dZ^i)^2, \tag{2.1}$$

where

$$R = \sqrt{X^2 + Y^2} \quad , \qquad \phi = \text{arc tg } (Y/X) \tag{2.2}$$

are cylindrical coordinates, but with the range

$$0 \le \phi \le 2\pi\alpha \quad , \quad \alpha = 1 - 4G\mu \tag{2.3}$$

$(dZ^i)^2$ is a flat (D-3)-dimensional space and Z^i, $3 \le i \le D-1$, are Cartesian coordinates. The spatial points (R, ϕ, Z^i) and $(R, \phi + 2\pi\alpha, Z^i)$ are identified. The space time is locally flat for $R \neq 0$ but has a conelike singularity at $R = 0$ with azimuthal deficit angle

$$\delta\Phi = 2\pi(1 - \alpha) = 8\pi G\mu \ . \tag{2.4}$$

This is the space-time geometry described by a straight cosmic string of zero thickness. This is a good approximation for very thin cosmic strings with large curvature radius. The space-time has a global non-trivial (multiply connected) topology.

Let us consider now the Neveu-Schwarz-Ramond (N.S.R.) model. In the conformal gauge, the covariant N.S.R. string action in flat space-time is given by [15]

$$S = -\frac{1}{4\pi\alpha'} \int d\sigma\, d\tau \left(\partial_A X^\mu \partial^A X_\mu - i\bar{\psi}^\mu \rho^A \partial_A \psi^\mu \right) \tag{2.5}$$

where $\mu = 0,1,\dots, D-1$, $A = 0,1$.

$X^\mu(\sigma,\tau)$ is a vector in the D-dimensional physical space-time and a scalar on the two-dimensional world sheet. $\psi^\mu(\sigma,\tau)$ is a vector in the D-dimensional space-time and a Majorana spinor on the world sheet [15]. Then $\psi^\mu(\sigma,\tau)$ can be written as

$$\psi^\mu = \begin{pmatrix} \psi^\mu_{(-)} \\ \psi^\mu_{(+)} \end{pmatrix} \quad , \quad \bar{\psi}^\mu = \psi^\mu \rho^0 \tag{2.6}$$

with the Dirac matrix representation (Sect. II.1.2).

The equations of motion, and the constraints, are the standard equations in flat space-time, but with the boundary conditions appropriated to the conical space-time. Equations of motion, constraints, and boundary conditions for the bosonic coordinates $X^{\mu}(\sigma,\tau)$ have been explicitly solved in ref. [4]. For the fermionic coordinates, the equations are

$$i\,\rho^{A}\,\partial_{A}\,\psi^{\mu} = 0$$

or (2.7)

$$(\partial_{\sigma} \pm \partial_{\tau})\ \psi^{\mu}_{(\mp)} = 0$$

i.e. $$\psi^{\mu}_{(-)} = \psi^{\mu}{}_{(-)}(\tau-\sigma) \quad , \quad \psi^{\mu}_{(+)} = \psi^{\mu}{}_{(+)}(\tau+\sigma) \tag{2.8}$$

In the Ramond model, open strings satisfy periodic conditions, namely

$$\begin{aligned} \psi^{\mu}_{(-)R}(0,\tau) &= \psi^{\mu}_{(+)R}(0,\tau) \\ \psi^{\mu}_{(-)R}(\pi,\tau) &= \psi^{\mu}_{(+)R}(\pi,\tau) \end{aligned} \tag{2.9}$$

and in the Neveu-Schwarz model, ψ^{μ} satisfy antiperiodic conditions:

$$\begin{aligned} \psi^{\mu}_{(-)NS}(0,\tau) &= \psi^{\mu}_{(+)NS}(0,\tau) \\ \psi^{\mu}_{(-)NS}(\pi,\tau) &= -\,\psi^{\mu}_{(+)NS}(\pi,\tau) \end{aligned} \tag{2.10}$$

As is known, the Ramond boundary conditions and integer modes are appropriate for the description of string states which are space-time fermions, whereas the Neveu-Schwarz boundary conditions and half-integer modes give bosonic states (which are different, of course, from those of the bosonic string theory).

In addition, the multiply connected topology of conical space-time imposes a non trivial periodicity condition on the spinors. Choosing the coordinates (6,7) perpendicular to the cosmic string, we have for the azimuthal angles $\phi_{\mp}$:

$$\begin{aligned} \phi_{\mp} &= \arctan\,[\psi^{6}_{(\mp)}\,/\,\psi^{7}_{(\mp)}] \\ \phi_{\mp} &\equiv \phi_{\mp} + 2\pi\alpha \end{aligned} \tag{2.11}$$

$\phi_{\mp}$ run in the range (0, $2\pi\alpha$) instead of the range (0, 2π). This amounts to having a deficit angle $\delta\Phi = 2\pi\,(1 - \alpha)$. The angles $\phi_{\mp}$ are periodic with period $2\pi\alpha$. From eqs (2.7) and open string boundary conditions (2.9), the Ramond string in terms of oscillator modes is given by

$$\psi^{\mu}_{(\mp)R}(\sigma, \tau) = \frac{1}{\sqrt{2}} \sum_{n \in Z} d^{\mu}_{n}\, e^{-in(\tau \mp \sigma)} \tag{2.12}$$

The Neveu-Schwarz string satisfying eqs. (2.7) and boundary condition (2.10) is written as

$$\psi^{\mu}_{(\mp)NS}(\sigma, \tau) = \frac{1}{\sqrt{2}} \sum_{r \in Z + 1/2} b^{\mu}_{r}\, e^{-ir(\tau \mp \sigma)} \tag{2.13}$$

Here

$$\{d^{\mu}_{n}, d^{\nu}_{m}\} = \eta^{\mu\nu}\, \delta_{n+m,0}\, 2\alpha' \text{ and } \{b^{\mu}_{r}, b^{\nu}_{r'}\} = \eta^{\mu\nu}\, \delta_{r+r',0}\, 2\alpha' \tag{2.14}$$

In both models due to the topologically non-trivial space-time, $\psi^{6}_{(\mp)}$, and $\psi^{7}_{(\mp)}$, have to satisfy eq. (2.11).

As we would like, eventually, to compare the N.S.R. model and the light-cone Green-Schwarz superstring formalism, we will consider the N.S.R. model in a light cone gauge.

The light-cone gauge conditions are given by

$$X^{+} = 2\alpha' p^{+} \tau \quad \text{and} \quad \psi^{+}_{(\mp)} = 0 \tag{2.15}$$

where

$$X^{+} = \frac{1}{\sqrt{2}} (X^{o} + X^{D-1}) \tag{2.16}$$

and

$$\psi^{+}_{(\mp)} = \frac{1}{\sqrt{2}} (\psi^{o}_{(\mp)} + \psi^{D-1}_{(\mp)}) \, . \tag{2.17}$$

The true dynamical variables are then X^{i} and $\psi^{i}_{(\mp)}$ (i = 1,..,8) where $\psi^{6}_{(\mp)}$ and $\psi^{7}_{(\mp)}$ satisfy the conical boundary conditions (2.11). X^{-} and $\psi^{-}_{(\mp)}$ are given through the usual constraint equations. As a consequence of condition (2.11), and as we have shown in ref. [9] for the bosonic coordinates, the string as a whole is deflected by an angle

$$\Delta_{\binom{R}{L}} = \pm\, \delta\phi = \pm\, 4\pi G\mu \tag{2.18}$$

where the (±) sign depends on whether the string passes to the right (R) or to the left (L) of the cosmic string.

The periodicity condition (2.11) is enforced by appropriately rotating free string solutions (see below). As a consequence, linear relations between ingoing ($\tau \to -\infty$) and outgoing ($\tau \to +\infty$) oscillators associated with the ψ^6 and ψ^7 coordinates will appear. Let us see this by studying the interaction between the superstring and the classical cosmic string (or conical space). As in the bosonic string case, we find essentially two different situations:

(i) The superstring does not touch the scattering center (here represented by the cosmic string); in this case the superstring only suffers a deflection at the origin. We refer to this situation as elastic scattering.

(ii) The superstring collides against the cosmic string. Then, in addition to being deflected, the superstring gets its normal modes excited. And this happens each time a point of the superstring collides with the center. We refer to this process as inelastic scattering.

Let us first consider the elastic scattering. The $\psi^i_{\pm}$ spinors naturally behave as the X^i coordinates: for $\tau \to -\infty$, the ingoing solution $\psi^i_{(\pm)<}$ is just the free solution without deflection. For $\tau \to +\infty$, the outgoing solution $\psi^i_{(\pm)>}$ is the free solution after deflection. We have

$$\begin{aligned} \psi^K_{(\pm)>}(\sigma,\tau) &= R^K_L(\pm\Delta)\, \psi^L_{(\pm)<}(\sigma,\tau) \\ \psi^\ell_{(\pm)>}(\sigma,\tau) &= \psi^\ell_{(\pm)<}(\sigma,\tau) \quad , \quad K,L=6,7;\ \ell \neq 6,7 \end{aligned} \tag{2.19}$$

where

$$R(\Delta) = \begin{pmatrix} \cos\Delta & -\sin\Delta \\ \sin\Delta & \cos\Delta \end{pmatrix} \tag{2.20}$$

$\pm\Delta$ being the deflection angles to the right (+) and left (-) of the cosmic string, given by eq. (2.18).

For the Ramond model, eqs. (2.19) imply

$$d^K_{n>} = R^K_L(\pm\Delta)\, d^L_{n<} \qquad K,L = 6,7$$
$$d^l_{n>} = d^l_{n<} \qquad l = 6,7 \tag{2.21}$$

and for the Neveu-Schwarz model we have

$$b^K_{r>} = R^K_L(\pm\Delta)\, b^L_{r<} \qquad K,L = 6,7$$
$$b^l_{r>} = b^l_{r<} \qquad l = 6,7 \tag{2.22}$$

Eqs. (2.21) and (2.22) amount to just a change of polarization of the outgoing string modes in the orthogonal plane to the cosmic string.

Let us discuss now the inelastic scattering [case (ii)].

The superstring evolution is described by the *free* equations *except* at the collision point (σ_o, τ_o) with the cosmic string.

Since the deflection angles to the right and to the left of the cosmic string are different, eqs. (2.21) and (2.22) do not hold anymore. We have now different matching conditions for $0 \leq \sigma < \sigma_o$ and $\sigma_o \leq \sigma \leq \pi$. In order to guarantee the continuity of the fermionic coordinates $\psi^\mu_{(\pm)}(\sigma,\tau)$ at $\tau = \tau_o$ for both the N.S. and the Ramond model, we have

$$\psi^K_{(\pm)>}(\sigma,\tau_o) = R^K_L(-\Delta)\, \psi^L_{(\pm)<}(\sigma,\tau_o) \quad , \quad 0 < \sigma < \sigma_o \tag{2.23}$$

$$\psi^K_{(\pm)>}(\sigma,\tau_o) = R^K_L(\Delta)\, \psi^L_{(\pm)<}(\sigma,\tau_o) \quad , \quad \sigma_o < \sigma < \pi \tag{2.24}$$

where $K,L \equiv 6,7$.

In addition, the string ends must obey the usual requirements (eq (2.9)). The Ramond string satisfies

$$\psi^i_{(-)R<}(0,\tau) = \psi^i_{(+)R<}(0,\tau) \qquad \text{for } \tau < \tau_o$$

$$\psi^i_{(-)R<}(\pi,\tau) = \psi^i_{(+)R<}(\pi,\tau) \tag{2.25}$$

and

$$\psi^i_{(-)R>}(0,\tau) = \psi^i_{(+)R>}(0,\tau) \qquad \text{for } \tau < \tau_o$$

$$\psi^i_{(-)R>}(\pi,\tau) = \psi^i_{(+)R>}(\pi,\tau) \tag{2.26}$$

For the Neveu-Schwarz string, one has (see eq. (2.10))

$$\psi^i_{(-)NS<}(0,\tau) = \psi^i_{(+)NS<}(0,\tau) \qquad \text{for } \tau < \tau_o$$

$$\psi^i_{(-)NS<}(\pi,\tau) = -\psi^i_{(+)NS<}(\pi,\tau) \tag{2.27a}$$

and

$$\psi^i_{(-)NS>}(0,\tau) = \psi^i_{(+)NS>}(0,\tau) \qquad \text{for } \tau > \tau_o$$

$$\psi^i_{(-)NS>}(\pi,\tau) = -\psi^i_{(+)NS>}(\pi,\tau) \tag{2.27b}$$

The equations of motion (2.7) and the Ramond boundary conditions (2.25) imply the usual expansion for the Fermi sector of the string solutions $\psi^i_{(\mp)R<}$ and $\psi^i_{(\mp)R>}$:

$$\psi^i_{(\pm)R<} = \frac{1}{\sqrt{2}} \sum_{n\in Z} d^i_{n<}\, e^{-in(\tau\mp\sigma)} \quad , \quad \tau < \tau_o \tag{2.28a}$$

$$\psi^i_{(\pm)R>} = \frac{1}{\sqrt{2}} \sum_{n\in Z} d^i_{n>}\, e^{-in(\tau\mp\sigma)} \quad , \quad \tau > \tau_o \tag{2.28b}$$

where $0 \leq \sigma \leq \pi$.

The Neveu-Schwarz boundary conditions (2.27) and eqs. (2.7) yield the mode expansion for the Bose sector $\psi^{i}_{(\mp)NS<}$ and $\psi^{i}_{(\mp)NS>}$:

$$\psi^{i}_{(\pm)NS<} = \frac{1}{\sqrt{2}} \sum_{r \in Z+1/2} b^{i}_{r<} \, e^{-ir(\tau \mp \sigma)} \quad , \quad \tau < \tau_0 \tag{2.29a}$$

$$\psi^{i}_{(\pm)NS>} = \frac{1}{\sqrt{2}} \sum_{r \in Z+1/2} b^{i}_{r>} \, e^{-ir(\tau \mp \sigma)} \quad , \quad \tau > \tau_0 \tag{2.29b}$$

where $0 \leq \sigma \leq \pi$.

Imposing Eqs. (2.23), (2.24) and (2.28) at $\tau = \tau_0$ yields linear relations between the ingoing and outgoing operators $d^K_{m<}$ and $d^K_{m>}$ on one hand, and $b^K_{r<}$ and $b^K_{r>}$ (K = 6,7), on the other hand (from eqs. (2.23), (2.24) and (2.29)). For the zero mode d^K_0 of the Ramond sector we have

$$d^{K}_{0>} = \frac{\sigma_0}{\pi} \left[R^{K}_{L}(-\Delta) - R^{K}_{L}(\Delta) \right] d^{L}_{0<} + \frac{1}{\pi} \sum_{n \neq 0} \left[R^{K}_{L}(-\Delta) - (-1)^{n} R^{K}_{L}(\Delta) \right] d^{L}_{n<} \, e^{-in\tau_0} \frac{\sin n\sigma_0}{n} \tag{2.30}$$

and for $m \neq 0$:

$$d^{K}_{-m>} = \frac{1}{\pi} \left[R^{K}_{L}(-\Delta) - (-1)^{m} R^{K}_{L}(\Delta) \right] e^{-im\tau_0} \frac{\sin m\sigma_0}{m} d^{L}_{0<} + \frac{1}{\pi} \sum_{n=1}^{\infty} \left(A^{K}_{mnL} d^{L}_{n<} + \mathcal{B}^{K}_{mnL} d^{L}_{-n<} \right) \tag{2.31}$$

where the Bogoliubov coefficients are given by

$$\mathcal{B}^{K}_{mnL} = \frac{1}{\pi} \left[R^{K}_{L}(-\Delta) - (-1)^{m-n} R^{K}_{L}(\Delta) \right] e^{-i(m-n)\tau_0} \frac{\sin (m-n)\sigma_0}{m-n} \tag{2.32a}$$

$$\mathcal{A}^{K}_{m,-n,L} = \mathcal{B}^{K}_{mnL} \tag{2.32b}$$

For the Neveu-Schwarz model there is no zero mode and we have

$$b^{K}_{-s>} = \sum_{r=1/2}^{\infty} \left(\mathcal{C}^{K}_{srL}\, b^{L}_{r<} + \mathcal{D}^{K}_{srL}\, b^{L}_{-r<} \right) \tag{2.33}$$

Here the Bogoliubov coefficients ($\mathcal{C}^{K}_{sr\,L}$ and $\mathcal{D}^{K}_{sr\,L}$) are

$$\mathcal{D}^{K}_{srL} = \frac{1}{\pi}\left[R^{K}_{L}(-\Delta) - (-1)^{s-r} R^{K}_{L}(\Delta) \right] e^{-i(s-r)\tau_0}\, \frac{\sin(s-r)\sigma_0}{s-r} \tag{2.34a}$$

$$\mathcal{C}^{K}_{s,-r,L} = \mathcal{D}^{K}_{srL} \tag{2.34b}$$

In this inelastic case -in which the superstring collides against the cosmic string- the modes become excited, in addition to having a change of polarization. The Bogoliubov transformations Eqs.(2.31) - (2.32) are interpreted as follows: if for $\tau < \tau_o$, the initial string is in a particle mode m polarized in a direction K ($d^{K}_{m<}$, K = 6,7), then for $\tau > \tau_o$, there will be: (a) an amplitude $\mathcal{A}^{K}_{mn\,L}(\Delta)$ for a mode n($d^{L}_{n>}$) polarized in a different direction L and (b) an amplitude $\mathcal{B}^{K}_{mn\,L}(\Delta)$ for an antiparticle mode ($d^{L}_{-n>}$) polarized in the L-direction. The same conclusions hold for the transverse modes d^{i}_{n} (i = 3,..., D-2), except that they do not suffer polarization changes. Identical conclusions hold also for the transformations (2.33) - (2.34) in terms of the Neveu-Schwarz operators $b_{\pm r>}$,$b_{\pm r<}$. It should be noticed that Eqs. (2.23)-(2.24) reproduce Eqs. (2.21) [and (2.22)] with signs $+\Delta$ or $-\Delta$ when $\sigma_o = 0$ or when $\sigma_o = \pi$, respectively, as it must be.

We have considered the cases where the string does not collide with the conical singularity or when it collides once. Depending upon its initial shape and velocity, the superstring may intersect two or more times with the cosmic string. The corresponding Bogoliubov transformation of the superstring-mode operators can be obtained by respeatedly using Eqs. (2.30) - (2.32) and (2.33) - (2.34), following on the same lines as for the bosonic string case [9].

II.2.2 Green-Schwarz (GS) light-cone model: elastic and inelastic scatterings

In the light-cone gauge, the Green-Schwarz action in flat space is given by eq. (1.30) (see Sect. II.1.2). In conical space-time, we choose, as before, the 6th and 7th coordinates to be perpendicular to the cosmic string. The spinor $S^a(\sigma,\tau)$ is periodic under a rotation $2\pi\alpha$ around the cosmic string, that is on the orthogonal plane (6.7).

Let us consider now the elastic scattering. The ingoing solution $S^{\ a}_{n<}$ (for $\tau \to -\infty$) is just the free solution without deflection. The outgoing solution $S^{\ a}_{n>}$ (for $\tau \to +\infty$) is the free solution but now rotated by an angle $+\Delta/2$ (instead of Δ) or $-\Delta/2$ (depending on whether the superstring passes to the right or to the left of the cosmic string respectively). The spinor components are rotated $S^{\ a}_{n>}$ as

$$S^{\ a}_{n>} = \mathcal{M}^{a}_{67b}\left(\Delta_{\binom{R}{L}}\right) S^{b}_{n<} \tag{2.35}$$

where Δ is given by eq. (2.18) and

$$\begin{aligned} \mathcal{M}^{a}_{67}(\Delta) &= \exp \frac{1}{4}[\gamma^6, \gamma^7]\Delta \\ &= \left(\mathbb{1}^{(2)} \otimes \mathbb{1}^{(2)} \otimes \mathbb{1}^{(2)}\right) \cos\Delta/2 - \left(\mathbb{1}^{(2)} \otimes \mathbb{1}^{(2)} \otimes i\sigma_2\right) \sin\Delta/2 \end{aligned} \tag{2.36}$$

i.e.

$$\mathcal{M}^{a}_{67b}(\Delta) = \mathcal{R}_{67}(\Delta/2) = R(\Delta/2) \otimes \mathbb{1}^{(4)} \tag{2.37}$$

Here according to the 8 x 8 real matrix representation:

$$\gamma^6 = i\sigma_2 \otimes \mathbb{1}^{(2)} \otimes \sigma_1 \quad , \quad \gamma^7 = i\sigma_2 \otimes \mathbb{1}^{(2)} \otimes \sigma_3 \tag{2.38}$$

$(\sigma^1, \sigma^2, \sigma^3)$ are Pauli matrices, $I^{(2)}$ and $I^{(4)}$ mean 2 x 2 and 4 x 4 unit matrices respectively and $R(\Delta/2)$ is the 2 x 2 rotation matrix (2.20).

In this elastic scattering, ingoing and outgoing particle and antiparticle solutions $S^a_{n<}$ and $S^a_{-n>}$ are not mixed. The outgoing solutions only suffer a polarization change. Notice that here in the Green-Schwarz model, <u>all</u> the outgoing components suffer a rotation, and with an angle $\pm\Delta/2$ (instead of $\pm\Delta$).

Let us consider now the inelastic scattering. The continuity of the S^a_n components at the collision time τ_o for $0 < \sigma < \sigma_o$ and for $\sigma_o < \sigma < \pi$ yield

$$
\begin{aligned}
S^a_>(\sigma,\tau_0) &= R^a_b(-\Delta/2)\, S^b_<(\sigma,\tau) \quad , \quad 0<\sigma<\sigma_0 \\
S^a_>(\sigma,\tau_0) &= R^a_b(\Delta/2)\, S^b_<(\sigma,\tau) \quad , \quad \sigma_0<\sigma<\pi
\end{aligned}
\tag{2.39}
$$

The string ends satisfy (eqs (1.34))

$$
S^{1a}_<(0,\tau) = S^{2a}_<(0,\tau) \quad , \quad S^{1a}_<(\pi,\tau) = S^{2a}_<(\pi,\tau) \tag{2.40a}
$$

and

$$
S^{1a}_>(0,\tau) = S^{2a}_>(0,\tau) \quad , \quad S^{1a}_>(\pi,\tau) = S^{2a}_>(\pi,\tau) \tag{2.40b}
$$

Therefore, the explicit free solutions satisfying eqs. (1.33) and conditions (2.40) are

$$
S^{1,2a}_<(\sigma,\tau) = \frac{1}{\sqrt{2}} \sum_{n\in Z} S^a_{n<}\, e^{-in(\tau\mp\sigma)} \tag{2.41a}
$$

$$
S^{1,2a}_>(\sigma,\tau) = \frac{1}{\sqrt{2}} \sum_{n\in Z} S^a_{n>}\, e^{-in(\tau\mp\sigma)} \tag{2.41b}
$$

From eqs. (2.39) and (2.41) we get for the operators $S^a_{n>}$, in terms of the operators $S^a_{n<}$, Bogoliubov transformations analogous to those given by eqs. (2.30) - (2.32) for the Ramond string, but with R^a_b $(\pm\Delta/2)$ instead of R^K_L $(\pm\Delta)$, that is to say:

$$S^a_{0>} = \frac{\sigma_0}{2}\left[\mathcal{R}^a_b(-\Delta/2) - \mathcal{R}^a_b(\Delta/2)\right] S^b_{0<}$$

$$+ \frac{1}{\pi}\sum_{n\neq 0}\left[\mathcal{R}^a_b(-\Delta/2) - (-1)^n \mathcal{R}^a_b(\Delta/2)\right] S^b_{n<}\, e^{-in\tau_0}\, \frac{\sin n\sigma_0}{n} \tag{2.42}$$

and

$$S^a_{-m>} = \frac{1}{\pi}\left[\mathcal{R}^a_b(-\Delta/2) - (-1)^m \mathcal{R}^a_b(\Delta/2)\right] S^b_{0<}\, e^{-im\tau_0}\, \frac{\sin m\sigma_0}{m}$$

$$+ \frac{1}{\pi}\sum_{n=1}^{\infty}\left(\mathcal{F}^a_{mnb}\, S^b_{n<} + \mathcal{G}_{mnb}\, S^b_{-n<}\right) \tag{2.43}$$

$$\mathcal{G}^a_{mnb} = \frac{1}{\pi}\left[\mathcal{R}^a_b(-\Delta/2) - (-1)^{m-n}\mathcal{R}^a_b(\Delta/2)\right] e^{-i(m-n)\tau_0}\, \frac{\sin(m-n)\sigma_0}{m-n} \tag{2.44}$$

$$\mathcal{F}^a_{m,-n,b} = \mathcal{G}^a_{mnb} \tag{2.45}$$

II.2.3 Mass and number operators in the RNS and GS light cone models

Flat space-time formal expressions hold for the square mass operator in the different ingoing (<) and outgoing (>) regions, for both the elastic and inelastic cases. In the light cone gauge, the mass formulas for the Ramond and Neveu-Schwarz models are given by

$$\alpha' M^2_{\lessgtr} = \sum_{n=1}^{\infty} \alpha^i_{-n\lessgtr}\, \alpha^i_{n\lessgtr} + \frac{1}{2\alpha'}\sum_{n=1}^{\infty} n\, d^i_{-n\lessgtr}\, d^i_{n\lessgtr} \tag{2.46}$$

$$\alpha' M^2_{\lessgtr} = \sum_{n=1}^{\infty} \alpha^i_{-n\lessgtr}\, \alpha^i_{n\lessgtr} + \frac{1}{2\alpha'} \sum_{r=1/2}^{\infty} r\, b^i_{-r\lessgtr}\, b^i_{r\lessgtr} - \frac{1}{16}(D-2) \tag{2.47}$$

where

$$N_{\alpha\lessgtr} = \sum_{n=1}^{\infty} \alpha^i_{-n\lessgtr}\, \alpha^i_{n\lessgtr} \tag{2.48}$$

$$N_{d\lessgtr} = \frac{1}{2\alpha'} \sum_{n=1}^{\infty} n\, d^i_{-n\lessgtr}\, d^i_{n\lessgtr} \tag{2.49}$$

$$N_{b\lessgtr} = \frac{1}{2\alpha'} \sum_{r=1/2}^{\infty} r\, b^i_{-r\lessgtr}\, b^i_{r\lessgtr} \tag{2.50}$$

are the corresponding number operators. From the masslessness of the vector states

$$b^i_{-1/2<} \,|0_{in}\rangle = 0$$

or (2.51)

$$b^i_{-1/2>} \,|0_{out}\rangle = 0$$

follows the critical dimension D = 10, the same as in flat space.

For the Green-Schwarz light-cone supersymmetric model we have

$$\alpha' M^2_{\lessgtr} = \sum_{n=1}^{\infty} \alpha^i_{-n\lessgtr}\, \alpha^i_{n\lessgtr} + \sum_{n=1}^{\infty} n\, S^a_{-n\lessgtr}\, S^a_{n\lessgtr} \tag{2.52}$$

Here $\alpha' M^2$ coincides with the number operator $(N_\alpha + N_S)$.

The two operators $M^2_<$ and $M^2_>$ have identical spectra which are simply the flat space-time mass spectrum. In the elastic case, they have also common particle eigenvectors (up to polarization changes). However, in the inelastic case, they do not have common particle eigenvectors: as discussed before, if the superstring collides with the conical singularity, a particle state in one of the regions (say, <) is an infinite superposition of particle states associated with the other region (>), and reciprocally.

If for $\tau < \tau_o$, the ingoing ground state $|0_{in}>$ for the R and N-S models, is defined by

$$\alpha_{n<} \ , \ d_{n<} \ , b_{r<} |0_{in}\rangle = 0 \qquad , \ n, r > 0 \tag{2.53}$$

then the number of physical, bosonic and fermionic, modes for $\tau > \tau_o$ created from the $\tau < \tau_o$ ground state is

$$\langle 0_{in} | N_{\alpha>} | 0_{in} \rangle = \sum_{n=1}^{\infty} \langle 0_{in} | \alpha^i_{-n>} \alpha^i_{n>} | 0_{in} \rangle \tag{2.54}$$

and

$$\langle 0_{in} | N_{d>} | 0_{in} \rangle = \frac{1}{2\alpha'} \sum_{n=1}^{\infty} n \langle 0_{in} | d^i_{-n>} d^i_{n>} | 0_{in} \rangle \tag{2.55}$$

$$\langle 0_{in} | N_{b>} | 0_{in} \rangle = \frac{1}{2\alpha'} \sum_{r=1/2}^{\infty} r \langle 0_{in} | b^i_{-r>} b^i_{r>} | 0_{in} \rangle \tag{2.56}$$

Since, in the inelastic case, $\alpha^K_{n>}$ contains $\alpha^L_{-n<}$ (equally for $d^K_{n>}$, $d^L_{-n<}$ and $b^K_{r<}$, $b^L_{-r<}$) all these expectation values are non-zero. These mode excitations yield final particle states different from the initial one, which is the phenomenon of particle transmutation described in the previous sections. With these excited modes one can construct a particle state of higher and well defined spin and mass. We have a single (test) string, and the final and initial states are one particle (but different) states. Here particle states are created at the classical (tree) level, as a consequence of the interaction with the geometry.

The same phenomenon happens for the superstring in the Green-Schwarz light cone model, where now the ingoing ground state for type I is defined by

$$\alpha^i_{n<} , S^a_{n<} | \Phi_{0\,in} \rangle = 0 \qquad , | \Phi_{0\,in} \rangle = \{ |i_{in}\rangle , |\dot{a}_{in}\rangle \} \tag{2.57}$$
$$(n > 0)$$

$|i_{in}>$ represent the eight physical transverse polarizations of the massless vector field, and $|a_{in}>$ its spinor partner. Similarly, the above argument is applied to

$$<\text{ground state}_{in}\, |N_{\alpha>} + N_{s>}\, |\text{ground state}_{in}> = 0 \tag{2.58}$$

II.2.4 Strings in spinning conical space-time

The space-time around a cosmic string possesing a spin S is described by the metric [16]

$$ds^2 = -\left(dX^0 + (G\,S/\alpha)\,d\phi\right)^2 + dR^2 + R^2 d\phi^2 + (dZ^i)^2 \tag{2.59}$$

The range of the azymuthal angle ϕ is given by eq. (2.3), and the points (X^0, R, ϕ, Z^i) and $(X^0 + 2\pi GS, R, \phi + 2\pi\alpha, Z^i)$ are identified. The Z^i coordinates are cartesian coordinates in a (D-3) dimensional flat space.

Let us consider a string propagating in this space time. The bosonic string equations are here

$$(\partial_\tau^2 - \partial_\sigma^2)\left[X^0(\sigma,\tau) + (GS)\,\phi(\sigma,\tau)\right] = 0 \tag{2.60a}$$

$$(\partial_\tau^2 - \partial_\sigma^2)R = R\left[(\partial_\tau\phi)^2 - (\partial_\sigma\phi)^2\right] \tag{2.60b}$$

$$(\partial_\tau^2 - \partial_\sigma^2)\phi = -(1/2R)\left[(\partial_\tau\phi)(\partial_\tau R) - (\partial_\sigma\phi)(\partial_\sigma R)\right] \tag{2.60c}$$

$$(\partial_\tau^2 - \partial_\sigma^2)Z^i = 0 \tag{2.60d}$$

Except for eq. (2.60a) for X^0, the spin S does not appear in these equations. Therefore, the spatial string coordinates are given by the solution found in ref. [9]. We call $X_<(\sigma,\tau)$, $Y_<(\sigma,\tau)$, and $X_>(\sigma,\tau)$, $Y_>(\sigma,\tau)$, , the ingoing and outgoing open string solution, (ϕ = arc tg Y/X). They are given by [9] ($X^{K,L} = X,Y$)

$$X^K_{\lessgtr} = q^k_{\lessgtr} + 2\alpha' p^k_{\lessgtr} \tau + i\sqrt{2\alpha'} \sum_{n\neq 0} \frac{\cos n\sigma}{n} \alpha^k_{n\lessgtr} e^{-in\tau} \tag{2.61}$$

$$q^k_> = R^K_L(\pm\Delta) q^L_< \quad , \quad p^k_> = R^K_L(\pm\Delta) p^L_< \tag{2.62}$$

$$\alpha^K_{n>} = R^K_L(\pm\Delta)\alpha^L_{n<} \quad , \quad (n\neq 0) \tag{2.63}$$

Here $R^K_L(\pm\Delta)$ is the rotation matrix eq. (2.20). Eqs. (2.62) and (2.63) refer to elastic scattering. In the inelastic case the solution is given by

$$q^K_> = R^K_L(\Delta) q^L_< + \frac{2\alpha'\sigma_0}{\pi} \mathcal{L}^K_L p^L_< \tag{2.64a}$$

$$p^K_> = R^K_L(\Delta) p^L_< + \frac{\sigma_0}{2\alpha'\pi} \mathcal{L}^K_L q^L_< \tag{2.64b}$$

$$\alpha^K_{n>} = \frac{\sin n\sigma_0}{n\pi}\sqrt{2\alpha'}\, \mathcal{L}^K_L p^L_< + \sum_{m=1}^{\infty} \left(A^K_{nmL}\alpha^L_{m<} + B^K_{nmL}\alpha^L_{-m<}\right) \tag{2.64c}$$

Here

$$\mathcal{L}^K_L = R^K_L(-\Delta) - R^K_L(\Delta) \tag{2.65}$$

$$A^k_{nmL} = R^k_L(\Delta)\,\delta_{nm} + \mathcal{L}^k_L \; \frac{\sin(n-m)\sigma_0}{(n-m)\pi}\; e^{-in\tau_0} \tag{2.66}$$

$$B^k_{nmL} = \mathcal{L}^k_L \; \frac{\sin(n+m)\sigma_0}{(n+m)\pi}\; e^{in\tau_0} \tag{2.67}$$

and σ_0 and τ_0 depend on the initial data of the string through the condition

$$q^k_< + 2\alpha' p^k_< \tau_0 + i\sqrt{2\alpha'} \sum_{n\neq 0} \frac{\cos n\sigma_0}{n}\; e^{-in\tau_0}\; \alpha^k_{n<} = 0 \tag{2.68}$$

The new features, when the cosmic string posseses spin, appear in the solution of eq. (2.60a) for $X^0(\sigma,\tau)$, which is given by:

$$\begin{aligned} X^0_{\lessgtr}(\sigma,\tau) = &- (G\,S/\alpha)\,\phi_{\lessgtr}(\sigma,\tau) + q^0 + 2\alpha' p^0 \tau \\ &+ i\sqrt{2\alpha'} \sum_{n\neq 0} \frac{\cos n\sigma}{n}\; \alpha^0_n\; e^{-in\tau} \end{aligned} \tag{2.69}$$

Here

$$\phi_{\lessgtr}(\sigma,\tau) = \arctan\left[Y_{\lessgtr}(\sigma,\tau)/X_{\lessgtr}(\sigma,\tau)\right] \tag{2.70}$$

In the elastic case:

$$X^0_>(\sigma,\tau) - X^0_<(\sigma,\tau) = \mp (G\,S/\alpha)\,\Delta \tag{2.71}$$

with the sign + (-) depending on whether the string passes to the right (+) or to the left (-) of the conical singularity.

In the inelastic case, we have

$$X^0_>(\sigma,\tau) - X^0_<(\sigma,\tau) = \begin{cases} +(GS/\alpha)\Delta & , \ 0<\sigma<\sigma_0 \\ -(GS/\alpha)\Delta & , \ \sigma_0<\sigma<\pi \end{cases} \tag{2.72}$$

We recall that $(0, \sigma_0)$ and (σ_0, π) are the pieces of the string passing to the left and to the right of the conical singularity respectively.

We can choose here the light-cone gauge

$$X^0(\sigma,\tau) + (GS/\alpha)\,\phi(\sigma,\tau) + Z^1(\sigma,\tau) = 2\alpha' p^0 \tau \tag{2.73}$$

Please recall that the space time around a cosmic string carrying spin has a *time-helical structure* in the coordinates $(\tilde{X}^0, R, \phi, Z^i)$ where $\tilde{X}^0 = X^0 + (GS\phi/\alpha)$. This means that the points

$$(\tilde{X}^0, R, \phi, Z^i) \ \text{ and } \ (\tilde{X}^0 + 2\pi GS, R, \phi + 2\pi\alpha, Z^i)$$

are identified.

The fermionic string propagation is not affected by the spin of the cosmic string. As stated above, the presence of the spin introduces a shift $2\pi GS/\alpha$ on $\tilde{X}^0$ besides the rotation $2\pi\alpha$ for two identical points. Since fermion coordinates are invariant under space-time translations (fermion string operators do not contribute to the string momentum), the fermion string solutions of Sections II.2.1 and II.2.2 also hold in the space-time (2.59) with $S \neq 0$.

The string constraint equations take here identical form as in the $S = 0$ case [9]. Therefore, the mass spectrum and the critical dimension are those of Minkowski space-time.

III. SUPERSTRINGS IN A BACKGROUND WITH "BOSONIC" AND "FERMIONIC" DEGREES OF FREEDOM

III.3 CONSTRUCTION OF THE TEN DIMENSIONAL SUPER-GRAVITY SHOCK-WAVE SPACETIME

We construct in this section a linearized supergravity solutions with a bosonic shock-wave background. In ten dimensions the N=1 supergravity multiplet contains :

a) a zehnbeine field e^m_μ ; μ = world index, m = tangent space index, $0 \leq \mu \leq 9$, $0 \leq m \leq 9$,

b) a Rarita-Schwinger field ψ^α_μ , $1 \leq \alpha \leq 32$

c) a 2-form potential $b_{\mu\nu}$

d) a Dirac field χ^α and a scalar ϕ .

ψ and χ are Majorana-Weyl spinors.

Since the supergravity lagrangian is rather involved [17] we will consider linearized equations except for the zehnbeine that will satisfy the usual non-linear pure gravitational field equations. This linearization corresponds physically to a weak matter source. Notice that for the pure gravitational field the linear approximation actually coincides with the exact shock-wave solution. The field equations result in this linear approximation

$$R_{\mu m}(e, \omega(e)) - \frac{1}{2} e_{\mu m} R(e, \omega(e)) = 0 \tag{3.1}$$

$$\Gamma^{\mu\rho\sigma} D_\rho(\omega(e)) \psi_\sigma = 0 \tag{3.2}$$

$$\partial_{[\rho} b_{\mu\nu]} = 0 \tag{3.3}$$

$$\Gamma^\mu D_\mu(\omega(e)) \chi = 0 \tag{3.4}$$

$$\partial_\mu \partial^\mu \phi = 0 \tag{3.5}$$

Here ω (e) stands for the spin connection (no torsion is considered).

$$\omega_{\mu mn}(e) = \frac{1}{2} e^{\nu}_{m} (\partial_{\mu} e_{n\nu} - \partial_{\nu} e_{n\mu}) - \frac{1}{2} e^{\rho}_{m} e^{\sigma}_{n} \partial_{\rho} e_{r\sigma} e^{r}_{\mu} - (m \leftrightarrow n) \tag{3.6}$$

and $\Gamma^{\mu\rho\sigma}$ is the antisymmetrized product of three Dirac matrices

$$\Gamma^{\mu\rho\sigma} \equiv \Gamma^{[\mu} \Gamma^{\rho} \Gamma^{\sigma]}$$

For further simplicity we choose trivial solutions for the spin 0, 1/2 and 1 fields, namely

$$b_{\mu\nu} = 0, \quad \chi = 0, \quad \phi = 1 \tag{3.7}$$

also in agreement with the constraint equations [17].

Although we are finally interested in a D = 10 solutions, we shall solve the Rarita-Schwinger equation (3.2) in D-dimensions for a shock-wave metric, as eqs. (3.1)-(3.5) are valid in any dimension. In order to deal with this problem it is particularly convenient to start by boosting the D-dimensional Schwarzschild metric [18] :

$$ds^2 = -g(r)\, dt^2 + (1/g(r))\, dr^2 + r^2 d\Omega^2_{D-2} \tag{3.8}$$

where $g(r) = 1 - 2BMr^{3-D}$ and

$$B = \frac{8\pi G}{(D-2)\Omega_{D-1}} \quad \text{and} \quad \Omega_D = \frac{2\pi^{D/2}}{\Gamma(D/2)}$$

Here M stands for the mass of the point source at r = 0 and G is the gravitational constant.

It is convenient first to transform to isotropic coordinates $(t, \bar{r}, \Omega_{D-2})$ where

$$r^{D-3} = \bar{r}^{D-3} \left(1 + \frac{BM}{2\bar{r}^{D-3}}\right)^2 \tag{3.9}$$

and

$$\bar{r}^2 = \sum_{i=1}^{D-1} (X_i)^2 \tag{3.10}$$

The line element can be written now as

$$ds^2 = -g(\bar{r})^2 dt^2 + h^2(\bar{r}) \sum_{i=1}^{D-1} (dX_i)^2 \tag{3.11}$$

where

$$g(\bar{r}) = 1 - \frac{2BM}{\bar{r}^{D-3}} \left(1 + \frac{BM}{2\bar{r}^{D-3}}\right)^{-2} \tag{3.12}$$

$$h(\bar{r}) = \left(1 + \frac{BM}{2\bar{r}^{D-3}}\right)^{2/D-3}$$

The curved shock-wave space-time is then obtained as the ultrarelativistic limit of the boosted Schwarzschild geometry [19]. Performing a boost in the X^{D-1} direction

$$\begin{aligned} t &= \gamma(t' - v\, x'^{D-1}) \\ X^{D-1} &= \gamma(x'^{D-1} - v t') \\ X^i &= X'^i \qquad , \quad 1 \leq i \leq D-2 \end{aligned} \tag{3.13}$$

where $\gamma = 1/\sqrt{1-v^2}$, the line element (3.11) becomes

$$\begin{aligned} ds^2 = {} & \gamma^2 [h^2(\bar{r}) - g(\bar{r})] (dt' - v\, dx'^{D-1})^2 \\ & + h^2(\bar{r}) \left[-dt'^2 + (dx'^{D-1})^2 + \sum_{i=1}^{D-2} (dx'^i)^2 \right] \end{aligned} \tag{3.14}$$

(we shall drop primes from now on).

The ultrarelativistic limit $v \to 1$ must be handled carefully [20]. We expand the line element in powers of the parameter

$$\lambda \equiv 1 - v^2 \tag{3.15}$$

In order to get a mathematically well defined and physically non-trivial ultra-relativistic limit we must keep the momentum

$$\tilde{p} \equiv \mathcal{M}/\sqrt{\lambda} \tag{3.16}$$

Fixed for $\lambda \to 0$. We find

$$\begin{aligned} \gamma^2[h^2(\bar{r}) - g(\bar{r})] &= (2\mathcal{H} + R)[1 + O(\sqrt{\lambda})] \\ h^2(\bar{r}) &= \left[1 + \frac{2\lambda\mathcal{H}}{D-2}\right][1 + O(\sqrt{\lambda})] \end{aligned} \tag{3.17}$$

where

$$\mathcal{H} = \frac{8\pi G}{(D-3)\Omega_{D-1}} \tilde{p} \frac{\lambda^{(D-4)/2}}{(\xi^2 + \lambda\rho^2)^{(D-3)/2}}$$

$$R = -\frac{4D^2 - 23D + 29}{2(D-2)^2} \lambda \mathcal{H}^2 \tag{3.18}$$

$$\xi \equiv X^{D-1} - tv$$

$$\rho = \sqrt{\sum_{i=1}^{D-2} (X^i)^2}$$

As we shall see below, higher order terms are irrelevant in the $\lambda = 0$ limit.

In the $\lambda \to 0$ limit, the metric takes the form

$$ds^2 = dU dV - (dX^i)^2 + f_D(\rho)\,\delta(U)(dU)^2 \tag{3.19}$$

where

$$f_D(\rho) = -8\pi^{2-D/2} G \tilde{p}\, \Gamma\left((D/2) - 1\right) \rho^{4-D}/D-4$$

Finally, the vielbein basis e_μ^m -showing the explicit λ dependence- is given by

$$e^{0}_{\mu} = \left(1-\left(1-\frac{\lambda}{D-2}\right)\mathcal{H}-\frac{R}{2},0,\ldots,0,\left(1-\frac{D-3}{D-2}\lambda\right)\mathcal{H}+\frac{R}{2}\right)$$

$$e^{i}_{\mu} = \left(0,\ldots\ldots,\overbrace{1+\frac{\lambda\mathcal{H}}{D-2}}^{ith},\ldots 0\right) \quad , \; 1 \leq i \leq D-2 \tag{3.20}$$

$$e^{D-1}_{\mu} = \left(-\left(1-\frac{\lambda}{D-2}\right)\mathcal{H}-\frac{R}{2},0,\ldots,0,1+\left(1-\frac{D-3}{D-2}\lambda\right)\mathcal{H}+\frac{R}{2}\right)$$

$e_{m\mu}$ follows trivially from e^{m}_{μ} by using the Minkowskian metric in tangent space $\eta^{oo} = -1$, $\eta^{mm} = 1$, $1 \leq m \leq D-1$.

In order to solve the linearized Rarita-Schwinger equation (3.2) in the shock-wave background we use the following method [21]. First notice that any pure gauge field identically fulfils eq. (3.2)

$$\psi^{gauge}_{\sigma} = D_{\sigma}\alpha \tag{3.21}$$

where $D_{\mu} = \partial_{\mu} + \frac{1}{4}\omega_{\mu mn}\Gamma^{mn}$ is the covariant derivative and

$$\Gamma^{mn} = \frac{1}{2}[\Gamma^{m},\Gamma^{n}]$$

We are of course interested in non-pure gauge solutions of (3.2). We start from a pure gauge solution for $\lambda \neq 0$

$$\psi_{\mu}(\lambda) = D_{\mu}(\lambda)\,\alpha(\lambda) \tag{3.22}$$

where

$$\alpha(\lambda) = \alpha_{c}/\lambda \tag{3.23}$$

and α_c is a covariantly constant spinor obeying

$$D_{\mu}(0)\,\alpha_{c} = 0 \tag{3.24}$$

Expanding $D_{\mu}(\lambda)$ in powers of λ [see (3.16)] yields

$$D_{\mu}(\lambda) = D_{\mu}(0) + \sqrt{\lambda}\,F^{(1/2)}_{\mu} + \lambda F^{(1)}_{\mu} + O(\lambda^{3/2}) \tag{3.25}$$

where $D_\mu(0)$, $F_\mu^{(1/2)}$ and $F_\mu^{(1)}$ have $\delta(U)$ singularities at $U = 0$.

Explicit calculation shows that

$$F_\mu^{(1/2)} \alpha_c = 0 \tag{3.26}$$

as it will become clear later on.

Combining eqs. (3.6), (3.20-21-22) we find in the $\lambda = 0$ limit

$$\Psi_\mu = F_\mu^{(1)} \alpha_c \tag{3.27}$$

Exactly at $\lambda = 0$ the only non-vanishing spin connections reduce to

$$\omega_{t0i} = \omega_{ti,D-1} = -\omega_{D-1,0i} = -\omega_{D-1,i,D-1} = \partial_i \mathcal{U}\Big|_{\lambda=0} \tag{3.28}$$

where

$$\partial_i \mathcal{U}\Big|_{\lambda=0} = -\frac{2\pi^{3/2}G}{\Omega_{D-1}} \tilde{p}\, X^i \delta(U) \frac{\Gamma((D/2)-1)}{\Gamma((D-1)/2)} \frac{1}{\rho^{D-2}} \tag{3.29}$$

and $U \equiv t - X^{0-1}$. With the help of eq. (3.28), eq. (3.24) can be written as

$$D_\mu(0)\alpha_c = \left[\partial_\mu + \frac{1}{2}\, \partial_i \mathcal{U}\Big|_{\lambda=0} \left(\Gamma^{0i} + \Gamma^{iD-1}\right)\right]\alpha_c = 0 \tag{3.30}$$

From now, we take D = 10 and the following 32x32 Dirac matrix representation

$$\Gamma^0 = \sigma_2 \otimes \mathbb{1}_{16} \;,\quad \Gamma^i = i\sigma_1 \otimes \gamma^i_{16} \;,\quad \Gamma^9 = i\sigma_3 \otimes \mathbb{1}_{16} \tag{3.31}$$

where $(\sigma_1 , \sigma_2 , \sigma_3)$ are Pauli matrices, $\mathbb{1}_{16}$ the unit 16x16 matrix and

$$\gamma^i_{16} = \begin{pmatrix} 0 & \gamma^i_{a\dot{a}} \\ \tilde{\gamma}^i_{\dot{b}b} & 0 \end{pmatrix} \;,\quad \tilde{\gamma}^i_{\dot{b}b} = \left(\gamma^i_{b\dot{b}}\right)^T \;,\quad 1 \le i \le 8 \tag{3.32}$$

provide a 16x16 reducible representation ($8_s + 8_c$).

Majorana-Weyl spinors α are (real) eigenvector of Γ_{11}

$$\Gamma_{11}\alpha = \eta\alpha \tag{3.33}$$

where $\eta = \pm 1$ stands for the chirality and

$$\Gamma_{11} \equiv -\Gamma_0 \Gamma_1 \Gamma_2 \ldots \Gamma_9 \tag{3.34}$$

In the above representation (3.31)-(3.32), the spinors solutions α from (3.33) take the form

$$\alpha = \begin{pmatrix} \lambda \\ \chi \\ -\eta\lambda \\ \eta\chi \end{pmatrix} \tag{3.35}$$

where λ and χ are real light-component spinors.

Let us now find the solutions of eq. (3.26) for D = 10. In this Γ-matrix representation, the four independent solutions are given by

$$\alpha_+ = \begin{pmatrix} \psi \\ 0 \\ -\psi \\ 0 \end{pmatrix} \qquad , \quad \alpha_- = \begin{pmatrix} 0 \\ \varphi \\ 0 \\ -\varphi \end{pmatrix} \tag{3.36}$$

$$\phi_+ = \begin{pmatrix} -2\theta(U)A(X^i/\rho^8)\gamma^i\varphi \\ \varphi \\ 2\theta(U)A(X^i/\rho^8)\gamma^i\varphi \\ \varphi \end{pmatrix}, \quad \phi_- = \begin{pmatrix} \psi \\ -2\theta(U)A(X^i/\rho^8)\tilde{\gamma}^i\psi \\ \psi \\ 2\theta(U)A(X^i/\rho^8)\tilde{\gamma}^i\psi \end{pmatrix}$$

where $A \equiv 3Gp^2/\pi^3$ and ψ, φ are arbitrary real and constant light component spinors. The indices + or - indicate the chirality of the respective solution.

Notice that for D = 4 there are half of these solutions, since Majorana and Weyl conditions cannot be satisfied simultaneously in such dimension.

The non-constant solutions $\phi_{\pm}(u,X^i)$ lead through eq. (3.27) to Rarita-Schwinger fields of infinite energy $\delta\,(U)^2$. Hence, we disregard these U-dependent solutions.

We come back to eqs. (3.22)-(3.23) in order to find the Rarita-Schwinger solutions. We choose for α_c either α_+ or α_-. We then find that eq. (3.26) holds thanks to the relation

$$\left(\Gamma^{0i} + \Gamma^{i9}\right)\alpha_{\pm} = 0 \tag{3.37}$$

which is identically satisfied. After some calculations from eq. (3.27), (3.6) and (3.20) we find for the ψ_μ components

$$\psi_u = \left[-\frac{3}{16}\,\partial_i \mathcal{U}\Big|_{\lambda=0} \Gamma^{i9} + \frac{5}{8}\,\frac{\partial \mathcal{U}}{\partial \xi}\Big|_{\lambda=0} \Gamma^{09} \right] \alpha_{\pm} \tag{3.38}$$

$$\psi_v = 0 \tag{3.39}$$

$$\psi_k = \left\{ \frac{1}{4}\left[\delta_{ik}\,\partial_j \mathcal{U}\Big|_{\lambda=0} - \delta_{jk}\,\partial_i \mathcal{U}\Big|_{\lambda=0} \right] \Gamma^{ij} - \frac{3}{16}\,\partial_k \mathcal{U}\Big|_{\lambda=0} \Gamma^{09} \right\} \alpha_{\pm} \tag{3.40}$$

$$1 \leq i, j, k \leq D-2$$

where

$$\frac{\partial \mathcal{U}}{\partial \xi}\Big|_{\lambda=0} = -\frac{G\tilde{p}}{\pi^3 \rho^6}\,\delta'(U) \tag{3.41}$$

The second term in eq. (3.38) is a pure gauge since it can be written as $D_u\beta$ with

$$\beta = \frac{5}{8\pi^3}\, \tilde{p}\, G\, \frac{\delta(U)}{\rho^6}\, \Gamma^{09}\, \alpha_{\pm} \tag{3.42}$$

(use of eq. (3.37) has been made). Dropping this gauge part, we have for positive chirality

$$\psi_u^{(+)} = \frac{9G}{8\pi^3}\, \tilde{p}\, \frac{\delta(U)}{\rho^8}\, X^i(\tilde{\gamma}^i \otimes \mathbb{1}_4) \begin{pmatrix} 0 \\ \psi \\ 0 \\ \psi \end{pmatrix}$$

$$\psi_k^{(+)} = -\frac{3G\tilde{p}}{\pi^3}\, \frac{\delta(U)}{\rho^8} \left[X^j(\gamma^{kj} \otimes \mathbb{1}_4) + \frac{3}{8} X^k \right] \alpha_+ \tag{3.43}$$

and for negative chirality

$$\psi_u^{(-)} = \frac{9G}{8\pi^3}\, \hat{p}\, \frac{\delta(U)}{\rho^8}\, X^i(\gamma^i \otimes \mathbb{1}_4) \begin{pmatrix} \varphi \\ 0 \\ \varphi \\ 0 \end{pmatrix}$$

$$\psi_k^{(-)} = -\frac{3G}{\pi^3}\, \hat{p}\, \frac{\delta(U)}{\rho^8} \left[X^j(\tilde{\gamma}^{kj} \otimes \mathbb{1}_4) + \frac{3}{8} X^k \right] \alpha_- \tag{3.44}$$

where $\mathbb{1}_4$ is the 4x4 unit matrix, $\gamma^{kj} = \frac{1}{2}[\gamma^k, \tilde{\gamma}^j]$ and $\tilde{\gamma}^{kj} = \frac{1}{2}[\tilde{\gamma}^k, \gamma^j]$.
For both chiralities, $\psi_v = 0$.

III.4 SUPERSTRINGS IN CURVED BACKGROUND SPACE-TIMES

III.4.1 General remarks

A non-linear supersymmetric sigma-model (a generalization of the Ramond-Neveu-Schwarz (NSR) model) in a bosonic gravitational shock-wave was analyzed in ref. [10]. It turned out that for that purely bosonic non-trivial background, it was possible to choose a light-cone gauge. The transverse equations of motion for the bosonic coordinates $X^i(\sigma,\tau)$ $(1 \leq i \leq 8)$ then decoupled from the world-sheet fermions $\psi^i(\sigma,\tau)$ and the fermion coordinates $\psi^i(\sigma,\tau)$ obeyed free equations. That is, the ψ^i were not affected by the bosonic shockwave background. The latter conclusion implied that the S matrix, which relates the in and out modes, did not involve fermionic degrees of freedom.

In curved spacetime admitting an 8_c gauge covariant constant spinor, the NSR formalism (where spinors carry a spacetime vector index) can be related to the GS formalism (where spinors carry a spacetime spinor index) [22]. Then, triality among the 8_v, 8_s, and 8_c representations of SO(8) enables to write the 8_v transverse components of the NSR spinors ψ^i in terms of the 8 GS spinor by

$$\psi^i = \bar{\alpha}\, \Gamma^i \Gamma^+ S \tag{4.1}$$

where α is a Majorana-Weyl 8_c gauge covariant constant spinor. Owing to triality, the above relation can be inverted. In flat spacetime, the triality property is well known [23]. In the bosonic-fermionic shockwave spacetime we propose, we find that the triality property is satisfied (due to the existence of a gauge covariant constant spinors) allowing us to relate the NSR and GS lightcone string variables. In the purely bosonic shock-wave background, both NSR and GS lightcone formalisms are equivalent.

Therefore, one could expect that a GS formalism in a purely bosonic background will be unable to provide us more interesting physics that the NSR model in such a background. In a superstring context everything mentioned so far lead us to consider a supersymmetric background including fermionic as well as bosonic fields.

In spite of the fact that superstrings in a background supergravity field have been extensively formulated in the literature [24 - 26], we would like to notice the following points :

(i) background fermions were set, sometimes equal to zero [27].

(ii) it turned out that the existence of a constant covariant spinor was necessary in order to establish a light-cone formulation. This fact implied non-trivial restrictions on the choice of background [22, 28].

(iii) only the right or left movers of the spinor string coordinates were considered as the lagrangians were obtained through superspace techniques.The model then investigated was called [28] "heterotic GS model in curved space". (Although the well known heterotic string [29] is different from it, as it has 26 bosonic degrees of freedom instead of 10).

III.4.2 Heterotic Green-Schwarz superstring propagation in a shockwave supergravity background

A general formulation of a superstring coupled to a supergravity background is already known in the framework of curved superspace ($Z=(x,\theta)$) [24 -26]. As a point Z in superspace has 10 bosonic coordinates (x) but only 16 fermionic coordinates (θ) one cannot accomodate both the right and left movers of the space-time Green-Schwarz spinor S, and the superspace formalism is only appropriate to a heterotic Green-Schwarz superstring.

Grisaru and Zanon [28], starting with a Green-Schwarz action in a full supergravity background proposed a light-cone action by fixing some gauges for the supergravity fields and imposing additional restrictions on the supergravity background. One of their aims was to link the NSR and the GS formulations for the case of non-trivial fermionic backgrounds.

In this section we propose a (heterotic) Green-Schwarz superstring coupled to the fermionic-bosonic background given in Section III.3.1. We will see that light -cone conditions are naturally satisfied in this Aichelburg-Sexl type metric.

Using a covariant expansion [30] to reduce the Green-Schwarz superaction to components, the lagrangian is given by [28] (up to second order in the spinor string variable S) by

$$\mathcal{L} = \frac{1}{2}\partial_A X^\mu \partial^A X^\nu e^m_\mu e_{\nu m} + (\eta^{AB} - \epsilon^{AB})\overline{(D_A S)}\Gamma^m S \partial_B X^\nu e_{\nu m}$$
$$+ 2(\eta^{AB} - \epsilon^{AB})\partial_A X^\mu (\bar{\psi}_\mu \Gamma^m S)\partial_B X^\nu e_{\nu m} \tag{4.2}$$

where $(D_A S)_\alpha = (\partial_A S_\alpha) + S^\beta \partial_A X^\mu e_\mu^m \; \omega_{m\beta\alpha}$ and A, B = 0, 1 (world sheet index); $\alpha = 1,\ldots,32$ spinor flat index. In Lagrangian (4.2) only the graviton (e_μ^m) and the gravitino (ψ_μ^α) supergravity fields have been considered. The Dirac field χ^α, the dilaton ϕ and $b_{\mu\nu}$ have taken trivial values (eq.3.7); also, the linear approximation for the Rarita-Schwinger fields has been used.

In light-cone coordinates the vielbeine for the AS metric (3.20) satisfy

$$e_+^- = 0, \quad e_\pm^i = 0, \quad e_r^\pm = 0, \qquad 1 \le i \le 8, 1 \le r \le 8 \tag{4.3}$$

where the lower indices are world indices and upper indices correspond to the tangent space. In addition the background is V independent and its fermionic part satisfies $\psi_V = 0$ [eq. (3.39)]. The set of equations (4.3) and (3.39) contain the conditions on the background set in ref. [28] in their way to obtain a light-cone gauge action. (Notice that our U coordinate corresponds to X- in ref.[26]).

The first term in the lagrangian (4.2) gives upon varying on V a free equation for $U(\sigma, \tau)$

$$(\partial_\tau^2 - \partial_\sigma^2)\, U(\sigma, \tau) = 0 \tag{4.4}$$

Since the background is V-independent we expect this equation to be exact to all orders. We can then choose the light-cone gauge.

$$U = 2\alpha' p^u \tau \tag{4.5}$$

We impose also a light-cone gauge condition on the spinor string variable

$$\Gamma^- S(\sigma, \tau) = 0 \tag{4.6}$$

where $\Gamma^- = \frac{1}{\sqrt{2}}(\Gamma^0 - \Gamma^9)$ (see eq. (3.31)).

The light-cone spinors

$$S_+ = \begin{pmatrix} S \\ 0 \\ -S \\ 0 \end{pmatrix}, \qquad S_- = \begin{pmatrix} 0 \\ \chi \\ 0 \\ -\chi \end{pmatrix} \tag{4.7}$$

fulfil eq. (4.6).

Here $\mathcal{S}_{\pm}$ has chirality ±1 and S and χ are real light-component spinors. Notice that the constant spinors $\alpha_{\pm}$ in (3.36) are light-cone spinors of the type (4.7). We choose from now on a positive chirality for our spinor string fields (That is $\mathcal{S} = \mathcal{S}_{+}$).

The shock-wave background we consider is Minkowski spacetime except at U = 0, that is $\tau = 0$, where the fermionic background and the non-Minkowski part of the bosonic background are concentrated [see eqs. (3.19), (3.43) and (3.44)]. Therefore, the string propagation will be free for $\tau < 0$ (before the collision) and for $\tau > 0$ (after the collision with the shock-wave).

Inserting now the metric (3.19), the connections (3.28), the Rarita-Schwinger field (eq.(3.43)), and the spin string variable $\mathcal{S}_{+}$ (eq.(4.7)) in the Lagrangian (eq.(4.2)), we have in the light-cone gauge (4.5)-(4.6)

$$\mathcal{L} = \frac{1}{2}\partial_A X^i \partial^A X^i + \alpha' p^U \delta(\tau) f_{10}(X^i) + 2\sqrt{2}\alpha' p^U \partial_+ S^T S$$

$$- \frac{6\sqrt{2}\, G\tilde{p}}{\pi^3 \rho^8}\, \delta(\tau)\left[X^j (\gamma^{ij}\psi)^T S + \frac{3}{8} X^i \psi^T S\right]\partial_+ X^i$$

$$+ \frac{9\, G\tilde{p}}{2\pi^3 \rho^8}\, \delta(\tau)\, X^i (\tilde{\gamma}^i \psi)^T (\tilde{\gamma}^j S)\, \partial_- X^j \qquad (4.8)$$

where ψ is an arbitrary real and constant 8_s spinor characterizing the Rarita Schwinger background and $f_{10} = -8Gp/\pi^3\rho^6$.

In the course of the calculations leading to eq. (4.8) we used the following relations

$$\Gamma^{i-}\Gamma^{j}\mathcal{S}^{\gamma} = 0$$

$$\Gamma^{i-}\Gamma^{+}\mathcal{S}^{\gamma} = -\Gamma^{i}\mathcal{S}^{\gamma}$$

$$\bar{\Psi}_u \Gamma^+ \mathcal{S} = 0$$
$$\bar{\Psi}_i \Gamma^j \mathcal{S} = 0$$
$$\bar{\mathcal{S}}^\alpha \Gamma^i_{\alpha\beta} \mathcal{S}^\beta = 0 \tag{4.9}$$

In order to have dimensionless fermionic string variables , we define

$$S' = 2^{3/4}\sqrt{p^U}\, S \qquad , \; \psi' = 2^{3/4}\sqrt{p^U}\,\psi \tag{4.10}$$

(we drop primes from now on).

The equations for $S_a(\sigma, \tau)$ result from eqs (4.8) and (4.10)

$$\partial_+ S_a = \frac{3G\tilde{p}}{\alpha' p^u \pi^3}\, \delta(\tau)\, \phi_a(\sigma) \tag{4.11}$$

where

$$\phi_a(\sigma) = \frac{1}{\rho^8}\left[X^j(\sigma,0)(\gamma^{ij}\psi)^T_a + \frac{3}{8} X^i(\sigma,0)\psi_a\right] \partial_+ X^i \Big|_{\tau=0}$$
$$- \frac{3X^i}{4\sqrt{2}\rho^8}\left[(\tilde{\gamma}^i\psi)^T \tilde{\gamma}^j\right]_a \partial_- X^j \Big|_{\tau=0} \tag{4.12}$$

The solution of this equation can be written as

$$S_a(\sigma,\tau) = \begin{cases} S_<(\sigma-\tau)_a & \tau < 0 \\ S_>(\sigma-\tau)_a & \tau > 0 \end{cases} \tag{4.13}$$

where $S_<(\sigma)$ is an arbitrary function and

$$S_>(\sigma)_a = S_<(\sigma)_a + \frac{3G\tilde{p}}{\alpha' p^U \pi^3}\, \phi_a(\sigma) \tag{4.14}$$

Eqs. (4.8) and (4.10) give the equation of motion for the string coordinates $X^i(\sigma, \tau)$

$$
(\partial_\tau^2 - \partial_\sigma^2) X^k = - p^U \alpha' \delta(\tau) \partial_k f_{10}(X^i) - \frac{3G\tilde{p}}{\pi^3 p^U} \frac{\delta'(\tau)}{\rho^8} X^j
$$

$$
\times \left[(\gamma^{kj}\psi)^T + \frac{3}{8} \delta^{kj} \psi^T + \frac{3}{4\sqrt{2}} (\tilde{\gamma}^j \psi)^T \tilde{\gamma}^k \right] S + \frac{3G\tilde{p}\,\delta(\tau)}{\pi^3 p^U \rho^8}
$$

$$
\times \left\{ - \left(\frac{8X^j}{\rho^2} [(\gamma^{ij} X^k + \gamma^{kj} X^i)\psi]^T + \frac{6X^i X^k}{\rho^2} \psi^T - 2(\gamma^{ik}\psi)^T \right) \partial_+ X^i S \right.
$$

$$
- \frac{3}{4\sqrt{2}} \left[\left(\delta^{jk} - \frac{8X^i X^k}{\rho^2} \right) \tilde{\gamma}^i - \left(\delta^{ij} - \frac{8X^i X^j}{\rho^2} \right) \tilde{\gamma}^k \right] (\tilde{\gamma}^j \psi)^T \partial_- X^j S
$$

$$
\left. + \frac{3}{4\sqrt{2}} X^i (\tilde{\gamma}^i \psi)^T (\tilde{\gamma}^k \partial_- S) \right\} \tag{4.15}
$$

For $\tau < 0$ and for $\tau > 0$ we can write for the string coordinates the usual Minkowski solutions $X^i_<(\sigma, \tau)$, $S^a_<(\sigma, \tau)$ and $X^i_>(\sigma, \tau)$, $S^a_>(\sigma, \tau)$ respectively (eqs. (1.12b) and (2.41a.b)).

Eqs. (4.14) and (4.15) determine the $>$ coefficients in terms of the $<$ coefficients. Let us start by analyzing the fermionic coordinates $S^a(\sigma, \tau)$. In eq. (4.14), ϕ_a is given by eq. (4.12) where $X^i = X^i_<(\sigma, \tau)$. Inserting eq. (1.12b) in (4.11) yields the Fourier expansion for $\phi_a(\sigma)$. We find

$$
\phi_a(\sigma) = M_a^{ij} \sum_{\ell \in Z} e^{i\ell\sigma} \beta^{ji}_{+,\ell} + N_a^{ij} \sum_{\ell \in Z} e^{i\ell\sigma} \beta^{ij}_{-,\ell} \tag{4.16}
$$

where

$$
M_a^{ij} \equiv (\gamma^{ij}\psi)_a^T + \frac{3}{8} \delta^{ij} \psi_a \, , \quad N_a^{ij} \equiv \frac{3}{4\sqrt{2}} \left[(\tilde{\gamma}^i \psi)^T \tilde{\gamma}^j \right]_a \tag{4.17}
$$

and

$$\beta^{ij}_{\pm,\ell} = \int_0^\pi \frac{d\sigma}{\pi} : \frac{X^i(\sigma,\tau)\,\partial_\pm X^j(\sigma,\tau)}{\rho^8(\sigma,\tau)}\Big|_{\tau=0} : e^{-i\ell\sigma} \tag{4.18}$$

Here, the normal order prescription in eq.(4.18) is explicitly implemented by

$$: \frac{X^i \partial_\pm X^j}{\rho^8(\sigma,\tau)}\Big|_{\tau=0} : = -\frac{1}{3}\int \frac{d^8K}{(4\pi)^4}\,\frac{K^i}{K^2}\, e^{iK\cdot q_<}\, A^j_\pm \tag{4.19}$$

where

$$A^j_\pm = e^{iK\cdot X^{(-)}(\sigma,0)}\, \partial_\pm X^j\Big|_{\tau=0}\, e^{iK\cdot X^{(+)}(\sigma,0)} \tag{4.20a}$$

$$X^{i\,(\pm)}_<(\sigma,0) = \pm i\sqrt{2\alpha'}\sum_{n=1}^{\infty}\frac{1}{n}\cos n\sigma\ \alpha^i_{\pm n,<} \tag{4.20b}$$

Thus, the fermion string operators $S^a{}_{n>}$ and $S^a{}_{n<}$ are related through

$$S^a_{\ell>} - S^a_{\ell<} = \frac{3\sqrt{2}\,G\tilde{p}}{\alpha' p^\upsilon \pi^3}\left[M^{ij}_a\, \beta^{ji}_{+,\ell} + N^{ij}_a\, \beta^{ij}_{-,\ell}\right] \tag{4.21}$$

We find a mixing between fermionic and bosonic oscillator modes.That is the fermion operators express in terms of the < fermions and the < bosonic oscillators. The expression is non-polynomial in the bosonic oscillators [(cf. the $\rho(\sigma,\tau)^{-8}$ in eq.(4.12)] and linear in the fermions in the approximation we consider. Notice that the spinor ψ is an anticommuting object.

It is easy to find the canonical transformation that generates (4.21). We find to this order

$$S^a_{n>} = e^{G_F}\, S^a_{n<}\, e^{-G_F} \simeq S^a_{n<} + [G_F, S^a_{n<}] \tag{4.22}$$

with

$$G_F = \frac{3\sqrt{2}\, G\tilde{p}}{\alpha' p^v n^3} \sum_{\ell\in Z} [M^{ij}_a\, \beta^{ji}_{+,\ell} + N^{ij}_a\, \beta^{ij}_{-,\ell}]\, S^a_{-\ell<} \tag{4.23}$$

By analogous steps one derives the relation between > and < bosonic oscillators $\alpha^i_{n>}$ and $\alpha^i_{n<}$ from eq.(4.15). Actually this transformation is generated by the same operator G_F (eq. (4.23)) plus the generator of the purely bosonic transformation [31]

$$G_B = -i\, G\tilde{p} \frac{8 p^v}{(2\pi)^8} \int_0^\pi d\sigma \int \frac{d^8 k}{k^2} :e^{i k\cdot X_<(\sigma,0)}: \tag{4.24}$$

$$\alpha^i_{n>} = e^{\mathcal{G}}\, \alpha^i_{n<}\, e^{-\mathcal{G}} \simeq \alpha^i_{n<} + [\mathcal{G}, \alpha^i_{n<}] \tag{4.25}$$

where $\mathcal{G} = G_F + G_B$ .The normal order prescription in eq. (4.24) is implemented by

$$:e^{i k\cdot X_<(\sigma,0)}: = e^{i k\cdot q_<}\, e^{i k\cdot X^{(-)}_<(\sigma,0)}\, e^{i k\cdot X^{(+)}_<(\sigma,0)} \tag{4.26}$$

III.4.3 Boson-fermion and fermion-boson transmutations

In the previous section, we showed the mixing between fermionic and bosonic oscillator modes for a superstring wich propagates in a supergravity background. Therefore, transitions from bosons to fermions (and fermions to bosons) are allowed, i.e., fermion number changing transitions appear. A particular consequence of this fact is the existence of non - zero transitions among the superstring ground states; this means we have transmutation of massless fermions into massless bosons and viceversa. (This massless super Yang - Mills N=1 , D=10 multiplet describes physical particles upon supersymmetry breaking at energies much lower than the Planck and grand unification scales).

In this section, we analyse and compute explicitely the more relevant and new particle transmutations that take place when an open superstring propagates through a supergravitational shock - wave in D=10 space - time dimensions.

The superstring ground state is

$$|\Phi_0\rangle = \{ |f, p\rangle , |\dot{a}, p\rangle \} \tag{4.27}$$

where **p** is the particle momentum, $|f\rangle$ and $|a\rangle$ ($1 \leq (a, f) \leq 8$ > transform under 8_V and 8_C SO(8) representation respectively, i.e. the super Yang-Mills multiplet N=1 , D=10. The ground states fulfil

$$\alpha^i_n |\Phi_0\rangle = 0 \quad (n>0) \quad , \quad S^a_\ell |\Phi_0\rangle = 0 \quad (\ell>0) \tag{4.28}$$

The fermionic and the bosonic ground states transform into each other as

$$S^a_0 |\dot{a}\rangle = \frac{1}{\sqrt{2}} \gamma^f_{a\dot{a}} |f\rangle \tag{4.29}$$

The transition amplitude between ingoing and outgoing states is given by

$$\langle out, p_1 | in, p_2 \rangle \simeq - \langle in, p_1 | G_{<} | in, p_2 \rangle \tag{4.30}$$

up to first order.

From the above equations, and with the help of eqs (4.23-4.24), we get the following amplitudes :

(i) Transitions among the ground states ($\alpha' M^2 = 0$) :

$$\langle f_{out}, p_1 | \dot{a}_{in}, p_2 \rangle = \frac{2iG\tilde{p}}{p^{\upsilon}\,\pi^3} \frac{1}{(4\pi)^4}\, p_2^j \frac{(p_2 - p_1)^i}{(p_2 - p_1)^2} \times \left(M_a^{ji} + N_a^{ij} \right) \gamma^f_{a\dot{a}} \tag{4.31}$$

$$\langle f_{1\,out}, p_1 | f_{2\,in}, p_2 \rangle = \delta_{f_1 f_2} Z \ , \quad \langle \dot{a}_{out}, p_1 | \dot{b}_{in}, p_2 \rangle = \delta_{\dot{a}\dot{b}} Z \tag{4.32a.b}$$

where

$$Z = \frac{8iG\tilde{p}\pi p^{\upsilon}}{(2\pi)^8} \frac{1}{(p_1 - p_2)^2}$$

We find a non - zero transition amplitude (eq. 4.31) from a massless **fermionic** initial state, with momentum $\mathbf{p}_1$, to a massless **bosonic** final state with momentum $\mathbf{p}_2$, and viceversa. This is especially interesting since these masssless states yield physical massive particles upon supersymmetry breaking at low energies. The non conservation of the particle fermion number proceeds from the fermionic gravitino background which *injects* half integer spin into the particle states.

It should be noticed that the transition amplitudes between the bosonic ground states themselves, and the fermionic ground states themselves, eqs. (4.32a,b) are diagonal. No rotation on the spin polarization takes place for the ground states.

(ii) Transitions between excited states:

We consider now the transitions involving excited states. The excited states of the first ($\alpha' M^2 = 1$) level are

bosonic states: $|\Phi^{ij}_{out}\rangle = \alpha^i_{-1\rangle}|j_{out}\rangle \quad , \quad S^a_{-1\rangle}|\dot{b}_{out}\rangle$

fermionic states: $|\Phi^{ia}_{out}\rangle = \alpha^i_{-1\rangle}|\dot{a}_{out}\rangle \quad , \quad S^a_{-1\rangle}|i_{out}\rangle$

The bosonic multiplet transforms under the $8_v \times 8_v = 1+28+35_v$ and the $8_s \times 8_c = 8_v + 56_v$ SO(8) representation . The fermionic multiplet transforms under $8_c \times 8_v = 8_s + 56_s$ and $8_v \times 8_s = 8_c + 56_c$.

In the linear approximation in which we are working we find the following three relevant non - zero transition amplitudes, due to the operator G_F (eq.4.23), between the first level of the excited states :

$$\langle f_{out}, p|\alpha^j_{1\rangle} S^a_{-1\langle}|f'_{in}, p'\rangle = \frac{iG\tilde{p}}{\sqrt{\alpha'}\, p^{\upsilon}\pi^3(4\pi)^4}\frac{(p'-p)^i}{(p'-p)^2}\delta_{ff'}$$

$$\times\left[N^{ij}_a - 2\alpha'(p'-p)^j p'_r\left(M^{ri}_a + N^{ir}_a\right)\right] \tag{4.33}$$

$$\langle \dot{a}_{out}, p|\alpha^j_{1\rangle} S^a_{-1\langle}|\dot{a}'_{in}, p'\rangle = \frac{iG\tilde{p}}{\sqrt{\alpha'}\, p^{\upsilon}\pi^3(4\pi)^4}\frac{(p'-p)^i}{(p'-p)^2}\delta_{\dot{a}\dot{a}'}$$

$$\times\left[N^{ij}_a - 2\alpha'(p'-p)^j p'_r\left(M^{ri}_a + N^{ir}_a\right)\right] \tag{4.34}$$

$$\langle f_{out}, p | \alpha^r_{1>} \alpha^j_{-1<} | \dot{a}_{in}, p' \rangle = \frac{-iG\tilde{p}}{p^{\upsilon}\pi^3(4\pi)^4} \frac{(p'-p)^i}{(p'-p)^2}$$

$$\times \Big\{ -\big[\alpha'(p'-p)^r (p'-p)^j p'^k (M^{ki}_a + N^{ik}_a) - \frac{1}{2}(p'-p)^r \cdot (M^{ji}_a + N^{ij}_a) + \frac{1}{2}(p'-p)^j (M^{ri}_a + N^{ir}_a) \big] \Big\} \gamma^f_{a\dot{a}} \tag{4.35}$$

Notice that the spinor (8) and the Rarita - Swinger (56) have a definite chirality (c or s) . Up to this order , there are no transitions between the vacuum and the first excited states.

The transitions between the first levels of the excited states due to the bosonic operator G_B (eq. (4.24)) are given by

$$\langle f_{out}, p | \alpha^i_{1>} \alpha^j_{-1<} | f_{in}, p' \rangle = \frac{8iG\tilde{p}\pi p^{\upsilon}\alpha'}{(2\pi)^8} \delta_{ff'} \times \frac{(p'-p)^i (p'-p)^j}{(p'-p)^2} \tag{4.36}$$

$$\langle \dot{a}_{out}, p | \alpha^i_{1>} \alpha^j_{-1<} | \dot{b}_{in}, p' \rangle = \frac{-8iG\tilde{p}\pi p^{\upsilon}\alpha'}{(2\pi)^8} \delta_{\dot{a}\dot{b}} \times \frac{(p'-p)^i (p'-p)^j}{(p'-p)^2} \tag{4.37}$$

For the second excited states, we find the following six non - zero transition amplitudes :

$$\langle f_{out}, p | \alpha^j_{2>} S^a_{-2<} | f'_{in}, p' \rangle = \frac{2iG\tilde{p}}{\sqrt{\alpha'}\, p^{\upsilon}\pi^3(4\pi)^4} \frac{(p'-p)^i}{(p'-p)^2} \delta_{ff'}$$

$$\times \big[N^{ij}_a - \alpha'(p'-p)^j p'_r (M^{ri}_a + N^{ir}_a) \big] \tag{4.38}$$

$$\langle \dot{a}_{out}, p | \alpha^j_{2>} S^a_{-2<} | \dot{a}'_{in}, p' \rangle = \frac{2iG\tilde{p}}{\sqrt{\alpha'}\, p^U \pi^3 (4\pi)^4} \frac{(p'-p)^i}{(p'-p)^2} \delta_{a\dot{a}}$$

$$\times [N^{ij}_a - \alpha'(p'-p)^j p'_r (M^{ri}_a + N^{ir}_a)]$$

(4.39)

$$\langle f_{out}, p | \alpha^r_{2>} \alpha^j_{-2<} | \dot{a}_{in}, p' \rangle = \frac{-iG\tilde{p}}{p^U \pi^3 (4\pi)^4} \frac{(p'-p)^i}{(p'-p)^2}$$

$$\times [\alpha'(p'-p)^r (p'-p)^j p'_k (M^{ki}_a + N^{ik}_a)$$

$$- (M^{ji}_a + N^{ij}_a)(p'-p)^r + (M^{ri}_a + N^{ir}_a)(p'-p)^j] \gamma^f_{a\dot{a}}$$

(4.40)

$$\langle f_{out}, p | \alpha^s_{-1<} S^a_{-1<} | f'_{in}, p' \rangle = \frac{-iG\tilde{p}}{\sqrt{\alpha'}\, p^U \pi^3 (4\pi)^4} \frac{(p'-p)^i}{(p'-p)^2} \delta_{ff'}$$

$$\times [M^{si}_a - 2\alpha'(p'-p)^s p'_j (M^{ji}_a + N^{ij}_a)]$$

(4.41)

$$\langle \dot{a}_{out}, p | \alpha^s_{-1<} S^a_{-1<} | \dot{b}_{in}, p' \rangle = \frac{-iG\tilde{p}}{\sqrt{\alpha'}\, p^U \pi^3 (4\pi)^4} \frac{(p'-p)^i}{(p'-p)^2} \delta_{\dot{a}\dot{b}}$$

$$\times [M^{si}_a - 2\alpha'(p'-p)^s p'_j (M^{ji}_a + N^{ij}_a)]$$

(4.42)

$$\langle f_{out}, p | \alpha^j_{-1<} \alpha^s_{-1<} | \dot{a}_{in}, p' \rangle = \frac{i G \tilde{p}}{p^U \pi^3 (4\pi)^4} \frac{(p'-p)^i (p'-p)^s}{(p'-p)^2}$$

$$\times \left(M^{ji}_a + N^{ij}_a \right) \gamma f_{a\dot{a}} \tag{4.43}$$

Eqs (4.38) - (4.43) describe the transitions from the vacuum to second excited states due to the operator G_F .

The transitions from the vacuum to second order excited states due to G_B are given by

$$\langle f_{out}, p | \alpha^i_{1>} \alpha^j_{-1<} | f'_{in}, p' \rangle = \frac{-4 i G \tilde{p} \pi p^U \alpha'}{(2\pi)^8} \delta_{ff'}$$

$$\times \frac{(p'-p)^i (p'-p)^j}{(p'-p)^2} \tag{4.44}$$

$$\langle \dot{a}_{out}, p | \alpha^i_{1>} \alpha^j_{-1<} | \dot{b}_{in}, p' \rangle = \frac{-4 i G \tilde{p} \pi p^U \alpha'}{(2\pi)^8} \delta_{\dot{a}\dot{b}}$$

$$\times \frac{(p'-p)^i (p'-p)^j}{(p'-p)^2} \tag{4.45}$$

Notice that up to this order there are no transitions between the vacuum and the following second excited states : $S^a_{-2} |\phi_o\rangle$, $S^a_{-1} S^a_{-1} |\phi_o\rangle$ and $\alpha^i_{-1} S^a_{-1} |\phi_o\rangle$.

It can be noticed that some of the transition amplitudes here are of anticommuting nature. This comes naturally from the gravitino background. The probabilities,which are the squared modulus of the transition amplitudes, are c- numbers as it must be. These square moduli are not identically zero since we have more than one (actually eight) independent anticommuting Grassman variables.

IV. CONCLUSIONS

We briefly sumarize some of the most important results we have shown, concerning particle creation and transmutation. Only the **non vanishing transition amplitudes** are mentioned.

(I) For a backgrond with purely "bosonic" degrees of freedom, we have the transitions:

I.A For bosonic strings:

(i) Ground state to ground state (pair creation; for both open and closed strings)

(ii) Ground state to highly massive state (Mass > M_{Planck}) generated by an **even** number of creation operators (for open and closed strings)

(iii) Ground state to graviton, and ground state to dilaton (for closed strings)

(iv) Dilaton to graviton (transmutation; for closed strings)

I.B For superstrings:

(i) Ground state to ground state elastic processes (this result is true to all orders in the perturbation scheme, due to the fact that the exact non-linear transformation between the ingoing and outgoing fermionic operators preserve the fermion number)

(ii) Ground state to highly massive state, generated by an even number of boson and/or fermion creation operators

(II) For a background with **both**, "bosonic" **and** "fermionic", degrees of freedom, we have:

(i) Ground state to ground state inelastic proceses: an ingoing massless fermion can transmute into an outgoing massless boson, and viceversa.

(ii) Transitions among excited states of the same level.

REFERENCES

[1] H.J.de Vega and N.Sánchez, Phys. Lett. **B197** (1987) 320.

[2] H.J.de Vega and N.Sánchez, Nucl. Phys. **B309** (1988) 552, 577.

[3] H.J.de Vega and N.Sánchez, Nucl. Phys. **B317** (1989) 706

[4] N.Sánchez and G.Veneziano, Nucl. Phys. **B333** (1990) 253

[5] M.Gasperini, N.Sánchez and G.Veneziano, Int.J.Mod.Phys. **A6** (1991) 3853; Nucl. Phys. **B364** (1991) 365

[6] H.J.de Vega and N.Sánchez, Phys. Lett. **B244** (1990) 215

[7] H.J.de Vega and N.Sánchez, Phys. Rev. Lett **C65** (1990) 1517

[8] M.Costa and H.J.de Vega, Ann. Phys. **211** (1991) 223, 235

[9] H.J.de Vega and N.Sánchez, Phys. Rev. **D42** (1990) 3969

[10] D.Amati and C.Klimcik, Phys. Lett. **B210** (1988) 92

[11] H.J.de Vega, M.Ramón - Medrano and N.Sánchez, Nucl. Phys. **B374** (1992) 425

[12] H.J.de Vega, M.Ramón - Medrano and N.Sánchez, Nucl. Phys. **B351** (1991) 277

[13] H.J.de Vega, M.Ramón - Medrano and N.Sánchez, Phys. Lett. **285** (1992) 206

[14] H.J.de Vega, M.Ramón - Medrano and N.Sánchez, Nucl. Phys. **B374** (1992) 405

[15] M.B.Green, J.H.Schwarz and E.Witten, " Superstring theory ", Vol. 1 (Cambridge Univ. Press, Cambridge, 1987)

[16] P.de Sousa Gerbert and R. Jackiw, Commun. Math. Phys. **124** (1989) 229

[17] E.Bergshoeff, M.de Roo, B.de Witt and P.van.Nieuwenhuizen, Nucl. Phys. **B195** (1982) 97; B.E. Wilson, Nucl. Phys. **B188** (1981) 176

[18] R.C.Myers and M.J.Perry, Ann. Phys. (N.Y) **172** (1986) 304

[19] C.O.Loustó and N.Sánchez, Int.J.Mod.Phys **A5** (1990) 915

[20] P.C.Aichelburg and R.U.Sexl , Lett. Nuovo Cimento **26** (1970) 1316: Gen.Rel.Grav. **2** (1971) 303

[21] P.C.Aichelburg and F.Embacher, Class. Quant. Grav. **1** (1984) 125

[22] M.D.Freeman, C.N.Pope, M.F.Sohnius and K.S.Stelle in Proc. Paris-Meudon Colloquium, ed. H.J.de Vega and N.Sánchez (World Scientific, Singapore 1986).

P.Candelas, G.Horowitz, A.Strominger and E.Witten, Nucl. Phys. **B258** (1985) 46

C.M.Hull, Phys. Lett. **B178** (1986) 357

A.Strominger, Nucl. Phys. **B274** (1986) 253

[23] R.I.Nepomechie, Phys. Lett. **B178** (1986) 207

[24] E.Witten, Nucl. Phys. **B266** (1986) 245

[25] J.J.Atick, A.Dhar and R.Ratra, Phys. Lett. **B169** (1986) 54
R.E.Kallosh, Phys. Scr. **T15** (1987) 118

[26] J.A.Shapiro and C.C.Taylor, Phys. Rep. **191** (1990) 221 and references therein.

[27] J.J.Atick and A.Dhar, Nucl.Phys. **B284** (1987) 131

[28] M.T.Grisaru and D.Zanon, Nucl. Phys. **B310** (1988) 57

[29] D.J.Gross, J.A.Harvey, E.Martinec and R.Rohm, Phys. Rev. Lett. **54** (1985) 502; Nucl. Phys. **B256** (1985) 253

[30] S.Mukhi, Nucl. Phys. **B264** (1986) 640
I.J.Atick and A.Dhar, Nucl. Phys. **B284** (1987) 131

[31] H.J.de Vega and N.Sánchez. Int.J.Mod.Phys. **A7** (1992) 3043

PARTICLE AND STRING SCATTERING AT THE PLANCK SCALE

C. O. LOUSTO[a]
IFAE - Grupo de Física Teórica, Universidad Autónoma de Barcelona,
E-08193 Bellaterra (Barcelona), Spain. E-mail: lousto@ifae.es

and

N. SÁNCHEZ
Observatoire de Paris DEMIRM, UA 336 Laboratoire associé au CNRS,
61, Av. de l'Observatoire, 75014 Paris, France.

ABSTRACT

The ultrarelativistic limit of the Kerr - Newman geometry is studied in detail. We find the corresponding gravitational shock wave background associated with this limit. Interestingly, this allows us to find the source of the Kerr - Newman geometry in the ultrarelativistic regime. We study the scattering of scalar fields in the gravitational shock wave geometries, and discuss the presence of the poles $iGs = n = 0, 1, 2, ...$, already present in the Aichelburg - Sexl metric. We compare this with the scattering by ultrarelativistic extended sources, for which such poles do not appear and with the scattering of fundamental strings.

We also study planckian energy string collisions in flat spacetime as the scattering of a string in the effective curved background produced by the others as the impact parameter b decreases. We find the effective energy density distribution $\sigma(\rho) \sim \exp\{-\rho^2/\Delta^2\}$, generated by these collisions. Two different regimes can be studied: intermediate impact parameters $x^d \ll b \ll \sqrt{\alpha' \ln s} \cong \Delta/2$, (x^d characterizing the string fluctuations) and large impact parameters, $b \gg \sqrt{\alpha' \ln s} \cong \Delta/2 \gg x^d$. The effective metric generated by these collisions is a gravitational shock wave of profile $f(\rho) \sim p\rho^{4-D}$, i.e. the Aichelburg - Sexl geometry for a point-like particle of momentum p for large b. For intermediate b, $f(\rho) \sim q\rho^2$, corresponding to an extended source of momentum q. The scattering matrix in this geometry and its implications for the string collision process are analysed. We show that the poles $iGs = n$, $n = 0, 1, 2...$, characteristic of the scattering by the A - S geometry are absent here, due to the extended nature of the effective source.

We finally study the emergence of string instabilities in D - dimensional black hole spacetimes (Schwarzschild and Reissner - Nordström), and De Sitter space (in static coordinates to allow a better comparison with the black hole case). We solve the first order string fluctuations around the center of mass motion at spatial infinity, near the horizon and at the spacetime singularity. We find that the time components are always well behaved in the three regions and in the three

[a]Permanent address: Instituto de Astronomía y Física del Espacio, Casilla de Correo 67 - Sucursal 28, 1428 Buenos Aires, Argentina. E-mail: lousto@iafe.edu.ar

backgrounds. The radial components are *unstable*: imaginary frequencies develop in the oscillatory modes near the horizon, and the evolution is like $(\tau - \tau_0)^{-P}$, $(P > 0)$, near the spacetime singularity, $r \to 0$, where the world - sheet time $(\tau - \tau_0) \to 0$, and the proper string length grows infinitely. In the Schwarzschild black hole, the angular components are always well - behaved, while in the Reissner - Nordström case they develop instabilities inside the horizon, near $r \to 0$ where the repulsive effects of the charge dominate over those of the mass. In general, whenever large enough repulsive effects in the gravitational background are present, string instabilities develop. In De Sitter space, all the spatial components exhibit instability. The infalling of the string to the black hole singularity is like the motion of a particle in a potential $\gamma(\tau - \tau_0)^{-2}$ where γ depends on the D spacetime dimensions and string angular momentum, with $\gamma > 0$ for Schwarzschild and $\gamma < 0$ for Reissner - Nordström black holes. For $(\tau - \tau_0) \to 0$ the string ends trapped by the black hole singularity.

1. Introduction

Recently, gravitational shock waves backgrounds have raised interest in the, context of both Field Theory and Strings[1,2,3,4,5,6,7,8,9,10,11,12,13] . Besides their role in classical gravitation, as an exact class of solutions, these geometries are relevant at energies of the order of the Planck scale.

At energies of the order of the Planck scale, the picture of particles propagating in flat space - time is no longer valid, and one must take into account the curved space - time geometry created by the particles themselves. In other words, gravitational interactions are at least as important as the rest, and can not be neglected anymore as it is usually the case in particle physics. In this context, gravitational shock wave backgrounds play a relevant role as they are the metrics generated by ultrarelativistic particles. The Aichelburg - Sexl (A - S) geometry[14] , is a typical example, as the ultrarelativistic limit of the Schwarzschild metric. The A - S metric was generalized to include the electromagnetic momentum of the particle, by obtaining the ultrarelativistic limit of the Reissner - Nordström solution[15] . In the paper 16 we find the ultrarelativistic limit of the Kerr - Newman geometry, to include the charge and spin of the particles.

In the elementary particle context in which we are studying these metrics, several interesting features are to be mentioned: Remarkable enough, the Klein - Gordon equation[1,2,15] and the string equations [3,6] have been exactly solved in these geometries. The scattering matrix in the A - S geometry exhibit poles at *iGs=n=0,1,2,...*, where s is the particle energy[1] . Such pole structure is also present in the two body scattering amplitude for the scalar ground state (tachyon) computed from strings.[17] It must be noticed that the presence of such poles in the S - matrix is associated to the point - like (structureless) character of the shock wave source. The contribution of small (zero) impact parameter yields in this case such poles. Recently, the present authors found the shock wave spacetimes generated by ultrarelativistic extended sources, as ultrarelativistic cosmic strings (with charge and spin), and other topological defects (as monopoles and domain walls). The scattering matrices of Klein - Gordon fields in this extended source geometries do not exhibit the Coulombian type poles characteristic of the point - like sources, (although, in the cosmic string case, the small s and small t behavior give the usual one - graviton exchange amplitude). In the case of the effective shock wave generated by Planckian energy string collisions,[6] the source is an extended one[7,18] , and the scattering matrix does not exhibit such poles neither.

Another feature in this context is that, for both, point - like as well as extended sources, the exact S - matrix of the relativistic geometry coincides with the S - matrix of the weak field limit geometry of the static sources, in the large impact parameter regime. In other words, the S - matrices produced by the ultrarelativistic limit and the weak field limit of a given geometry coincide.

In the ultrarelativistic limit, the sources travel at the speed of light (i.e. $\gamma = (1 - v^2)^{-1/2} \to 0$) and are massless. In order to describe the non trivial behavior of the boosted gravitational field in this limit, the different parameters of interest (mass, electric, magnetic or color charges, spin) must be chosen to be γ - dependent and vanish in a specific way. The mass goes to zero like $m = p\gamma^{-1}$, the charge like $Q = p_e\gamma^{-1/2}$ and the spin like $S = pa\gamma^{-1}$, keeping their respective momenta p, p_e, a fixed. It is the total spin (and not the spin per unit mass a) which becomes γ - dependent. The spin vanishes like the mass, whereas the charge vanishes slower.

Since an ultrarelativistic particle has only two possible spin polarizations, $\pm S$, in the direction of motion, we apply the Lorentz boost parallel to the spin. In order to apply a Lorentz boost, it is convenient to start with the metric expressed in Boyer - Lindquist type coordinates and then to take asymptotically Cartesian coordinates (as in eq (2) here). The limit $v \to c$ is subtle, although it is well defined in the sense of distributions functions. The method to compute this limit consists essentially of integrating over the null variable $u = z - t$ (the shock wave will be located at $u = 0$), then to take the limit $v \to c$ and finally derivate with respect to u. Thus, we obtain for the metric profile function $f(\rho)$ (see eq (9)), where we observe that the spin a^2 contribution couples to the kinematic momentum p only through even powers of $1/\rho$, whereas the coupling to the electromagnetic momentum p_e is through odd powers of $1/\rho$.

In the limit $v \to 1$, the terms corresponding to the crossed component $g_{t\varphi}$ go to zero as $\gamma^{-1}\delta(u)$. As can be seen from its multipole decomposition, boosting $g_{t\varphi}$ only adds a factor γ due to the boost in t (φ coordinates are not boosted). Thus, the limit will yield terms proportional to $\gamma^{-1}\delta(u)$.

Alternatively, metrics of this type can be found[19] by starting from flat space - time and generating a shock wave through an appropriate shift in the null coordinate $v = z+t$. In our problem here, since an explicit expression for the source of the Kerr geometry is not known, it is more convenient to take the problem all the way around. Thus, from the metric profile function $f(\rho)$ found here, we are able to obtain the source of the ultrarelativistic Kerr geometry (see eqs (14) - (15)). These results complete and enlarge our previous results presented in Ref. 20.

We have also studied the effect of the spin in the scattering of test particles and study the scattering of a Dirac field of spin 1/2 in this shock wave geometries.[16] The Dirac equation in this background can be diagonalized to have four uncoupled components of the spinor, each component fulfilling an equation similar to that of the scalar field. The S - matrix depends no only on the profile function $f(\rho)$, but also on its derivatives with respect to the transversal coordinates, due to the dependence on particle's spin. The poles appearing in the scattering amplitude of a scalar field in the Aichelburg - Sexl geometry survive, in some way, for the Dirac fields now considered. In the case of the ultrarelativistic Reissner - Nordström metric, the number of such poles is quadruplicated[15] and we expect that for the ultrarelativistic Kerr - Newman geometry this number will be still larger. Finally, we compared to and point out some features of the propagation and scattering of fundamental strings.

2. The Ultrarelativistic Kerr - Newman Geometry

The Kerr - Newman metric in Boyer - Lindquist coordinates reads[21]

$$ds^2 = -(\Delta - a^2 \sin^2 \theta) q^{-2} dt^2 - 2Mra \sin^2 \theta q^{-2} dt d\varphi + \frac{q^2}{\Delta} dr^2 + q^2 d\theta^2 +$$

$$+ \left[(r^2 + a^2)^2 - \Delta a^2 \sin^2 \theta \right] q^{-2} \sin^2 \theta d\varphi^2 \tag{1}$$

where

$$\Delta = r^2 - 2Mr + a^2 + Q^2 \quad , \quad q^2 = r^2 + a^2 \cos^2 \theta \ .$$

Having in mind to apply a Lorentz boost we write metric (1) in asymptotically Cartesian coordinates

$$\begin{aligned} x &= (r^2 + a^2)^{1/2} \sin\theta \cos\varphi \ , \\ y &= (r^2 + a^2)^{1/2} \sin\theta \sin\varphi \ , \\ z &= r \cos\theta \quad , \quad t = t \ . \end{aligned} \tag{2}$$

We now apply a Lorentz boost (with velocity v) in the z-direction (an ultrarelativistic particle have only two spin polarizations ($\pm S$), in the direction of motion)

$$\begin{aligned} \tilde{t} &= \gamma(t + vz) \quad , \quad \tilde{z} = \gamma(z + vt) \quad , \\ \tilde{y} &= y \quad , \quad \tilde{x} = x \quad , \\ \gamma &= (1 - v^2)^{-1/2} \end{aligned} \tag{3}$$

The limit $v \to 1$ in the resulting metric is a delicate point, but we can extract a meaningful expression from it by letting M , Q and S be γ - dependent and going to zero as

$$M = p\gamma^{-1} \quad , \quad Q^2 = p_e^2\gamma^{-1} \quad , \quad S = aM = ap\gamma^{-1} \ . \tag{4}$$

Thus, after use of eqs(2)-(4) in metric(1) we find

$$ds^2 = -d\tilde{t}^2 + d\tilde{z}^2 + d\tilde{y}^2 + d\tilde{x}^2 + (d\tilde{t} - vd\tilde{z})^2 \lim_{v\to 1} \left[\gamma^2(\Delta g_{\tilde{t}\tilde{t}} + \Delta g_{\tilde{z}\tilde{z}})\right] + O(\gamma^{-1}) \ , \tag{5}$$

where

$$\Delta g_{\tilde{t}\tilde{t}} = \frac{2M\tilde{r} - Q^2}{\tilde{r}^2 + a^2\cos^2\theta} \ ,$$

$$\Delta g_{\tilde{z}\tilde{z}} = \frac{(2M\tilde{r} - Q^2)(\tilde{r}^2 + a^2)\cos^2\theta}{(\tilde{r}^2 + a^2\cos^2\theta)(\tilde{r}^2 - 2M\tilde{r} + a^2 + Q^2)} \ , \tag{6}$$

and

$$\cos\theta = z(\tilde{z})/\tilde{r} \ ,$$

$$\tilde{r}^2 = \frac{1}{2}\left[z(\tilde{z})^2 + \rho^2 + \sqrt{(z^2(\tilde{z}) + \rho^2)^2 + 4a^2z^2(\tilde{z})}\right] \ , \tag{7}$$

$$\rho^2 = \tilde{x}^2 + \tilde{y}^2 - a^2 \ .$$

Let us notice that the terms corresponding to $g_{t\varphi}$ in the limit $v \to 1$ go to zero as $\gamma^{-1}\delta(\tilde{t} - \tilde{z})$. This is sketched in the appendix of Ref. 16.

The limits appearing in eq(5) are computed explicitly in the Appendix of Ref. 16 giving rise to a shock wave metric

$$ds^2 = du\ dv - f(\rho)\delta(u)du^2 + d\tilde{y}^2 + d\tilde{x}^2 \ , \tag{8}$$

where

$$f(\rho) = -8p\ln\rho - \frac{3}{2}\pi p_e^2\rho^{-1} + p\sum_{m=1}^{\infty} B_m(a^2/\rho^2)^m - \pi p_e^2\rho^{-1}\sum_{m=1}^{\infty} C_m(a^2/\rho^2)^m \tag{9}$$

where $u = \tilde{z} - \tilde{t}$ and $v = \tilde{z} + \tilde{t}$ are the usual null coordinates and the coefficients B_m and C_m are numerical factors which are found in the Appendix of Ref. 16

$$B_m = 4\sum_{k=0}^{m}(-)^{m-k}\sum_{j=0}^{m-1}(-)^j\binom{m-1}{j}\left\{\frac{A_{k,2m-2k+1/2}}{2j+2m+1} + \frac{A_{k,2m-2k+3/2}}{2j+2m+2}\right\} \ , \tag{10}$$

$$C_m = \sum_{k=0}^{m} (-)^{m-k} 4^{-m} \left\{ \frac{A_{k,2m-2k+1}}{(2m)!} (4m-4)!! \prod_{j=0}^{m-1} \frac{2m-2j-1}{2m+2j+1} + \right.$$
$$\left. + \frac{A_{k,2m-2k+2}}{2(2m+1)!} (4m+1)!! \prod_{j=0}^{m} \frac{2m-2j+1}{2m+2j+1} \right\}, \qquad (11)$$

The metric(5) represents the gravitational field of a boosted particle of mass M, charge Q and spin S in the ultrarelativistic limit $v \to 1$ when all these three quantities go to zero keeping their respective associated momenta (p, p_e , and a) fixed. The ρ dependence of each contribution is different. The spin contribution couples to the kinetic momentum only through even negative powers of ρ whereas the coupling to the charge is through odd negative powers of ρ.

The spacetime is flat everywhere except on the null plane orthogonal to the direction of propagation and has a $\delta(u)$-behaviour characteristic of such light-like contractions.

In the metric (8)-(9), the condition $\rho^2 = \tilde{x}^2 + \tilde{y}^2 - a^2 \geq 0$ represents that this boosted solution "remembers" the ring singularity of the original Kerr metric.

Let us notice that the profile function (9) remains invariant after a change in the orientation of the spin, i.e. $a \to -a$. This is due to the fact that in the ultrarelativistic limit $g_{t\varphi} \to 0$, which is the only spin - direction dependent term.

We should also point out that the boosted geometry (8)-(9), has $M^2 = p^2\gamma^{-2} < Q^2 + a^2 = p_e^2\gamma^{-1} + a^2$ when $\gamma \to \infty$. This means that metric(1) represents not a black hole, but a point-like particle at the irremovable singularity $r = 0$. This is in agreement with the interpretation and the elementary particle context in which we are using this geometry: the ratio $(Q^2 + a^2)/M^2$ is always bigger than one (This is not the case, of course, for astrophysical objects). The solutions representing black holes with $M^2 > Q^2 + a^2$ do not exist in the ultrarelativistic limit.

It is worth to remark that recently[22] , only the first term of eq(9) was obtained for the ultrarelativistic limit of the Kerr metric. The authors of [22] do not reach to find the additional powers of ρ proportional to B_m, because they set $u = 0$ before taking the singular limit $v \to 1$. If we had made this in eqs (47) and (53) all the additional terms would had gone to zero, but in this problem one has to take a double limit, $v \to 1$ and $u \to 0$, which must be treated in a distributional sense, that is, as we have done here, in the Appendix of Ref. 16. The final result will indeed be that given by eqs(9)-(11).

3. The Ultrarelativistic Energy - Momentum Tensor

The metric(8) can also be obtained from flat spacetime by generating a gravitational shock wave through a shift in the null coordinate v, following on the lines of the procedure developed by Dray and 't Hooft[19] . This procedure consist of solving the Einstein equations for a null source representing a particle travelling at the speed of light. This procedure can be generalized to D - dimensional curved backgrounds including sources and cosmological constant[23,24] .

This formalism will allow us to study the gravitational shock waves in a curved background:

$$ds^2 = 2A(u,v')dudv' + g(u,v')h_{ij}(x^k)dx^i dx^j \ ; \tag{12}$$

solution of Einstein equations without shock wave.

$$R_{\mu\nu} - \frac{1}{2}g_{\mu\nu}R + \Lambda g_{\mu\nu} = T_{\mu\nu} \tag{13}$$

Then, let us make the following ansätze for the metric solution of Einstein eqs., including the shock wave as source:

$$ds^2 = 2A(u,v)du[dv - f(x^k)\delta(u)du] + g(u,v)h_{ij}(x^k)dx^i dx^j \tag{14}$$

The source is given by

$$T_{\mu\nu} = T_{\mu\nu}^{(b)} + T_{\mu\nu}^{(p)} \ ,$$

where

$$T_{uu}^{(p)} = \sigma(x^k)\delta(u) \ , \tag{15}$$

The proposed metric (14) will be a solution of Einstein equations if:

$$\partial_v A(0,v) = 0 = \partial_v g(0,v) = 0 = T_{vv}(0,v) \ , \tag{16}$$

$$\frac{A}{g}\nabla^2_{D-2}f - (D-2)\frac{g,_{uv}}{2g}f = \sigma(x^k) \tag{17}$$

both evaluated at $u = 0$. Examples of application in a flat background, i.e. $A = \frac{1}{2}$ and $g = 1$, can be computed for an uncharged particle with energy-momentum tensor given by $T_{uu} = p\delta(\rho)\delta(u)$. Plugging this expression into(17):

$$\nabla^2_\rho f = 16\pi p\delta(\rho) \ . \tag{18}$$

we find the solution to be

$$f(\rho) = 8p\ln\rho \ . \tag{19}$$

i.e., the Aichelburg - Sexl metric. The cases of a charged and spinning particle is discussed in Refs. 15 and 16.

This method allows to obtain the metric function $f(\rho)$ given the $T_{\mu\nu}$ corresponding to the ultrarelativistic particle; in the Kerr-Newman case, however, it is more convenient to take the problem all the way around since an analytic expression for the exact source of the Kerr black hole remains unknown. Thus, given $f(\rho)$ by the expression (9) we can find $T_{\mu\nu}$, which in our case has only one component different from zero

$$T_{uu} = \delta(u)\nabla_{\perp}^{2} f(\rho) = \delta(u)\rho^{-1}\partial_{\rho}(\rho\partial_{\rho} f) \tag{20}$$

Plugging (9) into this equation we obtain

$$\begin{aligned} T_{uu} = \delta(u)\Big\{ & p\delta(\rho) + 4p\rho^{-2}\sum_{m=1}^{\infty} m^2 B_m (a^2/\rho^2)^m + \\ & + \frac{3}{2}\pi p_e^2 \rho^{-3} + \pi p_e^2 \rho^{-3} \sum_{m=1}^{\infty} (2m+1)^2 C_m (a^2/\rho^2)^m \Big\} \end{aligned} \tag{21}$$

Here we can identify the first addend as the source of an ultrarelativistic particle with momentum p. The third addend, as we have shown[15], corresponds to the contribution of the electromagnetic field generated by a point charge boosted to the speed of light. The other two contributions are the new terms which bring the effect of rotation. Clearly, we can distinguish between them: one is the effect of the Kerr metric (proportional to the kinetic momentum p) and the other is the effect of the electromagnetic field in rotation (proportional to p_e).

Let us observe that the term coming from the Kerr metric can be written as

$$T_{uu}^{Kerr} = \sigma(\rho)\delta(u) \tag{22}$$

Which allows to be interpreted as a "shell" of null fluid moving in the v direction with an axially symmetric surface energy density given by

$$\sigma(\rho) = p\left\{ \delta(\rho) + 4\rho^{-2}\sum_{m=1}^{\infty} m^2 B_m (a^2/\rho^2)^m \right\} \tag{23}$$

Of course, this expression for T_{uu} is the Kerr's source already boosted to the speed of light. We can not find uniquely from it the non-boosted $T_{\mu\nu}^{Kerr}$ (subject which would deserve the effort), because some information was lost in the process of taking the limit $v \to c$. Remarkably enough Balasin and Nachbagauer[25] in a series of papers have obtained the energy momentum tensor-distributions by using the Kerr-Schild decomposition of the metric successively for the Schwarzschild, Kerr and Kerr-Newman geometries, and have computed a closed form for the boosted Kerr energy momentum tensor and profile function $f(\rho)$ corresponding, presumably, to the resummation of the series appearing in eq. (21).

4. The Metric of Ultrarelativistic Particles and Extended Sources

As we have seen, the metric created by an ultrarelativistic source can be described by

$$ds^2 = dudv + f(x^i)\delta(u)du^2 + dx^i dx^j \quad , \quad i,j = 2,3,.....D-1 \tag{24}$$

where x^i are the coordinates transversal to the motion which takes place in the plane u, v.

$$u = x^1 - t \quad , \quad v = x^1 + t$$

This represents a gravitational shock wave located at the null plane $u = 0$. Outside the plane $u = 0$ this metric is flat. It is completely characterized by the function $f(x_\perp)$. For an uncharged and spinless point particle of momentum p in four dimensions

$$f(x_\perp) = 8Gp\ln\rho \quad , \quad \rho^2 = \sum_{i=2}^{3} x^i x^i \tag{25}$$

This is the Aichelburg - Sexl (A - S) metric[14] . This metric can be obtained by applying a Lorentz boost to the Schwarzschild geometry with the mass going to zero as $M = p\gamma^{-1}$, $\gamma^{-2} = (1 - v^2)$ and finally, taking the velocity tending to the speed of light ($v \to 1$). By the same method it can be obtained the form of the function f when one starts from a particle originally endowed of charge[15] (Reissner - Nordström metric) and spin[20,16] (Kerr - Newman metric)

$$f(\rho) = -8Gp\ln\rho - \frac{3}{2}\pi p_b^2 \frac{1}{\rho} + p\sum_{n=1}^{\infty} B_n \left(\frac{a^2}{\rho^2}\right)^n + \frac{\pi p_b^2}{\rho} \sum_{n=1}^{\infty} C_n \left(\frac{a^2}{\rho^2}\right)^n \tag{26}$$

where p_b is the impulse due to the electromagnetic field and a is the intrinsic angular momentum per unit mass associated to the particle. B_n and C_n are determined numerical constants.

There is an equivalent method[19,23] to obtain $f(\rho)$. This consists of writing down the Einstein equations for a metric like (24). This ansatz reduces the Einstein equations to only one relevant linear equation

$$\delta(u).\, \nabla^2_{x_\perp} f(x_\perp) = 2T_{uu} \tag{27}$$

T_{uu} being the only non zero component of the matter source in the ultrarelativistic limit. For an infinite thin shell of matter propagating at the speed of light: $T_{uu} = \sigma(x_\perp)\delta(u)$, $\sigma(x_\perp)$ being the matter density of the extended source.

Let us suppose now that instead of a point - like *particle* we proceed to boost a *cosmic string*. What we would obtain as result? The answer is[26] again a metric like (24), but now $f(x_\perp) = f(y)$. i.e. the function f will only depend on the distance y perpendicular to the string instead of the radial distance to the particle ρ, which is reasonable, if we take into account the symmetries of the problem. By replacing in eq (27) the expression for the $T_{\mu\nu}$ corresponding to cosmic strings, we find[26]

$$f(y) = 2p_l \mid y \mid \quad , \tag{28}$$

for *gauge* or *local* cosmic strings laying along the **z** axis. We have used that $T_t^t = T_z^z = \mu\delta(x)\delta(y)$, and when boosted in the **x** direction with $v \to 1$ this $T_{\mu\nu}$ reduces to

$$T_{uu} = p_l\delta(y)\delta(u) \quad . \tag{29}$$

It is worth to remark that for a local string in a D - dimensional spacetime, we had obtained:

$$f(y) = \frac{4Gq\Gamma[(D-5)/2]}{(2\pi)^{(D-5)/2}} \mid \vec{y}_{D-3} \mid^{5-D} , \tag{30}$$

where $\vec{y}_{D-3}$ is the $D-3$ dimensional vector perpendicular to the string and to the direction of motion. Eq(30) gives the same result than for the case of an ultrarelativistic particle in $D-1$ dimensional spacetime (Aichelburg - Sexl metric, see eq(26)).

The case of *global cosmic strings* with energy momentum tensor given by

$$T_t^t = T_z^z = T_r^r = T_\theta^\theta = -\frac{\mu}{2r^2} \tag{31}$$

when boosted at $v \to 1$ gives

$$T_{uu} = -p_g \mid y \mid^{-1} \delta(u) \quad , \quad y \neq 0 \; . \tag{32}$$

This, from eq (27) yields

$$f(y) = p_g \mid y \mid (ln \mid y \mid -1) \quad , \quad y \neq 0. \tag{33}$$

For the purposes of further discussion, it is interesting to see the metric generated by *Domain Walls.* Their formation is associated to a discrete symmetry breaking. For a wall located in the x - y plane, the energy momentum tensor can be represented by[26]

$$T_0^0 = T_x^x = T_y^y = \sigma\delta(z) \quad , \quad T_z^z = 0 \tag{34}$$

where σ is the energy scale of symmetry breaking corresponding to domain walls.

After boosting in the **z** - direction and by taking

$$\sigma = p_\sigma \gamma^{-1} \quad , \tag{35}$$

in the ultrarelativistic limit we have

$$T_{uu} = p_\sigma \delta(u) \quad , \quad u = z - t \quad , \tag{36}$$

and all other components vanishing. The metric function is given by

$$f(\rho) = \frac{1}{2} p_\sigma \rho^2 \tag{37}$$

5. The Scattering of Quantum Fields by these Gravitational Shock Waves

Let us study now the Klein-Gordon equation in the ultrarelativistic Reissner-Nordström geometry discussed before. This describes the following physical process: the scattering of two spinless particles (1 and 2) of masses m_1 , $m_2 = m$; both being $\ll m_{pl}$, and charges $e_1 = 0$, $e_2 = e$, at energies of the order or larger than Planck mass m_{pl}.

Following Ref. 1 we choose a frame where only one of the particles (particle 2) has an energy of the order of m_{pl}. Therefore the particle 2 is ultrarelativistic momenta p and p_e respectively and the space-time around it is described by generalized AS geometry discussed above. (The motion of particle 2 is taken here along the x axis with speed $+1$). Particle 1 is a test particle since its energy is much smaller than m_{pl}. Its dynamics is described by the Klein-Gordon equation in the geometry of Eq.(24).

By writing the wave function as

$$\Psi(u, v, x^k) = \exp(-i\omega v/4)\Phi(u, x^k) , \tag{38}$$

we have (since v is a cyclic variable)

$$\partial_u \Phi = \left[-i\omega/4f(x^k)\delta(u) + i/\omega(\nabla_\perp^2 - m^2)\right] \Phi . \tag{39}$$

Let us take an incoming plane wave for $u < 0$. Thus, for $u > 0$, the continuous regularization produces the following outcoming wave function[9]

$$\Phi_>(u, x^k) = \int dp_\perp^{D-2} \exp\left\{i\vec{p}_\perp \cdot \vec{x}_\perp - iu/\omega(p_\perp^2 + m^2)\right\} S(k_\perp, p_\perp, \omega) \tag{40}$$

where

$$S(k_\perp, p_\perp, \omega) = (2\pi)^{2-D} \int dx_\perp^{D-2} \exp\left[i(\vec{k}_\perp - \vec{p}_\perp) \cdot \vec{x}_\perp + i\varphi_D(x^k)\right] , \tag{41}$$

$S(k_\perp, p_\perp, \omega)$ is the S-matrix which transforms an incoming state with momentum $(k_\perp, \omega)$ into an outgoing state of momentum $(p_\perp, \omega)$. $\varphi_D(x^k)$ is the phase shift suffered by the incoming wave, that happens to be the only relevant effect since there is not particle creation.

$$\varphi_D(x^k) = \frac{\omega}{4} f(x^k) .$$

As examples of application of these results we can compute the S-matrix of a profile function $f(\rho)$ corresponding to the Aichelburg - Sexl metric. In this case we obtain[1]

$$S^{AS}(s,t) = \frac{1}{\pi}\left(\frac{4}{\mid t \mid}\right)^{iG(s-m^2)} \frac{\Gamma[1-iG(s-m^2)]}{\Gamma[iG(s-m^2)]} + \delta(\vec{k}_\perp - \vec{p}_\perp) \ . \tag{42}$$

where

$$Gs = \omega p + Gm^2 \quad , \quad t = - \mid \vec{k}_\perp - \vec{p}_\perp \mid \ .$$

It can be seen that there is no particle production[10] (the Bogoliubov coefficients β_{kp}=0). [This is not the case, however, when the fields interact with two colliding plane waves[27] .]

S^{cont} exhibits an infinite sequence of imaginary poles in

$$G(s-m^2) = -in \ , \ n = 1,2,3,.. \tag{43}$$

We can observe a similarity between eq (42) and the Veneziano amplitude for strings.

In the same way, we can compute the S-matrix for the profile corresponding to the Reissner - Nordström metric[15] .

This case contains the later and, in fact, it can be written as a superposition

$$S(s,t) = \sum_{n=0}^{\infty} \frac{1}{n!}\left[\frac{iG(s-m^2)}{4p}\right]^n S^{\alpha=0}\left[iG(s-m^2)+n/2,t\right] \ . \tag{44}$$

This series exhibits an infinite sequence of imaginary poles in

$$iG(s-m^2) = in/2 \quad , \quad n \text{ integer} \ . \tag{45}$$

We thus observe the very interesting feature that the electromagnetic momentum p_e, increases the number of poles by a factor four. From the residues of these poles, we see that for $p_e = 0$, the possible angular momenta for the bound state "resonances" are restricted by n. For $p_e \neq 0$, whatever the resonance be, all angular momenta are included (all spin are allowed).

6. Background Metric for the Scattering of Superstrings. Large and Intermediate Impact Parameters

Amati, Ciafaloni and Veneziano[4] have obtained an expression for the scattering amplitude of superstring collisions in flat space time, valid for high energies (i.e. energies much larger that $1/\sqrt{\alpha'}$). The high energy regime investigated is that of large effective coupling $g^2\alpha' s/\sqrt{\alpha'}^{D-4}$ and small loop expansion parameter $g^2/\sqrt{\alpha'}^{D-4}$, $(\alpha' g^2 = 16\pi G)$. By resumming in the eikonal approximation, the leading behaviour of the string loop diagrams in not compactified D-dimensional flat spacetime, they found an S-matrix:

$$S = \exp\left(2i\int_0^\pi : a(s, b + \hat{x}^u(\sigma_u, 0) - \hat{x}^d(\sigma_d, 0)) : \frac{d\sigma_u d\sigma_d}{\pi^2}\right) \tag{46}$$

where b is the impact parameter between the center of mass of the two colliding strings. x^u and x^d are the oscillations around the center of mass of the string at rest; s is the usual Mandelstam variable and a is the string tree amplitude

$$\begin{aligned} a(s,b) &= \frac{Gs}{2\pi^{D/2-2}} b^{4-D} \int_0^{g(b,s)} dt e^{-t} t^{D/2-3} \\ g(b,s) &= \frac{b^2}{4\alpha'[ln(s) - i\pi/2]} = \frac{b^2}{\Delta^2}, \end{aligned} \tag{47}$$

where the integral can be re-written in terms of the incomplete gamma function $\Gamma[D/2 - 2, g(b,s)]$ and α' is the usual string slope.

On the other hand, the scattering of free strings[6,2] in the shock wave backgrounds (24) is exactly solvable. For a free superstring in this background the S-matrix has the expression[2]

$$S = \exp\left(\frac{i}{2\pi} p^u \int_0^\pi d\sigma f\left(x(\sigma, 0)\right)\right) \tag{48}$$

where p^u is the impulse associated to the string, σ is the world sheet spatial coordinate and $x(\sigma, \tau = 0)$ is the projection of the D-dimensional superstring coordinate on the plane $u = p^u\tau = 0$ where the shock wave is located (perpendicular to the motion of the shock wave ultrarelativistic source).

Now, if we want both S-matrices describe the same phenomena, we equate expressions (48) and (46). One obtains in this way the form of the function f appearing in the metric(24)

$$f(\rho) = \frac{8q}{s}\int_0^\pi : a(s, \rho - \hat{x}^d(\sigma_d, 0)) : \frac{d\sigma_d}{\pi} \tag{49}$$

where q is the impulse associated to the shock wave and $s = 2p^u q$.

Plugging eq (47) into (49) we obtain for $f(\rho)$:

$$f(\rho) = C \int_0^\pi \frac{d\sigma}{\pi} : b^{4-D} \int_0^{g(\rho)} dt e^{-t} t^{\frac{D-6}{2}} : \tag{50}$$

where

$$g(\rho) = \frac{b^2}{4\alpha'(ln(s) - i\pi/2)} \quad , \quad C = \frac{4Gq}{2\pi^{D/2-2}} \quad , \quad b = \rho - x^d(\sigma, 0) \tag{51}$$

In the regime studied here, b never goes to zero ($b \gg x^d \sim$ string size). In the regime of large impact parameters described by

$$x^d \ll \sqrt{\alpha' \ln s} \ll b \quad , \tag{52}$$

where s is the usual Mandelstam variable and x^d the characteristic size of string fluctuations, the effective geometry generated by flat space string collisions is the Aichelburg - Sexl (A - S) geometry. This is the gravitational shock wave produced by an ultrarelativistic uncharged and spinless point particle in D dimensions.

We investigate here the regime

$$x^d \ll b \ll \sqrt{\alpha' \ln s} \tag{53}$$

In this regime (let us call it small-intermediate b), b is still much bigger than the characteristic size of string excitations, so that strings do not superpose each other. We find that it is possible to associate an effective curved metric to the string collision process in this regime. This is a gravitational shock wave, but with a profile function f given in $D = 4$ by

$$f(\rho) \simeq \frac{C}{(D-2)\Delta^{D-2}}\rho^2 \quad , \quad \Delta^2 = 4\alpha'(\ln(s) - i\pi/2) \ , \ C = \frac{4Gq}{2\pi^{D/2-2}} \ \text{(see below)} . \tag{54}$$

This metric is generated by an extended source with energy density distribution

$$\sigma(\rho) = \frac{2C}{\Delta^{D-2}} \exp\{-\rho^2/\Delta^2\} \ , \tag{55}$$

q being the kinematic momentum. This dependence on ρ in f and σ is characteristic of the string collision process at intermediate b, and reproduces at first order other simpler extended sources as domain walls[26] or "homogeneous beams"[28,7]. Notice also the difference with the point-like particle A - S geometry, for which $f(\rho) = 8Gp \ln \rho$ (in four dimensions) and $\sigma(\rho) = 8Gp\delta(\rho)$, ($p$ being the momentum of the particle). The scattering collision that generates the shock wave no longer behaves as an effective point-like source in the scattering at the small - intermediate b.

We study the scattering of test particles and fields by this geometry. The deflection angle is (in D=4)

$$\Theta = \begin{cases} 8Gp/b & \text{for large } b \\ 4Gqb/\Delta^2 & \text{for intermediate } b. \end{cases} \tag{56}$$

The S - matrix of Klein - Gordon fields in this geometry is analysed in order to describe some features of the ultra - high energy string - string scattering as one goes along decreasing impact parameters b. The whole picture is depicted in Figure 1. In this scattering matrix, the poles $iGs = n+1$, $n = 0,1,2,...$ do not appear. Such poles, appearing in the S - matrix of a particle (or string) in the A - S geometry, arise because of the extrapolated use of the A - S geometry down to small impact parameters $0 \leq b \leq x^d$ (including $b = 0$). For intermediate b, deviations from the A - S geometry are important and, on the other hand, for very small b $(0 \leq b \leq x^d)$, the use of the external metric approximation is not valid.

Let us study this function $f(\rho)$ in the case where

$$\rho \gg [\alpha' ln(s)]^{1/2} \gg x^d \quad . \tag{57}$$

This case corresponds to the scattering of superstrings at very high energies , but with large impact parameter. As ρ is large, the integral in the variable t in eq (50), can be taken between zero and infinity (due to the exponential in the integrand). Also, as $\rho \gg x^d(\sigma)$ we can take b^{4-D} out the integral. Then,

$$f(\rho) = C\Gamma\left(D/2 - 2\right)\rho^{4-D} \quad , \tag{58}$$

which corresponds to the A - S metric (uncharged and spinless particle) in D dimensions.

The other limit we want to study is

$$[\alpha' ln(s)]^{1/2} \gg \rho \gg x^d(\sigma) \quad , \tag{59}$$

which in some sense represents the opposite to the last limit studied, because it corresponds to the scattering of strings at very high energies and at smaller (intermediate, say) impact parameters (although not so small to superpose the superstrings each other). The regime given by eq (59) means that the impact parameters are smaller than those considered in the regime of eq (57), but still bigger than the characteristic size of fundamental string excitations.

Here, it is convenient to analyse the effective or equivalent energy - momentum tensor, T_{uu}, instead of the metric function $f(\rho)$. To do this we use eq (27)

$$T_{uu}(\rho) = \frac{1}{2}\delta(u)\frac{1}{\rho^{D-3}}\partial_\rho\left(\rho^{D-3}\partial_\rho f(\rho)\right) = \frac{2C\delta(u)\exp\{-\rho^2/\Delta^2\}}{\Delta^{D-2}} = R_{uu} \tag{60}$$

This expression is exact within the approximations posed to obtain (50). Here, can be observed clearly the meaning the regimes (57) and (59) that we have considered. In the limit (57), $T_{uu} \to 0$, which characterizes the Aichelburg - Sexl solution (represented by a $\delta(\rho)$). While in the limit (59), (for s fixed) $T_{uu} \sim$ constant (Let us observe that this is the case of domain walls).

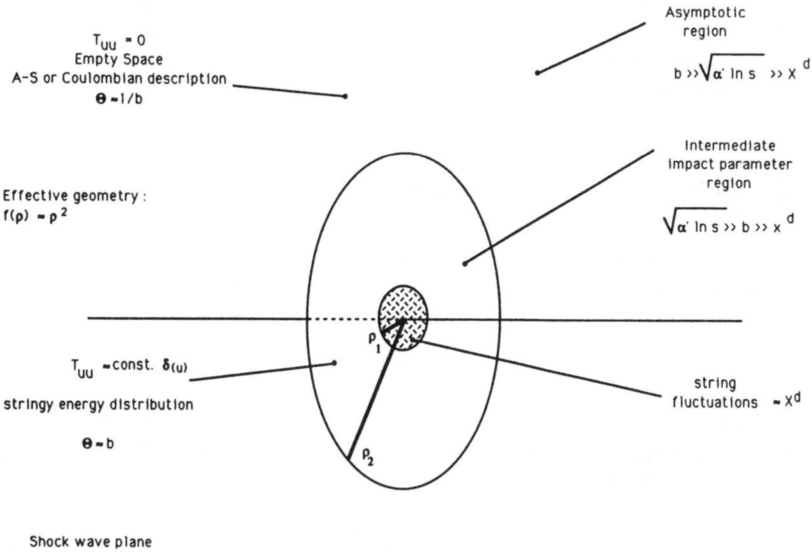

Figure 1: Schematic picture of the string collision process; the different impact parameter regimes and the associated effective geometry.

In terms of a metric to which we apply a Lorentz boost with $v \to 1$, $T_{uu} \sim$ constant would represent a particle that develops a non - localized interaction term. To this $T_{\mu\nu}$ corresponds a metric function

$$f(\rho) \simeq \frac{C}{(D-2)\Delta^{D-2}}\rho^2 \quad , \quad C = \frac{4Gq}{2\pi^{D/2-2}} \quad , \quad \rho < \Delta, \tag{61}$$

which differs from the Aichelburg - Sexl metric (see eq(58)). This new dependence arises as a consequence of the non-point-like interaction of strings in the small - intermediate impact parameter regime. One expects that the superstring collision that generates the effective shock wave geometry no longer behaves as an effective point like source when studying the scattering at smaller impact parameters.

It is worth to remark that the energy density (60) is proper of the scattering of fundamental strings. It contains the homogeneous beam approximation[28,7]

$$\sigma_h = \begin{cases} \sigma_0 & \text{if } \rho < \Delta \\ 0 & \text{if } \rho > \Delta. \end{cases}$$

It is also worth to point out that the presence of this energy density is enough to remove the so called 't Hooft poles[1,29] . This displays the "softener" character of fundamental strings, already noted in other computations within the theory of fundamental strings.

7. On the Universality of 't Hooft's Poles

Let us consider now the S - matrix of Klein - Gordon fields in the spacetime geometry of eq(24) in four dimensions. The S - matrix is given by[3]

$$S(\mathbf{k}_\perp, \mathbf{p}_\perp, \omega) = \int \frac{d\rho}{2\pi} \rho J_0(\mid \mathbf{k}_\perp - \mathbf{p}_\perp \mid \rho) e^{\frac{-i\omega}{4} f(\rho)} \quad , \tag{62}$$

In order to use this S - matrix to describe some of the features of the ultra high - energy string collisions, the integral in this context must be understood as the sum of at least three integrals

$$S = S_I + S_{II} + S_{III} = \int_0^{\rho_1} + \int_{\rho_1}^{\rho_2} + \int_{\rho_2}^{\infty} \tag{63}$$

In the last integral (large impact parameter regime), we can use the A - S metric, i.e. $f_{III}(\rho) = 8Gp\ln(\rho)$. In the second integral (intermediate region), $f_{II}(\rho) \simeq \frac{2Gq}{\Delta^2}\rho^2$ must be plugged in. Unfortunately, it seems not easy to find a closed expression for the second integral, while for the first region the use of the external effective metric approximation is not valid (the collision is so close that strings "touch" each other). The whole picture is depicted in Figure 1. The poles $iGs = n + 1$, $n = 0, 1, 2, ...$ appearing in the S - matrix of a particle (or string) in the A - S geometry arise because one usually extends the third integral down to impact parameters $\rho_2 = 0$.

This integral is given by

$$S_{III}(s, t, \rho_2) = \frac{1}{2\pi} \int_{\rho_2}^{\infty} d\rho \rho^{1-2iGs} J_0(\rho\sqrt{-t}) \tag{64}$$

where $t = - \mid \mathbf{k}_\perp - \mathbf{p}_\perp \mid^2$, $Gs = \omega p$ and we have taken $f_{III}(\rho) = 8Gp\ln\rho$. Then, we find

$$S_{III} = \frac{1}{2\pi}(-t)^{iGs-1}\{-2iGs\rho_2 J_0(\rho_2) S_{-2iGs,-1}(\rho_2) + \rho_2 J_1(\rho_2) S_{1-2iGs,0}(\rho_2)\} \tag{65}$$

where J_0 and J_1 are Bessel functions and

$$S_{\alpha,\beta}(z) = s_{\alpha,\beta}(z)+ \tag{66}$$

$$+2^{\alpha-1}\Gamma(\frac{\alpha-\beta+1}{2})\Gamma(\frac{\alpha+\beta+1}{2})\left[\sin(\frac{\alpha-\beta}{2}\pi)J_\alpha(z)-\cos(\frac{\alpha-\beta+1}{2}\pi)N_\alpha(z)\right]$$

$S_{\alpha,\beta}(z)$ and $s_{\alpha,\beta}(z)$ are Lommel's functions[30] and can be related to hypergeometric series

$$s_{\alpha,\beta}(z) = \frac{z^{\alpha+1}}{(\alpha+\beta+1)(\alpha-\beta+1)} \quad {}_1F_2\left(1;\frac{1}{2}(\alpha-\beta+3);\frac{1}{2}(\alpha+\beta+3);-z^2/4\right) \quad , \tag{67}$$

$$N_\alpha(z) = \csc(\alpha\pi)[J_\alpha(z)\cos(\alpha\pi) - J_{-\alpha}(z)] \quad .$$

For $\rho_2 \to 0$, eq(65) gives the A - S scattering amplitude[1,2] . For any ρ_2 finite and non zero, S_{III} do not have the poles at $iGs = n+1, \quad n = 0,1,....$ For the purposes of seeing this it is more convenient to make the decomposition: $\int_{\rho_2}^{\infty} = \int_0^{\infty} - \int_0^{\rho_2}$ in eq(64). The first term is given by[1,2]

$$\frac{1}{2\pi}2^{1-2iGs}(-t)^{iGs-1}\frac{\Gamma(1-iGs)}{\Gamma(iGs)} \tag{68}$$

while the second term yields

$$\frac{1}{2\pi}\int_0^{\rho_2} d\rho\rho^\mu J_0(\rho\sqrt{-t}) = \frac{1}{2\pi}(-t)^{-(\mu+1)/2}\sum_{k=0}^{\infty}\frac{(-1)^k}{4^k(k!)^2}\frac{\rho_2^{\mu+2k+1}}{\mu+2k+1} \quad , \quad \mu = 1-2iGs \tag{69}$$

where we have used

$$J_0(z) = \sum_{k=0}^{\infty}(-1)^k\frac{z^{2k}}{4^k(k!)^2} \tag{70}$$

Eq(69) have poles at $\mu = -(2k+1)$, i.e. $iGs = n+1, \quad n = 0,1,...$, the same than eq(68). Let us study if both expressions have the same residues. By multiplying eq(69) by $\mu + 2k + 1$ and taking the limit $\mu \to -(2n+1)$ we obtain:

$$Res\left\{\int_0^{\rho_2}\right\} = \frac{1}{2\pi}(-t)^{-n}\frac{(-1)^n}{4^n(n!)^2} \quad , \tag{71}$$

On the other hand, by using that

$$Res\{\Gamma(-n)\} = \frac{(-1)^n}{n!} \quad , \tag{72}$$

we obtain for the residue of eq(68) the same expression eq(71). Thus, in eq(64) the poles disappear.

8. String Instabilities in Black Hole Spacetimes

The next step in studying the scattering processes at Planckian scales is to consider, since the beginning, a theory which is considered well suited for dealing with such huge energies and tiny lengths. So far, the theory of strings provides a consistent frame where to study the very high energy scattering processes. In particular, we will discuss the collision (or better the infall) of a fundamental string in the gravitational background of a charged black hole.

The study of the string dynamics in curved space-times, reveals new insights with respect to string propagation in flat space-time (see for example refs [31, 32, 33, 34, 35, 36]).

The equations of motion and constraints for strings in curved spacetimes are highly non linear (and, in general, not exactly solvable). In ref [31], a method was proposed (the "strong field expansion") to study systematically (and approximately), the string dynamics in the strong curvature regime. In this method, one starts from an exact particular solution of the string equations in a given metric and then, one constructs a perturbative series around this solution. The space of solutions for the string coordinates is represented as

$$X^A(\sigma,\tau) = q^A(\sigma,\tau) + \eta^A(\sigma,\tau) + \xi^A(\sigma,\tau) + \ldots.. \quad , \tag{73}$$

$A = 0, ..., D-1$. Here $q^A(\sigma,\tau)$ is an exact solution of the string equations and $\eta^A(\sigma,\tau)$ obeys a linearized perturbation around $q^A(\sigma,\tau)$. $\xi^A(\sigma,\tau)$ is a solution of second perturbative order around $q^A(\sigma,\tau)$. Higher order perturbations can be considered systematically. A physically appealing starting solution is the center of mass motion of the string, $q^A(\tau)$, that is, the point particle (geodesic) motion. The world sheet time variable appears here naturally identified with the proper time of the center of mass trajectory. The space time geometry is treated *exactly*, and the string fluctuations around q^A are treated as perturbations. Even at the level of the zeroth order solution, gravitational effects including those of the singularities of the geometry are fully taken into account. This expansion corresponds to low energy excitations of the string as compared with the energy associated to the geometry. This corresponds to an expansion in powers of $(\alpha')^{1/2}$. Since $\alpha' = (l_{Planck})^2$, the expansion parameter turns out to be the dimensionless constant

$$g = l_{Planck}/R_c = 1/(l_{Planck}M) \quad , \tag{74}$$

where R_c characterizes the spacetime curvature and M is its associated mass (the black hole mass, or the mass of a closed universe in cosmological backgrounds). The expansion is well suited to describe strings in strong gravitational regimes (in most of the interesting situations one has clearly $g \ll 1$). The constraint equations are also expanded in perturbations. The classical $(\text{mass})^2$ of the string is defined through the center of mass motion (or Hamilton - Jacobi equation). The conformal generators (or world - sheet two dimensional energy - momentum tensor) are bilinear in the fields $\eta^A(\sigma,\tau)$. [If this method is applied to flat spacetime, the zeroth order plus the first order fluctuations provide the exact solution of the string equations].

This method was first applied to cosmological (De Sitter) spacetimes. One of the results was that for large enough Hubble constant, the frequency of the lower string modes, i.e. those with $|n| < \alpha' m H$, (α' being the string tension and m its mass), becomes imaginary. This was further analysed [33,34] as the onset of a physical instability, in which the proper string size starts to grow (precisely like the expansion factor of the universe). The string modes couple with the background geometry in such a way that the string inflates with the universe itself. The same happens for strings in singular gravitational plane waves[35,36] (see also ref[18]), and the results of paper 37 show that this is a generic feature of strings near spacetime singularities.

In black hole spacetimes, such unstable features have been explored in ref [37]. The string dynamics in black hole spacetimes is much more complicated to solve (even asymptotically and approximately). In ref [32], the study of string dynamics in a Schwarzschild black hole was started and the scattering problem was studied for large impact parameters. Stable oscillatory behavior of the string was found for the transversal (angular) components; scattering amplitudes, cross section and particle transmutation process were described, and explicitly computed in an expansion in $(R_s/b)^{D-3}$, R_s being the Schwarzschild radius and b the impact parameter. The aim of paper 37, was to find, and then to describe, the *unstable sector* of strings in black hole backgrounds. By unstable behavior, we mean here the following characteristic features: non oscillatory behavior in time, or the emergence of imaginary frequencies for some modes, accompanied of an infinite stretching of the proper string length. In addition, the spatial coordinates (some of its components) can become unbounded. Stable string behavior means the usual oscillatory propagation with real frequencies, (and the usual mode - particle interpretation), the fact that the proper string size does not blow up, and that the string modes remain well - behaved.

We express the first order string fluctuations η^μ, $(\mu = 0, ..., D-1)$ in D - dimensional Reissner - Nordström - De Sitter spacetime, as a Schrödinger type equation for the amplitudes $\Xi^\mu = q^R \eta^\mu$, q^R being the radial center of mass coordinate. We find the asymptotic behavior of the longitudinal and transverse string coordinates (Ξ^+, Ξ^-, Ξ^i) with $i = 2, ..., D-1$, at the spatial infinity, near the horizon and near the spacetime singularity. $+$ and $-$ stand for the longitudinal (temporal and radial) components respectively, and i for the transverse (angular) ones. We analyse first a head - on collision (angular momentum $L = 0$), that is, a radial infall of the string towards the black hole. Then, we analyse the full $L \neq 0$ situation. We consider Schwarzschild, Reissner - Nordström and De Sitter spacetimes (described here in static coordinates which allow a better comparison among the three cases). In all the situations (with and without angular momentum) and for the three cases we find the following results:

The time component Ξ^+ is *always stable* in the three regions (near infinity, the horizon and the singularity), and in the three cases (black holes and De Sitter spacetimes).

The radial component Ξ^- is *always unstable* in the three regions and in the three backgrounds. In the Schwarzschild case, the instability condition for the radial modes - which develop imaginary frequencies near the horizon - can be expressed as

$$n < \frac{\alpha' m\sqrt{D-3}}{R_s}\left[D-2-\left(\frac{D-3}{2}\right)\frac{m^2}{E^2}\right]^2 , \tag{75}$$

where α', m and E are the string tension, string mass and energy respectively. The quantity within the square brackets is always positive, thus the lower modes develop imaginary frequencies when the typical string size $\alpha' m\sqrt{D-3}$ is larger than the horizon radius. Notice the similarity with the instability condition in De Sitter space, $n < \alpha' m/r_H$, r_H being the horizon radius.

In the Schwarzschild black hole, the transverse modes Ξ^i are stable (well behaved) everywhere including the spacetime singularity at $q^R = 0$. In the Reissner - Nordström (R-N) black hole, the transverse modes Ξ^i are stable at infinity and outside the horizon. Imaginary frequencies appear, however, inside a region from $r_- < q^R < r_+$ to $q^R \to 0$, where $r_\pm = M \pm \sqrt{M^2 - Q^2}$, M and Q being the mass and charge of the black hole respectively. For the extreme black hole ($Q = M$), instabilities do not appear. There is a critical value of the electric charge of a Reissner - Nordström black hole, above which the string passing through the horizon passes from unstable to stable regime. In the De Sitter spacetime, the only stable mode is the temporal one (Ξ^+). All the spatial components exhibit instability, in agreement with the previous results in the cosmological context[31,33,34] . A summary of this analysis is given in Table 1.

Imaginary frequencies in the transverse string coordinates (Ξ^i) appear in the case in which the local gravity, i.e. $\partial_r a/2$, is negative (that is, repulsive effects). Here,

$$a(r) = 1 - (R_s/r)^{D-3} + (\tilde{Q}^2/r^2)^{D-3} + \frac{\Lambda}{3}r^2 , \tag{76}$$

where $\tilde{Q}^{2(D-3)} = \frac{8\pi G Q^2}{(D-2)(D-3)}$, and Λ is the cosmological constant. That is why the transverse modes (Ξ^i) are well behaved in the Schwarzschild case, and outside the Reissner - Nordström event horizon. But close to $q^R \to 0$, $a'_{RN} < 0$ (Reissner - Nordström has a repulsive inner horizon), and the gravitational effect of the charge overwhelms that of the mass; in this case instabilities develop. In the Reissner - Nordström - De Sitter spacetime, unstable string behavior appears far away from the black hole where De Sitter solution dominates, and inside the black hole where the Reissner - Nordström solution dominates. For $M = 0$ and $Q = 0$, we recover the instability criterion[31,33] $\alpha' m\Lambda/6 > 1$ for large enough Hubble constant (this is in agreement with the criterion given in ref[38].

We find that in the black hole spacetimes, the transversal first order fluctuations (Ξ^i) near the space time singularity $q^R = 0$, obey a Schrödinger type equation (with τ playing the role of a spatial coordinate), with a potential $\gamma(\tau-\tau_0)^{-2}$, (where τ_0 is the proper time of arrival to the singularity at $q^R = 0$). The dependence on D and L is concentrated in the coefficient γ. Thus, the approach to the black hole singularity is like the motion of a particle in a potential $\gamma(\tau - \tau_0)^{-\beta}$, with $\beta = 2$. And, then, like the case $\beta = 2$ of strings in singular gravitational waves[35,36] (in which case the spacetime is simpler and the exact full string equations become linear). Here $\gamma > 0$ for strings in the Schwarzschild spacetime, for which we have regular solutions Ξ^i ; while $\gamma < 0$ for Reissner - Nordström, that is, in the case we have a singular potential and an unbounded behavior (negative powers in $(\tau - \tau_0)$) for Ξ^i_{RN}. The fact that the angular coordinates Ξ^i_{RN} become unbounded means that the string makes infinite turns around the spacetime singularity and remains trapped by it.

For $(\tau - \tau_0) \to 0$, the string is trapped by the black hole singularity. In Kruskal coordinates $(u_k(\sigma,\tau), v_k(\sigma,\tau))$, for the Schwarzschild black hole we find $\lim_{(\tau-\tau_0)\to 0} u_k v_k = \exp 2KC(\sigma)(\tau - \tau_0)^P$, where $K = (D-3)/(2R_s)$ is the surface gravity, $P > 0$ is a determined coefficient that depends on the D dimensions, and $C(\sigma)$ is determined by the initial state of the string. Thus $u_k v_k \to 1$ for $(\tau-\tau_0) \to 0$. The proper spatial string length at fixed $(\tau - \tau_0) \to 0$ grows like $(\tau - \tau_0)^{-(D-1)P}$.

It must be noticed that in cosmological inflationary backgrounds, the unstable behavior manifests itself as non - oscillatory in $(\tau - \tau_0)$ (exponential for $(\tau - \tau_0) \to \infty$, power - like for $(\tau - \tau_0) \to 0$); the string coordinates η^i are constant (i.e. functions of σ only), while the proper amplitudes Ξ^i grow like the expansion factor of the universe. In the black hole cases, and more generally, in the presence of spacetime singularities, all the characteristic features of string instability appear, but in addition the spatial coordinates η^i (or some of its components) become unbounded. That is, not only the amplitudes Ξ^i diverge, but also the string coordinates η^i, what appears as a typical feature of strings near the black hole singularities.

It is also worth noting that we have considered m, the mass of the string, being different from zero, although to zeroth order the constraint equations are satisfied by a null geodesic rather than by a timelike one. To obtain the value of m one has to consider second order perturbations, however, since we were mainly concerned with the possibility of the string unstable behavior, and not its detailed trajectory, first order analysis is enough. Still the same comment can be applied to the fact that, as we know, only $D - 2$ string coordinate components are unconstrained. We could choose a covariant decomposition[39] that would leave, for example, untouched our analysis of unstable behaviour for the transversal coordinates. A full description of the string behavior near the black hole singularity will be reported elsewhere[40] .

9. Formulation of the problem, Non-collinear collision and Discussion

de Vega and Sánchez[31] have obtained the equations of motion of fundamental strings in curved backgrounds by expanding the fluctuations of the string around a given particular solution of the problem (for example, the center of mass motion). For the case of a black hole background[32] with mass M, charge Q and cosmological constant Λ,

$$ds^2 = -a(r)(dX^0)^2 + a^{-1}(r)dr^2 + r^2 d\Omega^2_{D-2} \ ,$$

$$a(r) = 1 - (R_s/r)^{D-3} + (\tilde{Q}^2/r^2)^{D-3} + \frac{\Lambda}{3}r^2 \ , \tag{77}$$

$$R_s^{D-3} = \frac{16\pi GM}{(D-2)\Omega_{D-1}} \ , \quad \Omega_D = \frac{2\pi^{D/2}}{\Gamma(D/2)} \ , \quad \tilde{Q}^{2(D-3)} = \frac{8\pi GQ^2}{(D-2)(D-3)} \ ,$$

the equations of motion of the first order fluctuations read

$$\left[\frac{d^2}{d\tau^2} + n^2\right]\eta_n^A(\tau) + 2A_B^A(\tau)\dot{\eta}_n^B(\tau) + B_B^A(\tau)\eta_n^B(\tau) = 0 \ . \tag{78}$$

Here we have expanded the first order string perturbations in a Fourier transform

$$\eta^A(\sigma,\tau) = \sum_n e^{in\sigma}\eta_n^A(\tau) \ . \tag{79}$$

η^A being the vector

$$\eta = \begin{pmatrix} \eta^0 \\ \eta^* \\ \eta^i \end{pmatrix} \quad , \quad i = 2, 3, \ldots\ldots, D-1 \tag{80}$$

and $A_B^A(\tau)$ and $B_B^A(\tau)$ are the components of the following matrices

$$A = \begin{pmatrix} -\frac{a'\dot{q}^R}{2a} & -\frac{\alpha' E a'}{2a} & \mathbf{O} \\ -\frac{\alpha' E a'}{2a} & -\frac{a'\dot{q}^R}{2a} & q^R\dot{q}^j \\ \mathbf{O} & \frac{\dot{q}^i}{q^R}a & \frac{\dot{q}^R}{q^R}\delta^{ij} + q^i\dot{q}^j \end{pmatrix} \ , \tag{81}$$

$$B = \begin{pmatrix} 0 & -\frac{\alpha' E a''\dot{q}^R}{a} & \mathbf{O} \\ 0 & -S & \mathbf{O} \\ \mathbf{O} & 2\dot{q}^i\dot{\left(\frac{a}{q^R}\right)} & \left(\frac{\alpha' L}{(q^R)^2}\right)^2\delta^{ij} \end{pmatrix} \ , \tag{82}$$

with

$$S = \frac{a''}{2a}\left[(q^R)^2 + \alpha'^2 E^2\right] - \left(\frac{\alpha' L}{(q^R)^2}\right)^2 a \ , \qquad (83)$$

Here the dot stands for $\partial/\partial\tau$ and $' = \partial/\partial q^R$. The solution for the string coordinates is given by the expansion

$$X^\mu = q^\mu + \eta^\mu + \xi^\mu + ... \ , \quad \mu = 0, 1, ..., D-1 \ . \qquad (84)$$

$q^\mu(\tau)$ being the center of mass coordinates (zeroth order solutions), which follow the geodesics of the background space-time, i.e.

$$\dot{q}^0 = \frac{\alpha' E}{a} \ , \ (\dot{q}^i)^2 = \left(\frac{\alpha' L}{(q^R)^2}\right)^2 \ , \ \frac{(\dot{q}^R)^2}{\alpha'^2 a} + \frac{L^2}{(q^R)^2} + m^2 - \frac{E^2}{a} = 0 \ , \qquad (85)$$

where we have identified the proper time of the geodesic with the world-sheet τ - coordinate.

To study the equation of motion of the first fluctuations, eq (78), it is convenient to apply a transformation to the vector η^A. Let us propose

$$\eta_n = G\Xi_n \ , \qquad (86)$$

where the matrix G is chosen to eliminate the term in the first derivative in eq (78), i.e.

$$G = P\exp\{-\int^\tau A(\tau')d\tau'\} \ , \qquad (87)$$

where P is a constant normalizing matrix. Thus, eq (78) transforms into

$$\ddot{\Xi}_n + G^{-1}(n^2 + B - \dot{A} - A^2)G\Xi_n = 0 \ , \qquad (88)$$

which is a Scrödinger-type equation with τ playing the role of the spatial coordinate.

It is simple to analyse the first order fluctuations in the transversal coordinates. In fact, for $i > 2$, the matrices A and B given by eqs (81) and (82) are diagonal,

$$A^{ij} = \frac{\dot{q}^R}{q^R}\delta^{ij} \ ; \ B^{ij} = \left(\frac{\alpha' L}{(q^R)^2}\right)^2 \delta^{ij} \ , \qquad (89)$$

then the equations for the first order fluctuations are

$$\ddot{\Xi}^i_n + (n^2 + \left(\frac{\alpha' L}{(q^R)^2}\right)^2 - \ddot{q}^R/q^R)\Xi^i_n = 0 \ . \qquad (90)$$

By use of the geodesic equations (85) we can rewrite eq (90) as

$$\ddot{\Xi}^i_n + \left[n^2 + \frac{(\alpha')^2}{2}\left(\frac{a'(q^R)m^2}{q^R} + 2\left(\frac{L}{(q^R)^2}\right)^2\left(1 - a + \frac{1}{2}a'q^R\right)\right)\right]\Xi^i_n = 0 \ . \quad (91)$$

We can still re-write the above equations as

$$\ddot{\Xi}^\mu_n + [n^2 + \lambda^\mu]\Xi^\mu_n = 0 \ , \quad (92)$$

where

$$\Xi^\mu_n = (\Xi^+_n, \Xi^-_n, \Xi^i_n) \ \ ; \ \ \lambda^\mu = (\lambda^+, \lambda^-, \lambda^i) \ , \quad (93)$$

and

$$\lambda^\pm = \lambda_\pm - n^2 \ \ , \ \ \lambda^i = \frac{(\alpha' m)^2}{2}\frac{a'(q^R)}{q^R} \ . \quad (94)$$

Thus we are able to analyse now the particular cases we are interested in:

9.1 Schwarzschild black hole

Plugging eq (77) into (91) we obtain

$$\lambda^i_{Sch} = \frac{(\alpha')^2}{2}R_s^{D-3}(q^R)^{1-D}\left[(D-3)m^2 + (D-1)\left(\frac{L}{q^R}\right)^2\right] \ . \quad (95)$$

We observe that for $D \geq 4$, $\lambda^i_{Sch} > 0$. Thus producing stable first order fluctuations. In fact, when we study the time evolution we obtain

$$\lim_{q^R \to 0} \lambda^i_{Sch} = \frac{2(D-1)}{(D+1)^2}(\tau - \tau_0)^{-2} \ , \quad (96)$$

where we have integrated the expression (85),

$$\alpha'(\tau - \tau_0) = \int_{q_0^R}^{q^R} \frac{dq^R}{\sqrt{E^2 - m^2 a(q^R) - a\left(\frac{L}{q^R}\right)^2}} \ , \quad (97)$$

in order to obtain the behavior

$$q_S^R(\tau) \to \left[\frac{(D+1)}{2}\alpha' R_s^{(D-3)/2}L(\tau - \tau_0)\right]^{2/(D+1)} \ . \quad (98)$$

Eq (95) has, then, a power-like solution in the limit $q^R \to 0$,

$$\Xi^i_{n,Sch}(\tau) \sim [n(\tau - \tau_0)]^P \ \ ; \ \ P^i_{Sch} = \frac{1}{2} \pm \sqrt{\frac{1}{4} - \frac{2(D-1)}{(D+1)^2}} \ . \quad (99)$$

That vanishes for $(\tau - \tau_0) \to 0$.

9.2 De Sitter spacetime

This case is very interesting because replacing expressions (77) into the first order fluctuations equations with $L \neq 0$, eq (91), we obtain that the L dependence disappears. Giving, in fact, the equation

$$\ddot{\Xi}^i_n + \left[n^2 - \frac{(\alpha' m)^2}{6}\Lambda\right] \Xi^i_n = 0 \,. \tag{100}$$

The solution to this equation are exponentials in the proper time, τ, suggesting the occurrence of instabilities.

The fact that the solutions should be L-independent, could have been guessed from the symmetries of the De Sitter space. This metric has not preferred point to refer the angular moment to as in the black hole cases (there is not singularity at $r = 0$). This allows us to say that the components Ξ^-, Ξ^+ and Ξ^i will behave as the ones already studied for the case $L = 0$.

9.3 Reissner-Nordström black hole

From the metric coefficients (77) and its first derivative we find that eq (91) reads

$$\ddot{\Xi}^i_n + \tag{101}$$

$$+\left\{n^2 + \frac{(\alpha')^2}{2}\left[\frac{(D-3)m^2}{(q^R)^2 x}(1 - \frac{2\beta}{x}) + \frac{L^2}{x(q^R)^4}\left((D-1) - \frac{2\beta}{x}(D-2)\right)\right]\right\}\Xi^i_n = 0 \,.$$

We observe that as $q^R \to \infty$, the term proportional to L^2 vanish faster than the other terms. Thus, in this limit we recover the $L = 0$ results.

As we approach to the black hole, and arrive to the horizon, we have for the Schwarzschild black hole

$$\lim_{q^R \to q^R_H} \lambda^i_{H,Sch} = \frac{(\alpha')^2}{2}\left[\frac{(D-3)m^2}{R_s^2} + (D-1)\frac{L^2}{R_s^4}\right] , \tag{102}$$

which is definite positive, and does not produce instabilities.

For the extreme Reissner-Nordström black hole we have,

$$\lim_{q^R \to q^R_H} \lambda^i_{H,ERN} = (\alpha')^2 2^{4/(D-3)} \frac{L^2}{R_s^4} \,. \tag{103}$$

Again, it is always positive. Thus, we see that the instabilities do not appear yet. The time-dependence of the solutions close to the horizon will be oscillatory with the squared frequency given by $\lambda^i_{H,ERN}$.

However, the picture changes when we go closer to the singularity. For the Reissner-Nordström black hole, when $q^R \to 0$, we have

$$\lim_{q^R \to 0} \lambda^i_{0,RN} = -(\alpha')^2 \frac{(D-2)\beta}{x^2} \frac{L^2}{(q^R)^4} \, . \tag{104}$$

As this squared frequency takes negative values allows the possibility that instabilities develop in the string transversal coordinates. In order to find the time dependence in the coordinates we integrate first the center of mass motion, eq (85). Then, as $q^R \to 0$, we obtain

$$\alpha'(\tau - \tau_0) \sim \frac{x}{L\sqrt{\beta}} \frac{(q^R)^2}{(D-1)} \, . \tag{105}$$

Plugging this expression into eq (104), we have

$$\lim_{q^R \to 0} \lambda^i_{0,RN} \to -\frac{(D-2)}{(D-1)^2} (\tau - \tau_0)^{-2} \, . \tag{106}$$

And the solution of the first order fluctuations is again a power-like one,

$$\Xi^i_{n,RN}(\tau) \sim [n(\tau - \tau_0)]^P \quad ; \quad P^i_{RN} = \frac{1}{2} \pm \sqrt{\frac{1}{4} + \frac{(D-2)}{(D-1)^2}} \, . \tag{107}$$

We have here that the solution with minus sign in front of the square root produces an unbounded solution as $(\tau - \tau_0) \to 0$, thus, suggesting the existence of instabilities.

It is worth to remark that the solutions (107) and (99) for the time dependence of the transversal coordinates for the Reissner-Nordström and Schwarzschild black holes respectively, are independent of L. They are, although, different from those of the case $L = 0$ (see Ref. 37). This is so because even if the L dependence cancels out from the final equations (107) and (108), the approach to the singularity, $q^R \to 0$, is different if $L \neq 0$, thus producing different final coefficients.

Another interesting feature of the equations for the transversal first order modes, is that for the black hole cases (Schwarzschild or Reissner-Nordström; orbit of the string center of mass with or without angular momentum), the time dependence of λ^i appears to be $(\tau - \tau_0)^{-2}$ as $q^R \to 0$. The behavior of λ^i as a function of q^R is different for each case, but the τ - dependence of the orbit in each case exactly compensates for such difference.

Thus, for the linearized string fluctuations, the approach to the black hole singularity corresponds to the case $\beta = 2$ of the motion of a particle in a potential $\gamma(\tau - \tau_0)^{-\beta}$. This is like the case of strings in singular plane - wave backgrounds.[35,36] In fact, the linearized first order string fluctuations produce a one - dimensional Schrödinger equation, with τ playing the role of a spatial coordinate. The potential term in eq (91), can be written fully τ - dependent, as we have seen, by plugging into it the center of mass trajectory, $q^R(\tau)$.(In the case of gravitational plane waves, the spacetime is simpler than in the black hole one, and the exact full string equations become linear).

The solution of eq (91) with a potential proportional to $\gamma(\tau-\tau_0)^{-2}$ can be given in terms of Bessel functions,

$$\Xi^i_n(\tau) = \sqrt{(\tau-\tau_0)}\left\{V^i_n J_\nu[n(\tau-\tau_0)] + W^i_n J_{-\nu}[n(\tau-\tau_0)]\right\} . \tag{108}$$

Where V^i_n and W^i_n are arbitrary constants coefficients and

$$\nu = \sqrt{\frac{1}{4}-\gamma} \quad \text{i.e. ,} \quad \nu = P^\mu - 1/2 , \tag{109}$$

Where $P^\mu(D,\beta)$'s are defined in Section IV of Ref. 37.

For $\gamma < 0$ we have Bessel functions (those with negative index) with a divergent behavior as $(\tau-\tau_0) \to 0$, indicating the existence of string instabilities. We would also to stress that for black holes, what determinates the possibility of instabilities is not the type of singularity (Reissner-Nordström or Schwarzschild), nor how it is approached ($L=0$ or $L \neq 0$), because we have seen that the time dependence of the potential is always $(\tau-\tau_0)^{-2}$, but the sign of the coefficient γ in front of it, i.e. the attractive character of the potential $(\tau-\tau_0)^{-2}$. Thus, we can conclude that whenever we have big enough repulsive effects in a gravitational background, instabilities in the propagation of strings on the spacetime background will appear. The coefficient γ is given by

$$\gamma_{Sch} = \begin{cases} \frac{2(D-3)}{(D-1)^2} , & L=0 \\ \frac{2(D-1)}{(D+1)^2} , & L \neq 0 \end{cases}$$

for the Schwarzschild black hole, and

$$\gamma_{RN} = \begin{cases} -\frac{2(D-3)}{(D-2)^2} , & L=0 \\ \frac{-2(D-2)}{(D-1)^2} , & L \neq 0 \end{cases}$$

for the Reissner-Nordström black hole.

Near the spacetime singularity, the dependence on the D spacetime dimensions is concentrated in γ. Notice the attractive singular character of the potential $\gamma(\tau-\tau_0)^{-2}$, for the Reissner-Nordström black hole, in agreement with the singular behavior of the string near $q^R = 0$; while for the Schwarzschild black hole γ is positive, and the string solutions Ξ^i are well behaved there.

The approach to the black hole singularity is better analysed in terms of the Kruskal coordinates (u_k, v_k)

$$u_k = e^{K u_{Sch}} \quad , \quad v_k = e^{K v_{Sch}}$$

u and v being null coordinates, and K the surface gravity of the black hole ($K = (D-3)/(2R_s)$ for Schwarzschild). We also have ($C^{\pm}(\sigma)$ being coefficients determined by the initial state of the string)

$$u_k(\sigma,\tau) = \exp K \left[C^+(\sigma)(\tau-\tau_0)^{P^+} - C^-(\sigma)(\tau-\tau_0)^{-|P^-|}\right]$$

$$v_k(\sigma,\tau) = \exp K \left[C^+(\sigma)(\tau-\tau_0)^{P^+} + C^-(\sigma)(\tau-\tau_0)^{-|P^-|}\right] \tag{110}$$

and thus, $u_k v_k \to 1$ for $(\tau-\tau_0) \to 0$. That is, for $(\tau-\tau_0) \to 0$, the string approaches the spacetime singularity $u_k v_k = 1$, and it is trapped by it. The proper spatial length element of the string at fixed $(\tau-\tau_0) \to 0$, between (σ,τ) and $(\sigma+d\sigma,\tau)$, stretches infinitely as

$$ds^2_{(\tau-\tau_0)\to 0} \to \left(C^-(\sigma)\right)'^2 d\sigma^2 (\tau-\tau_0)^{-(D-1)|P^-|} \,, \tag{111}$$

where P^- is given by eq (71) of Ref. 37. Here, τ_0 is the (finite) proper falling time of the string into the black hole singularity.

The fact that the angular coordinates Ξ^i become unbounded in the Reissner - Nordström case, means that the string makes infinite turns around the spacetime singularity and remains trapped by it.

It is also worth noting that we have considered m, the mass of the string being different from zero although to zeroth order the constraint equations are satisfied by a null geodesic rather than a timelike one. To obtain the value of m one has to consider second order perturbations, however, since we were mainly concerned with the possibility of the string unstable behavior, first order analysis is enough. The same conclusions can be drawn for the quantum propagation of strings. The τ dependence is the same because this is formally described by a Schrödinger equation with a potential $\gamma(\tau-\tau_0)^{-2}$, the coefficients of the solutions being quantum operators instead of C - numbers. The τ evolution of the string near the black hole singularity is fully determined by the spacetime geometry, while the σ - dependence (contained in the overall coefficients) is fixed by the state of the string.

For the modes Ξ^-, Ξ^+ and Ξ^i we can conclude that they should behave as in the case $L = 0$ far from the black hole, where the influence of the angular momentum vanishes. Then, Ξ^+ and Ξ^i will oscillate with bounded amplitude outside the horizon while Ξ^- will present an unbounded behavior. The approach to the singularity with $L \neq 0$ should not change qualitatively from the picture for $L = 0$. The analysis can be also made in terms of the geodesic orbit followed by the center of mass of the string. For a given energy E, there is a critical impact parameter b_c which determines whether the string will fall or not into the black hole. From our results here for $L = 0$ (see Table 1), and for $L \neq 0$, we can draw the following picture: For large impact parameters $b > b_c$, the transversal, Ξ^i, and temporal, Ξ^+, modes will be stable, while the radial modes, Ξ^-, begin to suffer instabilities. For small impact parameters, $b < b_c$, the string will fall into the black hole and, for a Reissner- Nordström background, also the transversal coordinates suffer instabilities.

Table 1 Regimes of string stability in black hole and De Sitter spacetimes: Here stable means well behaved string fluctuations and the usual oscillatory behavior with real frequencies. Unstable behavior corresponds to unbounded amplitudes $(\eta^{\pm}, \eta^i)$ with the emergence of non oscillatory behavior or imaginary frequencies, accompanied by the infinite string stretching of the proper string length. η^+,η^- and η^i, $(i = 2, ..., D-1)$, are the temporal, radial and angular (or transverse) string components respectively.

Region	Mode	Schwarzschild	Reissner-Nordström	De Sitter
	η^+	stable	stable	stable
$q^R \to 0$	η^-	unstable	unstable	unstable
	η^i	unstable	unstable	unstable
	η^+	stable	stable	stable
$q^R \to q^R_H$	η^-	unstable	unstable / stable	unstable
	η^i	stable	stable	unstable
	η^+	stable	stable	stable
$q^R \to \infty$	η^-	unstable	unstable	unstable
	η^i	stable	stable	unstable

It can be noticed that in cosmological inflationary backgrounds, for which unstable string behavior appears when $(\tau - \tau_0) \to 0$, the string coordinates η^i, remain bounded. In the black hole cases, all the characteristic features of string instabilities appear, but, in addition, the string coordinates η^i become unbounded near the $r = 0$ singularity. This happens to be a typical behavior of strings near spacetime singularities, describing the fact that the string is trapped by it.

10. Acknowledgements

C.O.L. was financially supported by the Dirección General de Investigación Científica y Técnica of the Ministerio de Educación y Ciencia de España and CICYT AEN93-0474. C.O.L. also thanks the DEMIRM - Observatoire de Paris, where part of this work was done, for kind hospitality and working facilities.

11. References

1. G.'t Hooft, *Phys. Lett. B* **198** (1987) 61.
2. H.J.de Vega and N.Sánchez, *Nucl. Phys. B* **317** (1989) 731.
3. H.J.de Vega and N.Sánchez, *Nucl. Phys. B* **317** (1989) 706.
4. D.Amati, M.Ciafaloni and G. Veneziano, *Int. J. Mod. Phys. A* **3** (1988) 1615. Ibid, *Nucl. Phys. B* **347** (1990) 550.
5. D.Gross and P.Mende, *Nucl. Phys. B* **303** (1988) 407.
6. D.Amati and C.Klimčik, *Phys. Lett. B* **210** (1988) 92.
7. G.Veneziano, "Strings and Gravitation", talk presented at the *5th Marcel Grossmann meeting* (Perth, Aug. 1988), D.G.Blair and M.J.Buckinghan eds. (World Sci., Singapore, 1989), p. 173.
8. G.T.Horowitz and A.R.Steif, *Phys. Rev. Lett.* **64** (1990) 260. Ibid *Phys. Rev. D* **42** (1990) 1950.
9. H.J. de Vega and N.Sánchez, *Phys. Lett. B* **244** (1990) 215. Ibid *Phys. Rev. Lett.* **65** (1990) 1517.
10. J.Garriga and E.Verdaguer, *Phys. Rev. D* **43** (1991) 391.
11. K. Hayashi and T. Samura, *Phys. Rev. D* **50** (1994) 3666. Ibid hep-th/940513.
12. P. C. Aichelburg and H. Balasin, *Class. Quantum Grav.* **11** (1994) L121.
13. S.Das and P.Majumdar, *Phys. Rev. Lett.* **72** (1994) 2524. Ibid, gr-qc/9411061. Ibid, hep-th/9412065.
14. P.C.Aichelburg and R.U.Sexl, *Gen. Relativ. Grav.* **2** (1971) 303.
15. C.O.Lousto and N.Sánchez, *Int. J. Mod. Phys. A* **5** (1990) 915.
16. C.O.Lousto and N.Sánchez, *Nucl. Phys. B* **383** (1992) 377.
17. M.E.V.Costa and H.J. de Vega, *Ann. Phys. (N.Y.)* **211** (1991) 223. Ibid, *Ann. Phys. (N.Y.)* **211** (1991) 235.
18. C. O. Lousto and N. Sánchez, *Phys. Rev. D* **46** (1992) 4520.
19. T.Dray and G.'t Hooft, *Nucl. Phys. B* **253** (1985) 173.
20. C.O.Lousto and N.Sánchez, *Phys. Lett. B* **232** (1989) 462.
21. C.W.Misner, K.S.Thorne and J.A.Wheeler, *Gravitation*, (W.H. Freeman and Co., San Francisco, 1973).
22. V.Ferrari and P.Pendenza, *Gen. Relativ. Grav.* **22** (1990) 1105.
23. C.O.Lousto and N.Sánchez, *Phys. Lett. B* **220** (1989) 55.
24. K.Sfetsos, *Nucl. Phys. B* **436** (1995) 721.
25. H. Balasin and H. Nachbagauer, *Phys. Lett. B* **315** (1993) 93. Ibid *Class. Quantum Grav.* **10** (1993) 2271. Ibid *Class. Quantum Grav.* **11** (1994) 1453. Ibid gr-qc/9405053.
26. C.O.Lousto and N.Sánchez, *Nucl. Phys. B* **355** (1991) 231.
27. M.Dorca and E.Verdaguer, *Nucl. Phys. B* **403** (1993) 770.

28. G. Veneziano, *DST Workshop on String Theory*, H. S. Mani and R. Ramachandran eds. (World Sci. Publ. Co. 1988) p. 1.
29. G.'t Hooft, *Nucl. Phys. B* **304** (1988) 867.
30. A. Erdelyi et al., *Tables of Integral Transforms*, Vol.II, (McGraw-Hill, New York, 1954).
31. H.J.de Vega and N.Sánchez, *Phys. Lett. B* **197** (1987) 320.
32. H.J.de Vega and N.Sánchez, *Nucl. Phys. B* **309** (1988) 552. Ibid *Nucl. Phys. B* **309** (1988) 577.
33. N.Sánchez and G.Veneziano, *Nucl. Phys. B* **333** (1990) 253.
34. M.Gasperini, N.Sánchez and G.Veneziano, *Int. J. Mod. Phys. A* **6** (1991) 3853. Ibid *Nucl. Phys. B* **364** (1991) 365.
35. H.J.de Vega and N.Sánchez, *Phys. Rev. D* **45** (1992) 2783.
36. H.J.de Vega, M.Ramón-Medrano and N.Sánchez, *Nucl. Phys. B* **374** (1992) 425.
37. C. O. Lousto and N. Sánchez, *Phys. Rev. D* **47** (1993) 4498.
38. M.Gasperini, *Gen. Relativ. Grav.* **24** (1992) 219.
39. A.L.Larsen and V.P.Frolov, *Nucl. Phys. B* **414** (1994) 129. A.L.Larsen, *Phys. Rev. D* **50** (1994) 2623.
40. C.O.Lousto and N.Sánchez, in preparation.

STRING IN CURVED SPACETIMES: THE NULL STRING APPROACH

H. J. de Vega

Laboratoire de Physique Théorique et Hautes Energies * †

A. Nicolaidis

Theoretical Physics Department, University of Thessaloniki. ‡

Abstract

The string propagation in curved spacetimes is investigated in the null string approach. In this approach the string equations of motion and constraints are systematically expanded in powers of c (the speed of light in the world-sheet). We obtain to zeroth order the null string (string with zero tension), while the first order correction incorporates the string dynamics.

We apply our formalism to null strings in de Sitter spacetime. After a reparametrization of the world-sheet coordinates, the equations of motion are solved. The quantum algebra generated by the constraints is considered, ordering the momentum operators to the right of the coordinate operators. No critical dimension appears. It is anticipated however that the conformal anomaly will appear when the first order corrections proportional to T_0, are introduced.

---

*Laboratoire Associé au CNRS UA 280.

†L.P.T.H.E., Tour 16 1er étage, Université Paris VI, 4 Place Jussieu, 75252, Paris cedex 05, FRANCE.

‡Thessaloniki 54006, GREECE

Classical and quantum string propagation in curved spacetimes is a very important subject. The investigations in this domain are relevant for the physics of quantum gravitation as well as for the understanding of the cosmic string models in cosmology.

Strings in curved spacetimes have been studied in different approaches: string oscillations around the center of mass [3,11,12], string solutions for symmetric ansatz (straight [4], circular [7,9]), soliton methods for de Sitter spacetime [5,6,8], semiclassical quantization of periodic solutions [16]. These works lead to understand in a large extent the string behaviour in curved spacetimes [1,2,4].

Strings are characterized by an energy scale $\sqrt{T_0}$ ($T_0 = \frac{1}{2\pi\alpha'}$ is the string tension). The frequencies of the string modes in the physical time are proportional to T_0 and the length of the string scales with $\frac{1}{\sqrt{T_0}}$. The background gravitational field provides another length scale R_c chosen as a typical curvature radius of the spacetime. For a string moving in a gravitational field a useful parameter is the dimensionless constant $g = R_c\sqrt{T_0}$. Large values of g correspond to weak gravitational field (the metric does not change appreciably over distances of the order of the string length). We may reach large values of g by letting $T_0 \to \infty$. In this limit the string shrinks to a point and a suitable expansion (small string oscillations around the center of mass of the string) has been proposed in ref. [3].

In the opposite limit, small values of g, we encounter strong gravitational fields. This corresponds to $T_0 \to 0$ and suggests that an expansion in powers of T_0 can be appropriate to investigate strings in strong gravitational fields.

In ref. [13] a systematic expansion in terms of the string tension has been presented. To zeroth order we obtain the null string [15], every point of the string moves independently following a null geodesic.

The limit $T_0 \to 0$ cannot be reached using the Nambu-Goto action for the strings. Following the analogous massless particle case we are led to a reformulated Lagrangian [13]

$$L = \frac{1}{4\lambda}[\dot{X}^\mu \dot{X}^\nu G_{\mu\nu}(X) - c^2 X'^\mu X'^\nu G_{\mu\nu}(X)] \tag{1}$$

where $c = 2\lambda T_0$ is the world-sheet 'speed of light' (a dot and a prime denote respectively differentiation with respect to the world-sheet time and space variables, τ and σ). Notice that c is *not* the speed of light in physical spacetime.

The string equations of motion read

$$\ddot{X}^\mu - c^2 X''^\mu + \Gamma^\mu_{\kappa\lambda}(\dot{X}^\kappa \dot{X}^\lambda - c^2 X'^\kappa X'^\lambda) = 0 \tag{2}$$

supplemented by the constraints

$$\dot{X}^\mu X'^\nu G_{\mu\nu} = 0 \qquad \dot{X}^\mu \dot{X}^\nu G_{\mu\nu} + c^2 X'^\mu X'^\nu G_{\mu\nu} = 0 \tag{3}$$

A systematic expansion in powers of c is feasible now [13]. If we write

$$X^\mu(\sigma,\tau) = A^\mu(\sigma,\tau) + c^2 B^\mu(\sigma,\tau) + \cdots$$

where the dots indicate higher powers of c^2 and substitute this form of X^μ in eq.(2) and (3) we find that to zeroth order in c the string dynamics is given by the equations of motion

$$\ddot{A}^\mu + \Gamma^\mu_{\kappa\nu}\dot{A}^\kappa \dot{A}^\nu = 0 \tag{4}$$

and the constraints

$$\begin{aligned} \dot{A}^\mu \dot{A}^\nu G_{\mu\nu} &= 0 \\ \dot{A}^\mu A'^\nu G_{\mu\nu} &= 0 \end{aligned} \tag{5}$$

$A^\mu(\sigma,\tau)$ represents a collection of massless particles moving independently along null geodesics. The only reminiscence from the string is the second constraint under eq.(5), which requires the velocity to be perpendicular to the string. We can rewrite this constraint as follows

$$A'^\nu\, \Pi_\nu = 0 \tag{6}$$

where Π_ν is the canonical momentum conjugated to A^ν. One can show that eq.(6) expresses the invariance under infinitesimal reparametrizations

$$\sigma \to \sigma + \epsilon\, f(\sigma) \tag{7}$$

where $f(\sigma)$ is an arbitrary smooth function and $\epsilon << 1$. This invariance is peculiar of null strings and does not hold for $T_0 \neq 0$. Let $F[A^\nu(.)]$ be a functional of $A^\nu(\sigma)$. Then, $F[A^\nu(.)]$ changes under the infinitesimal transformation (7) as follows,

$$\begin{aligned} \delta F &= \int d\sigma \, \frac{\delta F}{\delta A^\nu(\sigma)}\, A'^\nu(\sigma)\, f(\sigma) \\ &= \int d\sigma \, f(\sigma)\, \{\Pi_\nu(\sigma)\, A'^\nu(\sigma), F\} \end{aligned} \tag{8}$$

where $\{\,,\}$ stands for the canonical Poisson brackets. We see that eq.(6) guarantees that

$$\delta F = 0$$

This invariance generalizes to the quantum level replacing the Poisson brackets by quantum commutators.

The next order correction $B^\mu(\sigma,\tau)$ obeys the following equation of motion

$$\ddot{B}^\mu + \Gamma^\mu_{\kappa\lambda}(\dot{A}^\lambda \dot{B}^\kappa + \dot{A}^\kappa \dot{B}^\lambda) + \Gamma^\mu_{\kappa\lambda,\nu}\dot{A}^\kappa \dot{A}^\lambda B^\nu = A''^\mu + \Gamma^\mu_{\kappa\lambda} A'^\kappa A'^\lambda$$

supplemented with the constraints

$$\begin{aligned}
2\dot{A}^\mu \dot{B}^\nu G_{\mu\nu} + \dot{A}^\mu \dot{A}^\nu B^\rho G_{\mu\nu,\rho} + A'^\mu A'^\nu G_{\mu\nu} = 0 \\
\dot{B}^\mu A'^\nu G_{\mu\nu} + \dot{A}^\mu B'^\nu G_{\mu\nu} + \dot{A}^\mu A'^\nu B^\rho G_{\mu\nu,\rho} = 0
\end{aligned} \tag{9}$$

where $\Gamma^\mu_{\kappa\lambda,\rho}$ and $G_{\mu\nu,\rho}$ indicate the derivatives with respect to A^ρ.

We next apply this method to study the quantization of the null string in de Sitter spacetime [14]. The line element is defined as

$$ds^2 = C^2(X_0)(dX_0^2 - dX_1^2 - dX_2^2 - dX_3^2) \tag{10}$$

with

$$C(X_0) = \frac{R_0}{X_0} \tag{11}$$

where X_0 is the conformal time and R_0 is the scale factor. The Lagrangian for a null string propagating in de Sitter spacetime takes the form

$$L = C^2(A_0)(\dot{A}_0^2 - \dot{A}_1^2 - \dot{A}_2^2 - \dot{A}_3^2)$$

where we used eq.(10) The equations of motion, using also the constraints, provide

$$\begin{aligned}
C^2(A_0)\dot{A}_i = P_i(\sigma) \qquad i = 1, 2, 3 \\
C^2(A_0)\dot{A}_0 = P_0(\sigma)
\end{aligned} \tag{12}$$

Using eq. (11), we obtain from eq. (12)

$$A_0(\sigma, \tau) = \frac{\overline{A}_0 R_0^2}{R_0^2 - \overline{A}_0 P_0 \tau}$$

with $\overline{A}_0 = A_0(\sigma, \tau = 0)$. Eq. (12) gives then

$$A_i(\sigma, \tau) - \overline{A}_i = \frac{P_i \overline{A}_0^2 \tau}{R_0^2 - \overline{A}_0 P_0 \tau}$$

with $\overline{A}_i = A_i(\sigma, \tau = 0)$. The constraints (5) tie the initial shape and momentum of the string

$$\begin{aligned}
\overline{A}'_0 P_0 = \textstyle\sum_{i=1}^3 \overline{A}'_i P_i \\
P_0^2(\sigma) = \textstyle\sum_{i=1}^3 P_i^2(\sigma)
\end{aligned} \tag{13}$$

Notice that the relationship

$$A_i - \overline{A}_i = \frac{P_i}{P_0}(A_0 - \overline{A}_0)$$

holds, i.e in terms of the cosmic variables we obtain straight lines for the individual particles of the null string. The constraints retain their form under the reparametrization $\tau = f(\tilde{\tau}, \tilde{\sigma})$ and $\sigma = g(\tilde{\sigma})$ with f and g arbitrary functions. We may use this reparametrization freedom in order to simplify the equations of motion. Choosing

$$\tau = \frac{R_0^2 \tilde{\tau}}{\overline{A}_0 P_0 \tilde{\tau} + \overline{A}_0^2}$$

we obtain (after dropping tildes)

$$A_\mu(\sigma,\tau) = P_\mu(\sigma)\,\tau + \overline{A}_\mu(\sigma) \qquad \mu = 0,1,2,3$$

$$P_\mu(\sigma)^2 = 0 \tag{14}$$

Notice that $P_\mu(\sigma)^2 = 0$ is **necessary** in order to fulfil the equations of motion (4). The relation

$$\overline{A}'_\mu(\sigma) P^\mu(\sigma) = 0 \tag{15}$$

remains as a constraint [see eq.(5)].

Due to the periodicity in the σ-direction, since we deal with closed strings, we can expand $P^\mu(\sigma)$ and $\overline{A}^\mu(\sigma)$ in Fourier series as follows

$$P^\mu(\sigma) = \sum_n p_n^\mu \exp(in\sigma) \qquad \overline{A}^\mu(\sigma) = \sum_n x_n^\mu \exp(in\sigma)$$

Similarly the constraint can be expanded in Fourier series

$$P_\mu \overline{A}'^\mu = \sum_k G_k \exp(ik\sigma)$$

By substituting the expressions for the $P^\mu(\sigma)$ and $\overline{A}(\sigma)$ in the above relation we find

$$G_k = i \sum_n n\, p_{k-n}\, x_n$$

The string coordinate $\overline{A}^\mu(\sigma)$ and the conjugate momentum $P^\mu(\sigma)$ satisfy the following Poisson brackets

$$\{P^\mu(\sigma), \overline{A}^\nu(\sigma')\} = \delta(\sigma - \sigma')\, G^{\mu\nu}$$

while for their Fourier modes p_n^μ and x_n^μ we obtain

$$\{p_n^\mu, x_m^\nu\} = G^{\mu\nu}\delta_{m+n}$$

Using the above relations we can calculate the Poisson brackets of the moments G_k of the constraints. With a redefinition $G_k \to -G_k$ we find the following algebra (the analogue of the Virasoro algebra in the tensionful string)

$$\{G_n, G_m\} = i(n-m)G_{n+m}\ .$$

The constraints G_n generate reparametrizations in σ. The algebra of the constraints of the null string arises as the Inonu-Wigner contraction [17] ($T_0 \to 0$) of the algebra of the constraints of the tensile string. This can be seen by writing the constraint algebra of the tensile string in terms of the modes of the constraints in the null string.

We may now quantize the null string by replacing the Poisson bracket by the commutator

$$[p_n^\mu, x_m^\nu] = -i\, G^{\mu\nu} \delta_{m+n}$$

The passage from the classical domain to the quantum one offers the possibility for the emergence of central terms in the algebra. The algebra will take the form

$$[G_n, G_m] = (n-m)G_{n+m} + A(n)\, \delta_{n+m}.$$

In order to specify the form of the anomaly we make use of the Jacobi identity for the algebra. The most general form of the anomaly is given by

$$A(n) = \alpha\, n^3 + \beta\, n.$$

where α and β are constants which need to be determined. The actual calculation depends crucially upon the ordering of the operators [18–20].

Given our Lagrangian eq. (1) with $c = 0$, the vacuum $|0>$ satisfies the condition

$$p_n^\mu |0>= 0$$

We order then all the p_n to the right of the x_n. We arrive at similar conclusions by looking at the complete tensile theory and considering the limit $c \to 0$. The tensile string is quantized using harmonic oscillator operators, representing right and left movers. The operators which annihilate the vacuum state in the limit $c \to 0$ are reduced to momentum operators [20]. With the adopted ordering, it is relatively easy to evaluate the quantum algebra generated by the constraints. The quantum algebra **does not contain central terms**, namely we find that $\alpha = \beta = 0$ and therefore a quantum null string may exist in any dimension without anomalies. We anticipate however that a critical dimension will appear, as soon as the first order correction $B^\mu(\sigma, \tau)$, proportional to T_0, is introduced (work in progress).

Summarizing, we presented a consistent framework within which it is feasible to study string quantization in strong gravitational fields. We have shown that for the emergence of the conformal anomaly, responsible is the dimensionful string tension. Historical precedent (Higgs mechanism) may allow us in contemplation for a mechanism generating string tension and at the same time respecting the symmetries.

REFERENCES

[1] H.J. de Vega and N. Sánchez in "String Quantum Gravity and the Physics at the Planck Scale", Proceedings of the Erice Workshop held in June 1992. Edited by N. Sánchez, World Scientific, 1993. Pages 73-185, and references given therein.
[2] H.J. de Vega and N. Sánchez in 'Current Topics in Astrofundamental Physics', Proceedings Erice School, 4-16 September 1994, Edited by N. Sánchez.
[3] H. J. de Vega and N. Sánchez, Phys. Lett. **B 197**, 320 (1987).
[4] H. J. de Vega and N. Sánchez, Phys. Rev. **D50**, 7202 (1994).
[5] H. J. de Vega, A. V. Mikhailov and N. Sánchez, Teor. Mat. Fiz. **94** (1993) 232.
[6] F. Combes, H. J. de Vega, A. V. Mikhailov and N. Sánchez, Phys. Rev. **D50**, 2754 (1994).
[7] H. J. de Vega, A. L. Larsen and N. Sánchez, Nucl. Phys. **B 427**, 643 (1994).
[8] I. Krichever, Funct. Anal. and Appl. **28**, 21 (1994), [Funkts. Anal. Prilozhen. **28**, 26 (1994)].
[9] H. J. de Vega and I. L. Egusquiza, Phys. Rev. **D49**, 763 (1994).
[10] A. L. Larsen and N. Sánchez, Phys. Rev. **D50**, 7493 (1994).
[11] N. Sánchez and G. Veneziano, Nucl. Phys. **B333**, 253 (1990).
[12] M. Gasperini, N. Sánchez and G. Veneziano, Int. J. Mod. Phys. **A 6**, 3853 (1991) and Nucl. Phys. **B364**, 365 (1991).
[13] H. J. de Vega and A. Nicolaidis, Phys. Lett. **295B**(1992), 214.
[14] H. J. de Vega, I. Giannakis and A. Nicolaidis, LPTHE preprint, hep-th/9412081.
[15] A. Schild, Phys. Rev. **D16**(1977), 1722; A. Karlhede and U. Lindström, *Class. Quantum Grav.* **3** (1986), L73.
[16] H. J. de Vega, A. L. Larsen and N. Sánchez, 'Semi-Classical Quantization of Circular Strings in de Sitter and anti de Sitter Spacetimes', LPTHE preprint, hep-th/9410219, to appear in Phys. Rev. **D**.
[17] E. Inonu and E. Wigner, *Proc. Nat. Acad. Sci. (US)* **39** (1953), 510.
[18] F. Lizzi, B. Rai, G. Sparano and A. Srivastava, Phys. Lett. **182B**(1986), 326; F. Lizzi, Mod. Phys. Lett. **A9**(1994), 1465.
[19] J. Gamboa, C. Ramírez and M. Ruiz-Altaba, Nucl. Phys. **B338**(1990), 143.
[20] J. Isberg, U. Lindström, B. Sundborg and G. Theodoridis, Nucl. Phys. **B411**(1994), 122.

Strings and Multi-Strings in Black Hole and Cosmological Spacetimes

A.L. Larsen and N. Sánchez
Observatoire de Paris, DEMIRM. Laboratoire Associé au CNRS UA 336, Observatoire de Paris et École Normale Supérieure. 61, Avenue de l'Observatoire, 75014 Paris, France.

Abstract

Recent results on classical and quantum strings in a variety of black hole and cosmological spacetimes, in various dimensions, are presented. The curved backgrounds under consideration include the $2+1$ black hole anti de Sitter spacetime and its dual, the black string, the ordinary $D \geq 4$ black holes with or without a cosmological constant, the de Sitter and anti de Sitter spacetimes and static Robertson-Walker spacetimes. Exact solutions to the string equations of motion and constraints, representing circular strings, stationary open strings and dynamical straight strings, are obtained in these backgrounds and their physical properties (length, energy, pressure) are described. The existence of *multi-string* solutions, describing finitely or infinitely many strings, is shown to be a general feature of spacetimes with a positive or negative cosmological constant. Generic approximative solutions are obtained using the string perturbation series approach, and the question of the stability of the solutions is addressed.

Furthermore, using a canonical quantization procedure, we find the string mass spectrum in de Sitter and anti de Sitter spacetimes. New features as compared to the string spectrum in flat Minkowski spacetime appear, for instance the *fine-structure effect* at all levels beyond the graviton in both de Sitter and anti de Sitter spacetimes, and the *non-existence* of a Hagedorn temperature in anti de Sitter spacetime. We discuss the physical implications of these results. Finally, we consider the effect of spatial curvature on the string dynamics in Robertson-Walker spacetimes.

1 Introduction

The classical and quantum propagation of strings in curved spacetimes has attracted a great deal of interest in recent years. The main complication, as compared to the case of flat Minkowski spacetime, is related to the non-linearity of the equations of motion. It makes it possible to obtain the complete analytic solution only in a very few special cases like conical spacetime [1] and plane-wave/shock-wave backgrounds [2]. There are however also very general results concerning integrability and solvability for maximally symmetric spacetimes [3, 4] and gauged WZW models [5, 6]. These are the exceptional cases; generally the string equations of motion in curved spacetimes are not integrable and even if they are, it is usually an extremely difficult task to actually separate the equations, integrate them and finally write down the complete solution in closed form. Fortunately, there are several different ways to "attack" a system of coupled non-linear partial differential equations.

The systematic study of string dynamics in curved spacetimes and its associated physical phenomena was started in Refs.[7, 8]. Besides numerical methods, which will not be discussed here, approximative [7-10] and exact [11-14] methods for solving the string equations of motion and constraints in curved spacetimes, have been developed. Classical and quantum string dynamics have been investigated in black hole backgrounds [15-18], cosmological spacetimes [7, 11-14, 17-21], cosmic string spacetime [1], gravitational wave backgrounds [2], supergravity backgrounds (which are necessary for fermionic strings) [22], and near spacetime singularities [23]. Physical phenomena like the Hawking-Unruh effect in string theory [8, 24], horizon string stretching [8, 24], particle transmutation [15, 25], string scattering [1, 2, 15], mass spectrum and critical dimension [1, 7, 15, 27], string instability [7, 11-13, 16, 21] and multi-string solutions [11-13, 28] have been found.

In a generic D-dimensional curved spacetime with metric $g_{\mu\nu}$ and coordinates x^μ, $(\mu = 0, 1, ..., D-1)$, the string equations of motion and constraints are:

$$\ddot{x}^\mu - x''^\mu + \Gamma^\mu_{\rho\sigma}(\dot{x}^\rho \dot{x}^\sigma - x'^\rho x'^\sigma) = 0, \tag{1.1}$$

$$g_{\mu\nu}\dot{x}^\mu x'^\nu = g_{\mu\nu}(\dot{x}^\mu \dot{x}^\nu + x'^\mu x'^\nu) = 0, \tag{1.2}$$

where dot and prime stand for derivative with respect to the world-sheet coordinates τ and σ, respectively and $\Gamma^\mu_{\rho\sigma}$ are the Christoffel symbols with respect to the metric $g_{\mu\nu}$. In the following we present recent results [26-29] on solutions to Eqs.(1.1)-(1.2) in a variety of curved spacetimes from cosmology, gravitation and string theory. We obtain explicit (exact and/or approximate) mathematical solutions, discuss the corresponding physical properties, we quantize the solutions in different ways (canonical quantization, semi-classical quantization) and find the physical content: the mass spectrum.

The presentation is organized as follows: In Section 2, we discuss the classical string dynamics in the $2+1$ black hole anti de Sitter (BH-AdS) spacetime, recently found by Bañados et al [30]. We compare with the string dynamics in ordinary cosmological and black hole spacetimes. This clarifies the geometry (as seen by a string) of the BH-AdS spacetime. In Section 3, generalizing results from Section 2, we derive the quantum string mass spectrum in ordinary D-dimensional anti de Sitter spacetime. We discuss, in particular, the sectors of low and very high mass states. New physical phenomena arise like the fine-structure effect at all levels beyond the graviton and the non-existence of a Hagedorn critical temperature. The results are compared with corresponding results obtained in Minkowski and de Sitter spacetimes. Sections 4 and 5 are devoted to the investigation of the more general underlying structure of solutions to Eqs.(1.1)-(1.2). In Section 4, we find new classes of exact string and multi-string solutions in cosmological and black hole spacetimes, while in Section 5, we consider the effects of a non-zero spatial curvature on the classical and quantum string dynamics in Friedmann-Robertson-Walker (FRW) universes.

2 Classical String Dynamics in 2+1 BH-AdS

Anti de Sitter (AdS) spacetime is often considered to be a spacetime of minor importance in a cosmological context. However, first of all, it *is* a FRW universe and as such should not be neglected, secondly; it serves as a simple and convenient spacetime for comparison and understanding of results obtained in (say) Minkowski or de Sitter spacetimes and third, it has a tendency to show up (in disguise) as a solution in various models of dilaton-gravity and string theory. An example of the latter is represented by the $2+1$ BH-AdS spacetime of Bañados et al [30]. This spacetime background has arised much interest recently [31-35]. It describes a two-parameter family (mass M and angular momentum J) of black holes in 2+1 dimensional general relativity with metric:

$$ds^2 = (M - \frac{r^2}{l^2})dt^2 + (\frac{r^2}{l^2} - M + \frac{J^2}{4r^2})^{-1}dr^2 - Jdtd\phi + r^2 d\phi^2. \qquad (2.1)$$

It has two horizons $r_\pm = \sqrt{\frac{Ml^2}{2} \pm \frac{l}{2}\sqrt{M^2l^2 - J^2}}$ and a static limit $r_{\text{erg}} = \sqrt{M}l$, defining an ergosphere, as for ordinary Kerr black holes. The spacetime is not asymptotically flat; it approaches anti de Sitter spacetime asymptotically with cosmological constant $\Lambda = -1/l^2$. The curvature is constant $R_{\mu\nu} = -(2/l^2)g_{\mu\nu}$ everywhere, except probably at $r = 0$, where it has at most a delta-function singularity. Notice the weak nature of the singularity at $r = 0$ in 2+1 dimensions as compared with the power law divergence of curvature scalars in $D > 3$ (We will not discuss here the geometry near $r = 0$. For a discusion, see Refs.[33, 35]). The spacetime, Eq.(2.1), is also a solution of the low energy effective action of string theory with zero dilaton field $\Phi = 0$, anti-symmetric tensor field $H_{\mu\nu\rho} = (2/l^2)\epsilon_{\mu\nu\rho}$ (i.e. $B_{\phi t} = r^2/l^2$) and $k = l^2$ [31]. Moreover, it yields an exact solution of string theory in 2+1 dimensions, obtained by gauging the WZWN sigma model of the group

$SL(2,R)\times R$ at level k [31, 32] (for non-compact groups, k does not need to be an integer, so the central charge $c=3k/(k-2)=26$ when k=52/23). This solution is the black string background [36]:

$$d\tilde{s}^2=-(1-\tfrac{\mathcal{M}}{\tilde{r}})d\tilde{t}^2+(1-\tfrac{Q^2}{\mathcal{M}\tilde{r}})d\tilde{x}^2+(1-\tfrac{\mathcal{M}}{\tilde{r}})^{-1}(1-\tfrac{Q^2}{\mathcal{M}\tilde{r}})^{-1}\tfrac{l^2d\tilde{r}^2}{4\tilde{r}^2},$$

$$\tilde{B}_{\tilde{x}\tilde{t}}=\tfrac{Q}{r},\quad \tilde{\Phi}=-\tfrac{1}{2}\log\tilde{r}l, \tag{2.2}$$

which is related by duality [31, 37] to the 2+1 BH-AdS spacetime, Eq.(2.1). It has two horizons $\tilde{r}_\pm=r_\pm$, the same as the metric, Eq.(2.1), while the static limit is $\tilde{r}_{\text{erg}}=J/(2\sqrt{M})$.

We first investigate the string propagation in the $2+1$ BH-AdS background by considering the perturbation series around the exact center of mass of the string:

$$x^\mu(\tau,\sigma)=q^\mu(\tau)+\eta^\mu(\tau,\sigma)+\xi^\mu(\tau,\sigma)+... \tag{2.3}$$

The original method of Refs.[7, 8] can be conveniently formulated in covariant form. It is useful to introduce $D-1$ normal vectors n^μ_R $(R=1,..,D-1)$, (which can be chosen to be covariantly constant by gauge fixing), and consider comoving perturbations δx_R, i.e. those seen by an observer travelling with the center of mass, thus $\eta^\mu=\delta x^R n^\mu_R$. After Fourier expansion, $\delta x^R(\tau,\sigma)=\sum_n C^R_n(\tau)e^{-in\sigma}$, the first order perturbations satisfy the matrix Schrödinger-type equation in τ:

$$\ddot{C}_{nR}+(n^2\delta_{RS}-R_{\mu\rho\sigma\nu}n^\mu_R n^\nu_S\dot{q}^\rho\dot{q}^\sigma)C^S_n=0. \tag{2.4}$$

Second order perturbations ξ^μ and constraints are similarly covariantly treated, ξ^μ also satisfying Schrödinger-type equations with source terms, see Ref.[26].

For our purposes here it is enough to consider the non-rotating $(J=0)$ 2+1 BH-AdS background and a radially infalling string. We have solved completely the c.m. motion $q^\mu(\tau)$ and the first and second order perturbations $\eta^\mu(\tau,\sigma)$ and $\xi^\mu(\tau,\sigma)$ in this background. Eqs.(2.4) become:

$$\ddot{C}_{nR}+(n^2+\frac{m^2}{l^2})C_{nR}=0;\quad R=\perp,\parallel \tag{2.5}$$

The first order perturbations are independent of the black hole mass, only the anti de Sitter (AdS) part emerges. All oscillation frequencies $\omega_n=\sqrt{n^2+m^2/l^2}$ are real; there are no unstable modes in this case. The perturbations:

$$\delta x_R(\tau,\sigma)=\sum_n[A_{nR}e^{-i(n\sigma+\omega_n\tau)}+\tilde{A}_{nR}e^{-i(n\sigma-\omega_n\tau)}] \tag{2.6}$$

are completely finite and regular. This is also true for the second order perturbations which are bounded everywhere, even for $r\to 0$ $(\tau\to 0)$.

We have also computed the conformal generators L_n, (see Ref.[26]), and the string mass:

$$m^2 = 2\sum_n (2n^2 + \frac{m^2}{l^2})[A_{n\|}\tilde{A}_{-n\|} + A_{n\perp}\tilde{A}_{-n\perp}]. \tag{2.7}$$

The mass formula is modified (by the term m^2/l^2) with respect to the usual flat spacetime expression. This is due to the asymptotic (here AdS) behaviour of the spacetime. In ordinary $D \geq 4$ black hole spacetimes (without cosmological constant), which are asymptotically flat, the mass spectrum is the same as in flat Minkowski spacetime [15].

We compare with the string perturbations in the ordinary ($D \geq 4$) black hole anti de Sitter spacetime. In this case Eqs.(2.4) become:

$$\ddot{C}_{nS\perp} + (n^2 + m^2H^2 + \frac{Mm^2}{r^3})C_{nS\perp} = 0, \quad S = 1,2 \tag{2.8}$$

$$\ddot{C}_{n\|} + (n^2 + m^2H^2 - \frac{2Mm^2}{r^3})C_{n\|} = 0. \tag{2.9}$$

The transverse $\perp$-perturbations are oscillating with real frequencies and are bounded even for $r \rightarrow 0$. For longitudinal $\|$-perturbations, however, imaginary frequencies arise and instabilities develop. The ($\mid n \mid= 1$)-instability sets in at:

$$r_{\text{inst.}} = (\frac{2Mm^2}{1+m^2H^2})^{1/3} \tag{2.10}$$

Lower modes become unstable even outside the horizon, while higher modes develop instabilities at smaller r and eventually only for $r \approx 0$. For $r \rightarrow 0$ (which implies $\tau \rightarrow \tau_0$) we find $r(\tau) \approx (3m\sqrt{M/2})^{2/3}(\tau_0 - \tau)^{2/3}$ and:

$$\ddot{C}_{nS\perp} + \frac{2}{9(\tau-\tau_0)^2}C_{nS\perp} = 0, \quad S = 1,2 \tag{2.11}$$

$$\ddot{C}_{n\|} - \frac{4}{9(\tau-\tau_0)^2}C_{n\|} = 0. \tag{2.12}$$

For $\tau \rightarrow \tau_0$ the $\|$-perturbations blow up while the string ends trapped into the $r = 0$ singularity. We see the important difference between the string evolution in the 2+1 BH-AdS background and the ordinary 3+1 (or higher dimensional) black hole anti de Sitter spacetime.

We also compare with the string propagation in the 2+1 black string background, Eq.(2.2) (with $J = 0$). In this case, Eqs.(2.4) become:

$$\ddot{C}_{n\perp} + n^2C_{n\perp} = 0, \tag{2.13}$$

$$\ddot{C}_{n\|} + (n^2 - \frac{2m^2M}{lr})C_{n\|} = 0. \tag{2.14}$$

The $\perp$-modes are stable, while $C_{n\|}$ develop instabilities. For $r \to 0$ (which implies $\tau \to \tau_0$) we find $r(\tau) \approx \frac{m^2 M}{l}(\tau_0 - \tau)^2$ and:

$$\ddot{C}_{n\|} - \frac{2}{(\tau_0 - \tau)^2} C_{n\|} = 0, \tag{2.15}$$

with similar conclusions as for the ordinary 3+1 (or higher dimensional) black hole anti de Sitter spacetime.

In order to extract more information about the string evolution in these backgrounds, in particular exact properties, we consider the circular string ansatz:

$$t = t(\tau), \quad r = r(\tau), \quad \phi = \sigma + f(\tau), \tag{2.16}$$

in the equatorial plane ($\theta = \pi/2$) of the stationary axially symmetric backgrounds:

$$ds^2 = g_{tt}(r)dt^2 + g_{rr}(r)dr^2 + 2g_{t\phi}(r)dtd\phi + g_{\phi\phi}(r)d\phi^2. \tag{2.17}$$

This includes all the cases of interest here: The 2+1 BH-AdS spacetime, the black string, as well as the equatorial plane of ordinary Einstein black holes. The string dynamics, determined by Eqs.(1.1)-(1.2), is then reduced to a system of second order ordinary differential equations and constraints, also described as a Hamiltonian system:

$$\dot{r}^2 + V(r) = 0; \qquad V(r) = g^{rr}(g_{\phi\phi} + E^2 g^{tt}), \tag{2.18}$$

$$\dot{t} = -Eg^{tt}, \qquad \dot{f} = -Eg^{t\phi}; \quad E = -P_t = \text{const.}, \tag{2.19}$$

which in all backgrounds considered here are solved in terms of either elementary or elliptic functions. The dynamics of the circular strings takes place at the r-axis in the $(r, V(r))$-diagram and from the properties of the potential $V(r)$ (minima, zeroes, asymptotic behaviour for large r and the value $V(0)$), general knowledge about the string motion can be obtained. On the other hand, the line element of the circular string turns out to be:

$$ds^2 = g_{\phi\phi}(d\sigma^2 - d\tau^2), \quad \text{i.e.} \quad S(\tau) = \sqrt{g_{\phi\phi}(r(\tau))}, \tag{2.20}$$

$S(\tau)$ being the invariant string size. For all the static black hole AdS spacetimes (2+1 and higher dimensional): $S(\tau) = r(\tau)$.

For the rotating 2+1 BH-AdS spacetime:

$$V(r) = r^2(\frac{r^2}{l^2} - M) + \frac{J^2}{4} - E^2, \tag{2.21}$$

(see Fig.1). $V(r)$ has a global minimum $V_{\text{min}} < 0$ between the two horizons r_+, r_- (for $Ml^2 \geq J^2$, otherwise there are no horizons). The vanishing of $V(r)$ at $r = r_{01,2}$ (see Ref.[26]) determines three different types of solutions: (i) For $J^2 > 4E^2$, there are two positive zeroes $r_{01} < r_{02}$, the string never comes outside the static limit, never falls into $r = 0$ neither (there is a barrier between $r = r_{01}$ and $r = 0$). The mathematical solution oscillates between r_{01} and r_{02} with $0 < r_{01} < r_- < r_+ < r_{02} < r_{\rm erg}$. It may be interpreted as a string travelling between the different universes described by the maximal analytic extension of the manifold. (ii) For $J^2 < 4E^2$, there is only one positive zero r_0 outside the static limit and there is no barrier preventing the string from collapsing into $r = 0$. The string starts at $\tau = 0$ with maximal size $S_{\rm max}^{(ii)}$ outside the static limit, it then contracts through the ergosphere and the two horizons and eventually collapses into a point $r = 0$. For $J \neq 0$, it may be still possible to continue this solution into another universe like in the case (i). (iii) $J^2 = 4E^2$ is the limiting case where the maximal string size equals the static limit: $S_{\rm max}^{(iii)} = l\sqrt{M}$. In this case $V(0) = 0$, thus the string contracts through the two horizons and eventually collapses into a point $r = 0$.

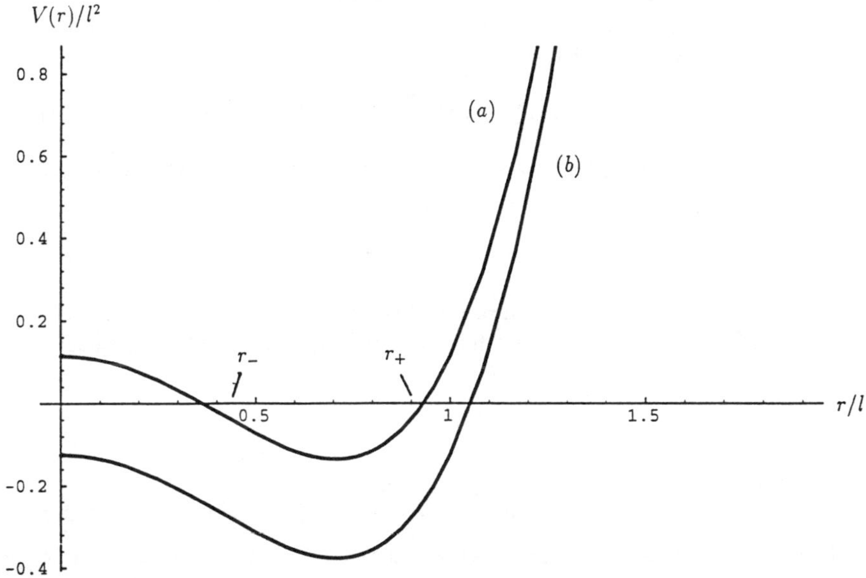

Figure 1. The potential $V(r)$, Eq.(2.21), for a circular string in the $2+1$ BH-AdS spacetime. In (a) we have $J^2 > 4E^2$ and a barrier between the inner horizon and $r = 0$, while (b) represents a case where $J^2 < 4E^2$ and a string will always fall into $r = 0$. The static limit is $r_{\rm erg} = l$.

The exact general solution in the three cases (i)-(iii) is given by:

$$r(\tau) = \mid r_m - \frac{1}{c_1\wp(\tau-\tau_0)+c_2} \mid, \tag{2.22}$$

where:

$$r_m = S_{\max} = \sqrt{\frac{Ml^2}{2}}\sqrt{1+\sqrt{1-\frac{4V(0)}{Ml^2}}}, \quad V(0) = \frac{J^2}{4} - E^2. \tag{2.23}$$

c_1, c_2 are constants in terms of (l, M, r_m), given in Ref.[26], and $\wp$ is the Weierstrass elliptic $\wp$-function with invariants (g_2, g_3), discriminant Δ and roots (e_1, e_2, e_3), also given in Ref.[26]. The three cases (i)-(iii) correspond to the cases $\Delta > 0$, $\Delta < 0$ and $\Delta = 0$, respectively. Notice that $S^{(ii)}_{\max} > S^{(iii)}_{\max} = l\sqrt{M} > S^{(i)}_{\max}$. In the case (i), $r(\tau)$ can be written in terms of the Jacobian elliptic function $\text{sn}[\tau^*, k]$, $\tau^* = \sqrt{e_1 - e_3}\,\tau$, $k = \sqrt{(e_2-e_3)/(e_1-e_3)}$. It follows that the solution (i) oscillates between the two zeroes r_{01} and r_{02} of $V(r)$, with period 2ω, where ω is the real semi-period of the Weierstrass function: $\omega = K(k)/\sqrt{e_1-e_3}$, in terms of the complete elliptic integral of the first kind $K(k)$. We have:

$$r(0) = r_m, \quad r(\omega) = \sqrt{Ml^2 - r_m^2}, \quad r(2\omega) = r_m, \ldots \tag{2.24}$$

In the case (ii) $(\Delta < 0)$ two roots (e_1, e_3) become complex, the string collapses into a point $r = 0$ and we have:

$$r(0) = r_m, \quad r(\frac{\omega_2}{2}) = 0, \quad r(\omega_2) = r_m, \ldots \tag{2.25}$$

where ω_2 is the real semi-period of the Weierstrass function for this case. In the case (iii) $(\Delta = 0)$ the elliptic functions reduce to hyperbolic functions:

$$r(\tau) = \frac{\sqrt{M}l}{\cosh(\sqrt{M}\tau)}, \tag{2.26}$$

so that:

$$r(-\infty) = 0, \quad r(0) = r_m = \sqrt{M}l, \quad r(+\infty) = 0. \tag{2.27}$$

Here, the string starts as a point, grows until $r = r_m$ (at $\tau = 0$), and then it contracts until it collapses again $(r = 0)$ at $\tau = +\infty$. In this case the string makes only one oscillation between $r = 0$ and $r = r_m$.

Notice that for the static background $(J = 0)$, the only allowed motion is (ii), i.e. $r_m > r_{\text{hor}} = \sqrt{M}l$ (there is no ergosphere and only one horizon in this case), with:

$$r_m = \sqrt{\frac{Ml^2}{2}}\sqrt{1+\sqrt{1+\frac{4E^2}{Ml^2}}}. \tag{2.28}$$

For $J = 0$, the string collapses into $r = 0$ and stops there. The Penrose diagram of the 2+1 BH-AdS spacetime for $J = 0$ is very similar to the Penrose diagram of the ordinary ($D \geq 4$) Schwarzschild spacetime, so the string motion outwards from $r = 0$ is unphysical because of the causal structure. The coordinate time $t(\tau)$ can be expressed in terms of the incomplete elliptic integral of the third kind Π, see Ref.[26]. The string has its maximal size r_m at $\tau = 0$, passes the horizon at $\tau = \tau_{\text{hor}}$ (expressed in terms of an incomplete elliptic integral of the first kind) and falls into $r = 0$ for $\tau = \omega_2/2$, ω_2 being the real semi-period of the Weierstrass function. That is, we have:

$$r(0) = r_m, \quad r(\tau_{\text{hor}}) = \sqrt{M}l, \quad r(\omega_2/2) = 0,$$
$$t(0) = 0, \quad t(\tau_{\text{hor}}) = \infty, \tag{2.29}$$

and $t(\omega_2/2)$ can be expressed in terms of the Jacobian zeta function Z, see Ref.[26].

We also study the circular strings in the ordinary $D \geq 4$ spacetimes. In the generic 3+1 Kerr anti de Sitter (or Kerr de Sitter) spacetime, the potential $V(r)$ is however quite complicated, see Ref.[26]. The general circular string solution involves higher genus elliptic functions and it is not necessary to go into details here. We will compare with the non-rotating cases, only.

It is instructive to recall [38] the circular string in Minkowski spacetime (Min), for which $V(r) = r^2 - E^2$ (Fig.2a), the string oscillates between its maximal size $r_m = E$, and $r = 0$ with the solution $r(\tau) = r_m \mid \cos\tau \mid$.

In the Schwarzschild black hole (S) $V(r) = r^2 - 2Mr - E^2$ (Fig.2b), the solution is remarkably simple: $r(\tau) = M + \sqrt{M^2 + E^2}\cos\tau$. The mathematical solution oscillates between $r_m = M + \sqrt{M^2 + E^2}$ and $M - \sqrt{M^2 + E^2} < 0$, but because of the causal structure and the curvature singularity the motion can not be continued after the string has collapsed into $r = 0$.

For anti de Sitter spacetime (AdS), we find: $V(r) = r^2(1 + H^2r^2) - E^2$ (Fig.2c). The string oscillates between $r_m = \frac{1}{\sqrt{2}H}\sqrt{-1 + \sqrt{1 + 4H^2E^2}}$ and $r = 0$ with the solution:

$$r(\tau) = r_m \mid \text{cn}[(1 + 4H^2E^2)^{1/4}, k] \mid, \tag{2.30}$$

which is periodic with period 2ω :

$$\omega = \frac{K(k)}{(1 + 4H^2E^2)^{1/4}}, \quad k = \sqrt{\frac{\sqrt{1 + 4H^2E^2} - 1}{2\sqrt{1 + 4H^2E^2}}}. \tag{2.31}$$

For Schwarzschild anti de Sitter spacetime (S-AdS), we find $V(r) = r^2(1 + H^2r^2) - 2Mr - E^2$ (Fig.2d) and:

$$r(\tau) = r_m - \frac{1}{d_1\wp(\tau) + d_2}, \quad r(0) = r_m \tag{2.32}$$

d_1, d_2 are constants given in terms of (M, H, r_m), see Ref.[26] (r_m is the root of the equation $V(r) = 0$, which has in this case exactly one positive solution). The invariants, the discriminant and the roots are also given explicitly in

Ref.[26]. The string starts with $r = r_m$ at $\tau = 0$, it then contracts and eventually collapses into the $r = 0$ singularity. The existence of elliptic function solutions for the string motion is characteristic of the presence of a cosmological constant. For $\Lambda = 0 = -3H^2$ the circular string motion reduces to simple trigonometric functions. From Fig.2 and our analysis we see that the circular string motion is qualitatively very similar in all these backgrounds (Min, S, AdS, S-AdS): the string has a maximal *bounded* size and then it contracts towards $r = 0$. There are however physical and quantitative differences: in Minkowski and pure anti de Sitter spacetimes, the string truly oscillates between r_m and $r = 0$, while in the black hole cases (S, S-AdS), there is only one half oscillation, the string motion stops at $r = 0$. This also holds for the 2+1 BH-AdS spacetime with $J = 0$ (Fig.1b). Notice also that in all these cases, $V(0) = -E^2 < 0$ and $V(r) \sim r^\alpha$; $(\alpha = 2, 4)$ for $r >> E$.

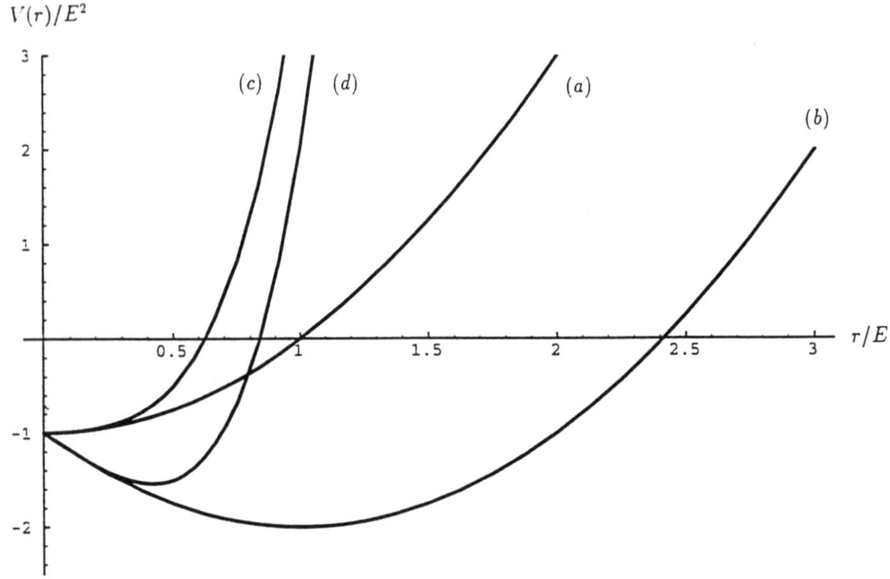

Figure 2. The potential $V(r)$, Eq.(2.18), for a circular string in the equatorial plane of the four $3+1$ dimensional spacetimes: (a) Minkowski (Min), (b) Schwarzschild black hole (S), (c) anti de Sitter (AdS), and (d) Schwarzschild anti de Sitter space (S-AdS).

The similarity can be pushed one step further by considering small perturbations around the circular strings. We find [18]:

$$\ddot{C}_n + (n^2 + \frac{r}{2}\frac{da(r)}{dr} + \frac{r^2}{2}\frac{d^2a(r)}{dr^2} - \frac{2E^2}{r^2})C_n = 0, \tag{2.33}$$

determining the Fourier components of the comoving perturbations. For the spacetimes of interest here, $a(r) = 1 - 2M/r + H^2r^2$ (Min, S, AdS, S-AdS), or $a(r) = r^2/l^2 - 1$ (2+1 BH-AdS), the comoving perturbations are regular except near $r = 0$, where we find (for all cases) $r(\tau) \approx -E(\tau - \tau_0)$ and:

$$\ddot{C}_n - \frac{2}{(\tau - \tau_0)^2}C_n = 0. \tag{2.34}$$

It follows that not only the unperturbed circular strings, but also the comoving perturbations around them behave in a similar way in all these non-rotating backgrounds (2+1 and higher dimensional). This should be contrasted with the string perturbations around the center of mass, which behave differently in these backgrounds. It must be noticed that for rotating ($J \neq 0$) spacetimes, the circular string behaviour is qualitatively different from the non-rotating ($J = 0$) spacetimes. For large J, both in the 2+1 BH-AdS as well as in the 3+1 ordinary Kerr-AdS spacetimes, non-collapsing circular string solutions exist. The potential $V(r) \to +\infty$ for $r \to 0$ and no collapse into $r = 0$ is possible.

The dynamics of circular strings in curved spacetimes is determined by the string tension, which tends to contract the string, and by the local gravity (which may be attractive or repulsive). In all the previous non-rotating backgrounds, the local gravity is attractive (i.e. $da(r)/dr > 0$), and it acts together with the string tension in the sense of contraction. But in spacetimes with regions in which repulsion (i.e. $da(r)/dr < 0$) dominates, the strings can expand with unbounded radius (unstable strings [11, 13]). It may also happen that the string tension and the local gravity be of the same order, i.e. the two opposite effects can balance, and the string is stationary. De Sitter spacetime provides an example in which all such type of solutions exist [11, 13]. In de Sitter spacetime, $V(r)$ is unbounded from below for $r \to \infty$ ($V(r) \sim -r^4$) and unbounded expanding circular strings are present. In addition, an interesting new feature appears in the presence of a positive cosmological constant: the existence of *multi-string* solutions [9-11]. The world-sheet time τ turns out to be a multi (finite or infinite) valued function of the physical time. That is, one single world-sheet where $-\infty \leq \tau \leq +\infty$, can describe many (even infinitely many [13]) different and independent strings (in flat spacetime, one single world-sheet describes only one string). In the S, AdS and S-AdS spacetimes, the multi-string feature for circular strings is absent.

The main conclusions of this section are given in Tables 1, 2. Further details can be found in Ref.[26]

Table 1. String motion described by the string perturbation series approach in the 2+1 BH-AdS, ordinary black hole-AdS, de Sitter (dS) and black string backgrounds.

String Perturbation Series Approach
$x^\mu(\tau,\sigma) = q^\mu(\tau) + \eta^\mu(\tau,\sigma) + \xi^\mu(\tau,\sigma) + \ldots$ $\eta^\mu = n^\mu_R \delta x^R, \quad \delta x^R(\tau,\sigma) = \sum_n C^R_n(\tau) e^{-in\sigma}; \; R = 1, \ldots, (D-1)$ $\ddot{C}_{nR} + (n^2 \delta_{RS} - R_{\mu\rho\sigma\nu} n^\mu_R n^\nu_S \dot{q}^\rho \dot{q}^\sigma) C^S_n = 0$

2+1 Black Hole-ADS	Ordinary ($D \geq 4$) Black Hole-ADS
$\ddot{C}_{n\perp} + (n^2 + \frac{m^2}{l^2}) C_{n\perp} = 0$ $\ddot{C}_{n\parallel} + (n^2 + \frac{m^2}{l^2}) C_{n\parallel} = 0$ No instability, $\omega_n = \sqrt{n^2 + \frac{m^2}{l^2}}$ δx^R, ξ^μ bounded everywhere. $\delta x_R = \sum_n [A_{nR} e^{-i(n\sigma + \omega_n \tau)} + \tilde{A}_{nR} e^{-i(n\sigma - \omega_n \tau)}]$ $m^2 = 2 \sum_n (2n^2 + \frac{m^2}{l^2}) \sum_R A_{nR} \tilde{A}_{-nR}$	$\ddot{C}_{n\perp} + (n^2 + m^2 H^2 + \frac{Mm^2}{r^3}) C_{n\perp} = 0$ $\ddot{C}_{n\parallel} + (n^2 + m^2 H^2 - \frac{2Mm^2}{r^3}) C_{n\parallel} = 0$ Instability, $r_{\text{inst}} = (\frac{2Mm^2}{1 + m^2 H^2})^{1/3}$ Unbounded perturbations $C_{n\parallel}$ for $r(\tau) \approx (3m\sqrt{M/2})^{2/3} (\tau_0 - \tau)^{2/3} \to 0$: $\ddot{C}_{n\parallel} - \frac{4}{9(\tau_0 - \tau)^2} C_{n\parallel} = 0$ $m^2(H = 0) = 4 \sum_n n^2 \sum_R A_{nR} \tilde{A}_{-nR}$

Ordinary ($D \geq 4$) de Sitter	Black String
$\ddot{C}_{nR} + (n^2 - m^2 H^2) C_{nR} = 0$ Unbounded modes for $\mid n \mid < mH$ Instability, $\omega_n = \sqrt{n^2 - m^2 H^2}$ $\delta x_R = \sum_n [A_{nR} e^{-i(n\sigma + \omega_n \tau)} + \tilde{A}_{nR} e^{-i(n\sigma - \omega_n \tau)}]$ $m^2 = 2 \sum_n (2n^2 - m^2 H^2) \sum_R A_{nR} \tilde{A}_{-nR}$	$\ddot{C}_{n\perp} + n^2 C_{n\perp} = 0$ $\ddot{C}_{n\parallel} + (n^2 - \frac{2Mm^2}{lr}) C_{n\parallel} = 0$ Instability, $r_{\text{inst}} = \frac{2Mm^2}{l}$ Unbounded perturbations $C_{n\parallel}$ for $r(\tau) \approx \frac{Mm^2}{l} (\tau_0 - \tau)^2 \to 0$: $\ddot{C}_{n\parallel} - \frac{2}{(\tau_0 - \tau)^2} C_{n\parallel} = 0$ $m^2 = 4 \sum_n n^2 \sum_R A_{nR} \tilde{A}_{-nR}$

Table 2. Circular exact string solutions in the indicated backgrounds. $S(\tau)$ is the invariant string size.

Exact Circular Strings

$$ds^2 = g_{tt}(r)dt^2 + g_{rr}(r)dr^2 + 2g_{t\phi}(r)dtd\phi + g_{\phi\phi}d\phi^2$$
$$t = t(\tau), \quad r = r(\tau), \quad \phi = \sigma + f(\tau), \quad (\theta = \pi/2)$$
$$\dot{r}^2 + V(r) = 0; \quad V(r) = g^{rr}(E^2 g^{tt} + g_{\phi\phi}), \quad \dot{t} = -Eg^{tt}, \quad \dot{f} = -Eg^{t\phi}$$
$$ds^2 = g_{\phi\phi}(-d\tau^2 + d\sigma^2), \quad \text{i.e.} \quad S(\tau) = \sqrt{g_{\phi\phi}(r(\tau))}$$

2+1 Black Hole ADS	Ordinary ($D \geq 4$) Black Hole ADS
$V(r) = r^2(r^2/l^2 - M) + J^2/4 - E^2$ $V(0) = J^2/4 - E^2, \quad V(r \to \infty) \propto r^4 \to \infty$ $S(\tau) = r(\tau) = r_m - (c_1 \wp(\tau - \tau_0) + c_2)^{-1}$ $S_{\text{max}} = r_m = \sqrt{\frac{Ml^2}{2}}\sqrt{1 + \sqrt{1 - \frac{4V(0)}{Ml^2}}}$ For $J^2 > 4E^2$, $S_{\text{min}} > 0$, no collapse. For $J^2 \leq 4E^2$, $S_{\text{min}} = 0$, collapse. No unbounded string size. No multi-string solutions.	$V(r) = r^2(1 + H^2 r^2) - 2Mr - E^2$ $V(0) = -E^2, \quad V(r \to \infty) \propto r^4 \to \infty$ $S(\tau) = r(\tau) = r_m - (c_1 \wp(\tau - \tau_0) + c_2)^{-1}$ $V(r_m) \equiv 0; \quad S_{\text{max}} = r_m = r_m(M, H, E)$ String contracts from $r(0) = r_m$ until it collapses into the $r = 0$ singularity. No unbounded string size. No multi-string solutions.
Ordinary ($D \geq 4$) Black Hole DS	**Black String**
$V(r) = r^2(1 - H^2 r^2) - 2Mr - E^2$ $V(0) = -E^2, \quad V(r \to \infty) \propto -r^4 \to -\infty$ Contracting, expanding or stationary solutions, depending on the balance between the string tension and the local gravity. $M = 0: \; S^2(\tau) = H^{-2}\wp(\tau - \tau_0) + H^{-2}/3$ Contracting, expanding or oscillating. Unbounded string size. Multi-string solutions.	$\tilde{V}(r) = J^2/(4r^2) - M/r^2 + 1/l^2 - E^2$ $\tilde{V}(0) = \infty \; (J \neq 0), \; \tilde{V}(0) = -\infty \; (J = 0)$ $\tilde{S}(\tau) = 1/r(\tau), \quad \tilde{V}(\infty) = 1/l^2 - E^2$ For $J \neq 0$, all solutions are bounded (finite $\tilde{S}$). For $J = 0$ unbounded size solutions exist as well. $J = 0, El = 1: \; \tilde{S}(\tau) = (\mid 2\sqrt{M}\tau \mid)^{-1/2}$ $\tilde{S}(-\infty) = 0, \; \tilde{S}(0) = \infty, \; \tilde{S}(\infty) = 0$ Unbounded string size. Multi-string solutions.

3 Quantum String Spectrum in Ordinary AdS

In the first part of the previous section, the classical string dynamics was studied in the $2+1$ BH-AdS spacetime. In this section we go further in the investigation of the physical properties of strings in AdS spacetime, by performing string quantization. Since the 2+1 BH-AdS spacetime is locally AdS, the results of Section 2 can be extended to classical strings in D-dimensional AdS spacetime as well. In AdS spacetime, the string motion is oscillatory in time and is *stable*; all fluctuations around the string center of mass are well behaved and bounded. Local gravity of AdS spacetime is always negative and string instabilities do not develop. The string perturbation series approach, considering fluctuations around the center of mass, is particularly appropriate in AdS spacetime, the natural dimensionless expansion parameter being $\lambda = \alpha'/l^2 > 0$, where α' is the string tension and the 'Hubble constant' $H = 1/l$. The negative cosmological constant of AdS spacetime is related to the 'Hubble constant' H by $\Lambda = -(D-1)(D-2)H^2/2$.

All the spatial ($\mu = 1, ..., D-1$) modes in D-dimensional AdS, oscillate with frequency $\omega_n = \sqrt{n^2 + m^2\alpha'\lambda}$, which are real for all n (m being the string mass). In this section, we perform a canonical quantization procedure. From the conformal generators L_n, $\tilde{L}_n$ and the constraints $L_0 = \tilde{L}_0 = 0$, imposed at the quantum level, we obtain the mass formula (the quantum version of Eq.(2.7) written in units where α' appears explicitly):

$$m^2\alpha' = (D-1)\sum_{n>0}\Omega_n(\lambda) + \sum_{n>0}\Omega_n(\lambda)\sum_{R=1}^{D-1}[(\alpha_n^R)^\dagger\alpha_n^R + (\tilde{\alpha}_n^R)^\dagger\tilde{\alpha}_n^R], \qquad (3.1)$$

where:

$$\Omega_n(\lambda) = \frac{2n^2 + m^2\alpha'\lambda}{\sqrt{n^2 + m^2\alpha'\lambda}}, \qquad (3.2)$$

and we have applied symmetric ordering of the operators. The operators α_n^R, $\tilde{\alpha}_n^R$ satisfy:

$$[\alpha_n^R, (\alpha_n^R)^\dagger] = [\tilde{\alpha}_n^R, (\tilde{\alpha}_n^R)^\dagger] = 1, \qquad \text{for all } n > 0. \qquad (3.3)$$

For $\lambda << 1$, which is clearly fulfilled in most interesting cases, we have found the lower mass states $m^2\alpha'\lambda << 1$ and the quantum mass spectrum [27]. Physical states are characterized by the eigenvalue of the number operator:

$$N = \frac{1}{2}\sum_{n>0} n \sum_{R=1}^{D-1}[(\alpha_n^R)^\dagger\alpha_n^R + (\tilde{\alpha}_n^R)^\dagger\tilde{\alpha}_n^R], \qquad (3.4)$$

and the ground state is defined by:

$$\alpha_n^R \mid 0 >= \tilde{\alpha}_n^R \mid 0 >= 0, \qquad \text{for all } n > 0. \qquad (3.5)$$

We find that $m^2\alpha' = 0$ is an *exact* solution of the mass formula in $D = 25$ and that there is a graviton at $D = 25$, which indicates, as in de Sitter space [7], that the critical dimension of AdS is 25 (although it should be stressed that the question whether these spacetimes are solutions to the full β-function equations remains open). As in Minkowski spacetime, the ground state is a tachyon. Remarkably enough, for $N \geq 2$ we find that a generic feature of all excited states beyond the graviton, is the presence of a *fine structure* effect: for a given eigenvalue $N \geq 2$, the corresponding states have different masses. For the lower mass states the expectation value of the mass operator in the corresponding states (generically labelled $\mid j >$) turns out to have the form (see Ref.[27]):

$$< j \mid m^2\alpha' \mid j >_{\text{AdS}} = a_j + b_j\lambda^2 + c_j\lambda^3 + \mathcal{O}(\lambda^4). \tag{3.6}$$

The collective index "j" generically labels the state $\mid j >$ and the coefficients a_j, b_j, c_j are all well computed numbers, different for each state, their precise values are given in Ref.[27]. The corrections to the mass in Minkowski spacetime appear to order λ^2. Therefore, the leading Regge trajectory for the lower mass states is:

$$J = 2 + \frac{1}{2}m^2\alpha' + \mathcal{O}(\lambda^2). \tag{3.7}$$

In Minkowski spacetime the mass and number operator of the string are related by $m^2\alpha' = -4 + 4N$. In AdS (as well as in de Sitter (dS)), there is no such simple relation between the mass and the number operators; the splitting of levels increases considerably for very large N. The *fine structure* effect we find here is also present in de Sitter space. Up to order λ^2, the lower mass states in dS and in AdS are the same, the differences appear to the order λ^3. The lower mass states in de Sitter spacetime are given by Eq.(3.6) but with the λ^3-term getting an opposite sign $(-c_j\lambda^3)$.

For the very high mass spectrum, we find more drastic effects. States with very large eigenvalue N, namely $N >> 1/\lambda$, have masses:

$$< j \mid m^2\alpha' \mid j >_{\text{AdS}} \approx d_j\lambda N^2 \tag{3.8}$$

and angular momentum:

$$J^2 \approx \frac{1}{\lambda}m^2\alpha', \tag{3.9}$$

where d_j are well computed numbers different for each state, their precise values are given in Ref.[27]. Since $\lambda = \alpha'/l^2$, we see from Eq.(3.8) that the masses of the high mass states are *independent* of α'. In Minkowski spacetime, very large N states all have the same mass $m^2\alpha' \approx 4N$, but here in AdS the masses of the high mass states with the same eigenvalue N are *different* by factors d_j. In addition, because of the fine structure effect, states with different N can get mixed up. For high mass states, the level spacing *grows* with N (instead of being constant as in Minkowski spacetime). As a consequence, the density of states $\rho(m)$ as a function of mass grows like $\text{Exp}[(m/\sqrt{\mid \Lambda \mid})^{1/2}]$ (instead of $\text{Exp}[m\sqrt{\alpha'}]$ as in Minkowski spacetime), and *independently* of α'. The partition function for a gas of strings at a temperature β^{-1} in AdS spacetime is well defined for all finite temperatures β^{-1}, discarding the existence of the Hagedorn temperature.

For the low mass states ($m^2\alpha'\lambda << 1$) in anti de Sitter spacetime, our results can be written as [27]:

$$< j \mid m^2(\alpha', l) \mid j >= \frac{4(N-1)}{\alpha'} + \frac{1}{l^2}\sum_{n=0}^{\infty} a_{jn}(\alpha'/l^2)^n, \qquad (3.10)$$

where "j" is a collective index labelling the state $\mid j >$. It is now important to notice that $a_{j0} = 0$ for *all* the low mass states [27], i.e. there is no "constant" term on the right hand side of Eq.(3.10). A non-zero a_{j0}-term would give rise to a α'-independent contribution to the string mass. Its absense, on the other hand, means that the first term on the right hand side of Eq.(3.10) is super-dominant (since, in all cases, $\alpha'/l^2 = \lambda << 1$) and that the string scale is therefore set by $1/\alpha'$.

For the high mass states ($m^2\alpha'\lambda >> 1$) we found instead [27]:

$$< j \mid m^2(\alpha', l) \mid j > \approx \frac{d_j}{l^2} N^2, \qquad \text{for } N >> l^2/\alpha' \qquad (3.11)$$

where the number d_j depends on the state. The right hand side of Eq.(3.11) is exactly like a non-zero dominant a_{j0}-term in Eq.(3.10). For the high mass states the scale is therefore set by $1/l^2$ which is equal to the absolute value of the cosmological constant Λ (up to a geometrical factor) and *independent* of α'. This suggests that for $\lambda << 1$, the masses of *all* string states can be represented by a formula of the form (3.10). For the low mass states $a_{j0} = 0$, while for the high mass states a_{j0} becomes a large positive number.

In the black string background, Eq.(2.2), we have calculated explicitly the first and second order string fluctuations around the center of mass [27]. We then determined the world-sheet energy-momentum tensor and we derived the mass formula in the asymptotic region. The mass spectrum is equal to the mass spectrum in flat Minkowski spacetime. Therefore, for a gas of strings at temperature β^{-1} in the asymptotic region of the black string background, the partition function goes like:

$$Z(\beta) \sim \int^{\infty} dm\; e^{-m(\beta-\sqrt{\alpha'})}, \qquad (3.12)$$

which is only defined for $\beta > \sqrt{\alpha'}$, i.e. there is a Hagedorn temperature:

$$T_{\rm Hg} = (\alpha')^{-1/2}. \qquad (3.13)$$

In higher dimensional ($D \geq 4$) black hole spacetimes the next step now would be to set up a scattering formalism, where a string from an asymptotic in-state interacts with the gravitational field of the black hole and reappears in an asymptotic out-state [15]. However, this is not possible in the black string background. In the uncharged black string background under consideration here, *all* null and timelike geodesics incoming from spatial infinity pass through the horizon and fall into the physical singularity [36]. No "angular momentum", as in the case of scattering off the ordinary Schwarzschild black hole, can prevent a point particle from falling into the singularity. The string solutions considered here are based on perturbations around the string center of mass which follows, at least approximately, a point particle geodesic. Therefore, a string incoming from spatial infinity inevitably falls into the singularity in the black string background.

Table 3. Characteristic features of the quantum string mass spectrum in anti de Sitter (AdS) and de Sitter (dS) spacetimes.

Quantum Strings	
Anti de Sitter spacetime (AdS)	de Sitter spacetime (dS)
Classical motion is stable and oscillatory in time with real frequencies $\omega_n = \sqrt{n^2 + m^2\alpha'^2 H^2}$	Classical motion is unstable with frequencies $\omega_n = \sqrt{n^2 - m^2\alpha'^2 H^2}$ Unbounded string size and energy for large de Sitter radius, $R \to \infty$.
The mass formula is well defined for all m. There is an infinite number of states with arbitrary high masses. $m^2\alpha' = 0$ is an exact solution at $D = 25$.	Real mass solutions only for $m < 1/(\alpha' H)$. Finite number of states, $N_{\max} \approx 0.15/(\alpha' H^2)$. $m^2\alpha' = 0$ is an exact solution at $D = 25$.
The coupling to the gravitational background produces a Fine structure effect at all levels in the mass spectrum. The number of levels considerably increases with respect to flat space.	Fine structure effect appears in low mass spectrum. Is similar to AdS; the differences appear to order $(\alpha' H^2)^3$
For the high mass states: $< m^2 > \sim \mid \Lambda \mid N^2, \quad J^2 \sim m^2 / \mid \Lambda \mid$ Both are *independent* of α' ! The level spacing grows with N. $\rho(m) \sim \text{Exp}[m/\sqrt{\mid \Lambda \mid})^{1/2}]$, high m. No Hagedorn temperature exists.	The similar region of high mass states does not exist in the de Sitter spacetime.

4 New Classes of Exact String and Multi-String Solutions in Curved Spacetimes

In this section we return to the string equations of motion and constraints, Eqs.(1.1)-(1.2), to look for new classes of exact solutions. In most spacetimes, quite general families of exact solutions can be found by making an appropriate ansatz, which exploits the symmetries of the underlying curved spacetime. In axially symmetric spacetimes, a convenient ansatz corresponds to circular strings, as we saw in Section 2. Such an ansatz effectively decouples the dependence on the spatial world-sheet coordinate σ, and the string equations of motion and constraints reduce to non-linear coupled ordinary differential equations, Eqs.(2.18)-(2.19). In this section we will make instead an ansatz which effectively decouples the dependence on the temporal world-sheet coordinate τ. This ansatz, which we call the "stationary string ansatz" is dual to the "circular string ansatz" in the sense that it corresponds to a formal interchange of the world-sheet coordinates (τ, σ), as well as of the azimuthal angle ϕ and the stationary time t in the target space:

$$\tau \leftrightarrow \sigma, \qquad t \leftrightarrow \phi. \tag{4.1}$$

The stationary string ansatz will describe stationary strings when t (and τ) are timelike, for instance in anti de Sitter spacetime (in static coordinates) and outside the horizon of a Schwarzschild black hole. On the other hand, if t (and τ) are spacelike, for instance inside the horizon of a Schwarzschild black hole or outside the horizon of de Sitter spacetime (in static coordinates), the stationary string ansatz will describe dynamical propagating strings. Considering for simplicity a static line element in the form:

$$ds^2 = -a(r)dt^2 + \frac{dr^2}{a(r)} + r^2(d\theta^2 + \sin^2\theta d\phi^2), \tag{4.2}$$

the stationary string ansatz reads explicitly:

$$t = \tau, \quad r = r(\sigma), \quad \phi = \phi(\sigma), \quad \theta = \pi/2. \tag{4.3}$$

The string equations of motion and constraints, Eqs.(1.1)-(1.2), reduce to two separated first order ordinary differential equations:

$$\phi' = \frac{L}{r^2}, \qquad r'^2 + V(r) = 0; \quad V(r) = -a(r)[a(r) - \frac{L^2}{r^2}], \tag{4.4}$$

where L is an integration constant. The qualitative features of the possible string configurations can be read off directly from the shape of the potential $V(r)$. Thereafter, the detailed analysis of the quantitative features can be performed by explicitly solving the (integrable) system of equations (4.4). The induced line element on the world-sheet is given by:

$$ds^2 = a(r)(-d\tau^2 + d\sigma^2). \tag{4.5}$$

Thus, if $a(r)$ is negative, the world-sheet coordinate τ becomes spacelike while σ becomes timelike and the stationary string ansatz (4.3) describes a dynamical string. If $a(r)$ is positive, the ansatz describes a stationary string.

In this section we solve explicitly Eqs.(4.4) and we analyze in detail the solutions and their physical interpretation in Minkowski, de Sitter, anti de Sitter, Schwarzschild and 2 + 1 black hole anti de Sitter spacetimes. In all these cases, the solutions are expressed in terms of elliptic (or elementary) functions. We furthermore analyze the physical properties, energy and pressure, of these solutions. The energy and pressure densities of the strings can be obtained from the 3 + 1 dimensional spacetime energy-momentum tensor:

$$\sqrt{-g}T^{\mu\nu} = \frac{1}{2\pi\alpha'}\int d\tau d\sigma(\dot{X}^\mu \dot{X}^\nu - X'^\mu X'^\nu)\delta^{(4)}(X - X(\tau,\sigma)). \tag{4.6}$$

After integration over a spatial volume that completely encloses the string [39], the energy-momentum tensor takes the form of a fluid:

$$T^\mu{}_\nu = \text{diag.}(-E, P_1, P_2, P_3). \tag{4.7}$$

In Minkowski spacetime (Min), the potential defined in Eqs.(4.4) is given by:

$$V_{\text{M}}(r) = \frac{L^2}{r^2} - 1, \tag{4.8}$$

and the solution of Eqs.(4.4) describes one infinitely long straight string with "impact-parameter" L. The equation of state, relating energy and pressure densities, takes the well-known form Ref.[28]:

$$dE = -dP_2 = \text{const.}, \quad P_1 = P_3 = 0. \tag{4.9}$$

In anti de Sitter spacetime (AdS), the potential is given by:

$$V_{\text{AdS}}(r) = -(1 + H^2 r^2)[1 + H^2 r^2 - \frac{L^2}{r^2}]. \tag{4.10}$$

The radial coordinate $r(\sigma)$ is periodic with finite period T_σ, which can be expressed in terms of a complete elliptic integral, see Ref.[28]. For $\sigma \in [0, T_\sigma]$, the solution describes an infinitely long stationary string in the wedge $\phi \in\ ]0, \Delta\phi[$, where:

$$\Delta\phi = 2k\sqrt{\frac{1-2k^2}{1-k^2}}[\Pi(1-k^2,\ k) - K(k)] \ \in\ ]0,\ \pi[\tag{4.11}$$

The elliptic modulus k, which is a function of HL [28], parametrizes the solutions, $k \in\ ]0, 1/\sqrt{2}[$. The azimuthal angle is generally not a periodic function of σ, thus when the spacelike world-sheet coordinate σ runs through the range $]-\infty, +\infty[$, the solution describes an *infinite number* of infinitely long stationary open strings. The general solution is therefore a *multi-string* solution. Until now multi-string solutions were only found in de Sitter spacetime [11-13]. Our results show that multi-string solutions are a general feature of spacetimes with a cosmological constant (positive or negative). The solution in anti de Sitter spacetime describes a *finite* number of strings if the following relation holds:

$$N\Delta\phi = 2\pi M. \tag{4.12}$$

Here N and M are integers, determining the number of strings and the winding in azimuthal angle, respectively, for the multi-string solution, see Fig.3. The equation of state for a full multi-string solution takes the form ($P_3 = 0$):

$$dP_1 = dP_2 = -\frac{1}{2}dE, \quad \text{for } r \to \infty \tag{4.13}$$

corresponding to extremely unstable strings [21].

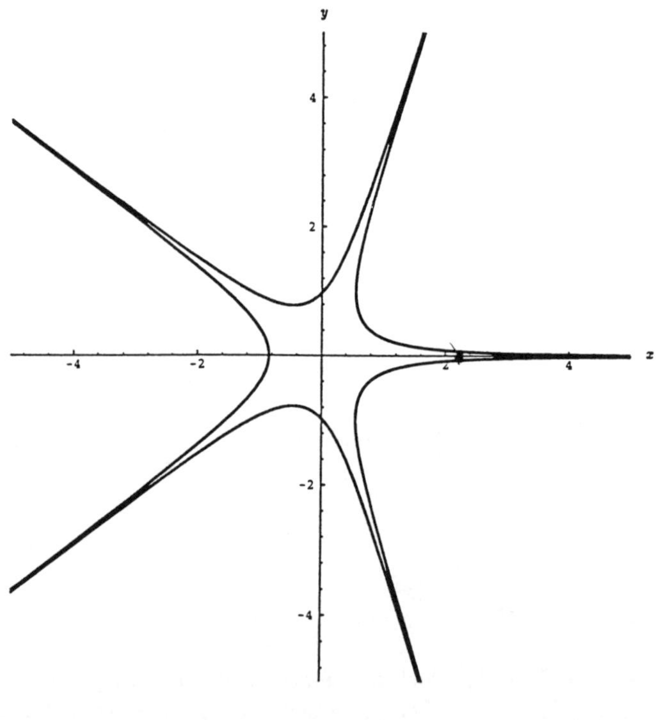

Figure 3. The $(N, M) = (5, 1)$ multi-string solution in anti de Sitter spacetime. The (N, M) multi-string solutions describe N stationary strings with M windings in the Azimuthal angle ϕ, in anti de Sitter spacetime.

In de Sitter spacetime (dS), the potential is given by:

$$V_{\rm dS}(r) = -(1 - H^2r^2)[1 - H^2r^2 - \frac{L^2}{r^2}]. \tag{4.14}$$

In this case we have to distinguish between solutions inside the horizon (where τ is timelike) and solutions outside the horizon (where τ is spacelike). Inside the horizon, the generic solution describes one infinitely long open stationary string winding around $r = 0$. For special values of the constants of motion, corresponding to a relation, which formally takes the same form as Eq.(4.12), the solution describes a closed string of finite length $l = N\pi/H$. The integer N in this case determines the number of "leaves", see Fig.4. The energy is positive and finite and grows with N. The pressure turns out to vanish identically, thus the equation of state corresponds to cold matter.

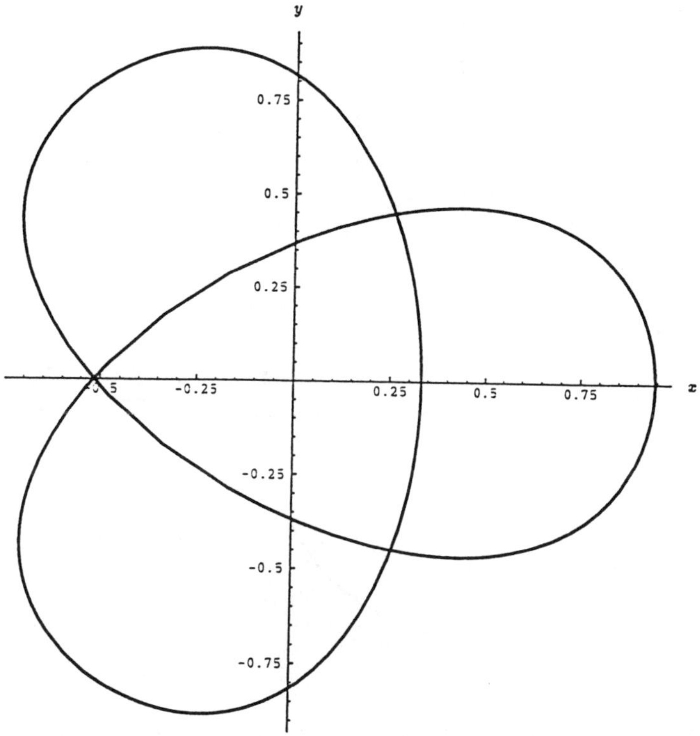

Figure 4. The $(N, M) = (3, 2)$ stationary string solution inside the horizon of de Sitter spacetime. Besides the circular string, this is the simplest stationary closed string configuration in de Sitter spacetime.

Outside the horizon, the world-sheet coordinate τ becomes spacelike while σ becomes timelike, thus we define:

$$\tilde{\tau} \equiv \sigma, \qquad \tilde{\sigma} \equiv \tau, \tag{4.15}$$

and the string solution is conveniently expressed in hyperboloid coordinates. The radial coordinate $r(\tilde{\tau})$ is periodic with a finite period $T_{\tilde{\tau}}$. For $\tilde{\tau} \in [0, T_{\tilde{\tau}}]$, the solution describes a straight string incoming non-radially from spatial infinity, scattering at the horizon and escaping towards infinity again, Fig.5.

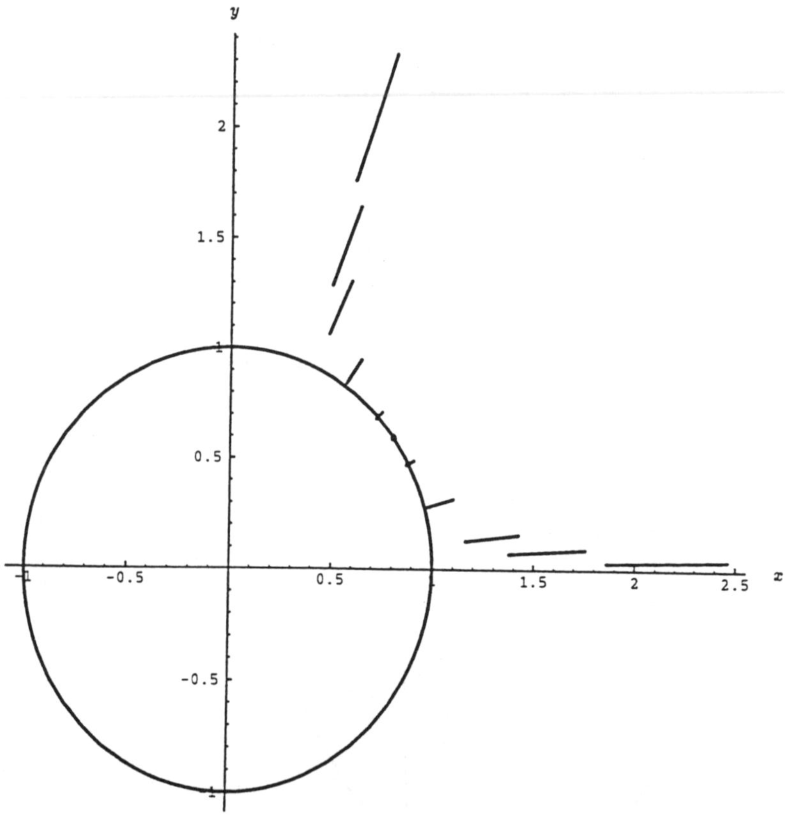

Figure 5. Schematic representation of the time evolution of the $(N, M) = (5, 1)$ dynamical multi-string solution, outside the horizon of de Sitter space-time. Only one of the 5 strings is shown; the others are obtained by rotating the figure by the angles $2\pi/5$, $4\pi/5$, $6\pi/5$ and $8\pi/5$. During the "scattering" at the horizon, the string collapses to a point and re-expands.

The string length is zero at the horizon and grows indefinetely in the asymptotic regions. As in the case of anti de Sitter spacetime, the azimuthal angle is generally not a periodic function, thus when the timelike world-sheet coordinate $\tilde{\tau}$ runs through the range $]-\infty, +\infty[$, the solution describes an *infinite* number of dynamical straight strings scattering at the horizon at different angles. The general solution is therefore a multi-string solution. In particular, a multi-string solution describing a *finite* number of strings is obtained if a relation of the form (4.12) is fulfilled. It turns out that the solution describes *at least* three strings. The energy and pressures of a full multi-string solution have also been computed. In the asymptotic region they fulfill an equation of state corresponding to extremely unstable strings, i.e. like Eq.(4.13).

In the Schwarzschild black hole background (S), the potential is given by:

$$V_{\text{S}}(r) = -(1 - 2m/r)[1 - 2m/r - \frac{L^2}{r^2}]. \tag{4.16}$$

No multi-string solutions are found in this case. Outside the horizon the solution of Eqs.(4.4) describes one infinitely long stationary open string. This solution was already derived in Ref.[40], and we shall not discuss it here. Inside the horizon, where τ becomes spacelike while σ becomes timelike, we make the redefinitions (4.15) and the solution is conveniently expressed in terms of Kruskal coordinates, see Ref.[28]. The solution describes one straight string infalling *non-radially* towards the singularity, see Fig.6. At the horizon, the string length is zero and it grows *indefinitely* when the string approaches the spacetime singularity. Thereafter, the solution can not be continued, see Ref.[28].

In the $2+1$ black hole anti de Sitter spacetime (BH-AdS) [30], the potential is given by:

$$V_{\text{BH-AdS}}(r) = -(\frac{r^2}{l^2} - 1)[\frac{r^2}{l^2} - 1 - \frac{L^2}{r^2}]. \tag{4.17}$$

Outside the horizon, the solutions "interpolate" between the solutions found in anti de Sitter spacetime and outside the horizon of the Schwarzschild black hole. The solutions thus describe infinitely long stationary open strings. As in anti de Sitter spacetime, the general solution is a *multi-string* describing *infinitely* many strings. In particular, for certain values of the constants of motion, corresponding to the condition of the form (4.12), the solution describes a finite number of strings. In the simplest version of the $2+1$ BH-AdS background ($M = 1$, $J = 0$), it turns out that the solution describes *at least* seven strings, see Ref.[28]. Inside the horizon, we make the redefinitions (4.15) and the solution is conveniently expressed in terms of Kruskal-like coordinates, see Ref.[28]. The solution is similar to the solution found inside the horizon of the Schwarzschild black hole, but there is one important difference at $r = 0$. As in the Schwarzschild black hole background, the solution describes one straight string infalling non-radially towards $r = 0$, and beyond this point the solution can not be continued because of the global structure of the spacetime. At the horizon the string size is zero and during the fall towards $r = 0$, the string size *grows* but stays *finite*. This should be compared with the straight string inside the horizon of the Schwarzschild black hole, where the string size grows indefinetely. The physical reason for this difference is that the point $r = 0$ is not a strong curvature singularity in the $2+1$ BH-AdS spacetime.

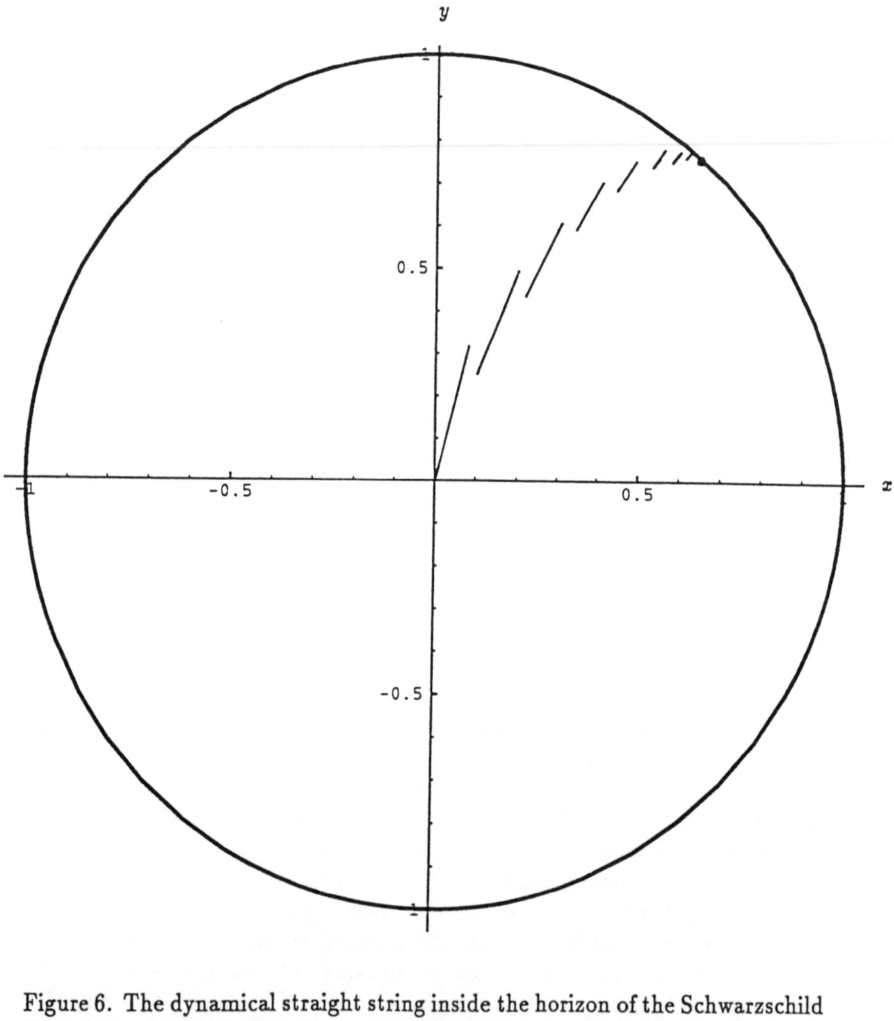

Figure 6. The dynamical straight string inside the horizon of the Schwarzschild black hole. The string infalls non-radially towards the singularity with indefinetely growing length.

We have considered also the Schwarzschild de Sitter and Schwarzschild anti de Sitter spacetimes. These spacetimes contain all the features of the spacetimes already discussed: singularities, horizons, positive or negative cosmological constants. All the various types of string solutions found in the other spacetimes (open, closed, straight, finitely and infinitely long, multi-strings) are therefore present in the different regions of the Schwarzschild-de Sitter and Schwarzschild-anti de Sitter spacetimes. The details are given in Ref.[28].

We close this section with a few remarks on the stability of the solutions. Generally, the question of stability must be addressed by considering small perturbations around the exact solutions. In Ref.[10], a covariant formalism describing physical perturbations propagating along an arbitrary string configuration embedded in an arbitrary curved spacetime, was developed. The resulting equations determining the evolution of the perturbations are however very complicated in the general case, although partial (analytical) results have been obtained in special cases for de Sitter [18, 43] and Schwarzschild black hole [10, 43, 44] spacetimes. The exact solutions found in this section fall essentially into two classes: dynamical and stationary. The dynamical string solutions outside the horizon of de Sitter (or S-dS) and inside the horizon of Schwarzschild (or S-dS, S-AdS) spacetimes, are already unstable at the zeroth order approximation (i.e. without including small perturbations), in the sense that their physical length grows indefinetely. For the stationary string solutions the situation is more delicate. The existence of the stationary configurations is based on an exact balance between the string tension and the local attractive or repulsive gravity. For that reason, it can be expected that the configurations are actually unstable for certain modes of perturbation, especially in strong curvature regions.

The main conclusions of this section are given in Table 4. Further details can be found in Ref.[28].

Table 4. Short summary of the features of the string solutions found in this section. In anti de Sitter spacetime and outside the horizon of de Sitter and $2+1$ BH-AdS spacetimes, the solutions describe a *finite* number of strings provided a condition of the form $N\Delta\phi = 2\pi M$ is fulfilled, where $\Delta\phi$ is the angle betwen the "arms" of the string and (N, M) are integers.

Line element:	$ds^2 = -a(r)dt^2 + \frac{dr^2}{a(r)} + r^2(d\theta^2 + \sin^2\theta d\phi^2)$
Ansatz:	$t = \tau, \quad r = r(\sigma) \quad \phi = \phi(\sigma), \quad \theta = \pi/2$
String solution:	$\phi' = \frac{L}{r^2} \quad r'^2 + V(r) = 0; \quad V(r) = -a(r)[a(r) - \frac{L^2}{r^2}]$
String length element:	$dl = \sqrt{a(r(\sigma))}\ d\sigma$

MINKOWSKI, $a(r) = 1, \quad V(r) = \frac{L^2}{r^2} - 1$

The solution describes one infinitely long stationary straight string.

$\frac{dE}{dl} = -\frac{dP_2}{dl} = \text{const.}, \qquad P_1 = P_3 = 0$

ANTI DE SITTER, $a(r) = 1 + H^2r^2, \quad V(r) = -(1 + H^2r^2)[1 + H^2r^2 - \frac{L^2}{r^2}]$

The solution describes a finite or infinite number of infinitely long stationary strings.

$\frac{dE}{dl} = -2\frac{dP_1}{dl} = -2\frac{dP_2}{dl}, \qquad P_3 = 0, \qquad$ (asymptotically for $r \to \infty$)

DE SITTER, $a(r) = 1 - H^2r^2, \quad V(r) = -(1 - H^2r^2)[1 - H^2r^2 - \frac{L^2}{r^2}]$

Inside the horizon, the solution describes one finitely or infinitely long stationary string winding around $r = 0$.

$P_1 = P_2 = P_3 = 0$, as cold matter, $\qquad E$ expressed in terms of elliptic functions.

Outside the horizon, the solution describes a finite or infinite number of dynamical straight strings "scattering at the horizon". The string length vanishes at the horizon, but stretches indefinetely at spatial infinity.

$\frac{dE}{dl} = -2\frac{dP_1}{dl} = -2\frac{dP_2}{dl}, \qquad P_3 = 0, \qquad$ (asymptotically for $r \to \infty$)

SCHWARZSCHILD, $a(r) = 1 - 2m/r, \quad V(r) = -(1 - 2m/r)[1 - 2m/r - \frac{L^2}{r^2}]$

Outside the horizon, the solution describes one infinitely long stationary string.

Inside the horizon, the solution describes one dynamical straight string. The string size is zero at the horizon, and grows *indefinetely* as the string falls towards $r = 0$.

2+1 BLACK HOLE-ADS, $a(r) = \frac{r^2}{l^2} - 1, \quad V(r) = -(\frac{r^2}{l^2} - 1)[\frac{r^2}{l^2} - 1 - \frac{L^2}{r^2}]$

Outside the horizon, the solution describes a finite or infinite number of infinitely long stationary string.

Inside the horizon, the solution describes one dynamical straight string. The string size is zero at the horizon, and grows *finitely* as the string falls towards $r = 0$.

5 Spatial Curvature Effects on the String Dynamics

The propagation of strings in Friedmann-Robertson-Walker (FRW) cosmologies has been investigated using both exact and approximative methods, see for example Refs.[7, 11-14, 17-20] (as well as numerical methods, which shall not be discussed here). Except for anti de Sitter spacetime, which has negative spatial curvature, the cosmologies that have been considered until now, have all been spatially flat. In this section we will consider the physical effects of a non-zero (positive or negative) curvature index on the classical and quantum strings. The non-vanishing components of the Riemann tensor for the generic D-dimensional FRW line element, in comoving coordinates:

$$ds^2 = -dt^2 + a^2(t)\frac{d\vec{x}d\vec{x}}{(1+\frac{K}{4}\vec{x}\vec{x})^2}, \tag{5.1}$$

are given by:

$$R_{itit} = \frac{-aa_{tt}}{(1+\frac{K}{4}\vec{x}\vec{x})^2}, \qquad R_{ijij} = \frac{a^2(K+a_t^2)}{(1+\frac{K}{4}\vec{x}\vec{x})^4}; \quad i \neq j \tag{5.2}$$

where $a = a(t)$ is the scale factor and K is the curvature index. Clearly, a non-zero curvature index introduces a non-zero spacetime curvature; the exceptional case provided by $K = -a_t^2 = \text{const.}$, corresponds to the Milne-Universe. From Eqs.(5.2), it is also seen that the curvature index has to compete with the first derivative of the scale factor. The effects of the curvature index are therefore most conveniently discussed in the family of FRW-universes with constant scale factor, the so-called static Robertson-Walker spacetimes. This is the point of view we take in the present section.

We consider both the closed ($K > 0$) and the hyperbolic ($K < 0$) static Robertson-Walker spacetimes, and all our results are compared with the already known results in the flat ($K = 0$) Minkowski spacetime. We determine the evolution of circular strings, derive the corresponding equations of state (using Eq.(4.6)), discuss the question of strings as self-consistent solutions to the Einstein equations [20], and we perform a semi-classical quantization. We also find all the stationary string configurations in these spacetimes.

The radius of a classical circular string in the spacetime (5.1), for $a = 1$, is determined by:

$$\dot{r}^2 + V(r) = 0; \qquad V(r) = (1-Kr^2)(r^2 - b\alpha'^2), \tag{5.3}$$

where b is an integration constant ($b\alpha'^2 = E$, in the notation of Eqs.(2.18)-(2.19)). This equation is solved in terms of elliptic functions and all solutions describe oscillating strings (Fig.7 shows the potential $V(r)$ for $K > 0$, $K = 0$, $K < 0$).

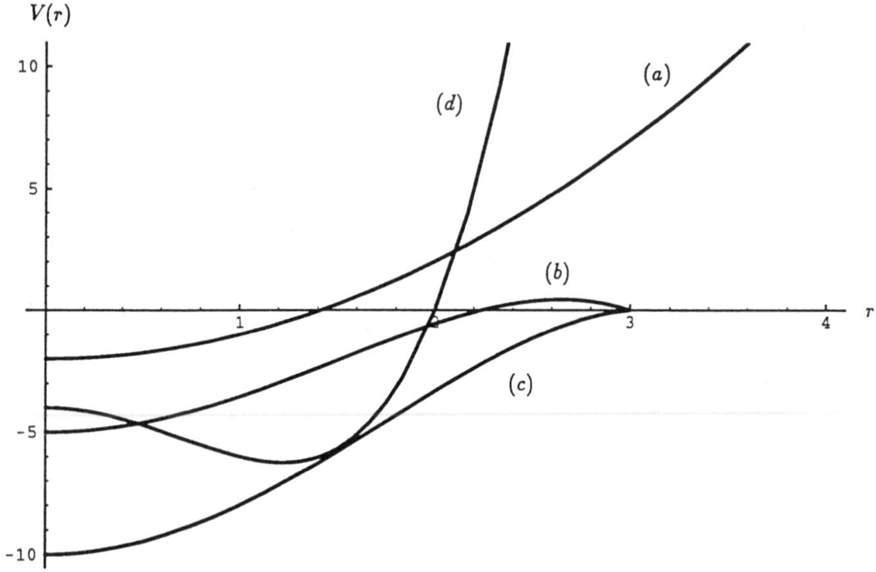

Figure 7. The potential $V(r)$ introduced in Eqs.(5.3) for a circular string in the static Robertson-Walker spacetimes: (a) flat ($K = 0$), (b) closed ($K > 0$ *and* $bK\alpha'^2 \leq 1$), (c) closed ($K > 0$ *and* $bK\alpha'^2 > 1$), (d) hyperbolic ($K < 0$).

For $K > 0$, when the spatial section is a hypersphere, the string either oscillates on one hemisphere or on the full hypersphere. The energy is always positive while the average pressure can be positive, negative or zero, depending on the precise value of an elliptic modulus; the equation of state is given explicitly in Table 5, in the different cases. Interestingly enough, we find that the circular strings provide a self-consistent solution to the Einstein equations with a selected value of the curvature index. Self-consistent solutions to the Einstein equations with string sources have been found previously in the form of power law expanding universes [20].

Table 5. Classical circular strings in the static Robertson-Walker spacetimes. Notice that a self-consistent solution to the Einstein equations, with the string back-reaction included, can be obtained only for $K > 0$.

CLASSICAL CIRCULAR STRING SOLUTIONS

Spacetime: $ds^2 = -dt^2 + \frac{dr^2}{1-Kr^2} + r^2 d\phi^2$

String solution: $\phi = \sigma, \quad t(\tau) = \sqrt{b}\alpha'\tau, \quad \dot{r}^2 + V(r) = 0; \; V(r) = (1 - Kr^2)(r^2 - b\alpha'^2)$

String length, energy and pressure: $l(\tau) = 2\pi|r(\tau)|, \quad T^{\mu}{}_{\nu}(\text{string}) = \text{diag.}(-\rho, P, P)$

MINKOWSKI ($K = 0$):

$r(\tau) = \sqrt{b}\alpha' \cos\tau$

$\rho = \sqrt{b}, \quad P = -\frac{\sqrt{b}}{2}\cos 2\tau, \quad < P >= 0,$ as cold matter.

Is not a self-consistent solution to the Einstein equations.

CLOSED ($K > 0$):

$$r(\tau) = \begin{cases} \sqrt{b}\alpha' \text{sn}[\tau, k_-] & \text{for } \sqrt{bK}\alpha' \leq 1, \quad k_- \equiv \sqrt{bK}\alpha' \in [0,1] \\ \frac{1}{\sqrt{K}}\text{sn}[\tau/k_+, k_+] & \text{for } \sqrt{bK}\alpha' > 1, \quad k_+ \equiv 1/(\sqrt{bK}\alpha') \in]0,1[\end{cases}$$

$$\rho = \sqrt{b}, \quad P = \begin{cases} \frac{k_-}{2\sqrt{K}\alpha'}(1 - 2\text{sn}^2[\tau, k_-]) & \text{for } \sqrt{bK}\alpha' \leq 1 \\ \frac{1}{2k_+\sqrt{K}\alpha'}(1 - 2k_+^2\text{sn}^2[\tau/k_+, k_+]) & \text{for } \sqrt{bK}\alpha' > 1 \end{cases}$$

$$< P >= \begin{cases} \frac{1}{2\sqrt{K}\alpha'}[\frac{k_-^2-2}{k_-} + \frac{2}{k_-}\frac{E(k_-)}{K(k_-)}] & \text{for } \sqrt{bK}\alpha' \leq 1, \quad \text{negative.} \\ \frac{1}{2k_+\sqrt{K}\alpha'}[\frac{2E(k_+)}{K(k_+)} - 1] & \text{for } \sqrt{bK}\alpha' > 1, \quad \text{positive or negative.} \end{cases}$$

A self-consistent solution to the Einstein equations is obtained for $k_+ = 0.9089...$

HYPERBOLIC ($K < 0$):

$r(\tau) = \frac{k}{\sqrt{-K}}\text{sd}[\mu\tau, k], \quad \mu \equiv \sqrt{1 - Kb\alpha'^2}, \quad k \equiv \sqrt{-Kb\alpha'^2}/\mu \in [0,1[$

$\rho = \sqrt{b}, \quad P = \frac{k - 2k(1-k^2)\text{sd}^2[\mu\tau,k]}{2\sqrt{1-k^2}\sqrt{-K}\alpha'}$

$< P >= \frac{1}{2\sqrt{1-k^2}\sqrt{-K}\alpha'}[-k + \frac{2}{k}(1 - \frac{E(k)}{K(k)})],$ positive.

Is not a self-consistent solution to the Einstein equations.

We have semi-classically quantized the circular string solutions of Eq.(5.3). We used an approach developed in field theory by Dashen et. al. [41], based on the stationary phase approximation of the partition function. The result of the stationary phase integration is expressed in terms of the function:

$$W(m) \equiv S_{\rm cl}(T(m)) + m\,T(m), \tag{5.4}$$

where $S_{\rm cl}$ is the action of the classical solution, m is the mass and the period $T(m)$ is implicitly given by:

$$\frac{dS_{\rm cl}}{dT} = -m. \tag{5.5}$$

The bound state quantization condition then becomes [41]:

$$W(m) = 2\pi n, \quad n \in N_0 \tag{5.6}$$

for n 'large'. Parametric plots of $K\alpha' W$ as a function of $K\alpha'^2 m^2$ in the different cases, are shown in Fig.8. The strings oscillating on one hemisphere give rise to a finite number N_- of states with the following mass-formula Ref.[29]:

$$m_-^2 \alpha' \approx \pi\, n, \qquad N_- \approx \frac{4}{\pi K \alpha'}. \tag{5.7}$$

As in flat Minkowski spacetime, the scale of these string states is set by α'. The strings oscillating on the full hypersphere give rise to an infinity of more and more massive states with the asymptotic mass-formula Ref.[29]:

$$m_+^2 \approx K\, n^2. \tag{5.8}$$

The masses of these states are independent of α', the scale is set by the curvature index K. Notice also that the level spacing grows with n. A similar result was found recently for strings in anti de Sitter spacetime [42].

For $K < 0$, when the spatial section is a hyperboloid, both the energy and the average pressure of the oscillating strings are positive. The equation of state is given in Table 5. In this case, the strings can not provide a self-consistent solution to the Einstein equations. After semi-classical quantization, we find an infinity of more and more massive states. The mass-formula is given by (see Ref.[29]):

$$\sqrt{-Km^2\alpha'^2}\, \log \sqrt{-Km^2\alpha'^2} \approx -\frac{\pi}{2} K\alpha'\, n \tag{5.9}$$

Notice that the level spacing grows faster than in Minkowski spacetime but slower than in the closed static Robertson-Walker spacetime.

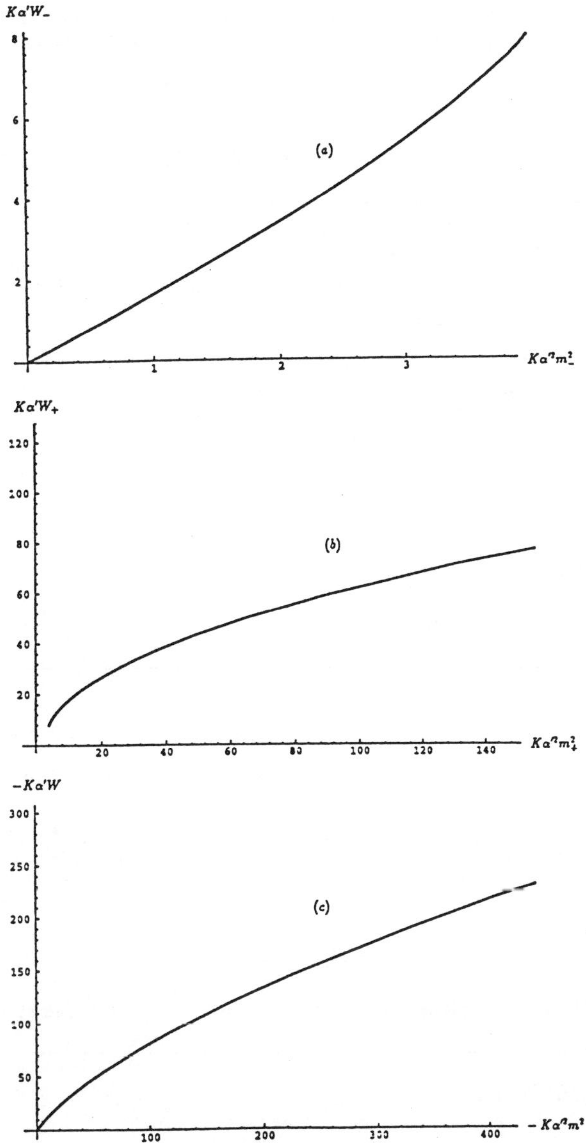

Figure 8. Parametric plot of $K\alpha' W$ as a function of $K\alpha'^2 m^2$ in the three cases: (a) $K > 0$, $bK\alpha'^2 \leq 1$ (strings oscillating on one hemisphere, only finitely many states), (b) $K > 0$, $bK\alpha'^2 > 1$ (strings oscillating on the full hypersphere, infinitely many states), (c) $K < 0$ (strings oscillating on the hyperboloid, infinitely many states).

Table 6. Semi-classical quantization of the circular strings in the static Robertson-Walker spacetimes. Notice in particular the different behaviour of the high mass spectrum of strings in the three cases.

SEMICLASSICAL QUANTIZATION OF CIRCULAR STRINGS

Stationary phase: $W(m) = S^{\text{cl}}(T(m)) + m\,T(m)$

Stationary phase point: $\frac{\partial S^{\text{cl}}}{\partial T} = -m,$ determines the period T

Quantization condition: $W(m) = 2\pi\, n, \quad n \in N_0$

MINKOWSKI ($K = 0$):

Mass: $m = 2\sqrt{b}$

Stationary phase: $W = 2\pi\sqrt{b}\alpha'$

Quantization condition: $\alpha' m^2 = 4\, n, \quad n \in N_0$

There are infinitely many states; $"m^2 \propto n"$

CLOSED ($K > 0$):

Mass: $$m = \begin{cases} \frac{2k_-}{\sqrt{K\alpha'}} & \text{for } \sqrt{bK}\alpha' \leq 1, \quad k_- \equiv \sqrt{bK}\alpha' \in [0,1] \\ \frac{2}{\sqrt{K\alpha'}k_+} & \text{for } \sqrt{bK}\alpha' > 1, \quad k_+ \equiv 1/(\sqrt{bK}\alpha') \in]0,1[\end{cases}$$

Stationary phase: $$W = \begin{cases} \frac{8}{K\alpha'}[E(k_-) - (1-k_-^2)K(k_-)] & \text{for } \sqrt{bK}\alpha' \leq 1 \\ \frac{8}{K\alpha'}\frac{E(k_+)}{k_+} & \text{for } \sqrt{bK}\alpha' > 1 \end{cases}$$

Quantization condition: $$\begin{cases} m_-^2\alpha' \approx \pi\, n & \text{for } \sqrt{bK}\alpha' \leq 1, \quad \text{on one hemisphere.} \\ m_+^2 \approx K\, n^2 & \text{for } \sqrt{bK}\alpha' > 1, \quad \text{on full hypersphere.} \end{cases}$$

Altogether there are infinitely many states; $"m \propto n"$, for high states

HYPERBOLIC ($K < 0$):

Mass: $m = \frac{2}{\sqrt{-K\alpha'}}\frac{k}{\sqrt{1-k^2}}, \quad k \equiv \sqrt{\frac{-Kb\alpha'^2}{1-Kb\alpha'^2}}$

Stationary phase: $W = \frac{8}{K\alpha'}\frac{E(k)-K(k)}{\sqrt{1-k^2}}$

Quantization condition: $\sqrt{-Km^2\alpha'^2}\log\sqrt{-Km^2\alpha'^2} \approx -\frac{\pi}{2}K\alpha'\, n$

There are infinitely many states; $"m \log m \propto n"$, for high states

On the other hand, the stationary strings are determined by Eqs.(4.4):

$$\phi' = \frac{L}{r^2}, \qquad r'^2 + V(r) = 0; \quad V(r) = (1 - Kr^2)(\frac{L^2}{r^2} - 1), \qquad (5.10)$$

where L is an integration constant (Fig.9 shows the potential $V(r)$ for $K > 0$, $K = 0$, $K < 0$).

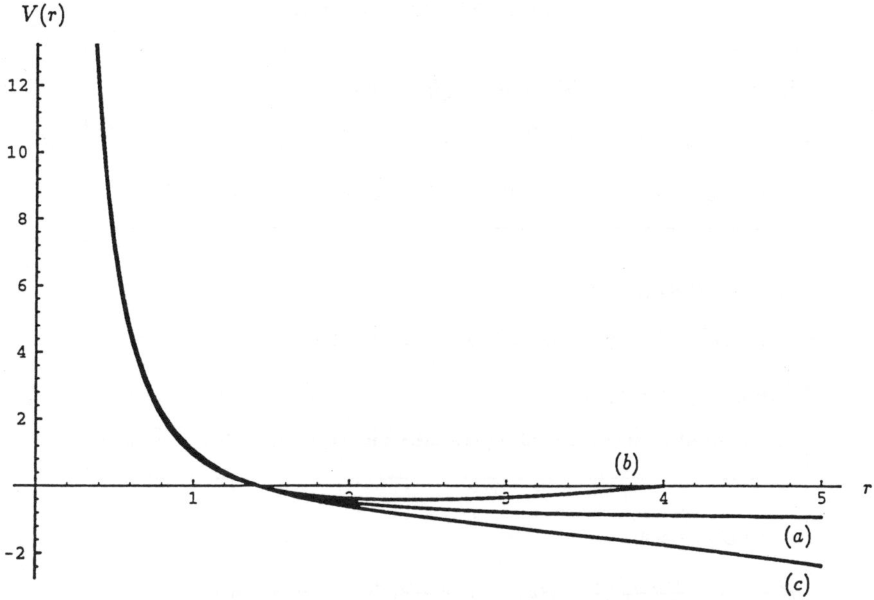

Figure 9. The potential $V(r)$ introduced in Eqs.(5.10) for a stationary string in the static Robertson-Walker spacetimes: (a) flat ($K = 0$), (b) closed ($K > 0$ *and* $KL^2 \leq 1$), (c) hyperbolic ($K < 0$).

For $K > 0$, all the stationary string solutions describe circular strings winding around the hypersphere. The equation of state is of the extremely unstable string type [21]. For $K < 0$, the stationary strings are represented by infinitely long open configurations with an angle between the two "arms" given by:

$$\Delta\phi = \pi - 2\arctan(\sqrt{-K}L). \qquad (5.11)$$

The energy density is positive while the pressure densities are negative. No simple equation of state is found for these solutions.

We also computed the first and second order fluctuations around a static string center of mass, using the string perturbation series approach [7] and its covariant versions [9, 26]. Up to second order, the mass-formula for arbitrary values of the curvature index (positive or negative) is identical to the well-known flat spacetime mass-formula; all dependence on K cancels out.

Table 7. Stationary strings in the static Robertson-Walker spacetimes. Notice that the pressure densities are always negative in all three cases.

STATIONARY STRING SOLUTIONS
Spacetime: $ds^2 = -dt^2 + \frac{dr^2}{1-Kr^2} + r^2 d\phi^2$ String solution: $t = \tau, \quad \phi' = \frac{L}{r^2}, \quad r'^2 + V(r) = 0; \quad V(r) = (1 - Kr^2)(\frac{L^2}{r^2} - 1)$ String length, energy and pressure: $dl = d\sigma, \quad T^{\mu}{}_{\nu}(\text{string}) = \text{diag.}(-\rho, P_i)$
MINKOWSKI $(K = 0)$: $r(\sigma) = \sqrt{\sigma^2 + L^2}, \quad \phi(\sigma) = \arctan(\sigma/L),$ straight strings. $\frac{d\rho}{d\sigma} = \frac{1}{2\pi\alpha'}, \quad \frac{dP_y}{d\sigma} = -\frac{1}{2\pi\alpha'}$ The equation of state corresponds to extremely unstable strings in 1+1 dimensions.
CLOSED $(K > 0)$: $r^2(\sigma) = \frac{1}{K} \frac{KL^2 + \tan^2(\sqrt{K}\sigma)}{1 + \tan^2(\sqrt{K}\sigma)}, \quad \phi(\sigma) = \arctan[\frac{\tan(\sqrt{K}\sigma)}{\sqrt{K}L}],$ circular strings. $\rho = \frac{1}{\sqrt{K}\alpha'}, \quad P_x = P_z = -\frac{1}{2\sqrt{K}\alpha'}$ The equation of state corresponds to extremely unstable strings in 2+1 dimensions.
HYPERBOLIC $(K < 0)$: $r^2(\sigma) = \frac{1}{K} \frac{KL^2 - \tanh^2(\sqrt{-K}\sigma)}{1 - \tanh^2(\sqrt{-K}\sigma)}, \quad \phi(\sigma) = \pm\left\{\frac{\pi}{2} - \arctan[\sqrt{-K}L \tanh^{-1}(\sqrt{-K}\sigma)]\right\}$ Infinitely long open strings, $\Delta\phi = \pi - 2\arctan(\sqrt{-K}L) \in]0, \pi]$ $\frac{d\rho}{d\sigma} = \frac{1}{2\pi\alpha'}, \quad \frac{dP_x}{d\sigma} = \frac{8KL^2}{\pi\alpha'(1-KL^2)^2} e^{-\sqrt{-K}\lvert\sigma\rvert}, \quad \frac{dP_y}{d\sigma} = \frac{-8}{\pi\alpha'(1-KL^2)^2} e^{-\sqrt{-K}\lvert\sigma\rvert}$ The pressure densities are negative but asymptotically go to zero exponentially. No simple equation of state in the general case.

6 Conclusion

We first studied the string propagation in the 2+1 BH-AdS background. We found the first and second order perturbations around the string center of mass as well as the mass formula, and compared with the ordinary black hole AdS spacetime and with the black string background as well. These results were then generalized to ordinary D-dimensional anti de Sitter spacetime. The classical string motion in anti de Sitter spacetime is stable in the sense that it is oscillatory in time with real frequencies and the string size and energy are bounded. Quantum mechanically, this reflects in the mass operator, which is well defined for any value of the wave number n, and arbitrary high mass states (and therefore an infinite number of states) can be constructed. This is to be contrasted with de Sitter spacetime, where string instabilities develop, in the sense that the string size and energy become unbounded for large de Sitter radius. For low mass states (the stable regime), the mass operator in de Sitter spacetime is given by Eq.(3.1) but with

$$\Omega_n(\lambda)_{\text{dS}} = \frac{2n^2 - m^2\alpha'\lambda}{\sqrt{n^2 - m^2\alpha'\lambda}}.$$

Real mass solutions can be defined only up to some *maximal mass* of the order $m^2\alpha' \approx 1/\lambda$ [7]. For $\lambda << 1$, real mass solutions can be defined only for $N \leq N_{\max} \sim 0.15/\lambda$ (where N is the eigenvalue of the number operator) and therefore there exists a *finite* number of states only. These features of strings in de Sitter spacetime have been recently confirmed within a different (semi-classical) quantization approach based on *exact* circular string solutions [42].

The presence of a cosmological constant Λ (positive or negative) increases considerably the number of levels of different eigenvalue of the mass operator (there is a splitting of levels) with respect to flat spacetime. That is, a non-zero cosmological constant *decreases* (although does not remove) the degeneracy of the string mass states, introducing a *fine structure effect*. For the low mass states the level spacing is approximately constant (up to corrections of the order λ^2). For the high mass states, the changes are more drastic and they depend crucially on the sign of Λ. A value $\Lambda < 0$ causes the *growing* of the level spacing linearly with N instead of being constant as in Minkowski space. Consequently, the density of states $\rho(m)$ grows with the exponential of $\sqrt{m}$ (instead of m as in Minkowski space) discarding the existence of a Hagedorn temperature in AdS spacetime, and the possibility of a phase transition. In addition, another important feature of the high mass string spectrum in AdS spacetime is that it becomes independent of α'. The string scale for the high mass states is given by $\mid \Lambda \mid$, instead of $1/\alpha'$ for the low mass states.

We have found new classes of exact string solutions obtained either by the circular string ansatz, Eq.(2.16), or by the stationary string ansatz, Eq.(4.3), in a variety of curved backgrounds including Schwarzschild, de Sitter and anti de Sitter spacetimes. Many different types of solutions have been found: oscillating circular strings, closed stationary strings, infinitely long stationary strings, dynamical straight strings and multi-string solutions describing finitely or infinitely many stationary or dynamical strings. In all the cases we have obtained the exact solutions in terms of either elementary or elliptic functions. Furthermore, we have analyzed the physical properties (length, energy, pressure) of the string solutions.

Finally, we solved the equations of motion and constraints for circular strings in static Robertson-Walker spacetimes. We computed the equations of state and found that there exists a self-consistent solution to the Einstein equations in the case of positive spatial curvature. The solutions were quantized semi-classically using the stationary phase approximation method, and the resulting spectra were analyzed and discussed. We have also found all stationary string configurations in these spacetimes and we computed the corresponding physical quantities, string length, energy and pressure.

Acknowledgements:

A.L. Larsen is supported by the Danish Natural Science Research Council under grant No. 11-1231-1SE

References

[1] H.J. de Vega and N. Sánchez, Phys. Rev. D42 (1990) 3969.
H.J. de Vega, M. Ramón-Medrano and N. Sánchez, Nucl. Phys. B374 (1992) 405.

[2] H.J. de Vega and N. Sánchez, Phys. Lett. B244 (1990) 215, Phys. Rev. Lett. 65C (1990) 1517, IJMP A7 (1992) 3043, Nucl. Phys. B317 (1989) 706 and ibid 731.
D. Amati and C. Klimcik, Phys. Lett. B210 (1988) 92.
M. Costa and H.J. de Vega, Ann. Phys. 211 (1991) 223 and ibid 235.
C. Loustó and N. Sánchez, Phys. Rev. D46 (1992) 4520.

[3] V.E. Zakharov and A.V. Mikhailov, JETP 47 (1979) 1017.
H. Eichenherr, in "Integrable Quantum Field Theories", ed. J. Hietarinta and C. Montonen (Springer, Berlin, 1982).

[4] H.J. de Vega and N. Sánchez, Phys. Rev. D47 (1993) 3394.

[5] I. Bars and K. Sfetsos, Mod. Phys. Lett. A7 (1992) 1091.

[6] H.J. de Vega, J.R. Mittelbrunn, M.R. Medrano and N. Sánchez, Phys. Lett. B232 (1994) 133 and "The Two-Dimensional Stringy Black Hole: a new Approach and a Pathology", PAR-LPTHE 93/14.

[7] H.J. de Vega and N. Sánchez, Phys. Lett. B197 (1987) 320.

[8] H.J. de Vega and N. Sánchez, Nucl. Phys. B299 (1988) 818.

[9] P.F. Mende, in "String Quantum Gravity and Physics at the Planck Energy Scale". Proceedings of the Erice Workshop held in June 1992, ed. N. Sánchez (World Scientific, 1993).

[10] A.L. Larsen and V.P. Frolov, Nucl. Phys. B414 (1994) 129.

[11] H.J. de Vega, A.V. Mikhailov and N. Sánchez, Teor. Mat. Fiz. 94 (1993) 232, Mod. Phys. Lett. A29 (1994) 2745.

[12] F. Combes, H.J. de Vega, A.V. Mikhailov and N. Sánchez, Phys. Rev. D50 (1994) 2754.

[13] H.J. de Vega, A.L. Larsen and N. Sánchez, Nucl. Phys. B427 (1994) 643.

[14] I. Krichever, "Two-Dimensional Algebraic-Geometrical Operators with Self-Consistent Potentials", Landau Institute Preprint, March, 1994, to appear in Funct. Analisis and Appl.

[15] H.J. de Vega and N. Sánchez, Nucl. Phys. B309 (1988) 552 and ibid, 577.

[16] C.O. Loustó and N. Sánchez, Phys. Rev. D47 (1993) 4498.

[17] H.J. de Vega and I.L. Egusquiza, Phys. Rev. D50 (1994) 763.

[18] A.L. Larsen, Phys. Rev. D50 (1994) 2623.

[19] A.L. Larsen and M. Axenides, Phys. Lett. B318 (1993) 47.

[20] H.J. de Vega and N. Sánchez, Phys. Rev. D50 (1994) 7202.

[21] N. Sánchez and G. Veneziano, Nucl. Phys. B333 (1990) 253.
M. Gasperini, N. Sánchez and G. Veneziano, IJMP A6 (1991) 3853, Nucl. Phys. B364 (1991) 365.

[22] H.J. de Vega, M. Ramón-Medrano and N. Sánchez, Nucl. Phys. B374 (1992) 425.

[23] H.J. de Vega and N. Sánchez, Phys. Rev. D45 (1992) 2783.
M. Ramón-Medrano and N. Sánchez, Class. Quantum Grav. 10 (1993) 2007.

[24] N. Sánchez, Phys. Lett. B195 (1987) 160.

[25] H.J. de Vega, M. Ramón-Medrano and N. Sánchez, Nucl. Phys. B351 (1991) 277.

[26] A.L. Larsen and N. Sánchez, Phys. Rev. D50 (1994) 7493.

[27] A.L. Larsen and N. Sánchez, "Mass Spectrum of Strings in Anti de Sitter Spacetime", DEMIRM-Paris-94048, to appear in Phys. Rev. D.

[28] A.L. Larsen and N. Sánchez, "New Classes of Exact Multi-String Solutions in Curved Spacetimes", DEMIRM-Paris-95003, to appear in Phys. Rev. D.

[29] A.L. Larsen and N. Sánchez, "The Effect of Spatial Curvature on the Classical and Quantum Strings", DEMIRM-Paris-95004.

[30] M. Bañados, C. Teitelboim and J. Zanelli, Phys. Rev. Lett. 69 (1992) 1849.

[31] G.T. Horowitz and D.L. Welch, Phys. Rev. Lett. 71 (1993) 328.

[32] N. Kaloper, Phys. Rev. D48 (1993) 2598.

[33] M. Bañados, M. Henneaux, C. Teitelboim and J. Zanelli, Phys. Rev. D48 (1993) 1506.

[34] C. Farina, J. Gamboa and A.J. Seguí-Santonja, Class. Quantum Grav. 10 (1993) L193.
N. Cruz, C. Martínez and L. Peña, Class. Quantum Grav. 11 (1994) 2731.

[35] A. Achúcarro and M.E. Ortiz, Phys. Rev. D48 (1993) 3600.
D. Cangemi, M. Leblanc and R.B. Mann, Phys. Rev. D48 (1993) 3606.

[36] J. Horne and G.T. Horowitz, Nucl. Phys. B368 (1992) 444.

[37] T. Buscher, Phys. Lett. B201 (1988) 466, B194 (1987) 59.

[38] A. Vilenkin, Phys. Rev. D24 (1981) 2082.

[39] H.J. de Vega and N. Sánchez, Int. J. Mod. Phys. A7 (1992) 3043.

[40] V.P. Frolov, V.D. Skarzhinsky, A.I. Zelnikov and O. Heinrich, Phys. Lett. B224 (1989) 255.

[41] R. Dashen, B. Hasslacher and A. Neveu, Phys. Rev. D11 (1975) 3424.

[42] H.J. de Vega, A.L. Larsen and N. Sánchez, "Semi-Classical Quantization of Circular Strings in de Sitter and anti de Sitter Spacetimes", DEMIRM-Paris-94049, submitted to Phys. Rev. D.

[43] A.L. Larsen, "Stable and Unstable Circular Strings in Inflationary Universes", NORDITA-94/14-P, to appear in Phys. Rev. D.

[44] A.L. Larsen, Nucl. Phys. B412 (1994) 372.

INTEGRABLE FIELD THEORIES

C. Destri

Dipartimento di Fisica, Università di Milano and INFN, sezione di Milano *

H. J. de Vega

Laboratoire de Physique Théorique et Hautes Energies † ‡

Abstract

Recent developments on integrable field theories and statistical models in two dimensions are reviewed.

The best way to investigate the continuum limit of lattice models is through the light-cone approach. This method allows to find the underlying quantum field theory for any integrable lattice model in its gapless regime. The relativistic spectrum and S-matrix follows straightforwardly in this way through the Bethe Ansatz. We show here how to derive the infinite number of local commuting and non-local and non-commuting conserved charges in integrable QFT, taking the massive Thirring model (sine-Gordon) as an example. They are generated by quantum monodromy operators and provide a representation of $q-$deformed affine Lie algebras $U_q(\hat{\mathcal{G}})$. Therefore, these models enjoy **infinite** dimensional **non-abelian** symmetries.

A new lattice local formulation of integrable field models is presented and applied to the massive Thirring model (sine-Gordon) as an example.

Finally, the new thermodynamic Bethe Ansatz (DDV equations) allowing an unified treatment of finite size and finite temperature integrable models is reviewed.

---

*Dipartimento di Fisica
Via Celoria 16, 20133 Milano, ITALIA.

†Laboratoire Associé au CNRS UA 280.

‡L.P.T.H.E., Tour 16 1er étage, Université Paris VI,
4 Place Jussieu, 75252, Paris cedex 05, FRANCE.

I. INTRODUCTION AND MOTIVATION

This paper covers recent progress on integrable field theories and statistical models in two dimensions. Integrable models are solvable in a large extent and thus provide **exact non-perturbative** information for **truly interacting** theories. As it is explained here, on-shell properties are explicitly calculable: mass, energy, momentum, and conserved charges spectra as well as the S-matrix.

Continuum field theories and lattice models are naturally related by scaling limits. For integrable models such correspondence is particularly powerful since it systematically provides continuum integrable field theories from solutions of the Yang-Baxter equations. Recall that each Yang-Baxter solution defines an integrable vertex or face model on the lattice.

The scaling limit of integrable models yields in addition conformal field theories (CFT). As is well known CFT and its infinite dimensional symmetry form the grounds of string theory. It could be that integrable models and its infinite dimensional non-abelian symmetry (Yang-Baxter symmetry, see section VI), provide the starting point for new mathematical constructions going beyond the present day string models.

II. INTEGRABLE LATTICE MODELS AND THEIR CONTINUUM LIMITS

How to take the continuum limit of integrable lattice models has always been a major problem.

We shall try to convince the reader in this review that the light-cone approach [1,2] is the best way to perform the continuum limit of lattice models. (For previous references to the light-cone approach see [3]).

We consider a N x M two dimensional square lattice whose links are labeled by an index $a = 1, \ldots, n$. The statistical weight of each vertex where the four links meet is defined by the $R-$matrix $R^{ab}_{cd}(\theta)$ where $1 \leq a, b, c, d \leq n$ (see fig.1). Here θ is a complex variable called spectral variable. In the present context θ can be considered as a sort of coupling constant. It must be noticed that universal magnitudes are θ-independent in integrable models [6].

It is convenient to introduce an operator $T_{ab}(\theta, \tilde{\omega})$ associated to horizontal lines (see fig. 2)

$$T_{ab}(\theta, \tilde{\omega}) = \sum_{a_1, \ldots, a_{N-1}} t_{a_1 b}(\theta + \omega_1) \otimes t_{a_2 a_1}(\theta + \omega_2) \otimes \ldots.. \otimes t_{a a_{N-1}}(\theta + \omega_N) \tag{2.1}$$

For fixed a, b, $T_{ab}(\theta, \tilde{\omega})$ acts on the vertical space $\mathcal{V} = \bigotimes_{1 \leq i \leq N} V_i$, $V_i \equiv \mathbb{C}^n$, and the local vertex operators are defined as $[t_{ab}(\theta)]_{cd} \equiv R^{bd}_{ca}(\theta)$. In eq.(2.1) we introduced arbitrary inhomogeneity parameters $\omega_1, \omega_2, \ldots., \omega_N$ associated to each site on a horizontal line.

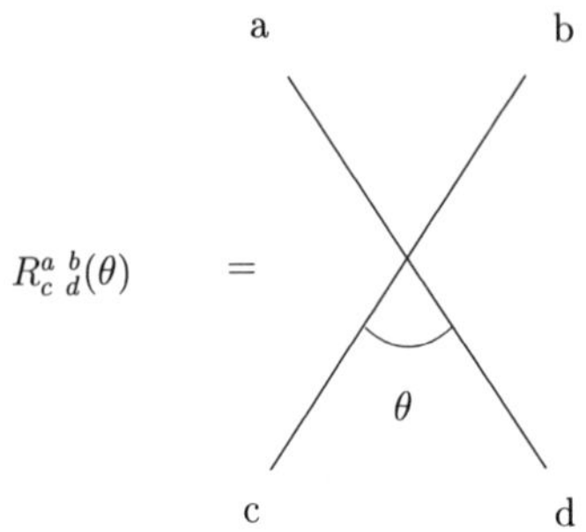

FIG. 1. The R-matrix elements $R_{cd}^{ab}(\theta)$ define the statistical weight of the depicted vertex configuration

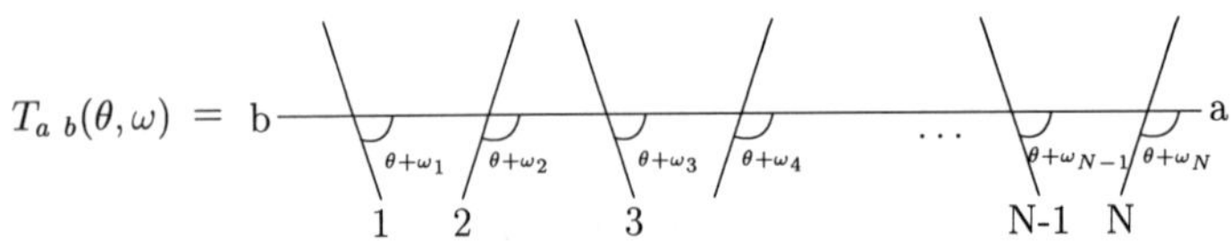

FIG. 2. Graphical representation of the inhomogeneous monodromy matrix. The angles between the horizontal and the vertical lines are site–dependent in an arbitrary way.

When periodic boundary conditions (PBC) are considered, it is useful to define the row-to-row transfer matrix as

$$t(\theta,\tilde{\omega}) = \sum_a T_{aa}(\theta,\tilde{\omega}) \tag{2.2}$$

For other types of boundary conditions, see refs. [6–8].

The first relevant physical problem is to compute the partition function $\mathcal{Z}$. It is defined as the sum over all possible configurations of the statistical weights for the whole lattice. For a N x M lattice with PBC in both directions, $\mathcal{Z}$ can be written as

$$\mathcal{Z} = Tr\left[t(\theta,\tilde{\omega})^M\right] \tag{2.3}$$

where Tr stands for the trace on the vertical space $\mathcal{V}$. The free energy is then given by

$$f(\theta,\tilde{\omega}) = -\lim_{(N,M)\to\infty}\frac{1}{NM}\log\mathcal{Z} \tag{2.4}$$

All considerations up to now are valid whether the model is integrable or not. We shall call integrable those models where the R-matrix $R(\theta)$ obeys the Yang-Baxter equations (YBE):

$$\begin{aligned}\sum_{1\le k,l,m\le n} & R^{kl}_{ba}(\theta-\theta')R^{dm}_{ck}(\theta)R^{ef}_{ml}(\theta') = \\ \sum_{1\le k,l,m\le n} & R^{lk}_{cb}(\theta')R^{mf}_{ka}(\theta)R^{de}_{lm}(\theta-\theta')\end{aligned} \tag{2.5}$$

or in tensor product notation

$$\begin{aligned}&[1\otimes R(\theta-\theta')]\,[R(\theta)\otimes 1]\,[1\otimes R(\theta')] = \\ &[R(\theta')\otimes 1]\,[1\otimes R(\theta)]\,[R(\theta-\theta')\otimes 1]\end{aligned} \tag{2.6}$$

It must be stressed that the YBE are a heavily overdetermined set of functional algebraic equations. They contain *a priori* n^4 unknowns [the elements of $R(\theta)$] and n^6 equations. Despite this fact a large set of solutions is known. All of them possess symmetries that reduce the number of independent equations and make possible the existence of solutions. The symmetries may be discrete as cyclic Z_n symmetries, continuous abelian symmetries as $U(1)^n$ and non-abelian as $GL(n)$ (see [6]).

The YBE (2.5-2.6) admit the natural graphical representation given in fig. 3. Graphically, the YBE express the freedom to push lines through intersections of pair of lines. This possibility of rigid line shifting can be interpreted as a zero curvature condition on the lattice.

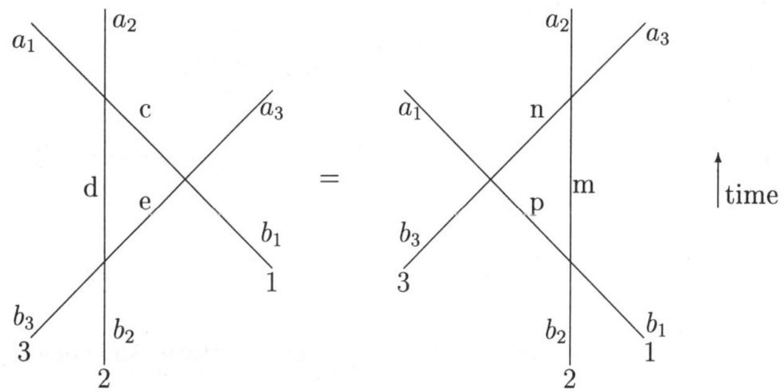

FIG. 3. The Yang-Baxter equation.

The YBE enjoy a powerful coproduct property. Namely, eqs.(2.5-2.6) implies that the operators $T_{ab}(\theta,\tilde\omega)$ fulfill the YB algebra

$$R(\lambda-\mu)\,[T(\lambda,\tilde\omega)\otimes T(\mu,\tilde\omega)] = [T(\mu,\tilde\omega)\otimes T(\lambda,\tilde\omega)]\,R(\lambda-\mu) \tag{2.7}$$

For one-site $(N = 1), T_{ab}(\theta, \tilde{\omega})$ reduces to $R(\theta)$ and eq.(2.7) becomes eq.(2.5). For N-sites, eq.(2.7) can be easily proved by repeatedly pushing lines through vertices (see [6]). That is, eq.(2.7) is the expression of the YBE for N-sites. The coproduct rule is here defined by eq.(2.1). Eq.(2.7) implies the commutativity of transfer matrices

$$[t(\theta, \tilde{\omega}), t(\theta', \tilde{\omega})] = 0 \tag{2.8}$$

That is, the transfer matrices *for fixed* $\omega_1, \omega_2, ..., \omega_N$ form a commuting family. Hence, one can expect to diagonalize it with θ-independent eigenvectors:

$$t(\theta, \tilde{\omega})\Psi(\tilde{\omega}) = \Lambda(\theta, \tilde{\omega})\ \Psi(\tilde{\omega}) \tag{2.9}$$

The Bethe Ansatz (BA) actually **does** this job [6]. Then, the free energy in the thermodynamic limit turns to be given by the largest eigenvalue $\Lambda(\theta, \tilde{\omega})_{max}$ of $t(\theta, \tilde{\omega})$. We find from eqs.(2.4) and (2.9)

$$f(\theta, \tilde{\omega}) = -\lim_{N\to\infty} \frac{1}{N} \log \Lambda(\theta, \tilde{\omega})_{max} \tag{2.10}$$

When $\theta = \theta'$, eq.(2.7) naturally suggest that $R(0)$ is a multiple of the unit matrix. This is usually the case. More precisely, a solution of the YBE (2.5) is called *regular* if

$$R(0) = c\ 1 \quad \text{that is} \quad R^{ab}_{cd}(0) = c\ \delta^a_c\ \delta^b_d \tag{2.11}$$

where c is a non-zero constant.

Setting $\theta = 0$ in eqs.(2.5) yields with the help of eq.(2.11)

$$M^{ef}_{ba}(\theta')\ \delta^d_c = \delta^f_a\ M^{de}_{cb}(\theta') \quad \text{where} \quad M^{ab}_{cb}(\theta) \equiv \sum_{1\leq k,l\leq n} R^{ab}_{kl}(-\theta) R^{kl}_{cd}(\theta)$$

We thus see that $M^{ab}_{cb}(\theta)$ must have the index structure $M^{ab}_{cb}(\theta) = \delta^a_c\ \delta^b_d\ \rho(\theta)$, where $\rho(\theta)$ is a c-number function. This can be written as

$$R(\theta)R(-\theta) = \rho(\theta) \ , \ \text{that is} \quad \sum_{1\leq c,d\leq n} R^{ab}_{cd}(\theta) R^{cd}_{ef}(-\theta) = \delta^a_e\ \delta^b_f\ \rho(\theta)$$

It follows that $\rho(\theta)$ is an even function. This property is usually called 'unitarity' although this may not be always the appropriate name.

From eqs.(2.1)-(2.2) at zero inhomogeneity $\omega_1 = \omega_2 = ... = \omega_N = 0$ and eq.(2.11), it follows that

$$t(0, \{\omega_k = 0\}) = c^N\ \Pi_s$$

where Π_s is the unit shift operator in the horizontal direction. The momentum operator is then given by

$$P \equiv -i \log \left[c^{-N} t(0, \{\omega_k = 0\})\right]$$

Moreover, it can be shown [9,6] that the operators

$$\mathcal{C}_m \equiv \frac{\partial^m}{\partial \theta^m} \log t(\theta, \{\omega_k = 0\})\Big|_{\theta=0}$$

couple $m+1$ neighbor sites on the horizontal line. Usually $\mathcal{C}_1$ is a quantum spin chain hamiltonian. The commutativity of the transfer matrices (2.8) implies

$$[\mathcal{C}_k, \mathcal{C}_l] = 0 \;\; \forall \; k, l$$

We thus find an infinite number of commuting magnitudes.

We are considering here vertex models where horizontal and vertical lines are of the same type. This is not necessary for integrability. It is possible to choose all vertical lines of a given kind (say with n_V states per vertical link) and all horizontal lines of a different kind (say with $n_H \neq n_V$ states per horizontal link) [6]. Moreover, it is possible to built integrable models mixing vertical (or horizontal) lines of different types [13].

It is also possible to construct integrable **face** models where the variables $(a, b, c, \ldots)$ are attached to the vertices. That is, to the dual lattice [10–12].

Up to now, we implicitly consider an euclidean two dimensional lattice. Let us now consider a diagonal-to-diagonal lattice (see fig.4) which represents a discretized **Minkowski** spacetime in light-cone coordinates. That is, the axis correspond to $x \pm t$, (x and t being the usual space and time variables).

In this approach we start from the discretized Minkowski 2D space–time formed by a regular diagonal lattice of right–oriented and left–oriented straight lines (see fig. 4). These represent true world–lines of "bare" objects (pseudo–particles) which are thus naturally divided in left– and right–movers. The right–movers have all the same positive rapidity Θ, while the left–movers have rapidity $-\Theta$. One can regard Θ as a cut–off rapidity, which will be appropriately taken to infinity in the continuum limit. Furthermore, we shall denote by V the Hilbert space of states of a pseudo–particle (we restrict here to the case in which V is the same for both left– and right–movers and has finite dimension n, although more general situations can be considered).

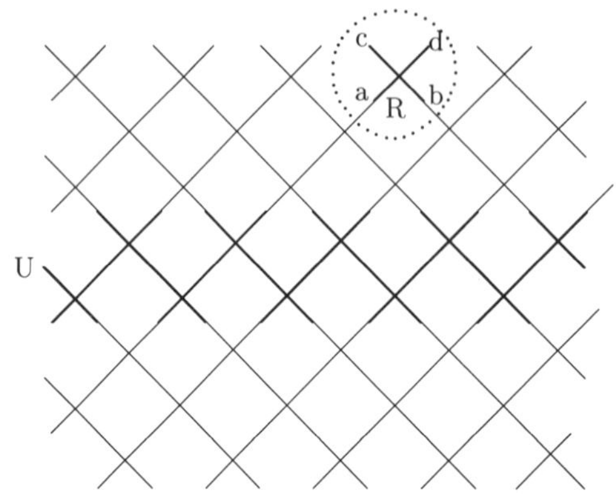

FIG. 4. Light–cone lattice representing a discretized portion of Minkowski space–time. A R–matrix of probability amplitudes is attached to each vertex. The bold lines correspond to the action, at a given time, of the one–step evolution operator U.

The dynamics of the model is fixed by the microscopic transition amplitudes attached to each intersection of a left– and a right–mover, that is to each vertex of the lattice. This amplitudes can be collected into linear operators R_{ij}, the local R–matrices, acting non–trivially only on the space $V_i \otimes V_j$ of ith and jth pseudo–particles. R_{ij} thus represent the relativistic scatterings of left–movers on right–movers and depend on the rapidity difference $\Theta - (-\Theta) = 2\Theta$, which is constant throughout the lattice. Moreover, by space–time translation invariance any other parametric dependence of R_{ij} must be the same for all vertices. We see therefore that attached to each vertex there is a matrix $R(2\Theta)_{ab}^{cd}$, where a, b, c, d are labels for the states of the pseudo–particles on the four links stemming out of the vertex, and take therefore n distinct values (see fig. 1). This is the general framework of a vertex model. The difference with the standard statistical interpretation is that the Boltzmann weights are in general complex, since we should require the unitarity of the matrix R. In any case, the integrability of the model is guaranteed whenever $R(\lambda)_{ab}^{cd}$ satisfy the Yang–Baxter equations (2.5).

For periodic boundary conditions, the one–step light–cone evolution operators $U_L(\Theta)$ and $U_R(\Theta)$, which act on the "bare" space of states $\mathcal{H}_N = (\otimes V)^{2N}$, (N is the number of sites on a row of the lattice, that is the number of diagonal lines), are built from the local R–matrices R_{ij} as [1].

$$\begin{aligned} U_R(\Theta) &= U(\Theta)V \,, \qquad U_L(\Theta) = U(\Theta)V^{-1} \\ U(\Theta) &= R_{12}R_{34}\dots R_{2N-1\,2N} \end{aligned} \tag{2.12}$$

where V is the one-step space translation to the right. U_R (U_L) evolves states by one step in right (left) light–cone direction. U_R and U_L commute and their product $U = U_R U_L$ is the unit time evolution operator. The graphical representation of U is given by the section of the diagonal lattice with fat lines in fig. 4. If a stands for the lattice spacing, the lattice hamiltonian H and total momentum P are naturally defined through

$$U = e^{-iaH} , \qquad U_R\, U_L^{-1} = e^{iaP} \tag{2.13}$$

The action of other fundamental operators is naturally defined on the same Hilbert space $\mathcal{H}_N$. These are the n^2 Yang-Baxter operators for $2N$ sites, which are conventionally grouped into the $n \times n$ monodromy matrix $T(\lambda) = \{T_{ab}(\lambda),\ a,b = 1,\ldots,n\}$. One usually regards the indices a, b of T_{ab} as *horizontal* indices fixing the out– and in–states of a reference pseudo–particle. Then $T(\lambda)$ is defined as horizontal coproduct of order $2N$ of the local vertex operators $L_j(\lambda) = R_{0j}(\lambda)P_{0j}$, where 0 label the reference space and P_{ij} is the transposition in $V_i \otimes V_j$. Explicitly

$$T(\lambda) = L_1(\lambda)L_2(\lambda)\ldots L_{2N}(\lambda)$$

The inhomogenuous generalization $T(\lambda, \vec{\omega}\,)$ then reads

$$T(\lambda, \vec{\omega}\,) = L_1(\lambda + \omega_1)L_2(\lambda + \omega_2)\ldots L_{2N}(\lambda + \omega_{2N})$$

and has the graphical representation of fig. 2. This expression is identical to eq.(2.1). $L_j(\lambda+\omega_j)$ can be regarded as the scattering matrix of the jth pseudo–particle carrying formal rapidity ω_j with the reference pseudo–particle carrying formal rapidity $-\lambda$.

In the case of our diagonal lattice of right– and left–moving pseudoparticles, there exists a specific, physically relevant choice of the inhomogeneities, namely

$$\omega_k = (-1)^k\Theta\ , \quad k = 1, 2, \ldots 2N$$

leading to the definition of the *alternating* monodromy matrix

$$T(\lambda, \Theta) \equiv T(\lambda, \{\omega_k = (-1)^k\Theta\})$$

In fact, the evolution operators $U_L(\Theta)$ and $U_R(\Theta)$ can be expressed in terms of the alternating transfer matrix

$$t(\lambda, \Theta) = t(\lambda, \{\omega_k = (-1)^k\Theta\}) \tag{2.14}$$

as [2]

$$U_R(\Theta) = t(\Theta, \Theta)\ , \quad U_L(\Theta) = t(-\Theta, \Theta)^{-1} \tag{2.15}$$

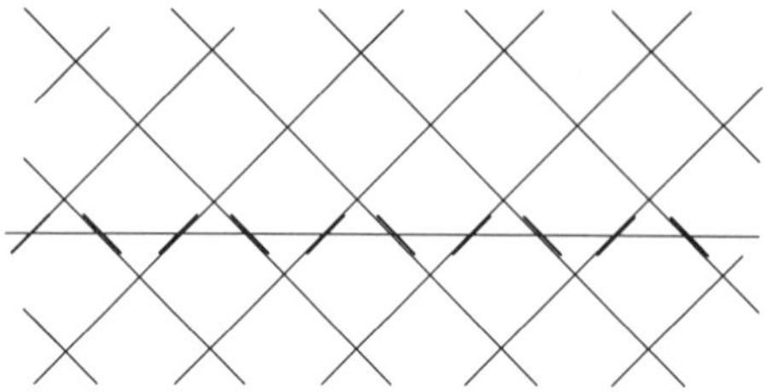

FIG. 5. Insertion of the alternating monodromy matrix in the light–cone lattice.

Notice that $T(\lambda,\Theta)$ fails to be conserved on the lattice only because of boundary effects. Indeed from fig. 5, which graphically represents the insertion of $T(\lambda,\Theta)$ in the lattice time evolution, one readily sees that U and $T(\lambda,\Theta)$ fail to commute only because of the free ends of the horizontal line. For all vertices in the bulk, the graphical interpretation of the YB equations (2.5), namely that lines can be freely pulled through vertices, allows to move $T(\lambda,\Theta)$ up or down, that is to freely commute it with the time evolution. The problem lays at the boundary: if periodic boundary conditions are assumed, then the free horizontal ends of $T(\lambda,\Theta)$ cannot be dragged along with the bulk, unless they are tied up, to form the transfer matrix $t(\lambda,\Theta)$. After all, for p.b.c., the boundary is actually equivalent to any point of the bulk and thus $t(\lambda,\Theta)$ commutes with U, as obvious also from eqs.(2.15) and the general fact that $[\,t(\lambda,\Theta),\, t(\mu,\Theta)\,] = 0$. One might think that the thermodynamic limit $N\to\infty$, by removing infinitely far away the troublesome free ends of $T(\lambda,\Theta)$, will allow for its conservation and thus for the existence of an exact YB symmetry with bare $R-$matrix. The situation however is not so simple: first of all one must fix the Fock sector of the $N\to\infty$ non–separable Hilbert space in which to take the thermodynamic limit. Different choices leads to different phases with dramatically different dynamics. Then the non–local structure of $T(\lambda,\Theta)$ must be taken into account. It is evident, for instance, that in the spin–wave Fock sector above ferromagnetic reference states $T(\lambda,\Theta)$ can never be conserved. Indeed, the working itself of the Quantum Inverse Scattering Method, where energy eigenstates are built applying non–diagonal elements of $T(\lambda,\Theta)$ on a specific ferromagnetic reference state, of course depends on $T(\lambda,\Theta)$ *not* commuting with the hamiltonian!

From the field–theoretic point of view, the most interesting phase is the antiferromagnetic one, in which the ground state plays the rôle of densely filled *interacting* Dirac sea (this holds for all known integrable lattice vertex models [1,2,5,6]. The corresponding Fock sector is formed by particle–like excitations which become relativistic massive particles within the scaling limit proper of the light–cone approach [2]. This consists in letting $a\to 0$ and $\Theta\to\infty$ in such a way that the physical mass scale

$$\mu = a^{-1}e^{-\kappa\Theta} \tag{2.16}$$

stays fixed. Here κ is a model–dependent parameter which for the integrable model where the R-matrix is a *rational* function of θ takes the general form [5]

$$\kappa = \frac{2\pi\, t}{h\, s} \tag{2.17}$$

where h is the dual Coxeter number of the underlying Lie algebra, s equals 1, 2 or 3 for simply, doubly and triply laced algebras, respectively, and $t = 1$ ($t = 2$) for non–twisted (twisted) algebras. For the class of model characterized by a trigonometric $R-$matrix (with anisotropy parameter γ) the expression (2.17) for κ is to be divided by γ [5].

The ground state or (physical vacuum) and the particle–like excitations of this antiferromagnetic phase are extremely more complicated than those of the ferromagnetic phase. It is therefore very hard to control, in the limit $N \to \infty$, the action of the alternating monodromy matrix $T(\lambda, \Theta)$ on the particle–like BA eigenstates of the alternating transfer matrix $t(\lambda, \Theta)$.

III. BOOTSTRAP CONSTRUCTION OF QUANTUM MONODROMY OPERATORS.

We briefly review in this section the work of refs. [14] where the exact (renormalized) matrix elements of a quantum monodromy matrix $\mathcal{T}_{ab}(u)$ (u is the generally complex spectral parameter) were derived using a bootstrap–like approach for a class of integrable local QFT's. In such theories there is no particle production and the $S-$matrix factorizes. The two–body $S-$matrix then satisfies the Yang–Baxter equations. Moreover, in the models considered in refs. [14] (the $O(N)$ nonlinear sigma model, the $SU(N)$ Thirring model and the $0(2N)$ $(\bar{\psi}\psi)^2$ model), thanks to scale invariance there exist classically conserved monodromy matrices. In general, the quantum $\mathcal{T}_{ab}(u)$ can be constructed by fixing its action on the Fock space of physical in and out many–particle states. The starting point are the following three general principles:

1. $\mathcal{T}_{ab}(u)$, $a, b = 1, 2, \ldots, n$, exist as quantum operators and are conserved.
2. $\mathcal{T}_{ab}(u)$ fulfill a quantum factorization principle.
3. $\mathcal{T}_{ab}(u)$ is invariant under P, T and the internal symmetries of the theory.

The quantum factorization principle referred above under 2. is nowadays called the "coproduct rule". This means that there exists the following relation between the action of $\mathcal{T}_{ab}(u)$ on $k-$particles states and its action on one–particle states

$$\begin{aligned}
\mathcal{T}_{ab}(u)\,|\theta_1\alpha_1, \theta_2\alpha_2, \ldots, \theta_k\alpha_k\rangle_{in} &= \textstyle\sum_{a_1 a_2 \ldots a_{k-1}} \mathcal{T}_{aa_1}(u)\,|\theta_1\alpha_1\rangle\, \mathcal{T}_{a_1 a_2}(u)\,|\theta_2\alpha_2\rangle \ldots \mathcal{T}_{a_{k-1} b}(u)\,|\theta_k\alpha_k\rangle \\
\mathcal{T}_{ab}(u)\,|\theta_1\alpha_1, \theta_2\alpha_2, \ldots, \theta_k\alpha_k\rangle_{out} &= \textstyle\sum_{a_1 a_2 \ldots a_{k-1}} \mathcal{T}_{a_1 b}(u)\,|\theta_1\alpha_1\rangle\, \mathcal{T}_{a_2 a_1}(u)\,|\theta_2\alpha_2\rangle \ldots \mathcal{T}_{a a_{k-1}}(u)\,|\theta_k\alpha_k\rangle
\end{aligned} \tag{3.1}$$

where θ_j and α_j $(1 \leq j \leq k)$ label the rapidities and the internal quantum numbers of the particles, respectively, in the asymptotic in and out states. Hence it is understood that $\theta_i > \theta_j$ for $i > j$.

Although $\mathcal{T}_{ab}(u)$ acts differently on in and out states, the assumption of conservation is nonetheless consistent. All the eigenvalues of a maximal commuting subset of $\{\mathcal{T}_{ab}(u),\ a, b = 1, 2, \ldots, n,\ u \in \mathbb{C}\}$ are identical for in and out states with given rapidities. Indeed the two in and out forms of the action on the internal quantum numbers are related by the unitary permutation $|\alpha_1, \alpha_2, \ldots, \alpha_k\rangle \to |\alpha_k, \alpha_{k-1}, \ldots, \alpha_1\rangle$.

Furthermore, principles 1. and 2. imply that $\mathcal{T}_{ab}(u)$ acts in a trivial way on the physical vacuum state $|0\rangle$:

$$\mathcal{T}_{ab}(u)\,|0\rangle = \delta_{ab}\,|0\rangle$$

This also fixes the normalization of $\mathcal{T}_{ab}(u)$ in agreement with the classical limit [14].

An immediate consequence of point 2. is that when $\mathcal{T}_{ab}(u)$ is expanded in powers of the spectral parameter u, it generates an infinite set of noncommuting and nonlocal conserved charges. This is the clue to the matching of the quantum monodromy matrix with its classical counterpart which is written nonlocally in terms of the local fields.

The main result in refs. [14] was to derive from 1. , 2. and 3. the explicit matrix elements of $\mathcal{T}_{ab}(u)$ on one–particle states. This result can be written as

$$\langle\theta\alpha|\,\mathcal{T}_{ab}(u)\,|\theta'\beta\rangle = \delta(\theta-\theta')S^{a\alpha}_{b\beta}(\kappa(u)+\theta) \tag{3.2}$$

where $S^{a\alpha}_{b\beta}(\theta-\theta')$ stands for the $S-$matrix of two–body scattering

$$|\theta b, \theta'\beta\rangle_{in} = \sum_{a\alpha} |\theta a, \theta'\alpha\rangle_{out}\, S^{a\alpha}_{b\beta}(\theta-\theta')$$

and $\kappa(u)$ is an odd function of u. Notice that this requires the presence in the model of particles with indices $a, b, \ldots$ as internal state labels. In the simplest situation these new labels coincide with those of the original particles. The appearance of a nontrivial "renormalization" $u \to \kappa(u)$ is to be expected when there exist a definition of the spectral parameter outside the bootstrap itself. This is the case of the models of refs. [14], which posses Lax pairs and auxiliary problems which fix the definition of u. Here we adopt the purely bootstrap viewpoint and fix the definition of u so that $\kappa(u) = u$. In principle, an extra $u-$ and $\theta-$dependent phase factor may appear in the r.h.s. of eq.(3.2). However, no phase showed up in the specific models of refs. [14], when nonperturbative checks were performed using the operator product expansion. Eq.(3.2) can be written in a more suggestive way as

$$\mathcal{T}_{ab}(u)\,|\theta\beta\rangle = \sum_{\alpha} |\theta\alpha\rangle\, S^{a\alpha}_{b\beta}(u+\theta) \tag{3.3}$$

This equation, when combined with eqs.(3.1), completely defines the quantum monodromy operators in the Fock space. From the YB equations satisfied by the $S-$matrix it then follows that $\mathcal{T}_{ab}(u)$ fulfills the YB algebra

$$\hat{R}(u-v)\,[\mathcal{T}(u)\otimes\mathcal{T}(v)] = [\mathcal{T}(u)\otimes\mathcal{T}(v)]\,\hat{R}(u-v) \tag{3.4}$$

where $\hat{R}^{a\alpha}_{b\beta}(u) = S^{\alpha a}_{b\beta}(u)$. It should be stressed that the conservation of $\mathcal{T}_{ab}(u)$ implies that this YB algebra is a true *non-abelian infinite symmetry algebra* of the relativistic local QFT. On the contrary the rôle of the YB algebra in integrable vertex and face models on finite lattices or in nonrelativistic quantum models is that of a dynamical symmetry underlying the Quantum Inverse Scattering Method. In these latter cases, only the transfer matrix, namely

$$\tau(u) = \sum_a \mathcal{T}_{aa}(u)$$

is conserved. Since $[\tau(u), \tau(v)] = 0$, the transfer matrix just generates an abelian symmetry.

The dynamical symmetry underlying the integrable QFT includes in addition non–conserved operators $Z_\alpha(\theta)$ which create the particle eigenstates out of the vacuum. In the bootstrap framework they can be introduced à la Zamolodchikov–Faddeev, by setting

$$|\theta_1\alpha_1, \theta_2\alpha_2, \ldots, \theta_k\alpha_k\rangle_{in} = Z_{\alpha_k}(\theta_k)Z_{\alpha_{k-1}}(\theta_{k-1})\ldots Z_{\alpha_1}(\theta_1)\,|0\rangle$$

$$|\theta_1\alpha_1, \theta_2\alpha_2, \ldots, \theta_k\alpha_k\rangle_{out} = Z_{\alpha_1}(\theta_1)Z_{\alpha_2}(\theta_2)\ldots Z_{\alpha_k}(\theta_k)\,|0\rangle$$

with the fundamental commutation rules

$$Z_{\alpha_2}(\theta_2)Z_{\alpha_1}(\theta_1) = \sum_{\beta_1\beta_2} S^{\beta_1\beta_2}_{\alpha_1\alpha_2}(\theta_1-\theta_2)Z_{\beta_1}(\theta_1)Z_{\beta_2}(\theta_2) \tag{3.5}$$

Combining now eqs.(3.1),(3.3) and (3.5), we obtain the algebraic relation between monodromy and Zamolodchikov–Faddeev operators:

$$\mathcal{T}_{ab}(u)Z_\beta(\theta) = \sum_{c\alpha} Z_\alpha(\theta)\mathcal{T}_{ac}(u)S^{c\alpha}_{b\beta}(u+\theta)$$

Together with eqs.(3.4) and (3.5), these relations close the complete dynamical algebra of an integrable QFT. For the XXZ spin chain in the regime $|q| < 1$, the ZF operators have been identified in ref. [42] with special vertex operators (or representation intertwiners of the relevant q–deformed affine Lie algebra). They are uniquely characterized by being solutions of the q–deformed Knizhnik–Zamolodchikov equation and by their normalization [21].

IV. LATTICE CONSTRUCTION OF QUANTUM MONODROMY OPERATORS AND BETHE ANSATZ

In order to study the infinite volume limit of $T(\lambda, \Theta)$ on the physical Fock space (that is, finite energy excitations around the antiferromagnetic vacuum), one needs to compute scalar products of Bethe Ansatz states to derive relations like (3.2) or (3.3) with $T(\lambda, \Theta)$ instead of

$\mathcal{T}(\lambda, \Theta)$ in the l.h.s. Since this kind of calculations are indeed possible but rather involved, we computed in ref. [49] the eigenvalues of $t(\lambda, \Theta)$ on a generic state of the physical Fock space. Then, we compare these eigenvalues with those of the bootstrap transfer matrix. This tell us whether the bare and the renormalized YB algebras have a common abelian subalgebra. Notice that this fact alone provides a microscopic basis for the TBA, which originally relies solely on the bootstrap.

We consider once more the sG model as example, although the same result would apply to any integrable QFT admitting a light–cone lattice regularization. This class of models contains also the O(N) nonlinear sigma model and the $SU(N)$ Thirring model considered from the bootstrap viewpoint in refs. [14].

The integrable light–cone lattice regularization of the sG–mT model is provided by the six-vertex model [1]. Therefore, the space V is $\mathbb{C}^2$ and the unitarized local $R-$matrices can be written

$$\begin{aligned} R_{jk}(\lambda) &= \tfrac{1+c}{2} + \tfrac{1-c}{2}\sigma_j^z\sigma_k^z + \tfrac{b}{2}(\sigma_j^x\sigma_k^x + \sigma_j^y\sigma_k^y) \\ b(\lambda) &= \tfrac{\sinh\lambda}{\sinh(i\gamma-\lambda)} \quad , \quad c(\lambda) = \tfrac{\sinh i\gamma}{\sinh(i\gamma-\lambda)} \end{aligned} \tag{4.1}$$

where γ is commonly known as anisotropy parameter and σ_j^x, σ_j^y and σ_j^z are Pauli matrices acting at the site j.

The standard Algebrized BA can be applied to the diagonalization of the alternating transfer matrix $t(\lambda, \Theta)$ with the following results [1,2,15]. The BA states are written

$$\Psi(\vec{\lambda}) = B(\lambda_1)....B(\lambda_M)\Omega$$

where $\vec{\lambda} \equiv (\lambda_1, \lambda_2, \ldots, \lambda_M)$, $B(\lambda_i) = T_{+-}(\lambda_i + i\gamma/2, \Theta)$ and Ω is the ferromagnetic ground-state (all spins up). They are eigenvectors of $t(\lambda, \Theta)$

$$t(\lambda, \Theta)\Psi(\vec{\lambda}) = \Lambda(\lambda; \vec{\lambda})\Psi(\vec{\lambda})$$

provided the λ_i are all distinct roots of the "bare" BA equations

$$\left(\frac{\sinh[i\gamma/2 + \lambda_j - \Theta]}{\sinh[i\gamma/2 - \lambda_j + \Theta]}\,\frac{\sinh[i\gamma/2 + \lambda_j + \Theta]}{\sinh[i\gamma/2 - \lambda_j - \Theta]}\right)^N = -\prod_{k=1}^{M} \frac{\sinh[+i\gamma + \lambda_j - \lambda_k]}{\sinh[-i\gamma + \lambda_j - \lambda_k]} \tag{4.2}$$

The eigenvalues $\Lambda(\lambda; \vec{\lambda})$ are the sum of a contribution coming from $A(\lambda) = T_{++}(\lambda, \Theta)$ and one coming from $D(\lambda) = T_{--}(\lambda, \Theta)$,

$$\Lambda(\lambda; \vec{\lambda}) = \Lambda_A(\lambda; \vec{\lambda}) + \Lambda_D(\lambda; \vec{\lambda}) \tag{4.3}$$

Here

$$\begin{aligned} \Lambda_A(\lambda; \vec{\lambda}) &= \exp\left[-iG(\lambda, \vec{\lambda})\right] \\ \Lambda_D(\lambda; \vec{\lambda}) &= e^{-iN[\phi(\lambda - i\gamma/2 - \Theta, \gamma/2) + \phi(\lambda - i\gamma/2 + \Theta, \gamma/2)]} \exp\left[iG(\lambda - i\gamma, \vec{\lambda})\right] \end{aligned}$$

and

$$G(\lambda,\vec{\lambda}) \equiv \sum_{j=1}^{M} \phi(\lambda-\lambda_j,\gamma/2)\ ,\quad \phi(\lambda,\gamma)\equiv i\log\frac{\sinh(i\gamma+\lambda)}{\sinh(i\gamma-\lambda)}$$

$G(\lambda,\vec{\lambda})$ is manifestly a periodic function of λ with period $i\pi$. Notice also that $\Lambda_D(\pm\Theta,\vec{\lambda})=0$. That is, only $\Lambda_A(\pm\Theta,\vec{\lambda})$ contributes to the energy and momentum eigenvalues:

$$\begin{aligned} E(\Theta) &= a^{-1}\sum_{j=1}^{M}[\phi(\Theta+\lambda_j,\gamma/2)+\phi(\Theta-\lambda_j,\gamma/2)-2\pi] \\ P(\Theta) &= a^{-1}\sum_{j=1}^{M}[\phi(\Theta+\lambda_j,\gamma/2)-\phi(\Theta-\lambda_j,\gamma/2)] \end{aligned} \tag{4.4}$$

The ground state and the particle–like excitations of the light–cone six–vertex model are well known [1,6]: the ground state corresponds to the unique solution of the BAE with $N/2$ consecutive real roots (notice that the energy in eq.(4.4) is negative definite, so that the ground state is obtained by filling the interacting Dirac sea). In the limit $N\to\infty$ this yields the antiferromagnetic vacuum. Holes in the sea appear as physical particles. A hole located at φ carries energy and momentum, relative to the vacuum,

$$e(\varphi)=2a^{-1}\arctan\left(\frac{\cosh\pi\varphi/\gamma}{\sinh\pi\Theta/\gamma}\right)\ ,\quad p(\varphi)=-2a^{-1}\arctan\left(\frac{\sinh\pi\varphi/\gamma}{\cosh\pi\Theta/\gamma}\right) \tag{4.5}$$

In the scaling limit $a\to 0,\ \Theta\to\infty$ with $e(0)$ held fixed, we then obtain $(e,p)=m(\cosh\theta,\sinh\theta)$ with

$$m\equiv 4a^{-1}\exp(-\pi\Theta/\gamma)\ ,\quad \theta\equiv -\pi\varphi/\gamma \tag{4.6}$$

We have thus proved that the continuum limit is **relativistic** with a finite non-zero mass. It provides a continuum relativistic massive field theory out of **any** gapless integrable model. Here, we have only considered the six-vertex model that yields the massive Thirring (mT) model [1].

We identify m as the physical mass and θ as the physical rapidity of a sG soliton (mT fermion) or antisoliton (antifermion). Complex roots of the BAE are also possible. They correspond to *magnons*, that is to different polarization states of several sG solitons (mT fermions), or to breather states (in the attractive regime $\gamma>\pi/2$).

Within the light-cone approach one can also perform the continuum limit at the bare level. For the six-vertex model we defined lattice fermion fields ψ_n and we found their equations of motion on the lattice Minkowski spacetime [1]:

$$\begin{aligned} U_R\psi_{2n-2}U_R^+ = U_L\psi_{2n}U_L^+ &= \bar{b}\psi_{2n}+\bar{c}\psi_{2n-1}+(c-\bar{c})\psi_{2n}^+\psi_{2n}\psi_{2n-1}-(b+\bar{b})\psi_{2n-1}^+\psi_{2n-1}\psi_{2n} \\ U_L\psi_{2n-1}U_L^+ = U_R\psi_{2n+1}U_R^+ &= \bar{b}\psi_{2n-1}+\bar{c}\psi_{2n}+(c-\bar{c})\psi_{2n-1}^+\psi_{2n-1}\psi_{2n}-(b+\bar{b})\psi_{2n}^+\psi_{2n}\psi_{2n-1} \\ &\text{where } b\equiv b(2\Theta) \text{ and } c\equiv c(2\Theta) \end{aligned} \tag{4.7}$$

These second quantized field equations are perfectly defined on the lattice. The bare scaling limit is not identical to the renormalized limit defined by eq.(4.6). The detailed proof in ref. [1] shows that one finds the bare continuum mTm if one takes in eq.(4.7)

$$\Theta \to \infty \quad , \quad a \to 0 \quad \text{with} \quad m_0 \equiv \frac{4}{a} \sin\gamma \; e^{-2\Theta} \quad \text{kept fixed}$$

We see that the **bare** mass m_0 scales as $e^{-2\Theta}$ while the **renormalized** mass scales as $e^{-\pi\Theta/\gamma}$ [eq.(4.6)].

After some calculations [1], the continuum limit of the momentum and hamiltonian defined by eq.(2.13) take the form

$$\begin{aligned} P &= -i \int dx \psi^+ \partial_x \psi \quad \text{and} \\ H &= \int dx \left[-i\psi^+ \left(\gamma^5 \partial_x + i m_0 \gamma^0 \right) \psi + \frac{g}{2} \left(\bar{\psi} \gamma_\mu \psi \right)^2 \right] \quad , \end{aligned} \tag{4.8}$$

where

$$\psi(x) = \begin{pmatrix} \psi_R(x) \\ \psi_L(x) \end{pmatrix} \; , \; \psi_{2n} = \sqrt{a}\, \psi_R(x + \xi a) \quad , \quad \psi_{2n-1} = \sqrt{a}\, \psi_L(x - \xi a)$$

$$\text{with } 0 < \xi < 1/2 \; , \; x = na \; , \; \text{ and } g = -2\cot\gamma \; , \; \gamma_1 = -i\sigma_y \; , \; \gamma_0 = \sigma_x \; , \; \gamma_5 = \sigma_z \; . \tag{4.9}$$

Notice that there is an *exact* and finite relation between the bare continuum (g), the lattice (γ) and the renormalized ($\tilde{\gamma} = \frac{\gamma}{1-\gamma/\pi}$) coupling.

For R-matrices acting on finite dimensional spaces V one gets fermion or parafermion field theories. In order to describe bosonic field theories one needs infinite dimensional representation spaces V in the framework of the light-cone approach. [Otherwise, bosons can appear as bound states of fermions as in the mTm-sG model].

Let us discuss briefly here the rational limit of the six-vertex R-matrix in its spin S representation [16]. That is, for $V = \mathbb{C}^{2S+1}$.

$$R_{jk}(\theta) = \frac{\Gamma(2S + 1 + i\theta)\Gamma(J + 1 - i\theta)}{\Gamma(2S + 1 - i\theta)\Gamma(J + 1 + i\theta)} \tag{4.10}$$

where the operator J is defined by

$$J(J+1) = 2S(S+1) + 2\vec{S}_j \otimes \vec{S}_k$$

where $\vec{S}_j$ and $\vec{S}_k$ are spin S operators acting on the spaces V_j and V_k respectively. [$(\vec{S}_j)^2 = (\vec{S}_k)^2 = S(S+1)$]. The hamiltonian and momentum (2.13) describe in the $S = \infty$ limit the principal chiral model (PCM) [17,18]. However, this is not the full hamiltonian. One finds in this way states which are left (or right) $SU(2)$-singlets. The lattice current construction (5.4)-(5.7) [see below] holds for the PCM. Notice that for large θ the R-matrix (4.10) possess an expansion like eq.(5.2). Then, the whole procedure works yielding a conserved a curvatureless current that we can identify either with the $SU(2)_L$ or with the $SU(2)_R$ current. This whole construction generalizes to the $SU(N)$ PCM. It also generalizes to PCM with one anisotropy axis (trigonometric R-matrices) [19].

The light-cone approach to the sine-Gordon-mTm model using bosonic fields is worked out in ref. [20].

V. VERTEX MODELS AND FIELD THEORIES ASSOCIATED TO Q-DEFORMED LIE ALGEBRAS

The R-matrices solutions of the YB eq.(2.5) can be classified according to

1. The Lie algebra (or q-deformed Lie algebra) to whom they are associated.
2. The couple of Lie algebra representations where they act: $V \otimes V'$, (Vmay coincide or not with V') .

The six-vertex R-matrix is associated to the $q - A_1$ Lie algebra in its fundamental (spin 1/2) representation. The R-matrix [6]

$$\begin{aligned} R^{ab}_{ab}(\theta) &= \tfrac{\sin\gamma}{\sin(\gamma-\theta)}\, e^{i\theta\,\mathrm{sign}(a-b)} \quad , \quad a \neq b \quad ; \\ R^{ab}_{ba}(\theta) &= \tfrac{\sin\theta}{\sin(\gamma-\theta)} \quad , \quad a \neq b \quad ; \\ R^{aa}_{aa}(\theta) &= 1 \\ & 1 \leq a, b \leq n \end{aligned} \tag{5.1}$$

corresponds to the $q - A_{n-1}$ Lie algebra in its n-dimensional (quark) representation. When $q = e^{i\gamma}$ is a root of unity representations which are unknown for $q = 1$ appear and hence new models can be constructed. They are better defined in face language. For the $q - A_1$ case they are called RSOS models [10]. R-matrices associated to other q-Lie algebras can be found in [23].

The eigenvectors of the transfer matrix can be obtained *via* the Bethe Ansatz (BA) in its various generalizations. (The BA can also be formulated in face language [12,11]). When the R-matrix corresponds to a Lie algebra of rank larger than one, the nested Bethe Ansatz (NBA)must be used.

The nested Bethe Ansatz (NBA) is probably the most sophisticated algebraic construction of eigenvectors for integrable lattice models. It consists of several levels of BA each one inside the previous one. This is the reason of its name. The NBA has been worked out for the A_{n-1} trigonometric and hyperbolic vertex model [6], for the $Sp(2n)$ symmetric vertex model [24] and for $O(2n)$ symmetric vertex model [25] (always in the fundamental representation). The structure of the NBA is closely related to the respective Dynkin diagram. For D_n , one starts by one end of the 'fork', goes till the end of the diagram and then back till the other end of the 'fork' [25].

The NBA equations (NBAE) for a class of vertex models associated to simple Lie Algebras has been proposed in ref. [26] and solved (in a large extent) in ref. [5]. Let us summarize the more relevant results for the field theory limit.

The structure of the NBAE for a given R-matrix is dictated by the associated Dynkin diagram. There are sets of NBA roots associated to each spot of the Dynkin diagram. The NBAE couple the roots associated to the same spot and to the roots associated to spots connected to it in the Dynkin diagram. The structure of the ground state may be antiferromagnetic (AF) or ferromagnetic (F) depending on the chosen regime (values of θ and γ).

The AF ground state yields the more interesting field theories. It is formed by filling all 'Dirac' seas with BA roots. There is a Dirac sea for each NBA level. The ground state roots are real for simply laced Lie algebras apart of a constant imaginary part that depends on the level in a simple way [6,5]. For non-simply laced cases [5] and for non-fundamental representations of all algebras, complex ('string' type) roots form the ground state.

On the top of the AF ground state (renormalizes or physical vacuum) there are excitations. One finds as many branches of excitations as the rank of the Lie algebra. They follow making holes and adding complex roots to each Dirac sea. The mass spectrum is usually q-independent except for non-compact q-Lie algebras [27]. The mass spectrum coincides (for simply laced cases) up to a general factor with the components of the Perron-Frobenius eigenvector of the Cartan matrix.

The field theories obtained for R-matrices in the fundamental representations are basically fermions or para-fermion models. In ref. [5] we computed the mass spectrum and the S-matrix in these models for most of the q-deformed simple Lie algebras.

Let us now briefly discuss the scaling limit of vertex models with rational R-matrices associated to a Lie algebra $\mathcal{G}$. These R-matrices have the asymptotic behaviour

$$R(\theta) \overset{\theta\to\infty}{=} P\left[1 + \frac{\Pi+\mu}{i\theta} + O\left(\frac{1}{\theta^2}\right)\right] \tag{5.2}$$

where μ is a numerical constant, $P^{ab}_{cd} = \delta^a_d\, \delta^b_c$ is the exchange operator and

$$\Pi = \sum_{\alpha=1}^{dim\mathcal{G}} T_\alpha \otimes T^\alpha \tag{5.3}$$

We then introduce the lattice operator

$$T^\alpha_n \equiv 1 \otimes \ldots \otimes \overbrace{T^\alpha}^{n^{th} \cdot site} \otimes \ldots \otimes 1 \tag{5.4}$$

Using eqs.(2.12),(5.2) and (5.3) and the Lie algebra commutators

$$\left[T^\alpha, T^\beta\right] = i f_\gamma^{\alpha\beta}\, T^\gamma$$

we can show that the operators T_n^α obey *local* equations of motion on the lattice [2]

$$\begin{aligned} U_R T_{2n-2}^\alpha U_R^+ = U_L T_{2n}^\alpha U_L^+ &= T_{2n}^\alpha + \tfrac{2i}{\theta} f_\gamma^{\alpha\beta} T_{2n-1}^\beta T_{2n}^\gamma + O(\tfrac{1}{\theta^2}), \\ U_R T_{2n-1}^\alpha U_R^+ = U_L T_{2n+1}^\alpha U_L^+ &= T_{2n-1}^\alpha - \tfrac{2i}{\theta} f_\gamma^{\alpha\beta} T_{2n-1}^\beta T_{2n}^\gamma + O(\tfrac{1}{\theta^2}). \end{aligned} \tag{5.5}$$

The bare scale limit is now defined as $a \to 0,\ \theta \to \infty,\ x = na$ fixed. We find

$$\partial^\mu J_\mu^\alpha(x) = 0 \quad , \quad \partial_0 J_1^\alpha - \partial_1 J_0^\alpha + ig f_\gamma^{\alpha\beta}\left[J_0^\beta, J_1^\gamma\right] = 0. \tag{5.6}$$

where $\mu = 0, 1$ and in light-cone coordinates

$$J_R^\alpha \equiv \frac{1}{ga\theta}\, T_{2n}^\alpha \quad , \quad J_L^\alpha \equiv \frac{1}{ga\theta}\, T_{2n-1}^\alpha \tag{5.7}$$

Therefore we have a lattice version of the $\mathcal{G}$-algebra currents $J_\mu^\alpha(x)$ associated to an exactly integrable discretization of the field theory model. Eqs.(5.6) characterize the currents in the non-abelian Thirring model associated to the Lie algebra $\mathcal{G}$. This model has as Lagrangian

$$\mathcal{L} = i\bar\psi\, \partial\!\!\!/ \psi - \frac{g}{4}\left(\bar\psi\gamma_\mu T^\alpha\psi\right)\left(\bar\psi\gamma^\mu T^\beta\psi\right) K_{\alpha\beta} \tag{5.8}$$

Here ψ transforms under an irreducible representation ρ of $\mathcal{G}$, T^α are the $\mathcal{G}$-generators in that representation and $K_{\alpha\beta}$ is proportional to the inverse of the Killing form. Actually the hamiltonian and momentum [H and P defined by eq.(2.13)] describe the zero-chirality (massive) sector of the model (5.8) (see ref. [2]) and we can identify

$$J_\mu^\alpha(x) = \bar\psi\gamma_\mu T^\alpha\psi \tag{5.9}$$

The renormalized scaling limit is discussed in [2,6] and through eqs.(2.16)-(2.17).

VI. THERMODYNAMIC LIMIT OF THE TRANSFER MATRIX FROM THE BETHE ANSATZ

In ref. [49] it is shown that the bootstrap construction (discussed in sec. III) of conserved $\mathcal{T}_{ab}(u)$ generalizes to integrable models with trigonometric $R-$matrices such as the sine-Gordon or massive Thirring model. In such cases the classical limit is abelian, as shown explicitly in ref. [49].

The main aim of ref. [49] was to investigate and clarify, from a microscopic point of view, the problem of unveiling the existence of the infinite YB symmetry of the sG–mT model. In other words, since lattice models provide regularized version of QFT, we seek an explicit connection between the lattice and the bootstrap YB algebras.

In order to investigate the operators present in such QFT, it is important to learn how the monodromy operators $T_{ab}(\lambda, \Theta)$ act on physical states. In ref. [49] we explicitly compute the eigenvalues of the alternating six–vertex transfer matrix $t(\lambda, \Theta)$, on a generic $n-$particle state in the thermodynamic limit.

The eigenvalues of $t(\lambda, \Theta)$ turned out to be $i\pi-$periodic and multi–valued functions of λ each determination of $t(\lambda, \Theta)$ being a meromorphic function of λ. We call $t^{II}(\lambda, \Theta)$ and $t^{I}(\lambda, \Theta)$ the determinations associated with the periodicity strips closer to the real axis . The ground-state contribution $\exp[-iG(\lambda)_V]$ is exponential on the lattice size, as expected, whereas the excited states contributions are finite and express always in terms of hyperbolic functions

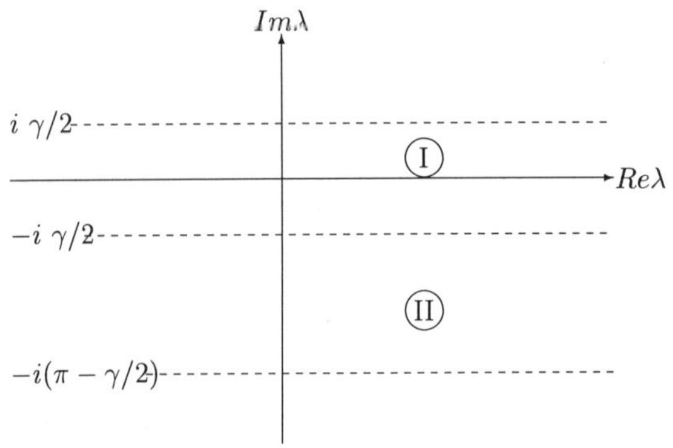

FIG. 6. The two main determinations, $G^{I}(\lambda)$ and $G^{II}(\lambda)$ are defined by $G(\lambda)$ with λ in strips I and II, respectively.

In strip I $|\text{Im}\,\lambda| < \gamma/2$, we define the renormalized type I transfer matrix

$$t^{I}(\lambda) = \lim_{N\to\infty} t(\lambda, \Theta) \exp[iG^{I}(\lambda)_V](-)^{J_z - N/2} \tag{6.1}$$

where $J_z = N/2 - M$ is to be identified with the soliton (or fermion) charge of the continuum sG–mT model. The last sign factor in eq.(6.1) corresponds to square–root branch choice suitable to obtain the relation

$$t^{I}(\pm\Theta) = \exp\{-ia[P_\pm - (P_\pm)_V]\} \tag{6.2}$$

where $P_\pm \equiv (H \pm P)/2$ [see eqs.(2.13), (2.15), (4.4)] and $(P_\pm)_V$ stands for the vacuum contribution. Notice that the $\Theta-$dependence of $t^{I}(\lambda)$ has been completely canceled out, since it is present only in the vacuum contribution. In fact, from the Bethe Ansatz calculations in ref [49], we read the eigenvalue $\Lambda^{I}(\lambda)$ of $t^{I}(\lambda)$ on a generic particle state:

$$\Lambda^I(\lambda) = \exp\left[-2i\sum_{n=1}^{k}\arctan\left(e^{\pi\lambda/\gamma+\theta_n}\right)\right] = \prod_{n=1}^{k}\coth\left(\frac{\pi\lambda}{2\gamma}+\frac{\theta_n}{2}+\frac{i\pi}{4}\right) \tag{6.3}$$

where $\theta_n \equiv -\pi\varphi_n/\gamma$ are the physical particle rapidities. Suppose now we expand $\log\Lambda^I(\lambda)$ in powers of $z = e^{-\pi|\lambda|/\gamma}$ around $\lambda = \pm\infty$,

$$\pm i\log\Lambda^I(\lambda) = \sum_{j=0}^{\infty} z^{2j+1}\frac{(-1)^j}{j+1/2}\sum_{n=1}^{k} e^{\pm(2j+1)\theta_n}$$

One has to regard the coefficients of the expansion parameter z as the eigenvalues of the conserved abelian charges generated by the transfer matrix. The additivity of the eigenvalues implies the locality of the charges. In terms of operators we can write, around $\lambda = \pm\infty$,

$$\pm i\log t^I(\lambda) = \sum_{j=0}^{\infty}\left[\frac{4z}{m}\right]^{2j+1} I_j^{\pm} \tag{6.4}$$

where $I_0^{\pm} = p_{\pm}$ is the continuum light–cone energy–momentum and the $I_j^{\pm}$, $j \geq 1$, are local conserved charges with dimension $2j+1$ and Lorentz spin $\pm(2j+1)$. Their eigenvalues

$$\frac{(-1)^j}{j+1/2}\sum_{n=1}^{k}\left[\frac{m}{4}e^{\pm\theta_n}\right]^{2j+1}$$

coincide with the values on multisoliton solutions of the higher integrals of motion of the sG equation [28]. It is remarkable that these eigenvalues are free of quantum corrections although the corresponding operators in terms of local fields certainly need renormalization. Let us stress that explicit expressions for these conserved charges can be obtained by writing the local R–matrices in terms of fermi operators, as in ref. [1]. Notice also that, combining eqs.(6.2) with (6.4), and recalling the scaling law (4.6), we can write

$$P_{\perp} - (P_{\perp})_V = p_{\perp} + \frac{m}{4}\sum_{j=1}^{\infty}\left(\frac{ma}{4}\right)^{2j} I_j^{\pm} \tag{6.5}$$

That is, the light-cone lattice hamiltonian and momentum can be expressed in a precise way as the continuum hamiltonian and momentum plus an infinite series of continuum higher conserved charges, playing the rôle of irrelevant operators.

The explicit Bethe Ansatz calculation in ref. [49] showed that in the strip II, the lattice transfer matrix eigenvalues match with the bootstrap transfer matrix eigenvalues.

We obtained as general form of the A and D contributions to the eigenvalue of $t(\lambda,\Theta)$ [see eq.(4.3)] on the $N\to\infty$ limit of the BA states for λ in strip II [49]:

$$\Lambda_A(\lambda) = -e^{-iG^{II}(\lambda)V}\left\{\prod_{n=1}^{k}S(x_n)\coth\frac{x_n}{2}\right\}\prod_{j=1}^{m}\frac{\sinh\hat{\gamma}[i/2+(\frac{\pi}{\gamma}(\lambda+i\gamma/2)-u_j)/\pi]}{\sinh\hat{\gamma}[i/2-(\frac{\pi}{\gamma}(\lambda+i\gamma/2)+u_j)/\pi]}$$

and

$$\Lambda_D(\lambda) = -e^{-iG^{II}(\lambda)_V} \left\{ \prod_{n=1}^{k} S(x_n)\hat{b}(x_n) \coth \frac{x_n}{2} \right\} \prod_{j=1}^{m} \frac{\sinh \hat{\gamma}[3i/2 + (\frac{\pi}{\gamma}(\lambda + i\gamma/2) - u_j)/\pi]}{\sinh \hat{\gamma}[-i/2 + (\frac{\pi}{\gamma}(\lambda + i\gamma/2) - u_j)/\pi]}$$

where for definiteness we chose the strip II , $-\pi + \gamma/2 < \mathrm{Im}\,\lambda < -\gamma/2$ and set $x_n = \frac{\pi}{\gamma}(\lambda + i\gamma/2) + \theta_n$. The distinct numbers $u_1, u_2, \ldots, u_m$ must satisfy the BA equations

$$\prod_{n=1}^{k} \frac{\sinh \hat{\gamma}[i/2 + (u_j + \theta_n)/\pi]}{\sinh \hat{\gamma}[i/2 - (u_j + \theta_n)/\pi]} = -\prod_{r=1}^{m} \frac{\sinh \hat{\gamma}[+i + (u_j - u_r)/\pi]}{\sinh \hat{\gamma}[-i + (u_j - u_r)/\pi]}$$

These last two expressions can be connected with that for the eigenvalues of the bootstrap transfer matrix $\tau(u)$ [49], provided we *identify* u with $\frac{\pi}{\gamma}(\lambda + i\gamma/2)$. We find indeed [49]:

$$\Lambda(\lambda) = -e^{-iG^{II}(\lambda)_V} \xi(u) \prod_{n=1}^{k} \coth\left(\frac{u + \theta_n}{2}\right) \tag{6.6}$$

where $\xi(u)$ is the eigenvalue of the bootstrap transfer matrix $\tau(u)$ and λ is in strip II . In analogy with eq.(6.1), we now define the type II renormalized transfer matrix

$$t^{II}(\lambda) = \lim_{N\to\infty} t(\lambda, \Theta) \exp[iG^{II}(\lambda)_V](-)^{J_z - N/2}$$

Then, taking into account eq.(6.3), eq.(6.6) can be rewritten

$$\xi(u) = \frac{\Lambda^{II}\left(\frac{\gamma}{\pi}u - i\frac{\gamma}{2}\right)}{\Lambda^{I}\left(\frac{\gamma}{\pi}u - i\frac{\gamma}{2}\right)} \tag{6.7}$$

Notice that the dependence on the cutoff rapidity Θ has completely disappeared from the r.h.s. of eq.(6.7). This holds true both for the explicit dependence in the vacuum function $G(\lambda)_V$ and for the implicit dependence through the bare BAE, which are now replaced by the $\Theta-$independent higher–level ones. In other words, the eigenvalues of the bootstrap transfer matrix can be recovered from the light–cone regularization already on the infinite diagonal lattice, with no need to take the continuum limit. This should cause no surprise, since after all a factorized scattering can be defined also on the infinite lattice, with physical rapidities replaced by lattice rapidities [see eq.(4.5)]. The bootstrap construction of the quantum monodromy operators $\mathcal{T}_{ab}(u)$ then proceeds just like on the continuum. In this case, some q_0-deformation of the two dimensional Lorentz algebra should act as a symmetry on the physical states. This q_0 becomes unit when $\Theta \to \infty$.

We then compare these Bethe Ansatz eigenvalues with the eigenvalues of the bootstrap transfer matrix $\tau(u)$. Remarkably enough, we find the following simple relation between the two results, for $0 < \gamma < \pi/2$ (repulsive regime),

$$\tau(u) = t^{II}(\frac{\gamma}{\pi}u - i\frac{\gamma}{2}, \Theta)\, t^{I}(\frac{\gamma}{\pi}u - i\frac{\gamma}{2}, \Theta)^{-1} \tag{6.8}$$

where $t^{II}(\lambda,\Theta)$ and $t^{I}(\lambda,\Theta)$ have been normalized to one on the ground state . Thus, we succeed in connecting the bootstrap transfer matrix $\tau(u)$ of the sG-mT model with the alternating transfer matrix $t(\lambda,\Theta)$ of the six vertex model. In the thermodynamic limit $\tau(u)$ coincide with the jump between the two main determinations of $t(\lambda,\Theta)$. Notice the renormalization of the rapidity by γ/π and the precise overall shift by $i\gamma/2$ in the argument in order the equality to hold.

We find in addition that $t(\lambda,\Theta)$, for $0 < \operatorname{Im}\lambda < \gamma/2$, generates the hamiltonian and momentum together with an infinite number of higher–dimension and higher–spin conserved abelian charges, through expansion in powers of $e^{\pm\pi\lambda/\gamma}$. We see therefore that the same bare operator generates two kinds of conserved quantities. Energy and momentum as well the higher–spin abelian charges are local in the basic fields which interpolate physical particles, whereas the infinite set of charges obtained from the jump from $t^{II}(\lambda,\Theta)$ to $t^{I}(\lambda,\Theta)$ are nonlocal in the same fields. The fact that local and nonlocal charges come from different sides of a natural boundary, clearly shows that they carry independent information. That is, one cannot produce the nonlocal charges from the sole knowledge of the local charges. We also recall that the monodromy matrix $T(\lambda,\Theta)$ can be written in terms of the lattice Fermi fields of the mT model [1], so that local and non local charges do admit explicit expressions in terms of local field operators.

We expect eqs.(6.8) and (6.5), and the discussion below eq.(6.8) , to be valid for many other integrable models provided the appropriate rapidity renormalization and imaginary shift are introduced.

The quantum monodromy operators $\mathcal{T}_{ab}(u)$ generate a Fock representation of the $q-$deformed affine Lie algebra $U_q(\hat{\mathcal{G}})$ corresponding to the given $R-$matrix. More precisely, by expanding $\mathcal{T}_{ab}(u)$ in powers of $z = e^u$ around $z = 0$ and $z = \infty$, one obtains non–abelian non-local conserved charges representing the algebra $U_q(\hat{\mathcal{G}})$ on the Fock space of in– and out–particles. This connects our approach based on the YB symmetry, to the $q-$deformed algebraic approach of ref. [29]. $U_q(\hat{\mathcal{G}})$ is a Hopf algebra endowed with an universal $R-$matrix, which reduces to the $R-$ explicitly entering the YB algebra, upon projection to the finite–dimensional vector space spanned by the indexes of $\mathcal{T}_{ab}(u)$ [21]. In particular, the two expansions around $z = 0$ and $z = \infty$ generate the two Borel subalgebras of $U_q(\hat{\mathcal{G}})$. A single monodromy matrix $\mathcal{T}(u)$ is sufficient for this purpose, since this field–theoretic representation has level zero . This fact receives a new explanation in the light–conc approach, since $U_q(\hat{\mathcal{G}})$ emerges as true symmetry only in the infinite–volume limit above the antiferromagnetic ground state (with no need to take the continuum limit), but its action is uniquely defined already on *finite* lattices, and all finite–dimensional representations have level zero.

VII. LATTICE LOCAL FORMULATION OF INTEGRABLE QFT: THE MASSIVE THIRRING MODEL

As one sees from eqs.(2.12,2.13) the light-cone Hamiltonian

$$H = \frac{i}{a}\log U(\Theta) \tag{7.1}$$

is a highly non-local operator that couple *all* lattice sites. On spite of that the equations of motion are local on the lattice [eqs.(4.7)] and moreover H becomes local in the continuum limit. [eqs.(4.8)]. An alternative formulation exists where the lattice hamiltonian is local [30]. By local we mean that it only couples a few neighboring sites (typically three). In this formulation we keep the time continuous and discretize only the space.

From the alternating transfer matrix $t(\lambda,\Theta)$ it is possible to obtain a full hierarchy of Θ–dependent local charges in involution. It suffices to take the logarithmic derivative of $t(\lambda|\,\Theta)$ [see eq.(2.14 with respect to the spectral parameter at $\lambda=\pm\Theta$:

$$H_n^{\pm}(\Theta) = i^{-1-n}\frac{\partial^n}{\partial\lambda^n}\log t(\lambda|\Theta)\Big|_{\lambda=\pm\Theta} . \tag{7.2}$$

By purely algebraic calculations one can verify that $H_n^{\pm}(\Theta)$ couples $2n+1$ neighboring sites. Notice that $H_n^{\pm}(\Theta)$ commutes with $U(\Theta)$ and is therefore a conserved charge also in the standard light–cone approach presented in sec. II.

The charges of level one read

$$H_1^+(\Theta) = \sum_{j=1}^{N} h_{2j-1}(2\Theta) , \quad H_1^-(\Theta) = \sum_{j=1}^{N} h_{2j}(-2\Theta) \tag{7.3}$$

in terms of the 'Hamiltonian density'

$$h_n(\lambda) = -R_{n\,n+1}(\lambda)^{-1}\left[\dot{R}_{n\,n+1}(\lambda) + \dot{R}_{n-1\,n}(0)R_{n\,n+1}(\lambda)\right] . \tag{7.4}$$

With them, one can now define the local Hamiltonian

$$\tilde{H} = \tfrac{1}{2}a^{-1}\left[H_1^+(\Theta) + H_1^-(\Theta)\right] , \tag{7.5}$$

which is indeed hermitean thanks to the unitarity of $R-$matrix. With this choice of hamiltonian, the evolution operator is $\tilde{U}(t) = e^{-it\tilde{H}}$, with the time t continuous, as required. Moreover, the locality of the Hamiltonian $\tilde{H}$ guarantees the addidivity of its spectrum, without any need to worry about logarithmic branches.

We shall now obtain the explicit form of the local hamiltonian H_1 in terms of the fermionic fields ψ_j, $j=1,2,\ldots,2N$, introduced in sec. IV. The best way is to write from the beginning the matrices $R_{jk}(\lambda)$ in terms of fermi fields as in [30] (this takes care of the problems with the boundary conditions that would appear by simply using the Jordan–Wigner transformation on the spin operators at the end). Then all one has to do is to perform rather long albeit trivial algebraic manipulations with the fermi fields. In particular we find

$$R_{jn}(\lambda)^\dagger \dot{R}_{jn}(\lambda) = (\bar{b}\dot{c} + \dot{b}\bar{c})K_{jn} + (\bar{b}\dot{b} + \dot{c}\bar{c})K_{jn}^2 \tag{7.6}$$

$$\begin{aligned} R_{jn}(\lambda)^\dagger \dot{R}_{ij}(0) R_{jn}(\lambda) = & \dot{b}_0[\psi_i^\dagger(b\psi_n + c\psi_j) + \text{h.c.}] \\ & + \dot{c}_0[b\bar{c}\psi_j\psi_n + \text{h.c.}) + Q_i + c\bar{c}Q_j + b\bar{b}Q_n] \\ & + \dot{b}_0[(b + \bar{b})Q_j K_{in} + (c - \bar{c})Q_n(\psi_j^\dagger\psi_i - \psi_i^\dagger\psi_j)] \\ & - 2\dot{c}_0[b\bar{c}Q_i(\psi_j^\dagger\psi_n - \psi_n^\dagger\psi_j) + Q_i(c\bar{c}Q_j + b\bar{b}Q_n)] \end{aligned} \tag{7.7}$$

where

$$K_{ij} = \psi_i^\dagger\psi_j + \psi_j^\dagger\psi_i\,, \quad Q_j = \psi_j^\dagger\psi_j = \tfrac{1}{2}(1 - \sigma_j^z)\,, \tag{7.8}$$

$b = b(\lambda)$, $c = c(\lambda)$ (see eq.(4.1)), $\dot{b}_0 = b'(0)$ and $\dot{c}_0 = c'(0)$. In the derivation of these results the unitarity relations $b\bar{b} + c\bar{c} = 1$ and $b\bar{c} + \bar{b}c = 0$ have been used. To obtain H_1 we must now set $(i, j, n) = (j-1, j, j+1)$, then put $\lambda = 2\Theta$ when j is odd and $\lambda = -2\Theta$ when j is even, and finally sum up over j. $\tilde{H}$ contains a piece quadratic in the fields, say H_D and a piece quartic in them, H_R. Having obtained the local Hamiltonian we may now take the thermodynamic limit $N \to \infty$.

The quadratic part H_D is diagonalized by the following Fourier transformation

$$\begin{pmatrix} \psi_{2j-1} \\ \psi_{2j} \end{pmatrix} = \int_{-\pi}^{\pi} dq \begin{pmatrix} \psi_+(q) \\ \psi_-(q) \end{pmatrix} e^{iqj} \tag{7.9}$$

One finds an equal amount of positive and negative energies, so that the perturbative filled Dirac sea (the perturbative vacuum state of the QFT) is obtained by filling all the negative levels. Thus $\langle Q_j \rangle = 1/2$ in this state, which is therefore an antiferromagnetic state in spin language. This justifies the following normal order prescription

$$Q_n =: Q_n : + \frac{1}{2} \tag{7.10}$$

which has a dramatic effect on the quadratic part of the Hamiltonian, leading to the perturbative one–particle energy spectrum

$$E = \frac{\sinh 4\Theta}{\sinh^2 2\Theta + \sin^2\gamma}\left[\sin^2 q + \frac{1}{2}(m_0 a)^2(1 + \cos q)\right]^{1/2} \tag{7.11}$$

where the $m_0 = 2a^{-1}\sin\gamma/\sinh(2\Theta)$ is the 'bare' mass of particles and holes.

It is clear that eq.(7.11) represent a lattice approximation to the relativistic spectrum of massive particles. It has a peculiarity though: it also describes massless particles at the boundaries of the Brillouin zone. This is inevitable, since we are working with a local lattice Hamiltonian which for $\Theta \to \infty$, that is in the massless limit $m_0 \to 0$, becomes chiral invariant. The Nielsen–Ninomiya theorem then implies the existence of the (in)famous 'fermion doublers'. In the model at hand, the left and right modes around $q = 0$ become massive for finite Θ, while the left and right doubler around $q = \pm\pi$ remain massless.

This observation provides the basis for the following representation of the Fermi fields in continuum limit $a \to 0$

$$\begin{aligned} \psi(2j) &= \sqrt{a}\left[\chi_L(ja) + (-)^j \eta_R(ja)\right] \\ \psi(2j+1) &= \sqrt{a}\left[\chi_R(ja) + (-)^j \eta_L(ja)\right] \end{aligned} \tag{7.12}$$

where χ and η are quantum relativistic Dirac fields. Then, taking into account the normal ordering and dropping all wildly oscillating terms, the continuum form of the Hamiltonian reads (here $\Theta \to \infty$ as $a \to 0$ so that the bare mass $m_0 \approx 4a^{-1} \sin\gamma\, e^{-2\Theta}$ stays fixed)

$$H \equiv \tilde{H}_{\text{cont}} = H_K + H_{\text{m}} + H_{\text{int}} \tag{7.13}$$

where H_K is the kinetic energy

$$H_K = -i \int dx \left(\chi_R^\dagger \partial_x \chi_R - \chi_L^\dagger \partial_x \chi_L + \eta_R^\dagger \partial_x \eta_R - \eta_L^\dagger \partial_x \eta_L\right) , \tag{7.14}$$

H_{m} is the mass term

$$H_{\text{m}} = m_0 \int dx \left(\chi_L^\dagger \chi_R + \chi_R^\dagger \chi_L\right) \tag{7.15}$$

and H_{int} the quartic interaction

$$H_{\text{int}} = g \int dx \,(J_R^\chi J_L^\chi - J_R^\eta J_L^\eta - J_R^\eta J_L^\chi - J_R^\chi J_L^\eta) \ . \tag{7.16}$$

Here $g = -2\cot\gamma$ is the same continuum coupling constant found above and the J's are free–field normal–ordered $U(1)$ currents:

$$J_\alpha^\chi = :\chi_\alpha^\dagger \chi_\alpha: \, , \quad J_\alpha^\chi = :\eta_\alpha^\dagger \eta_\alpha: \, , \quad A = R, L \ . \tag{7.17}$$

The nice feature of this result is that all terms surviving the naive continuum limit are manifestly Lorentz invariant. The more troublesome aspect is that η, the field describing the doublers, does not decouples from the putative massive Thirring field χ.

In order to find the the right decoupled description, we use abelian bosonization:

$$\begin{aligned} \chi_\alpha &= \mu :\exp(i\alpha\sqrt{4\pi}\, \phi_\alpha): \\ \eta_\alpha &= \mu :\exp(i\alpha\sqrt{4\pi}\, \sigma_\alpha): \end{aligned} \tag{7.18}$$

where $\alpha = \pm$ ($+ \equiv R$ and $- \equiv L$), μ is a normalization mass scale, and the fields

$$u_\alpha(x) = \int_{-\infty}^{x} dy\, J_\alpha^\chi(y)\ , \quad v_\alpha^\eta(x) = \int_{-\infty}^{x} dy\, J_\alpha^\eta(y) \tag{7.19}$$

can be identified with the chiral components of two free massless Bose fields. The symbols $:\ldots:$ in eq.(7.18) now stand for bosonic free–field normal ordering at the mass scale μ, so that the expressions (7.18) are effectively μ-independent.

With the standard rules of abelian bosonization, the Hamiltonian now takes the form, up to irrelevant constants,

$$\begin{aligned} H = \int dx\ \Big\{ & (\partial_x u_R)^2 + (\partial_x u_L)^2 + (\partial_x v_R)^2 + (\partial_x v_L)^2 \\ & + g\,[(\partial_x u_R)(\partial_x u_L) - (\partial_x v_R)(\partial_x v_L) \\ & \quad - (\partial_x u_R)(\partial_x v_L) + (\partial_x v_R)(\partial_x u_L)] \\ & + m_0 \mu^2 :\cos[\sqrt{4\pi}(u_R + u_L)]: \Big\} \end{aligned} \tag{7.20}$$

We now perform the canonical $O(2,2)$ transformation

$$\begin{pmatrix} \phi_R(x) \\ \xi_R(x) \\ \phi_L(x) \\ \xi_L(x) \end{pmatrix} \begin{pmatrix} a & c & -b & -c \\ c & -a & c & -b \\ -b & -c & a & c \\ c & -b & c & -a \end{pmatrix} \begin{pmatrix} u_R(x) \\ v_R(x) \\ u_L(x) \\ v_L(x) \end{pmatrix} \tag{7.21}$$

with

$$\begin{aligned} a &= \frac{\cosh^2\nu \cosh\lambda + \sinh^2\nu \sinh\lambda}{\sqrt{\cosh 2\nu}} \\ b &= \frac{\cosh^2\nu \sinh\lambda + \sinh^2\nu \cosh\lambda}{\sqrt{\cosh 2\nu}} \\ c &= \frac{\cosh\nu \sinh\nu\,(\cosh\lambda - \sinh\lambda)}{\sqrt{\cosh 2\nu}} \end{aligned}$$

and

$$\tanh(2\nu) = -\sinh(2\lambda)\ , \quad \tanh(2\lambda) = -\frac{g}{2\pi}\ . \tag{7.22}$$

In terms of the new fields the Hamiltonian reads

$$\begin{aligned} H = & \left[1 - \frac{g^2}{2\pi^2}\right]^{1/2} \int dx \left\{ (\partial_x\phi_R)^2 + (\partial_x\phi_L)^2 + (\partial_x\xi_R)^2 + (\partial_x\xi_L)^2 \right\} \\ & + m_0 \int dx\ :\cos[\sqrt{4\pi}\, e^\lambda(\phi_R + \phi_L)]: \end{aligned} \tag{7.23}$$

and we see that it correspond to a Sine–Gordon model plus a decoupled free massless field. We could now perform two identical canonical transformations separately on ϕ and ξ with parameter $\tanh(2\lambda') = \pm\frac{g}{\sqrt{2}\pi}$, and then an inverse bosonization trick, yielding at the end two decoupled Thirring models, one massive and the other massless.

VIII. NEW THERMODYNAMIC BETHE ANSATZ (DDV EQUATIONS)

The computation of thermodynamical functions for integrable models started with the seminal works of C.N. Yang and C.P. Yang [31], of M. Takahashi [32] and M. Gaudin [33]. In Refs. [32], [33] the free energy of the Heisenberg spin chain is written in terms of the solution of an infinite set of coupled nonlinear integral equations, derived on the basis of the so–called "string hypothesis".

These equation are commonly known as Thermodynamic Bethe Ansatz (TBA) equations. Nowadays, the TBA is mostly used as a nonperturbative tool to investigate Perturbed Conformal Field Theories and has been considerably generalized (see e.g. [34], [35]). In all cases some some kind of string hypothesis is used.

In ref. [36], we proposed a simpler way to solve the thermodynamics of the XXZ chain and of the sine–Gordon field theory by means of a single, rigorously derived, nonlinear integral equation. This method can be applied to a wide class of models solvable by Bethe Ansatz, and can provide also the excited–states scaling functions of the sG field theory [38]. Similar methods were used to also determine some correlation lengths at finite temperature in the XXZ chain [39].

The purpose of this approach as presented in ref. [37] is threefold. First we generalized the results of ref. [36] in the presence of external magnetic field for the XXZ Heisenberg model.

Second, we show that this approach provides an unified method to compute both at *finite* temperature and infinite size and at zero temperature and *finite* size. Third, we recast our NLIE in two alternative forms: a) as an infinite set of algebraic equations of BA type and b) as a Riemann-Hilbert problem. These forms are appropriate to compute the asymptotic behaviors around the conformal regime (low T for the magnetic chains and high T for field theory). In the recursive regime (the opposite to the conformal one), our NLIE can be easily solved by iteration.

Continuum quantum field theories at finite temperature T are obtained from periodic boundary conditions (PBC) in the imaginary time with period $\beta \equiv T^{-1}$. By Wick rotation, this leads to a periodic spatial box of finite size β. What we achieve here is the analogous to a Wick rotation on the exact Bethe Ansatz (BA) solution on the lattice.

Our starting point is the BA diagonalization of a row-to-row inhomogeneous and twisted transfer matrix $t(\lambda, \theta_1, \theta_2, \dots, \theta_N, \omega)$ where there is an inhomogeneity θ_n attached to each site and a twist ω in the boundary conditions ($\omega = 0$ corresponds to PBC).

Setting the inhomogeneities $\theta_n = 0$ yields the homogeneous six-vertex transfer matrix. Choosing the inhomogeneities alternating $\theta_n = (-1)^{n-1}\theta$ yields a diagonal-to-diagonal transfer matrix $T_N(\theta) = t_{2N}(\theta, \theta, 0)$ [2]. For $\theta \to 0$, $T_N(\theta)$ gives at order θ the XXZ hamiltonian.

$$T_N(\theta) \overset{\theta \to 0}{=} 1 - \frac{\theta}{2J \sin\gamma} H_{XXZ} + O(\theta^2)$$

and hence,

$$e^{-\beta H_{XXZ}} = \lim_{N\to\infty}\left[T_L\left(\frac{2\beta J\sin\gamma}{N}\right)\right]^N$$

This relation permits to write the XXZ partition function

$$\mathcal{Z} \equiv \mathrm{Tr}\left\{e^{-\beta H_{XXZ}}\right\}$$

in terms of the diagonal-to-diagonal transfer matrix $T_L(\theta)$. Moreover, using the crossing symmetry of the statistical weights one can prove that $\mathcal{Z}$ in the infinite volume limit is given just by the ground state eigenvalue of $T_N(\theta)$ in the $N \to \infty$ limit. This alternative way is much simpler than the usual TBA approach, where one has to sum over an infinite number of states with an arbitrary number of holes and an arbitrary number of strings of any length. What we do here is analog to duality in S-matrix theory, where the sum over an infinite number of resonances in the s-channel can be replaced by the exchange of one reggeon in the t-channel.

Setting $\theta = -i\Theta$ with Θ real, we get the lattice field theory evolution operators in the light-cone approach [1] , [2] [see eq.(2.15)]

$$U_R(\Theta) = a(-2i\Theta)^{-N}\; t_{2N}(-i\Theta, -i\Theta, 0) \quad , \quad U_L(\Theta) = a(2i\Theta)^{N}\; t_{2N}(+i\Theta, -i\Theta, 0)^{-1}$$

where $a(\pm 2i\Theta)^{\pm N}$ is a unitarization factor. That is, for imaginary θ, we interpret the lattice as a discretized **Minkowski** spacetime [6] and hence the operators U_R and U_L describe space and time translations in the light-cone directions of a discretized Minkowski spacetime.

The lattice hamiltonian (H) and momentum (P) operators. are defined by eq.(refevolu).

Moreover, the operator $t_{2N}(\lambda, -i\Theta, 0)$ is the generating functional of local conserved charges (including H and P) for $|\mathrm{Im}\,\lambda| < \gamma/2$ [49]. In the region $-(\pi - \gamma/2) < \mathrm{Im}\,\lambda < -\gamma/2$, $t_{2N}(\lambda, -i\Theta, 0)$ generates non-local conserved charges [49].

We diagonalize the transfer matrix $t(\lambda, \theta_1, \theta_2, \ldots, \theta_N, \omega)$ by algebraic Bethe Ansatz (BA). The BA the expresses eigenvectors and eigenvalues in terms of the roots of M coupled algebraic equations ($1 \le M \le N$), the so called Bethe Ansatz equations (BAE).

Using contour integral techniques, we rewrite these BAE as a non-linear integral equation. This equation may take different forms. A compact form is

$$Z(\mu) = \varphi(\mu) + 2\,\mathrm{Im}\int_{-\infty}^{+\infty} d\,\mu'\; G(\mu - \mu' - i\eta)\log\left[1 + e^{iZ(\mu'+i\eta)}\right] \tag{8.1}$$

where the kernel $G(\lambda)$ is explicitly given by

$$G(\lambda) = \int_{-\infty}^{+\infty}\frac{dk}{4\pi}\frac{\sinh(\pi/2-\gamma)k}{\sinh(\pi-\gamma)k/2\,\cosh(\gamma k/2)}\; e^{ik\lambda} \tag{8.2}$$

and

$$\varphi(\mu) = \begin{cases} 2N \arctan\left[\frac{\sinh(\pi\mu/\gamma)}{\cos(\pi\theta/[2\gamma])}\right] - \frac{\pi\omega}{\pi-\gamma} & \text{finite size vertex model with TBC} \\ -\frac{2\pi\beta J \sin\gamma}{\gamma\sinh[\pi\mu/\gamma]} + \frac{i\pi\beta h}{\pi-\gamma} & \text{XXZ thermodynamics in a magnetic field} \\ m\beta \sinh(\pi\mu/\gamma) & \text{sine-Gordon field theory} \end{cases} \tag{8.3}$$

$Z(\mu)$ is the unknown in eq.(8.1) . The real parameter η can be freely chosen in the interval $0 < \eta < \gamma/2$.

We have unified three different problems (finite size vertex model with TBC, the XXZ thermodynamics in a magnetic field and the sine-Gordon field theory in a finite volume) into the single NLIE (8.1) -(8.3) .

Once we have the solution $Z(\lambda)$ of eq.(8.1), we can compute the transfer matrix eigenvalues by quadratures. We consider three relevant applications

1. a) The finite size (homogeneous) six-vertex model with twisted boundary conditions (TBC).
2. b) The (infinite) XXZ chain at temperature $T = \beta^{-1}$ in an external magnetic field h.
3. c) The massive Thirring-sine Gordon model (mT-sG) on a periodic box of size β.

In the first case, we find for the six vertex partition function on a double periodic square lattice (T x $2N$)

$$\lim_{T\to\infty} T^{-1} \log \mathcal{Z}_{6V} = \log \tau_{max} = -2N f_0(\lambda,\omega) + L_{2N}(\lambda,\omega)$$

For large N, we obtain that L_N follows the conformal behaviour for TBC with unit central charge [see sec. 7.1 of ref. [37]]. Namely,

$$L_{2N} \overset{N\to\infty}{=} \frac{\pi}{6N} \tan\left(\frac{\pi\lambda}{\gamma}\right) \left[1 - \frac{6\omega^2}{\pi^2(1-\gamma/\pi)}\right]$$

In the case b) , we relate the XXZ free energy at temperature T and on an external magnetic field $2h$ in the z-direction with the maximal eigenvalue of the alternating transfer matrix $t_{2N}(\pm\theta,\theta,-i\beta h)$ with TBC. Notice that the twist, $-i\beta h$, is purely imaginary for real magnetic fields h. This new result allows us to express the XXZ free energy $f(\beta, h)$ in terms of the solution $Z(\lambda)$ of our non-linear integral equation (NLIE) (8.1) .

$$f(\beta,h) = h + 2J\cos\gamma + \frac{\sin\gamma}{\beta} \oint_\Gamma \frac{d\lambda}{2\pi i} \frac{\log\left[1+e^{-iZ(\lambda)}\right]}{\sinh\lambda \sinh(\lambda - i\gamma)}$$

Finally, we use the light-cone approach [1], [2] , [6] , [11], [49] to deal with c). We write the energy in the mT-sG model on a ring of length β in terms of the alternating six vertex transfer matrix $[t(\mp i\Theta, -i\Theta, 0)]$ eigenvalues. The basic relation is the the formula for the unit evolution operator for lattice field models,

$$U = e^{-iaH} = \left[\frac{a(2i\Theta)}{a(-2i\Theta)}\right]^N t(-i\Theta,-i\Theta,0)\; t(i\Theta,-i\Theta,0)^{-1}$$

where a is the lattice spacing, Θ plays the rôle of UV cutoff and $a(\pm 2i\Theta)$ are unitarizing c-number factors. In the continuum limit, $a \to 0,\ N \to \infty$ with $\beta \equiv Na$ fixed. At the same time Θ tends to infinite as

$$\Theta \simeq \frac{\gamma}{\pi} \log \frac{4N}{m\beta}$$

We find for the scaling function

$$\begin{aligned} E(\beta) &\equiv \lim_{N\to\infty} [E_N - E_c] \\ &= -\frac{m}{\gamma} \operatorname{Im} \int_{-\infty}^{+\infty} \frac{d\lambda}{\pi} \sinh[\pi(\lambda + i\eta)/\gamma] \log\left[1 + e^{iZ(\lambda+i\eta)}\right] \end{aligned} \tag{8.4}$$

where $Z(\lambda)$ satisfies [cfr. eq.(8.1)]

$$Z(\lambda) = m\beta \sinh(\pi\lambda/\gamma) + 2 \operatorname{Im} \int_{-\infty}^{+\infty} d\mu\, G(\lambda - \mu - i\eta) \log\left[1 + e^{iZ(\mu+i\eta)}\right] \tag{8.5}$$

provided we choose $0 < \eta < \frac{1}{2}\min(\gamma, \pi - \gamma)$.

Shifting λ by $i\gamma/2$, eq.(8.5) takes the form

$$\epsilon_f = m\beta \cosh\theta - G_0 * L_f + G_1 * L_{\bar{f}} \tag{8.6}$$

where

$$\begin{aligned} \epsilon_f(\theta) &= -iZ(\tfrac{\gamma}{\pi}\theta + i\gamma/2)\,, \qquad \epsilon_{\bar{f}} = \overline{\epsilon_f} \\ L_f &\equiv \log(1 + e^{\epsilon_f})\,, \qquad L_{\bar{f}} \equiv \log(1 + e^{\epsilon_{\bar{f}}}) = \overline{L_f} \\ G_0(\theta) &= \tfrac{\gamma}{\pi} G(\tfrac{\gamma}{\pi}\theta)\,, \quad G_1(\theta) = G_0(\theta + i\pi - i0) \end{aligned} \tag{8.7}$$

The kernels $G_0(\theta)$ and $G_1(\theta)$ are just the θ derivative of the fermion-fermion and fermion-antifermion physical phase-shifts. This transparent physical interpretation of eq.(8.6) suggests the form that it should take in other field theory models.

Now we face the problem of solving NLIE like (8.1) or (8.5) in order to get physical results. It must be noticed that we find just **one** NLIE for finite temperature and finite size situations. This situation contrasts with the standard TBA which for nonrational values of γ/π involves an infinite number of nonlinear integral equations for the infinitely many different types of "magnons" describing the energy degeneracies of the physical fermions and antifermions [51]. Therefore, **our equation effectively provides a resummation of the magnon degrees of freedom**.

There are essentially two regimes in our NLIE. We call them 'recursive' and 'conformal'. In the 'recursive' regime the NLIE can be simply solved by iterating the inhomogeneity. This happens for small β in the XXZ chain and for large β in field theory.

The 'conformal' regime is the opposite to the 'recursive' regime. In it, the calculations are more subtle. The dominant behaviour (large β for the XXZ chain and small β in field theory), follows in closed form with the help of the Lemma in sec. 7 of ref. [37]. This lemma avoids the detailed resolution of the NLIE for the dominant contributions. Higher corrections follow using the Riemann-Hilbert form of our NLIE. We find that the energy in the sG-mT model has a series of terms with $(m\beta)^{-1}$ times integer powers of $(m\beta)^{4\gamma/\pi}$ plus the central charge term proportional to $(m\beta)^{-1}$, in agreement with the predictions of Perturbed Conformal Field Theory.

Since the inhomogeneity in the NLIE for field theory is $m\beta\cosh\theta$, it is useful to define for small $m\beta$ kink solutions, $\epsilon_k(\theta)$ of the NLIE with inhomogeneity $e^{\pm\theta}$ [see eq.(7.19) in ref. [37]]. We obtain in this way for the energy $E(\beta)$ at high temperatures

$$E \overset{\beta\to 0}{=} -\frac{\pi}{6\beta} - \frac{m^2\beta}{4}\cot\left[\pi^2/2\gamma\right] - \frac{m}{2\pi}\int_{-\infty}^{+\infty} d\theta\,\cosh\theta\left[\ell(\theta)+\bar{\ell}(\theta)\right] \tag{8.8}$$

The first term is the universal conformal Casimir energy, from which we read the correct central charge $c = 1$ of the mT-sG model. The second term, linear in β, exactly coincides with minus the scaling bulk free energy [see eq.(5.7) of ref. [37]]. The last integral in eq.(8.8) represents the resummation of the perturbed conformal field theory around the massless Thirring model. The fact that it must contain non-integral powers of r, causing non-analyticity at $r = 0$, can be established directly from eq. (8.6), since this cannot be expanded in a Taylor series of r.

In the 'recursive' regime the high temperature expansion for the XXZ chain in the external field h is derived in ref. [37]. One **only needs the Cauchy theorem** to evaluate the high T expansion coefficients. The first five terms take the form [37]:

$$\begin{aligned} f(\beta) = &-\beta^{-1}\ln 2 + J\cos\gamma - \beta\left[J^2(1+\tfrac{1}{2}\cos^2\gamma)+\tfrac{1}{2}h^2\right] + \beta^2\cos\gamma(J^3+h^2) \\ &-\beta^3[(\tfrac{1}{4}+\tfrac{1}{6}\cos^2\gamma-\tfrac{1}{12}\cos^4\gamma)J^4 - J^2h^2 - \tfrac{1}{12}h^4] + O(\beta^4) \end{aligned} \tag{8.9}$$

which indeed agrees with the high T expansion.

We obtained also the ground state energy for the mT-sG model for small T [37]. It is non-analytic at $T = 0$. We find up to contributions $O(e^{-3m\beta})$

$$E(\beta) \overset{\beta\to\infty}{=} -\frac{2m}{\pi}K_1(m\beta) + e^{-2m\beta}\sqrt{\frac{m}{\pi\beta}}\left[\frac{1+\sqrt{2}}{2} + \sqrt{\frac{\pi}{m\beta}}\,K_\gamma + O(\frac{1}{\beta})\right] + O(e^{-3m\beta}) \tag{8.10}$$

where $K_1(x)$ stands for a modified Bessell function and the constant K_γ is defined by

$$K_\gamma = -\int_{-\infty}^{+\infty}\frac{dy}{\pi^2}\,\frac{\sinh(4-\frac{\pi}{\gamma})y + 3\sinh(\frac{\pi}{\gamma}-2)y}{2\,\cosh y\,\sinh(\frac{\pi}{\gamma}-1)y}$$

The first term in eq.(8.10) describes free massive bosons, the subsequent terms take into account the interactions.

In secs 4 and 5 of ref. [37], we show that our NLIE (8.1) and (8.5) are equivalent to an infinite set of algebraic equations of BA type. These equations differ from the usual BA equations [44], [6] in two essential aspects: i) they are infinite in number, ii) the roots are discretely spaced although they describe models with an infinite number of sites. The root spacing tends to zero only in the conformal limit.

We would like to remark that, contrary to the traditional thermodynamic Bethe Ansatz [32], [33], our approach does not rely on the string hypothesis on the structure of the solutions of the BA equations characteristic of the Heisenberg chain. This makes our approach definitely simpler. Most notably, the whole construction of the thermodynamics no longer depends on whether γ/π is a rational or not, unlike Takahashi approach [32]. This applys equally well to the problem of the ground state scaling function of the sine–Gordon field theory, since the standard TBA approach requires the string hypothesis.

Generalizations of the present TBA approach to higher spin chains as well as to magnetic chains and quantum field theories associated to Lie algebras other than A_1 (nested BA solutions) are relatively straightforward provided crossing symmetry holds.

Our NLIE has been recently used with success in ref. [52].

REFERENCES

[1] C. Destri and H.J. de Vega, Nucl. Phys. **B 290**, 363 (1987).

[2] C. Destri and H. J. de Vega, J. Phys. **A 22**, 1329 (1989).

[3] T. T. Truong, in Lecture Notes in Physics, vol. 226, Ed. by N. Sánchez, Springer, 1984.

[4] C. Destri and H. J. de Vega, Nucl. Phys. **B 406**, 566 (1993).

[5] H. J. de Vega and E. Lopes, Nucl. Phys. **B 362**, 261 (1991).

[6] H. J. de Vega, Int. J. Mod. Phys. **A 4**, 2371 (1989).

[7] E. K. Sklyanin, J. Phys. **A 21**, 2375 (1988).

[8] H. J. de Vega and A. González-Ruiz, Nucl. Phys **B 417**, 553 (1994).

[9] M. Lüscher, Nucl. Phys. **B 117**, 475 (1976).

[10] R. J. Baxter, 'Exactly solved models in Statistical Mechanics', Academic Press, 1982.

[11] H. J. de Vega, Int. J. Mod. Phys. **B 4**, 735 (1990).

[12] H. J. de Vega, Int. J. Mod. Phys. **A 5**, 1611 (1990).

[13] H. J. de Vega and F. Woynarovich, J. Phys. **A 25**, 4499 (1992).
H. J. de Vega, L. Mezincescu and R. I. Nepomechie, Phys. Rev. **B 49**, 13223 (1994).
S. R. Aladim and M. J. Martins, J. Phys. **A 26**, L529 and 7287 (1993).
M. J. Martins, J. Phys. **A 26**, 7301 (1993).

[14] H.J. de Vega, H. Eichenherr and J.M. Maillet, Nucl. Phys. **B 240**, 377 (1984), Comm. Math. Phys. **92**, 507 (1984) and Phys. Lett. **B 132**, 337 (1983).

[15] H.J. de Vega, Nucl. Phys. (*Proc. Suppl*) **18 A** (1990) 229.

[16] P. P. Kulish, N. Yu. Reshetikhin and E. K. Sklyanin, Lett. Math. Phys. **5**, 393 (1981).

[17] L. D. Faddeev and N. Yu. Reshetikhin, Ann. Phys. **167**, 227 (1986).

[18] C. Destri and H. J. de Vega, Phys. Lett. **B 201**, 245 (1988).

[19] A. N. Kirillov and N. Yu. Reshetikhin in Proc. of the Paris-Meudon Colloquiun, eds. H. J. de Vega and N. Sánchez, World Scientific, 1987.

[20] A. Yu. Volkov and L. D. Faddeev, Theor. Math. Phys. **92**, 837 (1993).

[21] I. B. Frenkel and N. Yu. Reshetikhin, Comm. Math. Phys. **146**, 1 (1992).

[22] B. Davies, O. Foda, M. Jimbo and A. Nakayashiki,
Comm. Math. Phys.**151**, 89 (1993).

[23] M. Jimbo in Lecture Notes in Physics, vol. 246,
edited by H. J. de Vega and N. Sánchez, Springer Verlag, 1985.

[24] N. Reshetikhin, Theor. Math. Phys. **63**, 555 (1986).

[25] H. J. de Vega and M. Karowski, Nucl. Phys. **B 280**, 225 (1987).

[26] N. Reshetikhin, Lett. Math. Phys. **14**, 235 (1987).

[27] H. J. de Vega and E. Lopes, Phys. Rev. Lett. **67**, 489 (1991).

[28] L. A. Takhtadzhyan and L. D. Faddeev, Theor. Mat. Phys. **21**, 1046 (1974).

[29] D Bernard and A LeClair, Comm. Math. Phys. **142** (1991) 99.

[30] C. Destri and T. Segalini, in preparation

[31] C.N. Yang and C.P. Yang, Jour. Math. Phys. 10 (1969) 1115.

[32] M. Takahashi, *Prog. Theor. Phys.* **46** (1971) 401.
M. Takahashi and M. Suzuki, *Prog. Theor. Phys.* **48** (1972) 2187.

[33] M. Gaudin, *Phys. Rev. Lett.* **26** (1971) 1302.

[34] Al.B. Zamolodchikov, *Nucl. Phys.* **B342** (1990) 695.

[35] Al.B. Zamolodchikov, *Nucl. Phys.* **B358** (1991) 524.
P. Christe and M. Martins, *Mod. Phys. Lett.* **A5** (1990) 2189.
M. Martins, *Phys. Rev. Lett.* **65** (1990) 2091.
T. Klassen and E. Melzer, *Nucl. Phys.* **B338** (1990) 485; *Nucl. Phys.* **B350** (1991) 635.
F. Ravanini, *Phys. Lett.* **282B** (1992) 73.
V. A. Fateev and Al. B. Zamolodchikov, *Phys. Lett.* **271B** (1991) 91.
P. Fendley, K. Intriligator *Nucl. Phys.* **B372** (1992) 533.
F.Ravanini, R. Tateo and A. Valleriani, *Int. J. Mod. Phys.* **A8** (1993) 1707.

[36] C. Destri and H. J. de Vega, *Phys. Rev. Lett.* **69** (1992) 2313 .

[37] C. Destri and H. J. de Vega, *Nucl. Phys.* **B438**, 413 (1995).

[38] C. Destri and H. J. de Vega, work in progress.

[39] A. Klumper, *Z. Phys.* **B 91** (1993) 507.

[40] R.J. Baxter, "Exactly solved models in Statistical Mechanics", Academic Press, 1982.

[41] M. Suzuki, *Prog. Theor. Phys.* **56** (1976) 1454.

[42] J. Suzuki, Y. Akutsu and M. Wadati, J. Phys. Soc. Jpn. **59** (1990) 2667.

[43] T. Koma, *Prog. Theor. Phys.* **81** (1988) 783. M. Takahashi , *Phys. Rev.* **B 43**, 5788 (1990).

[44] L. D. Faddeev and L. A. Takhtadzhyan, *Russian Math. Surveys* **34** (1979) 11.

[45] H.J. de Vega and F. Woynarovich, *Nucl. Phys.* **B251** (1985) 439.

[46] H.J. de Vega and M. Karowski, *Nucl. Phys.* **B280** (1987) 225.

[47] A. Klumper, M.T. Batchelor and P.A. Pearce, *J. Phys.* **A24** (1991) 3111.

[48] C. Destri and H.J. de Vega, *Nucl. Phys.* **B258** (1991) 251.

[49] C. Destri and H. J. de Vega, *Nucl. Phys.* **B406** (1993) 566.

[50] A. Polyakov and P. Wiegmann, *Phys. Lett.* **141B** (1984) 223.
N. Reshetikhin and P. Wiegmann, *Phys. Lett.* **189B** (1987) 125.
G. Japaridze, A. Nersesyan and P. Wiegmann, *Nucl. Phys.* **B230** (1984) 511.
P. Hasenfratz, M. Maggiore and F. Niedermayer, *Phys. Lett.* **245B** (1990) 522.
P. Hasenfratz and F. Niedermayer *Phys. Lett.* **245B** (1990) 529.

[51] M. Fowler and X. Zotos, *Phys. Rev.* **B 24** (1981) 2634 and **B25** (1982) 5806.

[52] Al. B. Zamolodchikov, *Phys. Lett.* **335B** (1994) 436 and *Nucl. Phys.* **B432**, 427 (1994).
A. LeClair, G. Mussardo, H. Saleur and S. Skorik, 'Boundary energy and boundary states in integrable QFT', hep-th/9503227.

VON NEUMANN AND SHANNON-WEHRL ENTROPY FOR SQUEEZED STATES AND COSMOLOGICAL PARTICLE PRODUCTION

M. Gasperini and M. Giovannini

Dipartimento di Fisica Teorica, Universitá di Torino,
Via P.Giuria 1, 10125 Turin, Italy

ABSTRACT

We show that the effective coarse graining of a two-mode squeezed density matrix, implicit in the Wehrl approaches to a semiclassical phase-space distribution, leads to results in agreement with previous different definitions of entropy for the process of pair production from the vacuum. We also present, in this context, a possible interpretation of the entropy growth as an amplification (due to the squeezing) of our lack of knowledge about the initial conditions, which gives rise to an effective decoherence of the squeezed density matrix.

Some years ago it was suggested that the in quantum process of pair production from the vacuum, induced by the action of a time-varying classical background, the growth of the average number of quanta of a given field should be naturally associated to the growth of entropy of that field [1,2]. More recently, such an entropy growth was quantified on the grounds of a squeezed state approach to the process of particle production [3-7]. With this approach one can easily compute, in particular, the total entropy S stored in the cosmological perturbations amplified by inflation, and one obtains [4,6]

$$S \simeq \left(\frac{H_f}{M_p}\right)^{3/2} S_{cmb} \tag{1}$$

where S_{cmb} is the thermal black-body entropy of the $3K^o$ photon microwave background, H_f the final curvature scale of inflation, and M_p the Planck mass.

The use of the squeezed state formalism to compute entropy production might seem to give rise to a puzzle, since the squeezed evolution is unitary. The solution of the puzzle is that, of course, there is no loss of information *in principle* associated to the evolution of the initial state into a final squeezed quantum state (and so no possible information paradox, like in the case of black-hole dynamics). The loss of information occurs however *in practice*, as a consequence of the process of measure of the observables characterizing the final squeezed state. There are indeed variables (related to the so-called "superfluctuant" operators [8]) which are amplified by the squeezed evolution and become thus available for macroscopic observations, while other variables (the "subfluctuant" ones [8]) are correspondingly "squeezed" and become unobservable.

A simple and intuitive way to illustrate this situation is to consider a semi-classical description of the squeezing process as a parametric amplification of the wave function, "hitting" the effective potential barrier of a Schroedinger-like equation (with time-like variable) [9,10]. The final amplitude of the wave scattered by the barrier may be larger or smaller than the initial amplitude, depending on the phase with which the wave enters the barrier. What is macroscopically detected, however, is the average amplification of the wave, out of an initial random distribution of phases. This point was clearly stressed already in the first paper on the cosmological amplification of tensor perturbations [11]. A macroscopic observation of the quantum fluctuations of the metric, amplified by the cosmic evolution, traces out over phases.

In order to compute the entropy associated to a squeezing process, this macroscopic loss of information has to be taken into account in the form of a suitable "coarse graining" of the density matrix of the final (pure) squeezed state. Such a coarse graining was originally performed in a Fock basis, i.e. with respect to the eigenstates of the number operator, by averaging over phases [3-5]. The number operator, however, is only a particular example of superfluctuant variable which grows in the case of a pair production process, while the conjugate variable (the phase) is correspondingly squeezed. This suggests the use, in general, of the eigenstates of the superfluctuant operator as the natural basis for the coarse graining of a squeezed density matrix [6,7]. A possible different basis, that of coherent states with real amplitude, has been suggested however in [12].

It is important to stress that all these different approaches [3-7,12] to the entropy of a squeezed state exactly lead to the same result when the entropy growth ΔS_k, for each mode k, is evaluated in the large squeezing regime, namely

$$\Delta S_k \simeq 2r_k, \qquad r_k >> 1 \tag{2}$$

where r_k is the modulus of the (complex) squeezing parameter [8]. In the previously quoted papers, however, the decoherence process which leads to a reduced density matrix ρ_{red} (from which to compute the quantum Von Neumann entropy) is imposed "by hand", by setting the off-diagonal terms of the matrix to zero in some chosen basis; at best, like in the case of the number basis, it is justified in a semiclassical way using a stochastic approach and a sort of random-phase approximation [3,4]. The problem is that it is formally impossible in the number basis, as well as in the more general superfluctuant basis, to compute the quantum von Neumann entropy $(-Tr\rho_{red}\ln\rho_{red})$ from a reduced matrix $(\rho_{red}\neq\rho_{red}^2)$ obtained by tracing out phases, or another subfluctuant variable. Indeed, a subfluctuant operator does not commute with the conjugate superfluctuant one, so that both operators cannot be simultaneously included in the complete set of observables characterizing a pure squeezed state.

In view of these preliminary remarks, we can say that the aim of this paper is twofold. First of all we want to compute the semiclassical Shannon-Wehrl entropy [13,14] for a two-mode squeezed vacuum state, in order to show that also in that approach one recovers the result (2) for the large squeezing limit. This confirms that the entropy of a highly squeezed state is very robust to the particular scheme of coarse graining implemented (as discussed also in [15,16]).

In the second place we want to present a possible justification to the reduction of the density matrix in a superfluctuant basis [6,7], by interpreting such a reduction as the consequence of our lack of knowledge about the initial configuration which is the starting point of the squeezed evolution. To this purpose we shall represent the initial state, in general, as a statistical mixture of coherent states. This representation automatically defines a quantum von Neumann entropy associated to the initial mixture. This entropy however is constant, as the squeezed evolution is unitary, and then it does not contribute to the difference ΔS between final and initial entropy. However, if the statistical distribution satisfies (at least approximately) a principle of equipartition of the probabilities among the states of the mixture, then the off-diagonal terms of the density matrix turn out to be suppressed. An effective decoherence is thus produced, whose associated entropy in the superfluctuant basis is eventually amplified by the squeezed evolution.

The arguments presented in this paper are intended to apply to any process of pair production, described by a Bogoliubov transformation connecting the particle $\{b,b^\dagger\}$ and anti-particle $\{\tilde{b},\tilde{b}^\dagger\}$ annihilation and creation $|in\rangle$ operators to the $|out\rangle$ ones, $\{a,a^\dagger,\tilde{a},\tilde{a}^\dagger\}$. Such a transformation reads, for each mode k,

$$\begin{aligned} a_k &= c_+(k)b_k + c_-^*(k)\tilde{b}^\dagger_{-k} \\ \tilde{a}^\dagger_{-k} &= c_-(k)b_k + c_+^*(k)\tilde{b}^\dagger_{-k} \end{aligned} \tag{3}$$

where the Bogoliubov coefficients $c_{\pm}(k)$ satisfy $|c_+|^2 - |c_-|^2 = 1$. By putting

$$c_+(k) = \cosh r_k, \qquad c_-^*(k) = e^{2i\theta_k} \sinh r_k \tag{4}$$

the transformation (3) can be rewritten as a unitary transformation,

$$a_k = \Sigma_k b_k \Sigma_k^\dagger, \qquad \tilde{a}_{-k}^\dagger = \Sigma_k \tilde{b}_{-k}^\dagger \Sigma_k^\dagger \tag{5}$$

generated by the two-mode squeezing operator [8]

$$\Sigma_k = \exp\left(z_k^* b_k \tilde{b}_{-k} - z_k b_k^\dagger \tilde{b}_{-k}^\dagger\right), \qquad z_k = r_k e^{2i\theta_k} \tag{6}$$

Our discussion can thus be applied in general to any dynamic situation in which the initial vacuum state $|0\rangle$ evolves towards a final two-mode squeezed vacuum state, $|z\rangle = \Sigma_k|0\rangle$, characterized by a non-vanishing value of the squeezing parameter $r_k = |z_k|$ (related to the expectation value of the number of produced particles by $\langle n_k\rangle = \langle z_k|b_k^\dagger b_k|z_k\rangle = \langle 0|a_k^\dagger a_k|0\rangle = |c_-(k)|^2 = \sinh^2 r_k$).

It is important to note that in the context of such a dynamic evolution one can always define two operators x and $\tilde{x}$, called superfluctuant, whose variances $(\Delta x)_z$, $(\Delta \tilde{x})_z$ are amplified with respect to their vacuum value $(\Delta x)_0$, $(\Delta \tilde{x})_0$, namely [8]

$$(\Delta x)_0 \to (\Delta x)_z = (\Delta x)_0 e^r, \qquad (\Delta \tilde{x})_0 \to (\Delta \tilde{x})_z = (\Delta \tilde{x})_0 e^r \tag{7}$$

where $(\Delta x)_z^2 = \langle z|x^2|z\rangle - (\langle z|x|z\rangle)^2$ (and the same for $\tilde{x}$). The variance of the canonically conjugate operators $y, \tilde{y}$ is correspondingly squeezed, $(\Delta y)_z = (\Delta y)_0 e^{-r} = (\Delta \tilde{y})_z$. In terms of the superfluctuant variables the operators b and $b^\dagger$ have the differential representation [6,7,8]

$$b = \frac{i}{2} e^{i\theta}(x - i\tilde{x} + \partial_x - i\partial_{\tilde{x}})$$

$$\tilde{b} = \frac{i}{2} e^{i\theta}(x + i\tilde{x} + \partial_x + i\partial_{\tilde{x}}) \tag{8}$$

(henceforth the mode index k is to be understood, if not explicitly written, and any correlation among modes with different $|k|$ will be neglected, following the coarse graining approach of [2]). The normalized wave function ψ_z for the two-mode squeezed vacuum (such that $a\psi_z = 0 = \tilde{a}\psi_z$) becomes, in the $(x, \tilde{x})$-space representation [6,7],

$$\psi_z(x, \tilde{x}) = \langle x\tilde{x}|\Sigma|0\rangle \equiv \langle x\tilde{x}|z\rangle = \left(\frac{\sigma}{\pi}\right)^{1/2} e^{-\frac{\sigma}{2}(x^2+\tilde{x}^2)},$$

$$\sigma = e^{-2r}, \qquad \langle z|z\rangle = 1 \tag{9}$$

The squeezed evolution $|0\rangle \to |z\rangle = \Sigma|0\rangle$ may thus represented, in the superfluctuant basis, as a scaling transformation $x \to \sqrt{\sigma}x$, $\tilde{x} \to \sqrt{\sigma}\tilde{x}$, with real positive parameter $\sigma \leq 1$, related to the squeezing parameter r according to eq.(9).

In order to compute the Shannon-Wehrl entropy for a squeezed state we recall that the classical entropy is defined, for any given phase-space distribution $f(q,p)$, as

$$S_{cl}(f) = -\int dqdpf(q,p)\ln f(q,p) \tag{10}$$

while the quantum (or von Neumann) entropy is defined in terms of the density operator ρ as

$$S_q(\rho) = -Tr\rho\ln\rho \tag{11}$$

For a pure state, in particular, $\rho^2 = \rho$ and $S_q = 0$. Suppose now that we want to compute, for a given quantum state represented by ρ, its entropy in the "semiclassical" limit. It is well known that a possible procedure to obtain the semiclassical limit of a quantum observable is to compute the expectation value of that observable in the basis of the coherent states $|\alpha\rangle$. According to Wehrl [14], a natural semiclassical phase-space distribution associated to ρ is thus given by

$$f(\alpha) = \frac{1}{\pi}\langle\alpha|\rho|\alpha\rangle \tag{12}$$

(the so-called Glauber distribution function [17]), which leads to define the semiclassical (Wehrl) entropy S_W as

$$S_W = S_{cl}(f(\alpha)) \tag{13}$$

For a two-mode squeezed state $|z\rangle$ (see [18,19] for the one-mode case) we have, in particular,

$$f(\alpha,\tilde{\alpha}) = \frac{1}{\pi^2}\langle\alpha\tilde{\alpha}|z\rangle\langle z|\alpha\tilde{\alpha}\rangle \tag{14}$$

where the coherent states $|\alpha\tilde{\alpha}\rangle$ satisfy

$$b|\alpha\tilde{\alpha}\rangle = \alpha|\alpha\tilde{\alpha}\rangle, \qquad \tilde{b}|\alpha\tilde{\alpha}\rangle = \tilde{\alpha}|\alpha\tilde{\alpha}\rangle \tag{15}$$

(the operators $b, \tilde{b}$ are defined in eq.(8); their corresponding eigenvalues are complex numbers, $\alpha = \alpha_1 + i\alpha_2$, $\tilde{\alpha} = \tilde{\alpha}_1 + i\tilde{\alpha}_2$). By expanding in the superfluctuant basis $|x\tilde{x}\rangle$ one easily obtains

$$\langle x\tilde{x}|\alpha\tilde{\alpha}\rangle = \frac{1}{\sqrt{\pi}}\exp\left[-\frac{1}{2}(\Re k)^2 - \frac{1}{2}(\Re\tilde{k})^2 + kx + \tilde{k}\tilde{x} - \frac{1}{2}(x^2+\tilde{x}^2)\right],$$

$$\langle\alpha\tilde{\alpha}|\alpha\tilde{\alpha}\rangle = 1 \tag{16}$$

where $\Re z$ denotes the real part of z, and

$$k = -ie^{-i\theta}(\alpha + \tilde{\alpha}), \qquad \tilde{k} = -e^{-i\theta}(\tilde{\alpha} - \alpha) \tag{17}$$

Moreover

$$f(\alpha, \tilde{\alpha}) = \frac{1}{\pi^2} \int dx d\tilde{x} dx' d\tilde{x}' \psi_z(x\tilde{x})\psi_z^*(x'\tilde{x}')\langle\alpha\tilde{\alpha}|x\tilde{x}\rangle\langle x'\tilde{x}'|\alpha\tilde{\alpha}\rangle =$$

$$= \frac{1}{\pi^2 \cosh^2 r} \exp\left[-\tanh r(\alpha\tilde{\alpha}e^{-2i\theta} + \alpha^*\tilde{\alpha}^*e^{2i\theta}) - |\alpha|^2 - |\tilde{\alpha}|^2\right] \tag{18}$$

With this Glauber distribution, the semiclassical Wehrl entropy (13) for the two-mode squeezed vacuum is then

$$S_W = -\int d^2\alpha d^2\tilde{\alpha} f(\alpha, \tilde{\alpha}) \ln f(\alpha, \tilde{\alpha}) = 2 + 2\ln\pi + 2\ln\cosh r \tag{19}$$

The same result holds for a squeezed-coherent state $|z\alpha_0\tilde{\alpha}_0\rangle = \Sigma|\alpha_0\tilde{\alpha}_0\rangle$, obtained by applying a squeezing transformation to the "displaced" vacuum $|\alpha_0\tilde{\alpha}_0\rangle = D(\alpha_0)D(\tilde{\alpha}_0)|0\rangle$, where $D(\alpha) = \exp(\alpha b^\dagger - \alpha^* b)$ is the Glauber displacement operator.

It is important to note that for $r >> 1$ this semiclassical entropy reproduces the large squeezing behavior of eq.(2), previously obtained [1,3-7,12] with different approaches to the entropy of a particle production process. We also note that the entropy (19) is exactly the sum of two "marginal" entropies,

$$S_W = S^W_{\alpha_1\tilde{\alpha}_1} + S^W_{\alpha_2\tilde{\alpha}_2} \tag{20}$$

obtained by decomposing the probability distribution $f(\alpha, \tilde{\alpha})$ in the complex $\alpha, \tilde{\alpha}$ planes, and by tracing out one of the two cartesian variables,

$$S^W_{\alpha_1\tilde{\alpha}_1} = -\int d\alpha_1 d\tilde{\alpha}_1 f(\alpha_1, \tilde{\alpha}_1) \ln f(\alpha_1, \tilde{\alpha}_1), \quad f(\alpha_1, \tilde{\alpha}_1) = \int d\alpha_2 d\tilde{\alpha}_2 f(\alpha, \tilde{\alpha})$$

$$S^W_{\alpha_2\tilde{\alpha}_2} = -\int d\alpha_2 d\tilde{\alpha}_2 f(\alpha_2, \tilde{\alpha}_2) \ln f(\alpha_2, \tilde{\alpha}_2), \quad f(\alpha_2, \tilde{\alpha}_2) = \int d\alpha_1 d\tilde{\alpha}_1 f(\alpha, \tilde{\alpha}) \tag{21}$$

This is a non-trivial result, as it implies that the information-theoretical Araki-Lieb inequality [20], $S \leq S_{\alpha_1\tilde{\alpha}_1} + S_{\alpha_2\tilde{\alpha}_2}$, is maximized both by the two-mode squeezed vacuum and by a squeezed-coherent state, in agreement with the properties of one-mode squeezed states [18].

According to eq.(19), the semiclassical entropy associated to a pure squeezed state $|z\rangle$ is thus non-vanishing, in spite of the fact that $\rho = |z\rangle\langle z| = \rho^2$ and $S_q(\rho) = 0$. This occurs because an effective reduction of the quantum density matrix is implicit in the semiclassical limit, as even a pure state is seen, classically, as a disordered one. However, the coarse graining associated to the semiclassical limit cannot give us explicit information about the physical origin of the decoherence mechanism, which can only be described in the context of some dynamic model for the quantum to classical transition (see also [21]). For a squeezed state such a model must account, in particular, for the fact that some degree of freedom of the system is made dynamically less relevant than others for what concerns macroscopic observations.

Let us consider now the possibility of defining also a von Neumann entropy for the squeezing process, in terms of a properly reduced density operator. In the approach of Refs.[5-7] the reduction was performed by neglecting the off-diagonal elements of the density matrix in a superfluctuant representation. The use of such a representation is justified because the squeezed density operator can only be reconstructed, observationally, by measuring superfluctuant variables and their momenta. But is the reduction a formally justified procedure?

The answer is certainly not for the pure squeezed vacuum state (9), whose density matrix in the superfluctuant representation

$$\rho_z(x,x',\tilde{x},\tilde{x}') = \langle x\tilde{x}|z\rangle\langle z|x'\tilde{x}'\rangle = \psi_z(x,\tilde{x})\psi_z^*(x',\tilde{x}') =$$

$$= \frac{\sigma}{\pi}\exp\left[-\frac{\sigma}{2}(x^2+x'^2+\tilde{x}^2+\tilde{x}'^2)\right] \qquad (22)$$

is perfectly symmetric in the (x,x') and $(\tilde{x},\tilde{x}')$ planes. The same answer holds for a pure squeezed number and squeezed-coherent state. On the contrary, the reduction would be naturally justified if the off-diagonal matrix elements would be suppressed, $|\rho_z(x,x)| >> |\rho_z(x,x'\neq x)|$ (and the same for $\tilde{x}$), for any given value of the squeezing parameter. It is thus interesting to note that such requirement can be satisfied when the initial state is not pure, but is represented by a mixture, whose statistical weights tend to be equally distributed among all the states of the mixture.

Suppose to start, in fact, from an initial configuration more general than the vacuum, represented by an ensemble of states which, like the vacuum, minimize however the quantum fluctuations Δx, $\Delta\tilde{x}$ (and their conjugate Δy, $\Delta\tilde{y}$). This corresponds to an initial density operator which can be expressed in terms of the coherent states (15), (16) as

$$\rho_i = \int d^2\alpha d^2\tilde{\alpha} P(\alpha, \tilde{\alpha}) |\alpha\tilde{\alpha}\rangle\langle\alpha\tilde{\alpha}| \tag{23}$$

where the statistical weights $P \geq 0$ satisfy the normalization condition

$$\int d^2\alpha d^2\tilde{\alpha} P(\alpha, \tilde{\alpha}) = 1 \tag{24}$$

The squeezed evolution leads then to a final density operator, $\rho_f = \Sigma\rho_i\Sigma^\dagger$, which in the superfluctuant basis is explicitly represented by

$$\rho_f(x, x', \tilde{x}, \tilde{x}') = \int d^2\alpha d^2\tilde{\alpha} P(\alpha, \tilde{\alpha}) \psi_{z\alpha\tilde{\alpha}}(x, \tilde{x}) \psi^*_{z\alpha\tilde{\alpha}}(x', \tilde{x}') \tag{25}$$

Here $\psi_{z\alpha\tilde{\alpha}}(x, \tilde{x})$ is the squeezed-coherent wave function, representing in the $(x, \tilde{x})$ space the eigenfunctions of $\{a, \tilde{a}\}$ with eigenvalues $\{\alpha, \tilde{\alpha}\}$, namely

$$\psi_{z\alpha\tilde{\alpha}}(x, \tilde{x}) = \langle x\tilde{x}|z\alpha\tilde{\alpha}\rangle = \langle x\tilde{x}|\Sigma|\alpha\tilde{\alpha}\rangle =$$

$$= \sqrt{\frac{\sigma}{\pi}} \exp\left[-\frac{1}{2}(\Re k)^2 - \frac{1}{2}(\Re\tilde{k})^2 + \sqrt{\sigma}(k\sigma + \tilde{k}\tilde{x}) - \frac{\sigma}{2}(x^2 + \tilde{x}^2)\right] \tag{26}$$

($k, \tilde{k}$ are defined in eq.(17)). Note that, for $\rho_i \neq \rho_i^2$, there is a non-vanishing quantum entropy, $S_q = -\int d^2\alpha d^2\tilde{\alpha} P(\alpha, \tilde{\alpha}) \ln P(\alpha\tilde{\alpha})$, naturally associated to the initial mixture. Such an entropy, however, is constant throughout the whole process of squeezing $\rho_i \to \rho_f$ (which describes a unitary evolution), so that it cannot contribute to the overall entropy difference $\Delta S = S_f = S_i$.

The choice of the vacuum as initial condition corresponds to the assumption $P(\alpha, \tilde{\alpha}) = \delta^2(\alpha)\delta^2(\tilde{\alpha})$ in eq.(25). We shall consider here an initial distribution which, instead of being infinitely peaked upon the vacuum, tends to be spread rather uniformly over a wide range of states, so as to reflect our lack of knowledge about the initial conditions. We shall assume, however, that such a distribution is always centered around the vacuum, and we shall conveniently translate the $(x, \tilde{x})$ frame in such a way that $\langle z\alpha\tilde{\alpha}|x|z\alpha\tilde{\alpha}\rangle = 0 = \langle z\alpha\tilde{\alpha}|\tilde{x}|z\alpha\tilde{\alpha}\rangle$ (which implies $\Re k = 0 = \Re\tilde{k}$, see eq.(26)). The simplest example satisfying the above requirements is a normalized square-like distribution, in which the probability of a state $|\alpha\tilde{\alpha}\rangle$ is constant and equal to $1/L$ for $|\alpha|, |\tilde{\alpha}| < L/2$, and vanishing for $|\alpha|, |\tilde{\alpha}| > L/2$, namely

$$\int d^2\alpha d^2\tilde{\alpha} P(\alpha, \tilde{\alpha}) = d\alpha_1 d\alpha_2 d\tilde{\alpha}_1 d\tilde{\alpha}_2 \delta(\alpha_1 - \tilde{\alpha}_1)\delta(\alpha_2 + \tilde{\alpha}_2)\times$$

$$\times \frac{1}{L^2}\left[\theta(\alpha_1 + L/2) - \theta(\alpha_1 - L/2)\right]\left[\theta(\alpha_2 + L/2) - \theta(\alpha_2 - L/2)\right] \tag{27}$$

Here θ is the Heaviside step function, and the two delta distributions have been inserted to implement the conditions $\langle x\rangle = 0 = \langle\tilde{x}\rangle$; we have also rescaled the variables $\alpha, \tilde{\alpha}$ so as to absorb the overall phase $e^{-i\theta}$ which does not contribute to the integration over the whole complex plane.

With this probability distribution the squeezed matrix (25) becomes

$$\rho_f(x, x', \tilde{x}, \tilde{x}') = \frac{\sigma}{\pi}\frac{\sin\left[L\sqrt{\sigma}(x'-x)\right]}{L\sqrt{\sigma}(x'-x)}\frac{\sin\left[L\sqrt{\sigma}(\tilde{x}-\tilde{x}')\right]}{L\sqrt{\sigma}(\tilde{x}-\tilde{x}')}e^{-\frac{\sigma}{2}(x^2+x'^2+\tilde{x}^2+\tilde{x}'^2)} \tag{28}$$

and it is immediately evident that, for large enough L, the off-diagonal terms $x \neq x'$, $\tilde{x} \neq \tilde{x}'$ are suppressed with respect to the diagonal ones. Such a suppression grows with L, namely with the number of states which are considered equiprobable in the initial distribution, i.e. with the growth of our "ignorance" about the initial configuration. For large L we may thus approximate the squeezed matrix with the diagonal elements only, and in particular we can formally define in this basis a (normalized) reduced density matrix, in the limit $L \to \infty$ (at finite σ), as

$$\rho_{red}(x, x', \tilde{x}, \tilde{x}') = \lim_{L\to\infty}\{L^2\sigma\rho_f(x, x', \tilde{x}, \tilde{x}')\} =$$

$$= \frac{\sigma}{\pi}e^{-\sigma(x^2++\tilde{x}^2)}\delta(x-x')\delta(\tilde{x}-\tilde{x}') \tag{29}$$

This is exactly the representation in the $(x, \tilde{x})$ basis of the reduced operator

$$\rho_{red} = \int dx d\tilde{x}|\psi_z(x, \tilde{x})|^2|x\tilde{x}\rangle\langle x\tilde{x}| \tag{30}$$

previously used in Refs.[6,7].

Note that $Tr\rho_{red} = 1$, but $\rho_{red}^2 \neq \rho_{red}$. This reduction leads then to a quantum entropy $S_q(\rho_{red})$ which, unlike the entropy of the mixture $S_q(\rho_f)$, depends on σ, and in particular grows with the degree of squeezing. By taking the difference between final and initial entropy we find indeed the result of eq.(2),

$$\Delta S = (Tr\rho_{red}\ln\rho_{red})_{\sigma=1} - (Tr\rho_{red}\ln\rho_{red})_{\sigma} = 2r \tag{31}$$

The accuracy of this estimate in our context becomes larger and larger as the initial probability distribution becomes flatter, so as to share probabilities among a larger number of states. It is also important to note that if, for a given L, the reduced matrix represent a good approximation to the exact matrix (28) at large squeezing ($\sigma >> 1$), then the approximation is even better as we go back in time towards the initial configuration, where $\sigma = 1$.

We may thus conclude that the effective reduction of the squeezed density operator, introduced "ad hoc" in previous papers [6,7], can be formally justified through a limiting procedure based on an appropriate mixture of coherent states, assumed to represent the initial configuration. The corresponding entropy growth may thus physically interpreted as an amplification, due to the squeezed evolution, of our ignorance about the initial conditions, in a context in which we have no motivation to single out some preferred initial state, and we cannot reconstruct the initial state through final measurements of superfluctuant variables only. Such an interpretation seems to be particularly appropriate in the context of cosmological perturbation theory, where the amplification of the background fluctuations is computed starting from an initial state which is left unspecified except for the requirement of minimizing the quantum fluctuations, and which can thus be properly represented as a general mixture of coherent states.

Acknowledgments

One of us (M. Gasperini) wishes to thank D. Boyanovsky for rising stimulating questions that motivated in part the work reported here. We are also grateful to G. Veneziano for many helpful discussions.

References

1. B. L. Hu and D. Pavon, Phys. Lett. B 180, 329 (1986)
2. B. L. Hu and H. E. Kandrup, Phys. Rev. D 35, 1776 (1987);
   H. E. Kandrup, Phys. Rev. D 37, 3505 (1988);
   for a recent review on cosmological particle production see also B. L. Hu, G. Kang and A. Matacz, Int. J. Mod. Phys. A 9, 991 (1994)
3. R. Brandenberger, V. Mukhanov and T. Prokopec, Phys. Rev. Lett. 69, 3606 (1992)
4. R. Brandenberger, V. Mukhanov and T. Prokopec, Phys. Rev. D 48, 2443 (1993)

5. T. Prokopec, Class. Quantum Grav. 10, 2295 (1993)
6. M. Gasperini and M. Giovannini, Phys. Lett. B 301, 334 (1993)
7. M. Gasperini and M. Giovannini, Class. Quantum Grav. 10, L133 (1993)
8. B. L. Shumaker, Phys. Rep. 135, 317 (1986);
J. Grochmalicki and M. Lewenstein, Phys. Rep. 208, 189 (1991)
9. B. L. Hu, Phys. Rev. D 9, 3263 (1974)
10. L. P. Grishchuk and Y. V. Sidorov, Phys. Rev. D 42, 3413 (1990);
L. P. Grishchuk and L. Solokhin, Phys. Rev. D 43, 2566 (1991)
11. L. P. Grishchuk, Zh. Eksp. Teor. Fiz. 67, 825 (1975)
12. A. Matacz, Phys. Rev. D 49, 788 (1994)
13. C. E. Shannon and W. Weaver, "The mathematical theory of communication" (Univ. of Illinois Press, Urbana, IL, 1963)
14. A. Wehrl, Rev. Mod. Phys. 50, 221 (1978)
15. E. Keski-Vakkuri, Phys. Rev. D49, 2122, (1994)
16. M. Kruczenski, L. E. Oxman and M. Zaldarriaga, "Large squeezing behavior of cosmological entropy generation", Preprint gr-qc/9403024
17. R. London and P. L. Knight, J. Mod. Opt. 34, 709 (1987)
18. C. Keitel and K. Wodkiewicz, Phys. Lett. A 167, 151 (1992)
19. H. Rosu and M. Reyes, "Shannon-Wehrl entropy for cosmological and black-hole squeezing", Preprint gr-qc/9406001
20. H. Araki and E. Lieb, Commun. Math. Phys. 18, 160 (1970)
21. W. H. Zurek, S. Habib and J. P. Paz, Phys. Rev. Lett. 70, 1187 (1993)

PREHEATING AND REHEATING IN INFLATIONARY COSMOLOGY: A PEDAGOGICAL SURVEY

D. Boyanovsky[(a)], **H.J. de Vega**[(b)], **R. Holman**[(c)] **and J. F. J. Salgado**[(b)]

(a) Department of Physics and Astronomy, University of Pittsburgh, Pittsburgh, PA. 15260, U.S.A.

*(b) LPTHE, * Université Pierre et Marie Curie (Paris VI) et Denis Diderot (Paris VII), Tour 16, 1er. étage, 4, Place Jussieu 75252 Paris, Cedex 05, France*

(c) Department of Physics, Carnegie Mellon University, Pittsburgh, PA. 15213, U. S. A.

(August 1996)

Recent progress in the preheating phenomena for inflationary cosmology is reviewed. We first discuss estimates of the preheating time scale and particle production at the early stages of parametric amplification within the Mathieu and Lamé approximations and we analyze their precision and limitations. The necessity of self-consistent calculations including the non-linearity of the field theory equations in an energy conserving scheme is stressed. The large N calculations including the field back-reaction are reviewed. For spontaneously broken theories the issue of symmetry restoration is analyzed. A discussion of the possibility and criterion for symmetry restoration is presented. (To appear in the Proceedings of the Paris Euronetwork Meeting 'String Gravity', Observatoire de Paris, June 1996).

I. INTRODUCTION

Research activity on inflationary cosmologies has continued steadily since the concept of inflationary cosmology was first proposed in 1981 [1].

It was recognized that in order to merge an inflationary scenario with standard Big Bang cosmology a mechanism to reheat the universe was needed. Such a mechanism must be present in any inflationary model to raise the temperature of the Universe at the end of inflation, thus the problem of reheating acquired further importance deserving more careful investigation. The original version of reheating envisaged that during the last stages of inflation when the accelerated universe expansion slows down, the energy stored in the oscillations of the inflaton zero mode transforms into particles via single particle decay. Such particle production reheats the universe whose temperature reduced enormously due to the inflationary expansion [2].

It was realized recently [4,15,16,18], that in fact, the elementary theory of reheating [2] does not describe accurately the quantum dynamics of the fields.

Our programme on non-equilibrium dynamics of quantum field theory, started in 1992 [3], is naturally poised to provide a framework to study these problems. The larger goal of the program is to study the dynamics of non-equilibrium processes from a fundamental field-theoretical description, by solving the dynamical equations of motion of the underlying four dimensional quantum field theory for physically relevant problems: phase transitions out of equilibrium, particle production out of equilibrium, symmetry breaking and dissipative processes.

The focus of our work is to describe the quantum field dynamics when the energy density is **high**. That is, a large number of particles per volume m^{-3}, where m is the typical mass scale in the theory. Usual S-matrix calculations apply in the opposite limit of low energy density and since they only provide information on *in* $\rightarrow$ *out* matrix elements, are unsuitable for calculations of expectation values. Our methods were naturally applied to different physical problems like pion condensates [5,7], supercooled phase transitions [3,6] and inflationary cosmology [4,6,8–10].

An analogous program has been pursued by the Los Alamos group, whose research focuses on non-linear effects in scalar QED linked to the pair production in strong electric fields [11], the Schwinger-Keldysh non-equilibrium formalism in the large N expansion [12], and the dynamics of chiral condensates in such framework [13].

II. PREHEATING IN INFLATIONARY UNIVERSES

As usual in inflationary cosmology, matter is described in an effective way by a self-coupled scalar field $\Phi(x)$ called the inflaton. The spacetime geometry is a cosmological spacetime with metric $ds^2 = (dt)^2 - a(t)^2\,(d\vec{x})^2$, where $a(t)$ is the scale factor.

The evolution equations for the k-modes of the inflaton field as considered by different groups can be summarized as follows,

$$\ddot{\chi}_k + 3H(t)\,\dot{\chi}_k + \left(\frac{k^2}{a^2(t)} + M^2[\Phi(.)]\right)\chi_k(t) = 0 \quad (2.1)$$

where $H(t) = \dot{a}(t)/a(t)$, and $M^2[\Phi(.)]$ is the effective mass felt by the modes. The expression considered depends on the model (here the $\lambda\ \Phi^4$ model) and the approximations made. The value of λ is scenario-dependent but it is usually very small. $M^2[\Phi(.)]$ depends on the scale factor and on the physical state. Therefore, it depends on the modes $\chi_k(t)$ themselves in a complicated way. One is definitely faced with a complicated non-linear problem. We call 'back-reaction' the effect of the modes $\chi_k(t)$ back on themselves through $M^2[\Phi(.)]$.

In the initial stage, all the energy is assumed in the zero mode of the field $\phi(t)$ [3,6–8,15–18]. That is, the field expectation value $\phi(t) \equiv <\Phi(x)>$, where $<\cdots>$ stays for the expectation value in the translational invariant but non-equilibrium quantum state. For very weakly coupled theories, and early times, such that the back-reaction effects of the non-equilibrium quantum fluctuations can be neglected, one can approximate,

$$M^2[\Phi(.)] \simeq m^2 + \frac{\lambda}{2}\phi(t)^2 \tag{2.2}$$

At this moment, the scale factor is set to be a constant in ref. [15,18]. Refs. [3,6–8] consider Minkowski spacetime from the start. In ref. [16], the scale factor is set to be a constant for a model without $\lambda\Phi^4$ inflaton self-coupling. That is, for the classical potential [16]

$$V = \frac{1}{2}m^2\Phi^2 + g\sigma^2\Phi^2 \ , \tag{2.3}$$

one can consider as classical solution $\Phi(t) = \Phi_0\ \cos(mt)$, $\sigma = 0$. In such case, the Mathieu equation approximation is exact in Minkowski spacetime. However, the potential (2.3) is unstable under renormalization (a Φ^4 counterterm is needed from the one-loop level). Hence, the $\lambda = 0$ choice is a fine-tuning not protected by any symmetry. In a second case a massless selfcoupled $\lambda\Phi^4$ field in a radiation dominated universe $a(t) \propto \sqrt{t}$ is considered in conformal time [16]. In such specific case the classical field equations take the Minkowski form for $\sqrt{t}\Phi$.

In a way or another, using the classical oscillating behaviour of $\phi(t)$, one is led by eq.(2.1) to an effective mass that oscillates in time. In this approximation (which indeed, may be very good for small coupling, see [8]), eq.(2.1) exhibits **parametric resonance** as noticed first in ref [15]. Namely, there are allowed and forbidden bands in k^2. The modes within the forbidden bands grow exponentially whereas those in the allowed bands stay with bounded modulus. The growth of the modes in the forbidden bands translates into profuse particle production, the particles being created with these particular unstable momenta. The rate of particle production is determined by the imaginary part of the Floquet index in these unstable bands. Notice that the approximation (2.2) breaks down as soon as many particles are produced. Namely, when the energy of the produced particles becomes of the order of the zero-mode energy and eq.(2.2) is no more valid.

Now, in order to compute quantitatively the particles produced one needs the form of $\phi(t)$. In ref. [15,16,18] $\phi(t)$ is approximated by a cosinus function in the calculations. The mode equations become a Mathieu equation (the scale factor is set to be a constant). In ref. [15] the Bogoliubov-Krylov approximation is used to compute estimates. In ref. [16,17], estimates are obtained using asymptotic formulas for the Mathieu equation. In ref. [8] the exact classical solution is used (a cn Jacobi function) to compute estimates.

Let us now compare the results from the exact mode solutions obtained in ref. [8] with the Mathieu equation approximation to it. In units where $m^2 = 1$ and setting $\eta(t) \equiv \sqrt{\frac{\lambda}{2}}\ \phi(t)$, one finds

$$\eta(t) = \eta_0\ \mathrm{cn}\left(t\sqrt{1+\eta_0^2}, \bar{k}\right)$$
$$\bar{k} = \frac{\eta_0}{\sqrt{2(1+\eta_0^2)}} \ , \tag{2.4}$$

where cn stands for the Jacobi cosinus and we choose for initial conditions $\eta(0) = \eta_0$, $\dot{\eta}(0) = 0$.

Here $\eta(t)$ has period $4\omega \equiv 4K(\bar{k})/\sqrt{1+\eta_0^2}$, where $K(\bar{k})$ is the complete elliptic integral of first kind, and $\eta(t)^2$ has period 2ω.

Inserting this form for $\eta(\tau)$ in eqs.(2.2) and (2.1) yields

$$\left[\frac{d^2}{dt^2} + k^2 + 1 + \eta_0^2\ \mathrm{cn}^2\left(t\sqrt{1+\eta_0^2}, \bar{k}\right)\right]\chi_k(t) = 0 \ . \tag{2.5}$$

This is the Lamé equation for a particular value of the coefficients that make it solvable in terms of Jacobi functions [24]. As shown in ref. [8], this equation has only one forbidden band for positive k^2 going from $k^2 = 0$ to $k^2 = \frac{\eta_0^2}{2}$. One can choose Floquet solutions of eq.(2.5) that fullfil the relation

$$U_k(t+2\omega) = e^{iF(k)}\ U_k(t), \tag{2.6}$$

where the Floquet indices $F(k)$ are independent of t. In the forbidden band the $F(k)$ posses an imaginary part. The production rate is determined by the imaginary part of the Floquet index.

The exact form of $F(k)$ results [8],

$$F(k) = -2iK(\bar{k})\ Z(2K(\bar{k})\,v) + \pi$$

where $Z(u)$ is the Jacobi zeta function [25] and v is a function of k in the forbidden band defined by

$$k = \frac{\eta_0}{\sqrt{2}}\,\mathrm{cn}(2K(\bar{k})\,v, k) \ , \ 0 \le v \le \frac{1}{2}. \tag{2.7}$$

All these elliptic functions posses fastly convergent expansions in powers of the elliptic nome

$$q \equiv e^{-\pi K'(\bar{k})/K(\bar{k})} \ .$$

Since $0 \le \bar{k} \le 1/\sqrt{2}$ [see eq.(2.4)], we have

$$0 \le q \le e^{-\pi} = 0.0432139\ldots \,. \tag{2.8}$$

Then,

$$F(k) = 4i\,\pi\, q\, \sin(2\pi v)\, \left[1 + 2\,q\, \cos 2\pi v + O(q^2)\right] + \pi\,. \tag{2.9}$$

The imaginary part of this function has a maximum at $k = k_1 = \frac{1}{2}\,\eta_0\,(1-q) + O(q^2)$ where [8]

$$\mathcal{F} \equiv Im F(k_1) = 4\,\pi\, q + O(q^3)\,. \tag{2.10}$$

This simple formula gives the maximum of the imaginary part of the Floquet index in the forbidden band with a precision better than $8.\,10^{-5}$. q can be expressed in terms of η_0 as follows [8]

$$q = \frac{1}{2}\,\frac{(1+\eta_0^2)^{1/4} - (1+\eta_0^2/2)^{1/4}}{(1+\eta_0^2)^{1/4} + (1+\eta_0^2/2)^{1/4}}\,.$$

with an error smaller than $\sim 10^{-7}$.

Let us now proceed to the Mathieu equation analysis of this problem. The cn Jacobi function can be expanded as [26]

$$\mathrm{cn}(z,\bar{k}) = (1-q)\cos(1-4q)z + q\cos 3z + O(q^2)\,.$$

To zeroth order in q we have

$$\eta(t)^2 = \frac{\eta_0^2}{2}\left[1 + \cos(2t\sqrt{1+\eta_0^2})\right] + O(q)\,.$$

and $2\omega = \pi/\sqrt{1+\eta_0^2} + O(q)$. Under such approximations eq.(2.5) becomes the Mathieu equation [27]

$$\frac{d^2y}{dz^2} + (a - 2\bar{q}\cos 2z)\,y(z) = 0\,, \tag{2.11}$$

where

$$a = 1 + \frac{k^2 - \frac{\eta_0^2}{2}}{\eta_0^2+1}\,,\; \bar{q} = \frac{\eta_0^2}{4(\eta_0^2+1)}$$

and $z = \sqrt{\eta_0^2+1}\;t$. Notice that $0 \le \bar{q} \le 1/4$ in the present case. Eq.(2.11) posses an infinite number of forbidden bands for $k^2 > 0$. The lower and upper edge of the first band are respectively [27]

$$k^2_{inf} = \frac{\eta_0^2}{4}\left[1 - \frac{\eta_0^2}{2^5(\eta_0^2+1)} + \frac{\eta_0^4}{2^{10}(\eta_0^2+1)} + \ldots\right]\,,$$

till

$$k^2_{sup} = \frac{\eta_0^2}{4}\left[3 - \frac{\eta_0^2}{2^5(\eta_0^2+1)} - \frac{\eta_0^4}{2^{10}(\eta_0^2+1)} + \ldots\right]\,.$$

These values must be compared with the exact result for the Lamé equation (2.5) : $k^2_{inf} = 0$, $k^2_{sup} = \frac{\eta_0^2}{2}$. The width of the band is well approximated but not its absolute position. The numerical values of the maximum of the imaginary part of the Floquet index are given in Table I and compared with the exact values from eq.(2.10).

We see that the Mathieu approximation **underestimates** the exact result by a fraction ranging from 13% to 39%. The second forbidden band in the Mathieu equation yields $\mathcal{F}_{Mathieu} = 0.086\ldots$ for $\eta_0 \to \infty$. This must be compared with $\mathcal{F}_{Lame} = 0$ (no further forbidden bands).

In ref. [17], an even larger discrepancy between Lame and Mathieu Floquet indices has been reported within a different approximation scheme.

It must be noticed that the Floquet indices entering in exponents produces a dangerous error propagation. For example, the number of particles produced during reheating is of the order of the exponential of $2\mathcal{F}$ times the reheating time in units of $\pi/\sqrt{1+\eta_0^2}$. An error of 25% in $\mathcal{F}$ means an error of 25% in the exponent. (For instance, one would find 10^9 instead of 10^{12}).

The mode equations (2.1) apply to the self-coupled $\lambda\,\Phi^4$ scalar field. Models for reheating usually contain at least two fields: the inflaton and a lighter field $\sigma(x)$ in which the inflaton decays. For a $g\sigma^2\Phi^2$ coupling, the mode equations for the σ field take the form [7,16,18]

$$\ddot{V}_k + 3H(t)\dot{V}_k + \left(\frac{k^2}{a^2(t)} + m_\sigma^2 + \frac{g}{\lambda}F[\Phi(.),\sigma(.)]\right)V_k(t) = 0 \tag{2.12}$$

A new dimensionless parameter $\frac{g}{\lambda}$ appears here. Neglecting the σ and Φ backreaction, we have

$$F[\Phi(.),\sigma(.)] \simeq \eta^2(t)\,. \tag{2.13}$$

In ref. [6,8], it is shown that abundant particle production (appropriate for reheating) shows up even for $g = \lambda$.

η_0	$\mathcal{F}_{Lame}$	$\mathcal{F}_{Mathieu}$	$\%error$
1	$0.2258\ldots$	$0.20\ldots$	13%
4	$0.4985\ldots$	$0.37\ldots$	35%
$\eta_0 \to \infty$	$4\pi e^{-\pi} = 0.5430\ldots$	$0.39\ldots$	39%

TABLE I. The maximum of the imaginary part of the Floquet index $\mathcal{F}$ for the Lamé equation and for its Mathieu approximation.

Eqs.(2.12)-(2.13) becomes a Lamé equation when $\eta(t)$ is approximated by the classical solution in Minkowski spacetime given by (2.4). Such Lamé equation is solvable in closed form when the couplings g and λ are related as follows [8]

$$\frac{2g}{\lambda} = n(n+1) \ , \ n = 1, 2, 3, \ldots$$

In those cases there are n forbidden bands for $k^2 \geq 0$. The Lamé equation exhibits an infinite number of forbidden bands for generic values of $\frac{g}{\lambda}$. The Mathieu and WKB approximations has been applied in the non-exactly solvable cases [16–18]. However, as the above analysis shows (see Table I) such estimations cannot be trusted quantitatively. The only available precise method consists on accurate numerical calculations as those of ref. [6–8] (where the precision is at least 10^{-6}).

Estimates in the cosinus approximation for FRW-de Sitter backgrounds and open universes using the Bogoliubov-Krylov approximation are given in ref. [21]. In ref. [22] the Bogoliubov-Krylov approximation is applied to the large N equations.

Applications of preheating to various relevant aspects of the early cosmology are considered in ref. [23].

As soon as the quantum fluctuations grow and cease to be negligible compared with the the classical piece (2.13), all the approximations discussed so far (Lamé, Mathieu, etc.) break down. This time is the so-called preheating time t_{reh} [8]. One can estimate t_{reh} by equating the zero mode energy (2.13) with the estimation of the quantum fluctuations derived from the unstable Floquet modes [16,8]. Such estimation yields when the Lamé Floquet indices are used [8],

$$t_{reh} \approx \frac{1}{B} \log \frac{N(1+\eta_0^2/2)}{g\sqrt{B}} \ , \tag{2.14}$$

where

$$B = 8\sqrt{1+\eta_0^2}\, q\,(1-4q) + O(q^3) \ ,$$

$$N = \frac{4}{\sqrt{\pi}} \sqrt{q}\, \frac{(4+3\,\eta_0^2)\,\sqrt{4+5\,\eta_0^2}}{\eta_0^3\,(1+\eta_0^2)^{3/4}} \left[1+O(q)\right] \ . \tag{2.15}$$

Where B is determined by the maximum of the imaginary part of the Floquet index in the band. It is now clear that a few percent correction to the Floquet indices will result in a large error in the estimate of the preheating time scale.

III. THE END OF PREHEATING AND BEYOND: SELF-CONSISTENT LARGE N CALCULATIONS

In order to compute physical magnitudes beyond t_{reh}, one **must** solve self-consistently the field equations including the back reaction. In ref. [6,8] this is done for the $N \to \infty$ limit and in ref. [7] to one-loop order.

In the large N limit the zero mode and k-modes renormalized equations take the form,

$$\ddot{\eta} + \eta + \eta^3 + g\,\eta(t)\,\Sigma(t) = 0 \ , \tag{3.1}$$

$$\left[\frac{d^2}{dt^2} + k^2 + 1 + \eta(t)^2 + g\,\Sigma(t)\right] \varphi_k(t) = 0 \ , \tag{3.2}$$

where $g\Sigma(t)$ is given by

$$g\Sigma(t) = g\int_0^\infty k^2 dk \left\{ \mid \varphi_k(t) \mid^2 - \frac{1}{\Omega_k} + \frac{\theta(k-\kappa)}{2k^3}\left[-\eta_0^2 + \eta^2(t) + g\,\Sigma(t)\right] \right\} \ ,$$

$$\Omega_k = \sqrt{k^2+1+\eta_0^2} \ . \tag{3.3}$$

We choose the initial state such that at $t=0$ the quantum fluctuations are in the ground state of the oscillators. That is,

$$\varphi_k(0) = \frac{1}{\sqrt{\Omega_k}} \quad , \quad \dot{\varphi}_k(0) = -i\,\sqrt{\Omega_k} \ ,$$

$$\eta(0) = \eta_0 \quad , \quad \dot{\eta}(0) = 0 \ .$$

In the one-loop approximation the term $g\,\Sigma(t)$ is absent in eq.(3.2).

Eqs.(3.2)-(3.3) were generalized to cosmological spacetimes (including the renormalization aspects) in ref. [4].

Eqs.(3.2) is an infinite set of coupled non-linear differential equations in the k-modes $\varphi_k(t)$ and in the zero mode $\eta(t)$. We have numerically solved eqs.(3.2) for a variety of couplings and initial conditions [6–8]. In figs. 1-3 we display $\eta(t)$, $\Sigma(t)$ and the total number of created particles $N(t)$ as a function of t for $\eta_0 = 4$ and $g = 10^{-12}$ [8]. One sees that the quantum fluctuations are indeed very small till we approach the reheating time t_{reh}. As one can see from the figures, eq.(2.14) gives a very good approximation for t_{reh} and for the behaviour of the quantum fluctuations for $t \leq t_{reh}$. For times earlier than t_{reh}, the Lamé estimate for the envelope of $\Sigma(t)$ and $N(t)$ take the form [8],

$$\Sigma_{est-env}(t) = \frac{1}{N\sqrt{t}}\, e^{Bt} \ ,$$

$$N_{est-env}(t) = \frac{1}{8\pi^2}\, \frac{4+\frac{9}{2}\eta_0^2}{\sqrt{4+5\eta_0^2}}\, \Sigma_{est-env}(t) \ . \tag{3.4}$$

where B and N are given by eq.(2.15).

This analysis presents a very clear physical picture of the properties of the "gas" of particles created during the preheating stage. This "gas" turns to be **out** of thermodynamic equilibrium as shown by the energy spectrum obtained, but it still fulfills a radiation-like equation of state $p \simeq e/3$ [8].

In approximations that do not include the backreaction effects there is infinite particle production, since these

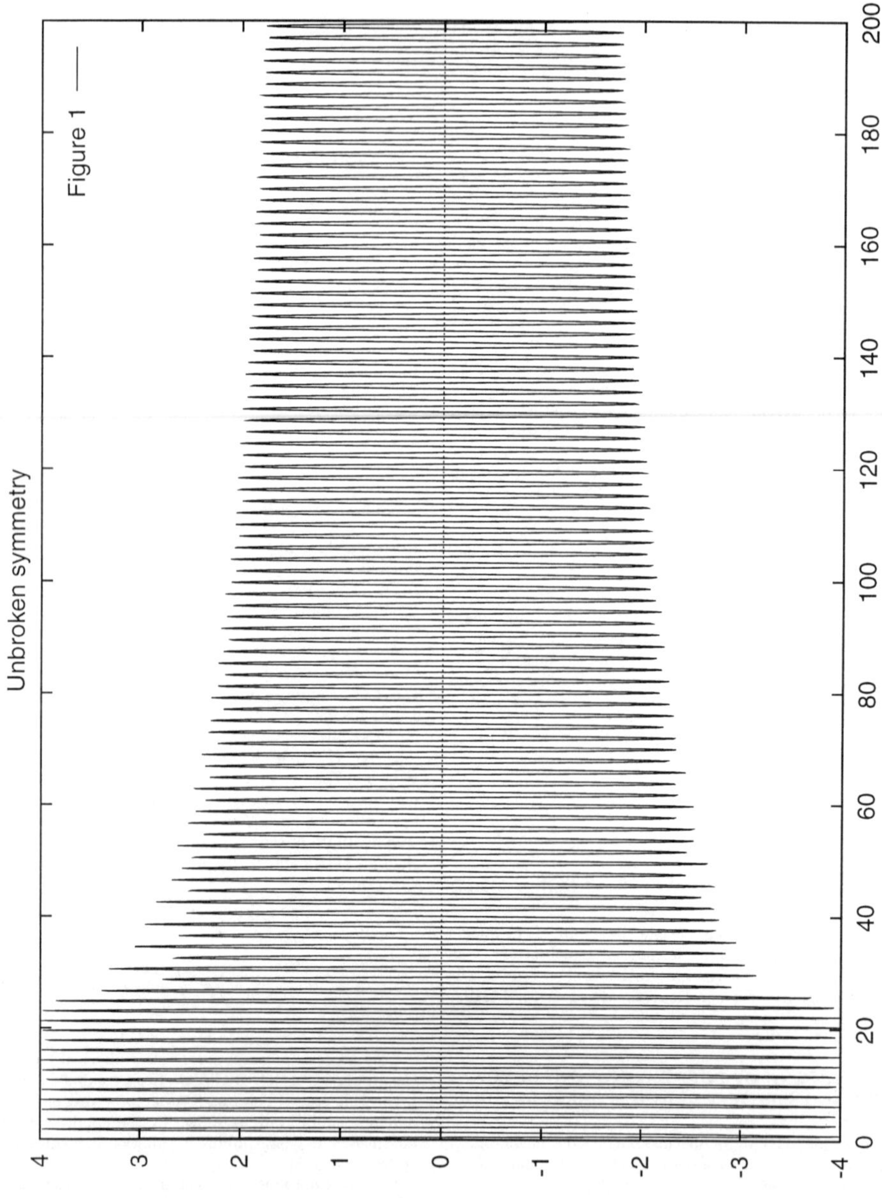

Figure 1: $\eta(\tau)$ vs. τ for the unbroken symmetry case with $\eta_0 = 4$, $g = 10^{-12}$.

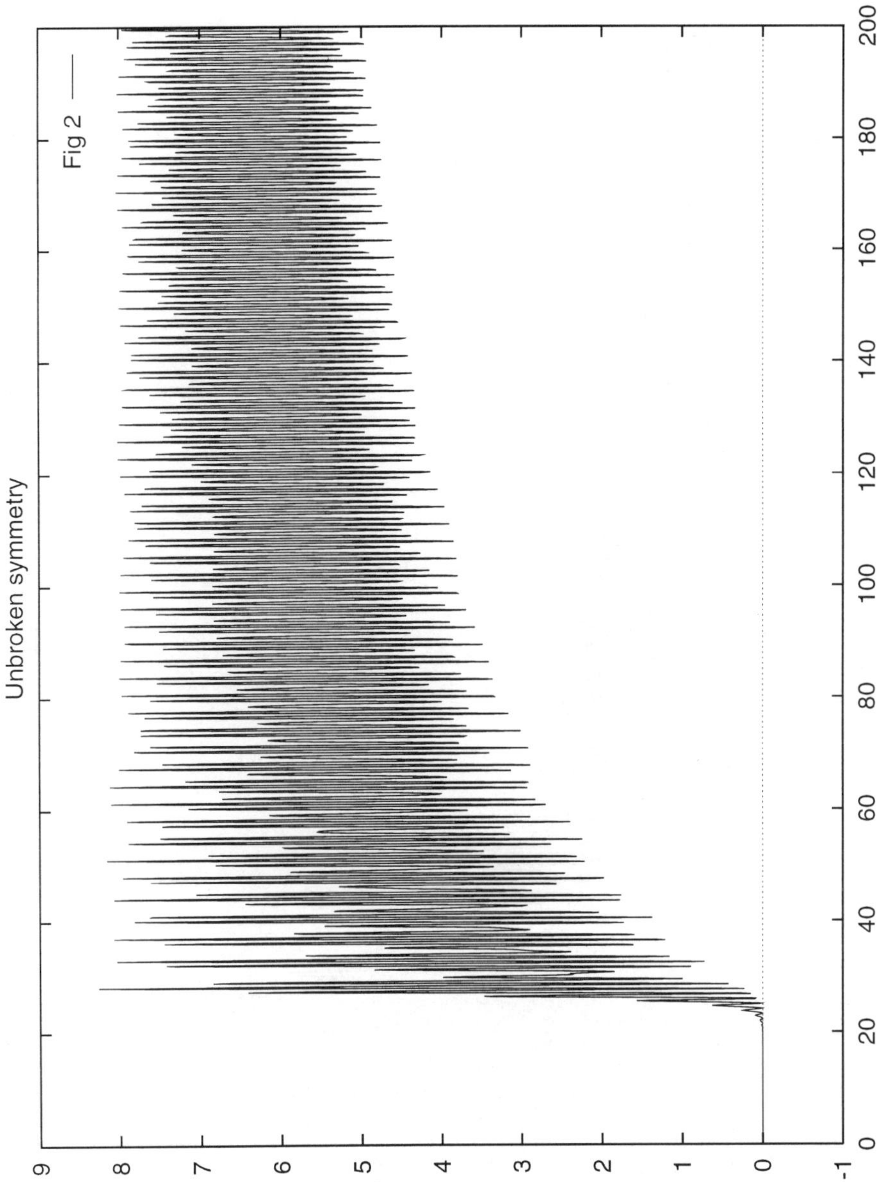

Figure 2: $g\Sigma(\tau)$ vs. τ for the same values of the parameters as in Fig. 1.

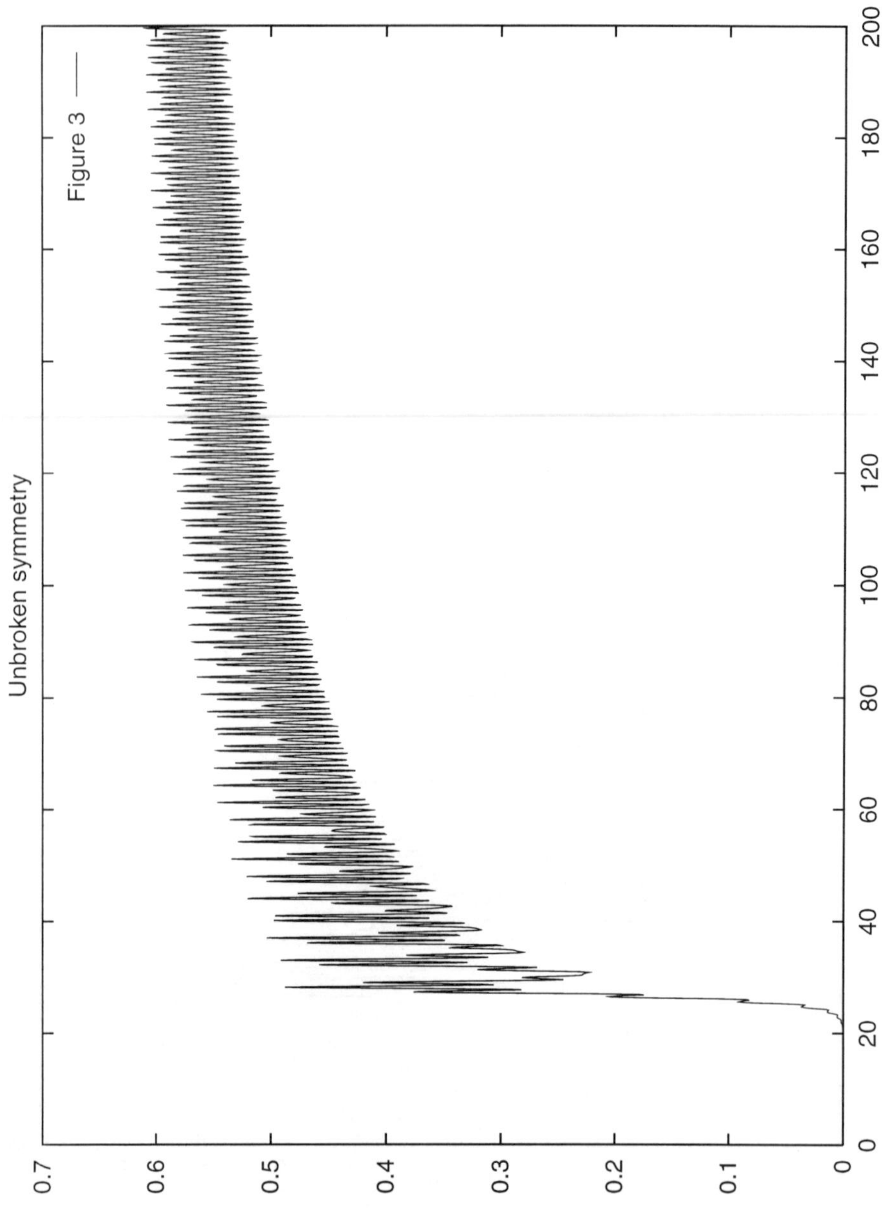

Figure 3: $g\mathcal{N}(\tau)$ for the same parameters as in fig. 1.

do not maintain energy conservation [15,18]. The full backreaction problem and the approximation used in our work exactly conserves the energy. As a result, particle production shuts-off when the back-reaction becomes important ending the preheating stage.

In ref. [14] results of ref. [6] are rederived with a different renormalization scheme.

IV. BROKEN SYMMETRY AND ITS QUANTUM RESTORATION

Up to now we discussed the unbroken symmetry case $m^2 = 1 > 0$. In the spontaneously broken symmetry case $m^2 = -1 < 0$, the classical potential takes the form

$$V(\eta) = \frac{1}{4}(\eta^2 - 1)^2 \tag{4.1}$$

In refs. [6,8] the field evolution is numerically solved in the large N limit for $m^2 < 0$. Analytic estimations are given in ref. [8] for early times.

In ref. [6] it is shown that for small couplings and small η_0, the order parameter damps very quickly to a constant value close to the origin. That is, the order parameter ends close to the classical false vacuum and far from the classical minimum $\eta = 1$. The fast damping is explained, in this spontaneously broken symmetry case by the presence of Goldstone bosons. The massless Goldstone bosons dissipate very efficiently the energy from the zero mode. We plot in figs. 4 and 5, $\eta(t)$ and $\Sigma(t)$ for $\eta_0 = 10^{-5}$ and $g = 10^{-12}$ [8]. The profuse particle production here is due to spinodal unstabilities.

In ref. [19], it is claimed that quantum fluctuations have restored the symmetry in the situations considered in ref. [6]. However, the presence of massless Goldstone bosons and the non-zero limiting value of the order parameter clearly show that the symmetry **is** always broken in the cases considered in ref. [6,8]. These results rule out the claim in ref. [19]. In particular, the asymptotic value of the order parameter results from a very detailed dynamical evolution that conserves energy.

In the situation of 'chaotic initial conditions' but with a broken symmetry tree level potential, the issue of symmetry breaking is more subtle. In this case the zero mode is initially displaced with a large amplitude and very high in the potential hill. The total energy *density* is non-perturbatively large. Classically the zero mode will undergo oscillatory behavior between the two classical turning points, of very large amplitude and the dynamics will probe both broken symmetry states. Even at the classical level the symmetry is respected by the dynamics in the sense that the time evolution of the zero mode samples equally both vacua. This is not the situation that is envisaged in usual symmetry breaking scenarios. For broken symmetry situations there are no finite energy field configurations that can sample both vacua. In the case under consideration with the zero mode of the scalar field with very large amplitude and with an energy density much larger than the top of the potential hill, there is enough energy in the system to sample both vacua. (The energy is proportional to the spatial volume). Parametric amplification transfers energy from the zero mode to the quantum fluctuations. Even when only a fraction of the energy of the zero mode is transferred thus creating a non-perturbatively large number of particles, the energy in the fluctuations is very large, and the equal time two-point correlation function is non-perturbatively large and the field fluctuations are large enough to sample both vacua. The evolution of the zero mode is damped because of this transfer of energy, but in most generic situations it does not reach an asymptotic time-independent value, but oscillates around zero, sampling the tree level minima with equal probability. This situation is reminiscent of finite temperature in which case the energy density is finite and above a critical temperature the ensemble averages sample both tree level vacua with equal probability thus restoring the symmetry. In the dynamical case, the "symmetry restoration" is just a consequence of the fact that there is a very large energy density in the initial state, much larger than the top of the tree level potential, thus under the dynamical evolution the system samples both vacua equally. This statement is simply the dynamical equivalent of the equilibrium finite temperature statement that the energy in the quantum fluctuations is large enough that the fluctuations can actually sample both vacua with equal probability.

Thus the criterion for symmetry restoration when the tree level potential allows for broken symmetry states is that the energy density in the initial state be larger than the top of the tree level potential. That is when the amplitude of the zero mode is such that $V(\eta_0) > V(0)$. In this case the dynamics will be very similar to the unbroken symmetry case, the amplitude of the zero mode will damp out, transferring energy to the quantum fluctuations via parametric amplification, but asymptotically oscillating around zero with a fairly large amplitude.

To illustrate this point clearly, we plot in figs. 6 and 7, $\eta(t)$ and $\Sigma(t)$ for $\eta_0 = 1.6 > \sqrt{2}$ [and hence $V(\eta_0) > V(0)$, see eq.(4.1)] and $g = 10^{-3}$. We find the typical behaviour of unbroken symmetry. Notice again that the effective or tree level potential is an irrelevant quantity for the dynamics, the asymptotic amplitude of oscillation of the zero mode is $\eta \approx 0.5$, which is smaller than the minimum of the tree level potential $\eta = 1$ but the oscillations are symmetric around $\eta = 0$.

Since the dynamical evolution sampled both vacua symmetrically from the beginning, there never was a symmetry breaking in the first place, and "symmetry restoration" is just the statement that the initial state has enough energy such that the *dynamics* probes both vacua symmetrically despite the fact that the tree level potential allows for broken symmetry ground states.

In their comment (hep-ph/9608341), KLS seem to agree with our conclusion that in the situations studied in our articles [6,7], when the expectation value is

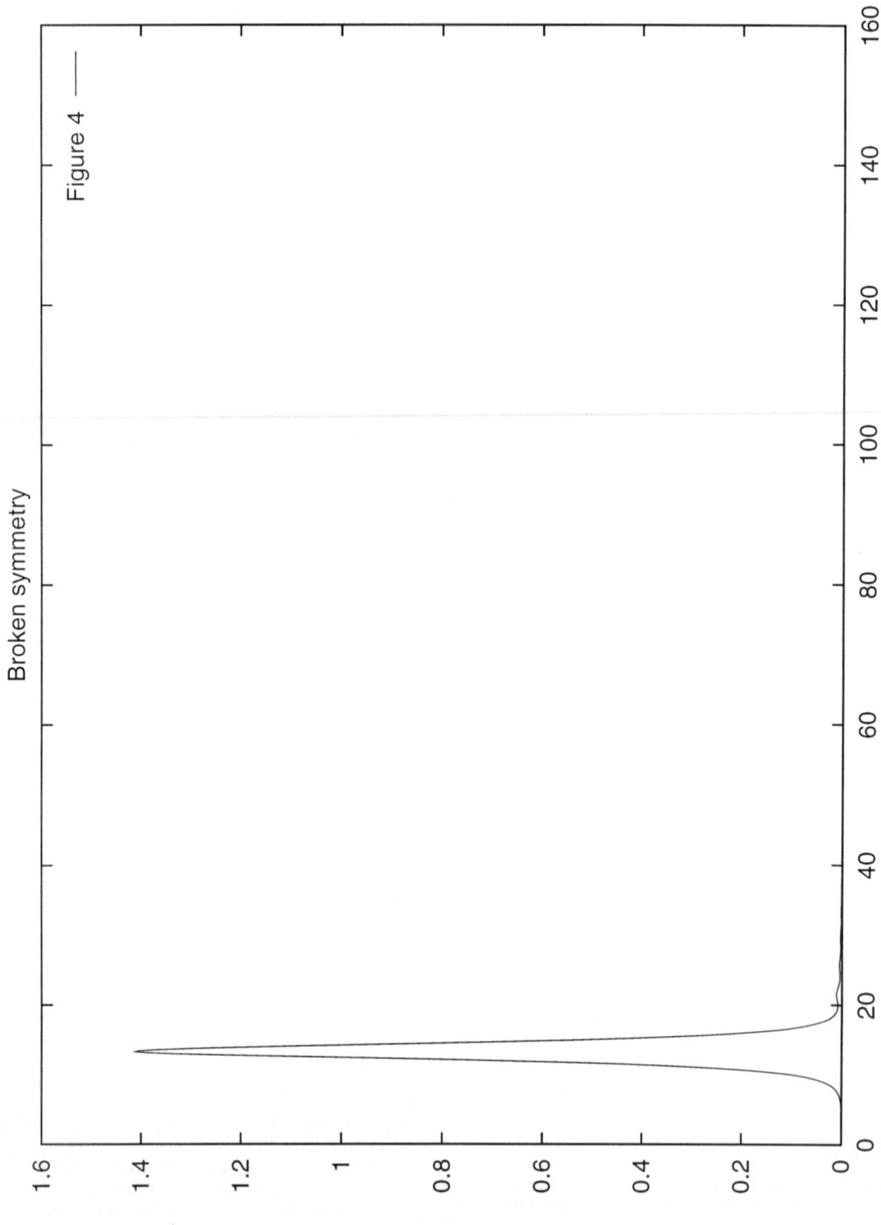

Figure 4: $\eta(\tau)$ vs. τ for the broken symmetry case with $\eta_0 = 10^{-5}$, $g = 10^{-12}$.

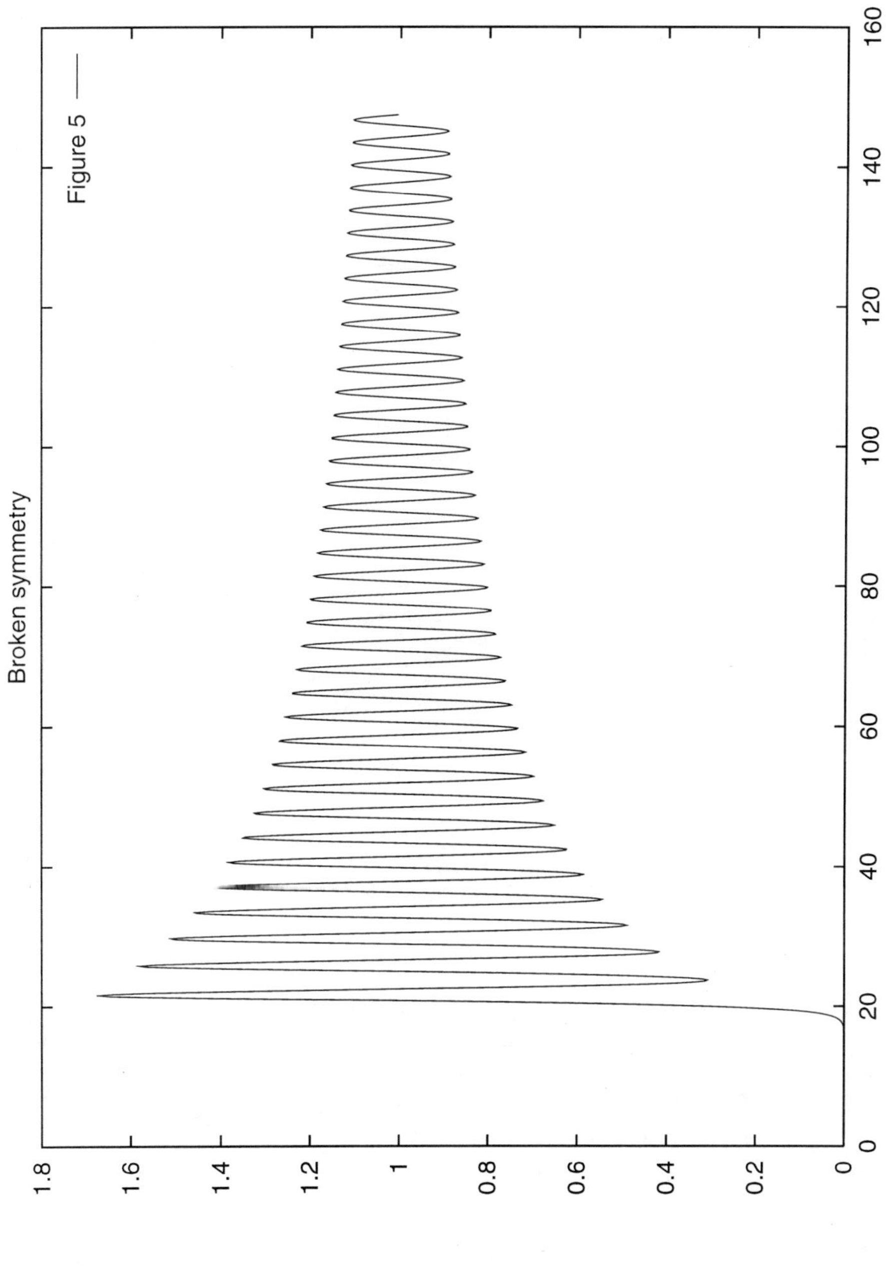

Figure 5: $g\Sigma(\tau)$ for the same values of the parameters as in fig. 4.

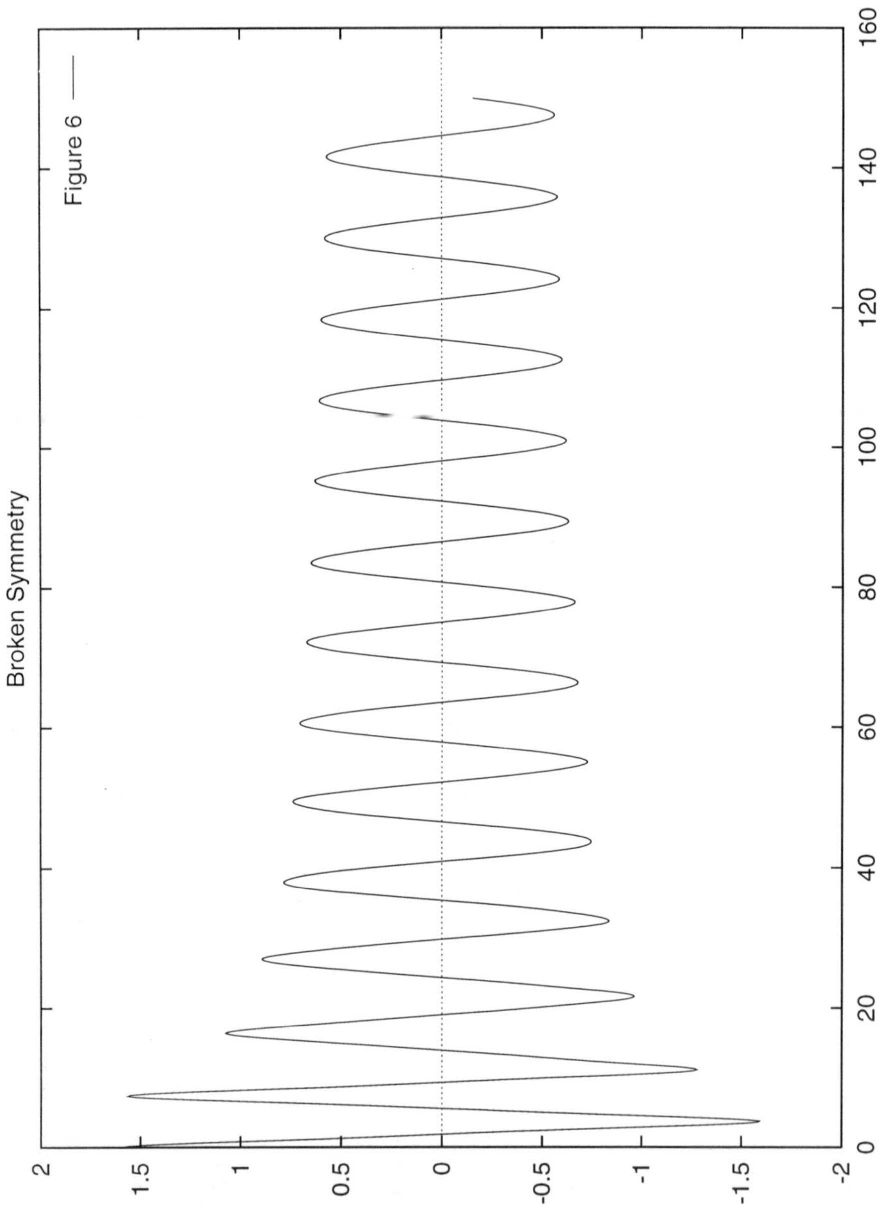

Figure 6: $\eta(\tau)$ vs. τ for the broken symmetry case with $\eta_0 = 1.6 > \sqrt{2}$, $g = 10^{-3}$.

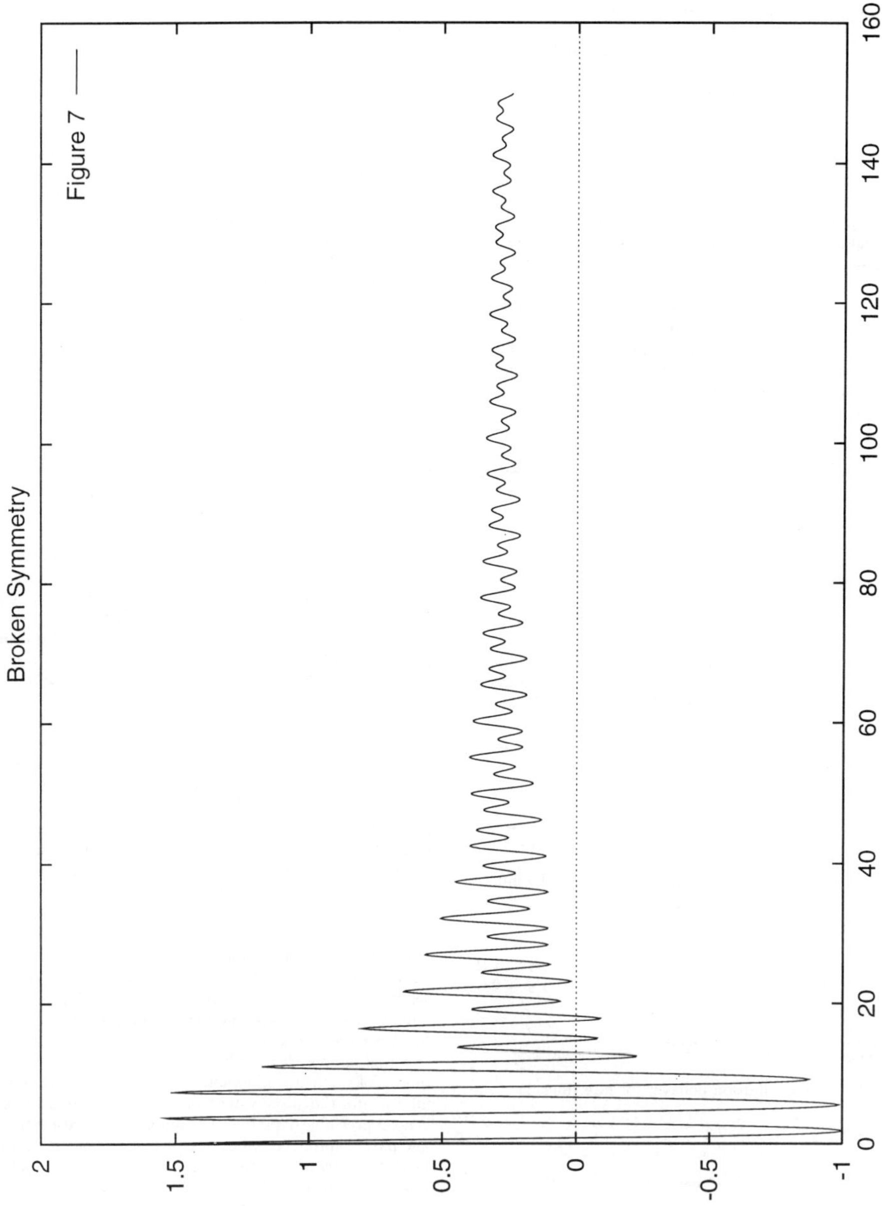

Figure 7: The effective mass squared vs. τ for the same values of the parameters as in fig. 6.

released below the potential hill, symmetry restoration does not occur.

We will present a deeper analytical and numerical study of the subtle aspects of the dynamics in this case in a forthcoming article [9].

V. LINEAR VS. NONLINEAR DISSIPATION (THROUGH PARTICLE CREATION)

As already stressed, the field theory dynamics is unavoidable nonlinear for processes like preheating and reheating. It is however interesting to study such processes in the amplitude expansion. This is done in detail in refs. [6,7]. To dominant order, the amplitude expansion means to linearize the zero mode evolution equations. This approach permits an analytic resolution of the evolution in closed form by Laplace transform. Explicit integral representations for $\eta(t)$ follow as functions of the initial data [6,7]. Moreover, the results can be clearly described in terms of S-matrix concepts (particle poles, production thresholds, resonances, etc.).

Let us consider the simplest model where the inflaton Φ couples to a another scalar σ and to a fermion field ψ, and potential [7]

$$V = \frac{1}{2}\left[m_\Phi^2\Phi^2 + m_\sigma^2\sigma^2 + g\,\sigma^2\Phi^2\right] + \frac{\lambda_\Phi}{4!}\,\Phi^4 + \frac{\lambda_\sigma}{4!}\,\sigma^4 + \bar{\psi}(m_\psi + y\Phi)\psi\ .$$

In the unbroken symmetry case ($m_\Phi^2 > 0$) the inflaton is always stable and we found for the order parameter (expectation value of Φ) evolution in the amplitude expansion [7],

$$\eta(t) = \frac{\eta_i}{1-\frac{\partial\Sigma(im_\Phi)}{\partial m_\Phi^2}}\cos[m_\Phi t] + \frac{2\,\eta_i}{\pi}\int_{m_\Phi+2m_\sigma}^{\infty}\frac{\omega\Sigma_I(\omega)\cos\omega t\;d\omega}{[\omega^2-m_\Phi^2-\Sigma_R(\omega)]^2+\Sigma_I(\omega)^2}\ . \tag{5.1}$$

where $\Sigma_{\rm physical}(i\omega\pm 0^+) = \Sigma_R(\omega)\pm i\Sigma_I(\omega)$ is the inflaton self-energy in the physical sheet, $\eta_i = \eta(0)$ and $\dot\eta(0) = 0$. The first term is the contribution of the one-particle pole (at the physical inflaton mass). This terms oscillates forever with constant amplitude. The second term is the cut contribution $\eta(t)_{cut}$ corresponding to $\Phi\to\Phi+2\sigma$.

In general, when

$$\Sigma_I(\omega\to\omega_{\rm threshold}) \overset{\omega\to\omega_{\rm threshold}}{=} B\,(\omega-\omega_{\rm threshold})^\alpha\ ,$$

the the cut contribution behaves for late times as

$$\eta(t)_{cut} \simeq \frac{2\,\eta_i}{\pi}\frac{B\,\omega_{\rm threshold}\,\Gamma(1+\alpha)}{[\omega^2-m_\Phi^2-\Sigma_R(\omega_{\rm threshold})]^2}\, t^{-1-\alpha}\cos\left[\omega_{\rm threshold}t+\frac{\pi}{2}(1+\alpha)\right]\ . \tag{5.2}$$

Here, $\omega_{\rm threshold} = m_\Phi + 2M_\sigma$ and $\alpha = 2$ since to two-loops, [7]

$$\Sigma_I(\omega) \overset{\omega\to m_\Phi+2M_\sigma}{=} \frac{2g^2\pi^2}{(4\pi)^4}\frac{M_\sigma\sqrt{m_\Phi}}{(m_\Phi+2M_\sigma)^{7/2}}[\omega^2-(m_\Phi+2M_\sigma)^2]^2$$

In the broken symmetry case ($m_\Phi^2 < 0$) we may have either $M < 2m_\sigma$ or $M > m_\sigma$, where M is the physical inflaton mass. ($M = |m_\Phi|\sqrt{2}$ at the tree level). In the first case the inflaton is stable and eq.(5.1) holds. However, the self-energy starts now at one-loop and vanishes at threshold with a power $\alpha = 1/2$. For $M > m_\sigma$ the inflaton decomes a resonance (an unstable particle) with width (inverse lifetime)

$$\Gamma = \frac{g^2\Phi_0^2}{8\pi M}\sqrt{1-\frac{4m_\sigma^2}{M^2}}\ .$$

This pole dominates $\eta(t)$ for non asymptotic times

$$\delta(t) \simeq \delta_i\,A\,e^{-\Gamma t/2}\,\cos(Mt+\gamma)\ , \tag{5.3}$$

where

$$A = 1+\frac{\partial\Sigma_R(M)}{\partial M^2}\ ,\qquad \gamma = -\frac{\partial\Sigma_I(M)}{\partial M^2}\ . \tag{5.4}$$

In summary, eq.(5.3) holds provided: a) the inflaton is a resonance and b) $t \le \Gamma^{-1}\ln(\Gamma/M_\sigma)$. For later times the fall off is with a power law $t^{-3/2}$ determined by the spectral density at threshold as before [7].

In ref. [7] the selfconsistent nonlinear evolution is computed to one-loop level for the model (5.1). In fig. 8 $\eta(t)$ is plotted as a function of time for $\lambda = g = 1.6\pi^2$, $y = 0$, $m_\sigma = 0.2m_\Phi$, $\eta(0) = 1$ and $\dot\eta(0) = 0$.

Figure 8 shows a very rapid, non-exponential damping within few oscillations of the expectation value and a saturation effect when the amplitude of the oscillation is rather small (about 0.1 in this case), the amplitude remains almost constant at the latest times tested. Figures 8 and 9 clearly show that the time scale for dissipation (from fig. 8) is that for which the particle production mechanism is more efficient (fig. 9). Notice that the total number of particles produced rises on the same time scale as that of damping in fig. 8 and eventually when the expectation value oscillates with (almost) constant amplitude the average number of particles produced remains constant. This behaviour is a close analog to the selfcoupled inflaton for unbroken symmetry (fig.1). The amplitude expansion predictions are in qualitative agreement with both results.

These figures clearly show that damping is a consequence of particle production. At times larger than about 40 m_Φ^{-1} (for the initial values and couplings chosen) there is no appreciable damping. The amplitude is rather small and particle production has practically shut off. If we had used the *classical* evolution of the expectation value in the mode equations, particle production would not shut off (parametric resonant amplification), and thus we clearly

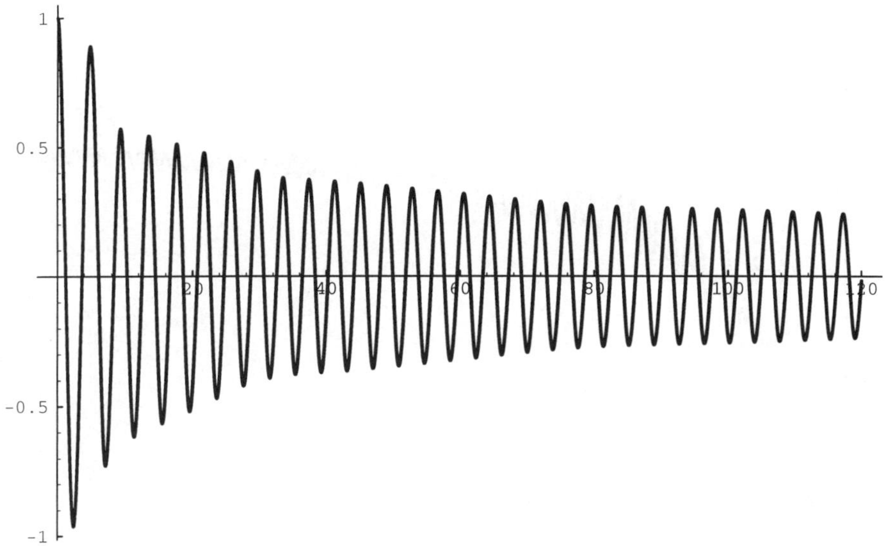

Figure 8: The inflaton coupled to a lighter scalar field σ: $\eta(\tau)$ vs τ for the values of the parameters $y = 0$; $\lambda/8\pi^2 = 0.2$; $g = \lambda$; $m_\sigma = 0.2\, m_\phi$; $\eta(0) = 1.0$; $\dot{\eta}(0) = 0$.

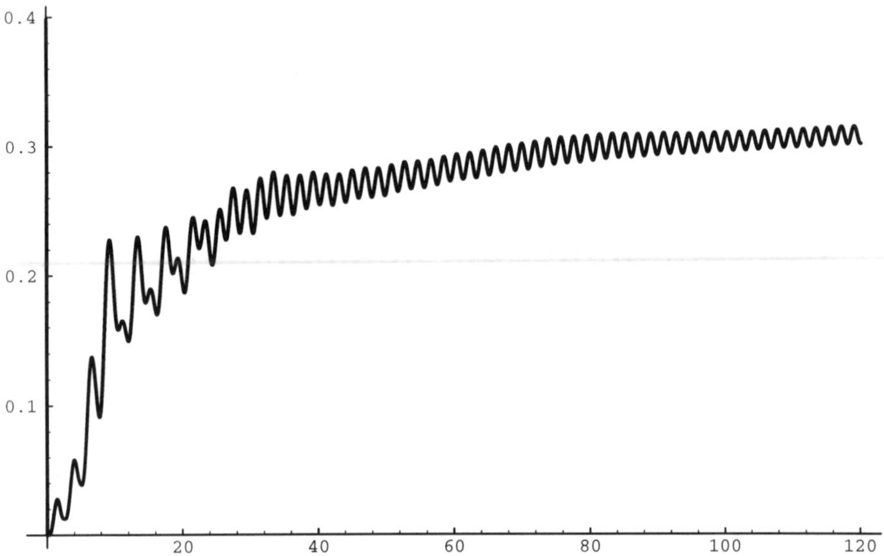

Figure 9: $\mathcal{N}_\sigma(\tau)$ vs. τ for the same value of the parameters as figure 8.

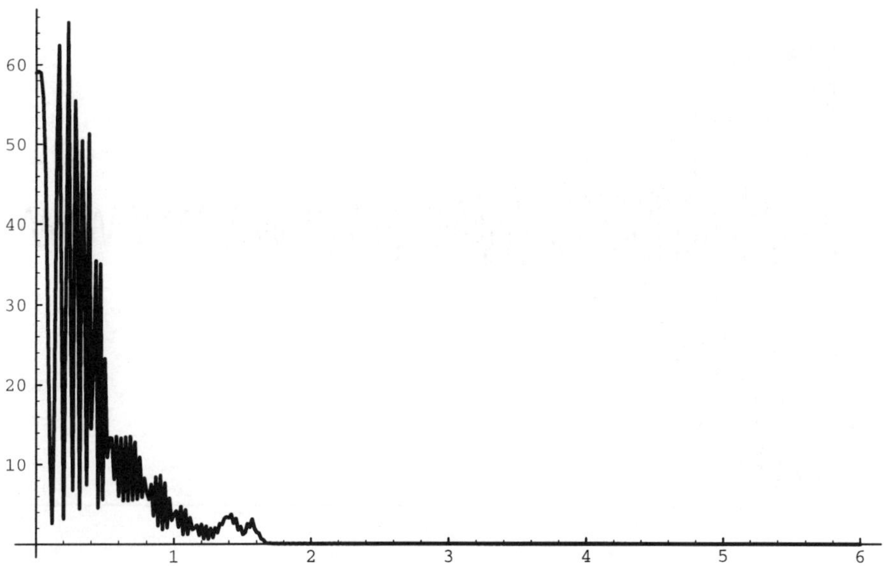

Figure 10: $\mathcal{N}_{q,\sigma}(\tau = 120)$ vs. q for the same values as in fig. 8.

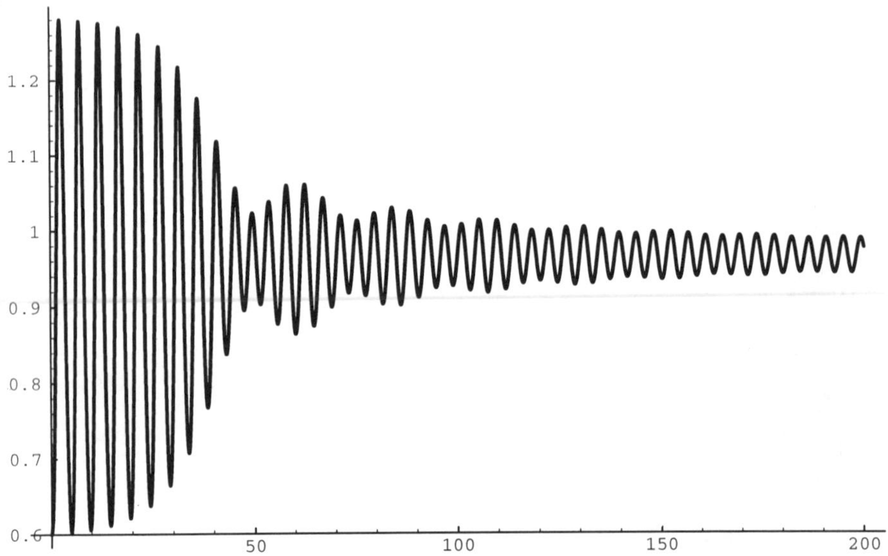

Figure 11: The inflaton in the broken symmetry case coupled to a lighter scalar σ. $\eta(\tau)$ vs τ for the values of the parameters $y = 0$; $\lambda/8\pi^2 = 0.2$; $g = \lambda$; $m_\sigma = 0.2\,|m_\phi|$; $\eta(0) = 0.6$; $\dot{\eta}(0) = 0$.

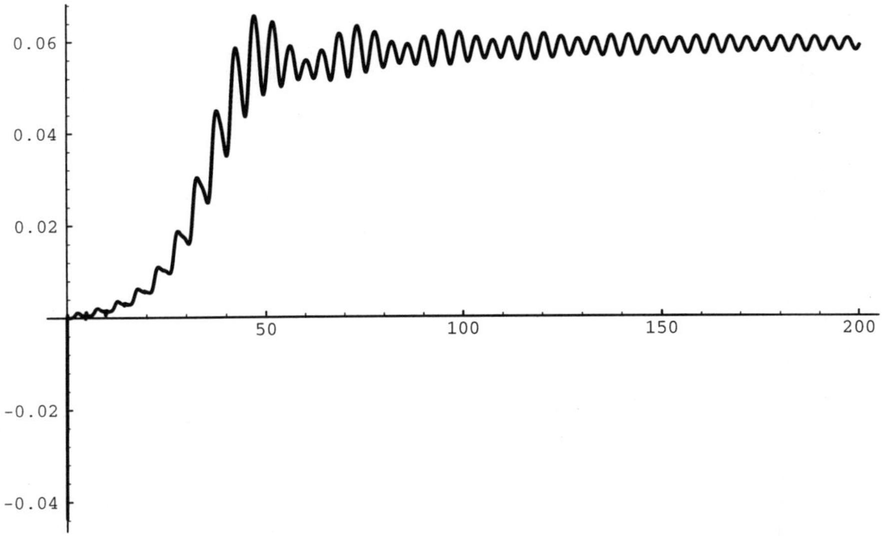

Figure 12: $\mathcal{N}_\sigma(\tau)$ vs. τ for the same value of the parameters as fig. 11.

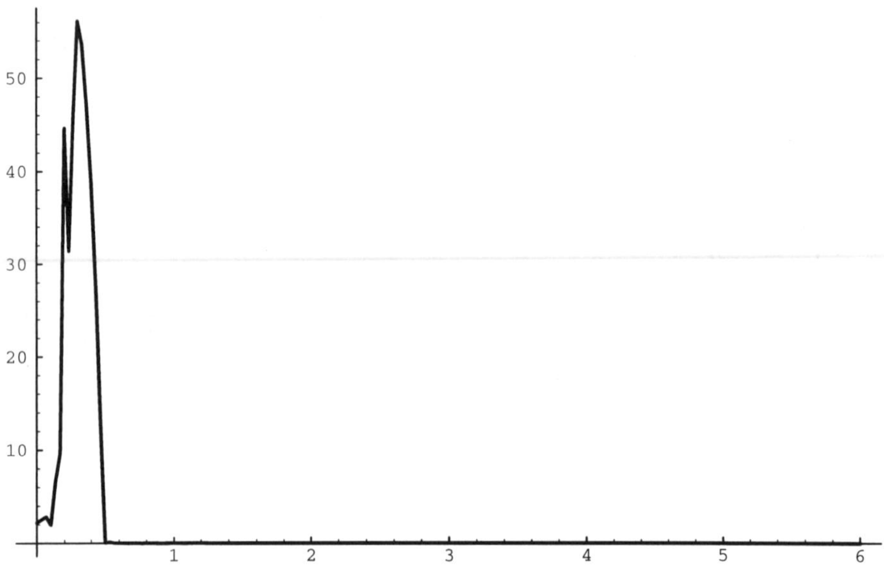

Figure 13: $\mathcal{N}_{q,\sigma}(\tau = 200)$ vs. q for the same values as in fig. 11.

see the dramatic effects of the inclusion of the back reaction.

In ref. [7] the broken symmetry case $m_\Phi^2 < 0$ is then studied. Figures 11-13 show $\eta(\tau)$ vs τ, $\mathcal{N}_\sigma(\tau)$ vs τ and $\mathcal{N}_{q,\sigma}(\tau = 200)$ vs q respectively, for $\lambda/8\pi^2 = 0.2$; $g/\lambda = 0.05$; $m_\sigma = 0.2\,|m_\Phi|$; $\eta(0) = 0.6$; $\dot\eta(0) = 0$. Notice that the mass for the linearized perturbations of the Φ field at the broken symmetry ground state is $\sqrt{2}\,|m_\Phi| > 2m_\sigma$. Therefore, for the values used in the numerical analysis, the two-particle decay channel is open. For these values of the parameters, linear relaxation predicts exponential decay with a time scale $\tau_{rel} \approx 300$ (in the units used). Figure 11 shows very rapid non-exponential damping on time scales about *six times shorter* than that predicted by linear relaxation. The expectation value reaches very rapidly a small amplitude regime, once this happens its amplitude relaxes very slowly. In the non-linear regime relaxation is clearly *not* exponential but extremely fast. The amplitude at long times seems to relax to the expected value, shifted slightly from the minimum of the tree level potential at $\eta = 1$. This is as expected from the fact that there are quantum corrections. Figure 12 shows that particle production occurs during the time scale for which dissipation is most effective, giving direct proof that dissipation is a consequence of particle production. Asymptotically, when the amplitude of the expectation value is small, particle production shuts off. We point out again that this is a consequence of the back-reaction in the evolution equations. Without this back-reaction, as argued above, particle production would continue without indefinitely. Figure 13 shows that the distribution of produced particles is very far from thermal and concentrated at low momentum modes $k \leq |m_\Phi|$. This distribution is qualitatively similar to that in the unbroken symmetry case, and points out that the excited state obtained asymptotically is far from thermal.

In ref. [7] the case where the inflaton is only coupled to fermions is studied ($g = 0$, $y \neq 0$). The damping of the zero mode is very inefficient in such case due to Pauli blocking. Namely, the Pauli exclusion principle forbids the creation of more than 2 fermions per momentum state. Pauli blocking shuts off particle production and dissipation very early on.

VI. FUTURE PERSPECTIVES

The preheating and reheating theory in inflationary cosmology is currently a very active area of research in fast development, with the potential for dramatically modifying the picture of the late stages of inflationary phase transitions.

As remarked before, estimates and field theory calculations have been done mostly assuming Minkowski spacetime. Results in de Sitter [10] and FRW backgrounds [28] are just beginning to emerge. A further important step will be to consider the background dynamics. Namely, the coupled gravitational and matter field dynamics. The matter state equations obtained in Minkowski [8] and de Sitter backgrounds [9] give an indication, through the Einstein-Friedmann equation, of the scale factor behaviour.

---

[1] A. H. Guth, Phys. Rev. **D23**, 347 (1981).

[2] See for reviews,
E. W. Kolb and M. S. Turner, "The Early Universe", Addison Wesley (Frontiers in Physics) (1990),
A. Linde, Particle Physics and Inflationary Cosmology, Harwood Academic Publishers (1990), and references therein.

[3] D. Boyanovsky and H. J. de Vega, Phys. Rev. **D47**, 2343 (1993).

[4] D. Boyanovsky, H. J. de Vega and R. Holman, Phys. Rev. **D49**, 2769 (1994).

[5] D. Boyanovsky, H. J. de Vega and R. Holman, Phys. Rev. **D 51**, 734 (1995).

[6] D. Boyanovsky, H. J. de Vega, R. Holman, D.-S. Lee and A. Singh, Phys. Rev. **D51**, 4419 (1995);
See for reviews:
D. Boyanovsky, H. J. de Vega and R. Holman, 'Non-equilibrium dynamics of phase transitions in the early universe', in 'Advances in Astrofundamental Physics' 1995, N. Sánchez and A. Zichichi Editors, p. 343, World Scientific, 1995.
D. Boyanovsky, M. D'Attanasio, H. J. de Vega, R. Holman and D.-S. Lee, 'New aspects of reheating', in the Proceedings of the Erice Chalonge School, 'String Gravity and Physics at the Planck Energy Scale', NATO Advanced Study Institute, N. Sánchez and A. Zichichi Editors, Kluwer 1996, p. 451-492.

[7] D. Boyanovsky, M. D'Attanasio, H. J. de Vega, R. Holman and D. S. Lee, Phys. Rev. **D52**, 6805 (1995);
D. Boyanovsky, M. D'Attanasio, H. J. de Vega and R. Holman, Phys. Rev. **D54**, 1748 (1996).

[8] D. Boyanovsky, H. J. de Vega, R. Holman and J. F. J. Salgado, 'Analytic and Numerical Study of Preheating Dynamics', hep-ph/9608205.

[9] D. Boyanovsky, H. J. de Vega, R. Holman and J. F. J. Salgado, in preparation.

[10] 'Out of equilibrium dynamics of an inflationary phase transition'
D. Boyanovsky, D. Cormier, H. J. de Vega and R. Holman, to appear in hep-ph.
'Preheating in FRW Universes',
D. Boyanovsky, D. Cormier, H.J. de Vega, R. Holman, A. Singh and M. Srednicki, to appear in hep-ph.

[11] F. Cooper, J. M . Eisenberg, Y. Kluger, E. Mottola, B. Svetitsky, Phys. Rev. Lett. 67, 2427 (1991);
F. Cooper, J. M. Eisenberg, Y, Kluger, E. Mottola, B. Svetitsky, Phys. Rev. D48, 190 (1993);

[12] F. Cooper and E. Mottola,

Mod. Phys. Lett. A 2, 635 (1987);
F. Cooper, S. Habib, Y. Kluger, E. Mottola, J. P. Paz, P. R. Anderson, Phys. Rev. D50, 2848 (1994).

[13] F. Cooper, Y. Kluger, E. Mottola, J. P. Paz, Phys. Rev. **D 51**, 2377 (1995).

[14] J. Baacke, K. Heitmann and C. Paetzold, hep-th/9608006.

[15] J. Traschen, and R. Brandenberger, Phys. Rev. D **42**, 2491 (1990).
Y. Shtanov, J. Traschen, and R. Brandenberger, Phys. Rev. D **51**, 5438 (1995).

[16] L. A. Kofman, A. D. Linde and A. A. Starobinsky, Phys. Rev. Lett. **73**, 3195 (1994).

[17] L. A. Kofman, astro-ph/9605155.

[18] M. Yoshimura, Prog. Theor. Phys. **94**, 873 (1995); hep-th/9506176; H. Fujisaki, K. Kumekawa, M. Yamaguchi, M. Yoshimura, Phys. Rev. D **53**, 6805 (1996), hep-ph/9508378. H. Fujisaki, K. Kumekawa, M. Yamaguchi, and M. Yoshimura, TU/95/493, hep-ph/9511381; M. Yoshimura, TU-96-500 preprint, hep-ph/9605246. S. Kasuya and M. Kawasaki, preprint ICRR-Report-360-96-11, hep-ph/9603317.

[19] L. A. Kofman, A. D. Linde and A. A. Starobinsky, Phys. Rev. Lett. **76**, 1011 (1996).

[20] I. Tkachev, Phys. Lett. **B376**, 35 (1996).

[21] D. Kaiser, Phys. Rev. **53**, 1776 (1995) and astro-ph/9608025.

[22] D. T. Son, preprint UW-PT-96-05, hep-ph/9604340.

[23] E.W. Kolb, A.D. Linde and A. Riotto, FERMILAB-Pub-133-A, hep-ph/9606260.
G. W. Anderson, A. D. Linde and A. Riotto, FERMILAB-PUB-96-078-A, hep-ph/9606416.
E. W. Kolb and A. Riotto, astro-ph/9602095, FERMILAB-Pub-96/036-A.
A. Riotto and I. Tkachev, hep-ph/9604444.
R. Allahverdi and B.A. Campbell, Alberta-THY-18/96, hep-ph/9606463.
G. Dvali and A. Riotto, hep-ph/9606431.

[24] C. Hermite, Notes de Cours à l'École Polytechnique, 1872 (unpublished). Whittaker and Watson, A Course on Modern Analysis. (AMS, Press 1979).

[25] Bateman Manuscript Project, chapter XIII, vol. II, A. Erdély Editor, McGraw-Hill, 1953.

[26] I. S. Gradshteyn and I. M. Ryshik, Table of Integrals, Series and Products, Academic Press.

[27] Handbook of Mathematical functions, Edited by M. Abramowitz and I. A. Stegun, NBS, 1965.

[28] S. Yu. Khlebnikov, I. I. Tkachev, hep-ph/9608458.

Circular Strings and Multi-Strings in de Sitter and Anti de Sitter Spacetimes*

H.J. de Vega

Laboratoire de Physique Théorique et Hautes Energies. Laboratoire Associé au CNRS UA280, Université de Paris VI et VII, Tour 16, 1er étage, 4, Place Jussieu, 75252 Paris, France.

and

A.L. Larsen, N. Sánchez

Observatoire de Paris, DEMIRM. Laboratoire Associé au CNRS UA 336, Observatoire de Paris et École Normale Supérieure. 61, Avenue de l'Observatoire, 75014 Paris, France.

Abstract

The exact general solution of circular strings in $2+1$ dimensional de Sitter spacetime is described completely in terms of elliptic functions. The novel feature here is that one single world-sheet generically describes *infinitely many* (different and independent) strings. This has no analogue in flat spacetime. The circular strings are either oscillating ("stable") or indefinitely expanding ("unstable"). We then compute the *exact* equation of state of circular strings in the $2+1$ dimensional de Sitter (dS) and anti de Sitter (AdS) spacetimes, and analyze its properties for the different (oscillating, contracting and expanding) strings. We finally perform a semi-classical quantization of the oscillating circular strings. The string mass is $m = \sqrt{C}/(\pi H \alpha')$, C being the Casimir operator, $C = -L_{\mu\nu}L^{\mu\nu}$, of the $O(3,1)$-dS [$O(2,2)$-AdS] group, and H is the Hubble constant. We find the mass formula $\alpha' m^2_{\text{dS}} \approx 4n - 5H^2\alpha' n^2$, ($n \in N_0$), and a *finite* number of states $N_{\text{dS}} \approx 0.34/(H^2\alpha')$ in de Sitter spacetime; $m^2_{\text{AdS}} \approx H^2 n^2$ (large $n \in N$) and $N_{\text{AdS}} = \infty$ in anti de Sitter spacetime. The level spacing grows with n in AdS spacetime, while is approximately constant (although smaller than in Minkowski spacetime and slightly decreasing) in dS spacetime.

*Research report to appear in "New developments in string gravity and physics at the Planck energy scale" Edited by N. Sánchez (World Scientific, Singapore, 1995)

1 Introduction

The study of string dynamics in curved spacetimes reveals new insights and new physical phenomena with respect to string propagation in flat spacetime (and with respect to quantum fields in curved spacetimes) [1, 2, 3]. The results of this programme are relevant both for fundamental (quantum) strings and for cosmic strings, which behave essentially in a classical way.

Among the cosmological backgrounds of interest, de Sitter spacetime occupies a special place. It is on one hand relevant for inflation, and on the other hand, string propagation turns out to be specially interesting there. String-instability, in the sense that the string proper length grows indefinitely (proportional to the expansion factor of the universe) is particularly present in de Sitter spacetime [1-8]. Moreover, a novel feature for strings in curved spacetimes was first found in de Sitter spacetime: Exact *multi-string* solutions [6, 7]. That is, one single world-sheet generically describes two strings [7], several strings [6] or even *infinitely many* (different and independent) strings [8]

Circular strings are specially suited for detailed investigation. Since the string equations of motion become separable, one has to deal with non-linear ordinary differential equations instead of non-linear partial differential equations. In order to obtain generic non-circular string solutions the full power of the inverse scattering method is needed in de Sitter spacetime [6].

Here we will present some more recent results [8, 9] on exact circular string solutions in de Sitter spacetime and, for comparison, in anti de Sitter spacetime too. That is, we consider the circular string solutions of the equations of motion and constraints:

$$\ddot{x}^\mu - x''^\mu + \Gamma^\mu_{\rho\sigma}(\dot{x}^\rho \dot{x}^\sigma - x'^\rho x'^\sigma) = 0, \tag{1.1}$$

$$g_{\mu\nu}\dot{x}^\mu x'^\nu = g_{\mu\nu}(\dot{x}^\mu \dot{x}^\nu + x'^\mu x'^\nu) = 0, \tag{1.2}$$

where dot and prime stand for derivative with respect to the world-sheet coordinates τ and σ, respectively and $\Gamma^\mu_{\rho\sigma}$ are the Christoffel symbols with respect to the metric $g_{\mu\nu}$. The exact general solution of circular strings in $2+1$ dimensional de Sitter spacetime is described closely and completely in terms of elliptic functions (Section 2).

By computing the string energy and pressure, we obtain the corresponding equations of state, providing the physical interpretation of the solutions in a cosmological context. We analyze the equations of state for the different (oscillating, contracting and expanding) strings. The string equation of state turns out to have the perfect fluid form $P = (\gamma - 1)E$, with the instantaneous coefficient γ depending on an elliptic modulus (Section 3).

Cosmological backgrounds like de Sitter and anti de Sitter spacetimes are not Ricci flat and hence they are not string vacua even at first order in α'. Strings are there non-critical and quantization will presumably lead to features like ghost states. No definite answer is available by now to such conformal anomaly effects. It is important in this context to investigate the quantum aspects in the semi-classical regime, where anomaly effects are practically irrelevant. Semi-classical, in this context, means the regime in which $H^2\alpha' << 1$, where H is the Hubble constant. We semi-classically quantize the time-periodic string solutions in de Sitter and anti de Sitter spacetimes. Time-periodic string solutions here include all the circular string solutions in anti de Sitter spacetime, as well as the oscillating string solutions in de Sitter spacetime (Section 4).

2 Circular Strings in de Sitter Spacetime, Multi-Strings

Recently, several progresses in the understanding of string propagation in de Sitter spacetime have been obtained [6-11]. The classical string equations of motion (plus the string constraints) were shown to be integrable in D-dimensional de Sitter spacetime [10, 11]. They are equivalent to a non-linear sigma model on the Grassmannian $SO(D,1)/O(D)$ with periodic boundary conditions (for a closed string). In addition, the string constraints imply a zero world-sheet energy-momentum tensor, and these constraints are compatible with the integrability. Moreover, the exact string dynamics in de Sitter spacetime is equivalent to a generalized sh-Gordon model with a potential unbounded from below [11]. The sh-Gordon function $\alpha(\sigma,\tau)$ has here a clear physical meaning: $H^{-1}\exp[\alpha(\sigma,\tau)/2]$ determines the proper size of the string (H is the Hubble constant). In $2+1$ dimensions, the string dynamics is exactly described by the standard sh-Gordon equation.

More recently, a novel feature for strings in de Sitter spacetime was found: Exact multi-string solutions [6, 7]. Exact circular string solutions were found describing two different strings [7]. One string is stable (the proper size is bounded), and the other one is unstable (the proper size blows up) for large de Sitter radius. Soliton methods, (the so-called "dressing method" in soliton theory) were implemented using the linear problem (Lax pair) of this system, in order to construct systematically exact string solutions [6]. The one-soliton string solution constructed in this way, generically describe five different and independent strings: one stable string and four unstable strings. These solutions (even the stable string) do not oscillate in time.

In this section, we go further in the investigation of exact string solutions in de Sitter spacetime. We find exact string solutions describing *infinitely many* different and independent strings. The novel feature here is that we have one single world-sheet but multiple (infinitely many) strings. The world-sheet time τ turns out to be an infinite-valued function of the target space time X^0 (which can be the hyperboloid time q^0, the comoving time T or the static coordinate time t). Each branch of τ as a function of q^0 corresponds to a different string. In flat spacetime, multiple string solutions are necessarily described my multiple world-sheets. Here, a single world-sheet describes infinitely many different and simultaneous strings as a consequence of the coupling to the spacetime geometry. These strings do not interact among themselves; all the interaction is with the curved spacetime.

We apply the circular string *Ansatz*:

$$t = t(\tau), \quad \phi = \sigma, \quad r = r(\tau), \tag{2.1}$$

in $2+1$ dimensional de Sitter spacetime, particularly convenient in terms of the static de Sitter coordinates (t, r, ϕ) (we also describe the solutions in the hyperboloid and comoving parametrizations). The string equations of motion and constraints, Eqs.(1.1)-(1.2), can be solved directly and completely in terms of elliptic functions. They reduce to two decoupled first order differential equations for the time component $t(\tau)$ and the string radius $r(\tau)$:

$$\dot{t} = \frac{\sqrt{b}\alpha'}{1 - H^2 r^2} \tag{2.2}$$

$$\dot{r}^2 + V(r^2) = b\alpha'^2; \quad V(r^2) = r^2(1 - H^2 r^2) \tag{2.3}$$

Notice that we are using the notation of Ref.[9], which is slightly different from the notation of Ref.[8]. The $\dot{r}$-equation is solved by:

$$H^2 r^2(\tau) = \wp(\tau - \tau_o) + 1/3, \tag{2.4}$$

where $\wp$ is the Weierstrass elliptic function [12] with discriminant $\Delta = 16b^2 H^4 \alpha'^4 (1 - 4bH^2\alpha'^2)$, and b and τ_o are integration constants (τ_o is generally complex and must be chosen such that $r(\tau)$ is real for real τ). The solutions depend on one constant parameter b (for fixed H, α') related to the string energy, see Section 3, and fall into three classes, depending on whether:

$$bH^2\alpha'^2 < 1/4 \; (\Delta > 0),$$
$$bH^2\alpha'^2 = 1/4 \; (\Delta = 0),$$
$$bH^2\alpha'^2 > 1/4 \; (\Delta < 0).$$

As can be seen in the diagram $(r^2, V(r^2))$, Fig.1., in which the full string dynamics takes place, these cases correspond to either oscillatory motion or infinite (unbounded) motion.

The proper string size S of the circular strings are given for all r by:

$$S(\tau) = r(\tau), \tag{2.5}$$

as can be seen from the induced line element on the string world-sheet:

$$ds^2 = r^2(\tau)(-d\tau^2 + d\sigma^2). \tag{2.6}$$

In the $bH^2\alpha'^2 = 1/4$ case, the Weierstrass elliptic function degenerates into a hyperbolic function:

$$H^2 r^2(\tau) = \frac{1}{2}[1 + \sinh^{-2}(\frac{\tau - \tau_o}{\sqrt{2}})]. \tag{2.7}$$

Two real independent solutions appear for the choices $\tau_o = i\pi/2$ and $\tau_o = 0$, respectively:

$$H^2 r_\pm^2(\tau) = \frac{1}{2}[\tanh(\frac{\tau}{\sqrt{2}})]^{\pm 2}. \qquad (2.8)$$

(They were first found in Ref.[7]). We have also the solution $H^2 r_0^2 = 1/2$, corresponding to a stable string with constant proper size $S_0 = 1/(\sqrt{2}H)$ (i.e., sh-Gordon function $\alpha = 0$). This solution was found in Ref.[7] and we shall not discuss it here.

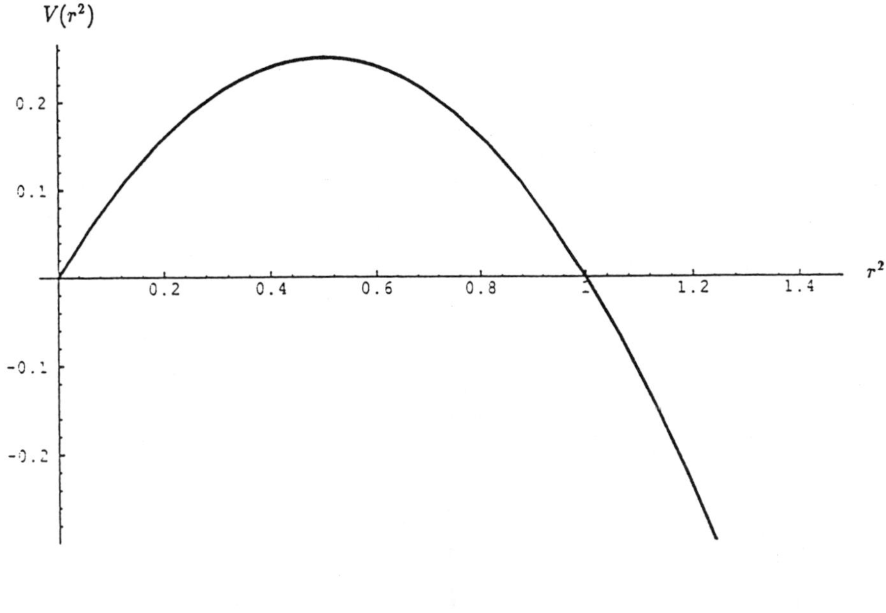

Figure 1. The potential $V(r^2) = r^2(1 - H^2 r^2)$, defined in Eq.(2.3). For $bH^2\alpha'^2 < 1/4$, it acts effectively as a barrier. The horizon corresponds to $H^2 r^2 = 1$.

The solution r_- describes two different strings, I and II, as it can be seen from the hyperboloid time $q_-^0(\tau)$, Fig.2. Here τ is a two-valued function of q_-^0: String I corresponds to $-\infty < \tau < 0$ and string II to $0 < \tau < \infty$. The proper size S_- for both strings is given by Eq.(2.5), using Eq.(2.8). For $q_-^0 \to \infty$, string I is unstable, while string II is stable. The proper size $S_-(q_-^0 \to \infty)$ blows up for string I (for which $q_-^0 \to \infty$ corresponds to $\tau \to 0_-$), while tends to a constant value for string II (for which $q_-^0 \to \infty$ corresponds to $\tau \to \infty$). String I starts with minimal size $S_- = 1/(\sqrt{2}H)$ at $\tau = -\infty$ and blows up at $\tau = 0$. String II starts with infinite size at $\tau = 0$ but approaches $S_- = 1/(\sqrt{2}H)$ for $\tau \to \infty$.

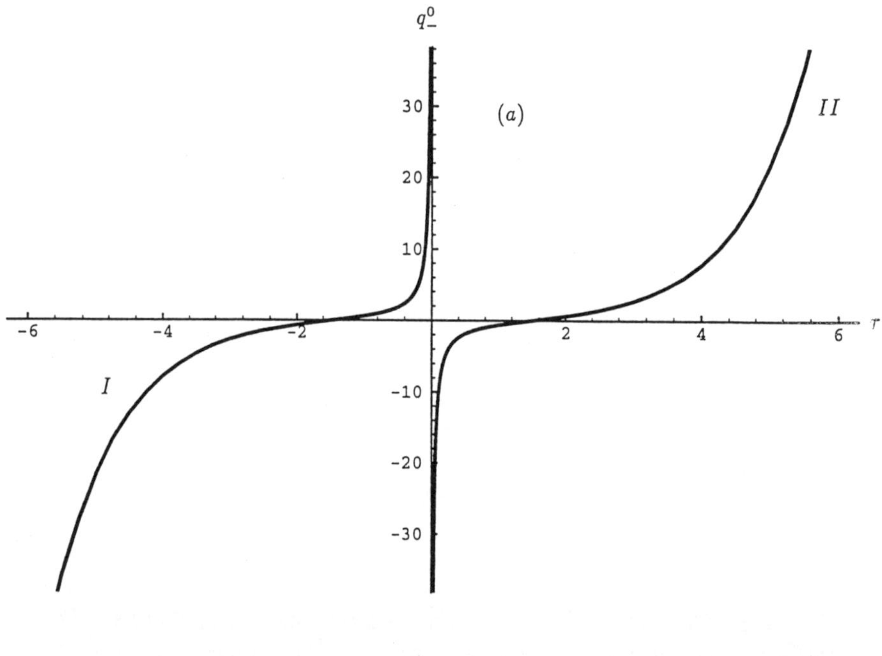

Figure 2. The hyperboloid time q_-^0, in the $bH^2\alpha'^2 = 1/4$ case, as a function of τ. Notice that τ is a two-valued function of q_-^0.

The solution r_+ of this $bH^2\alpha'^2 = 1/4$ case describes only one stable string. The hyperboloid time $q_-^0(\tau)$, is a monotonically increasing function of τ and the corresponding string has bounded proper size S_+. The string starts with $S_+ = 1/(\sqrt{2}H)$, at $q_+^0 = -\infty$, it contracts until it collapses ($S_+ = 0$), then it expands until it reaches the original size for $q_+^0 = \infty$. For $bH^2\alpha'^2 = 1/4$ the evolution is always non-oscillatory. Even the stable string does not oscillate in time.

For $bH^2\alpha'^2 < 1/4$ there exist two real independent solutions for the choices $\tau_o = 0$ and $\tau_o = \omega'$, where ω' is the imaginary semi-period of the Weierstrass function:

$$H^2 r_-^2(\tau) = \wp(\tau) + 1/3, \tag{2.9}$$

$$H^2 r_+^2(\tau) = \wp(\tau + \omega') + 1/3. \tag{2.10}$$

r_- and r_+ are oscillating solutions as functions of τ. The solution r_- describes infinitely many strings; r_- has infinitely many branches $[0, 2\omega]$, $[2\omega, 4\omega]$, ..., each of which corresponds to a different string (ω is the real semi-period, of the Weierstrass function). This can be seen from the hyperboloid time $q_-^0(\tau)$, Fig.3: The world-sheet time τ is an infinite-valued function of q_-^0. The hyperboloid time q_-^0 blows up at the boundaries of the branches $\tau = \pm 2N\omega$ (N being an integer):

$$\mid q_-^0(\tau) \mid \sim \frac{1}{\mid 2N\omega - \tau \mid}. \tag{2.11}$$

Further insight is obtained by considering the comoving time T_- and the static coordinate time t_-. Closed expressions for them are given in Ref.[8], in terms of Weierstrass ζ and σ-functions, and also rewritten in terms of elliptic theta-functions. The cosmic time T_- is singular at $\tau = 0$, $\tau = x/\mu$, $\tau = 2\omega$ and similarly in the other branches. (x, μ are two real constants. x is expressed as an incomplete elliptic integral of first kind while $\mu = \sqrt{(1 + \sqrt{1 - 4bH^2\alpha'^2})/2}$). The static coordinate time t_-, on the other hand, is regular at the boundaries of the branches, but is singular at $\mu\tau = 2KN \pm x$:

$$t_-(\tau) \sim \frac{1}{2\pi} \log \mid \mu\tau - 2KN \mp x \mid, \tag{2.12}$$

where K is a complete elliptic integral of the first kind. It must be noticed that although r_- is periodic in τ, the comoving time is not, i.e. $T_-(\tau) \neq T_-(\tau + 2\omega)$. This implies that the infinitely many strings are different. The difference in proper size of the n'th and the $(n+1)$'th string for a given comoving time T_- is given by:

$$\Delta S_- = \frac{\pi}{H} \frac{\theta_1'}{\theta_1}\left(\frac{\pi x}{2K}\right) e^{HT_-} e^{-n\pi \frac{\theta_1'}{\theta_1}\left(\frac{\pi x}{2K}\right)}. \tag{2.13}$$

However, *all* the strings are of the same type: *unstable*. For instance, in the

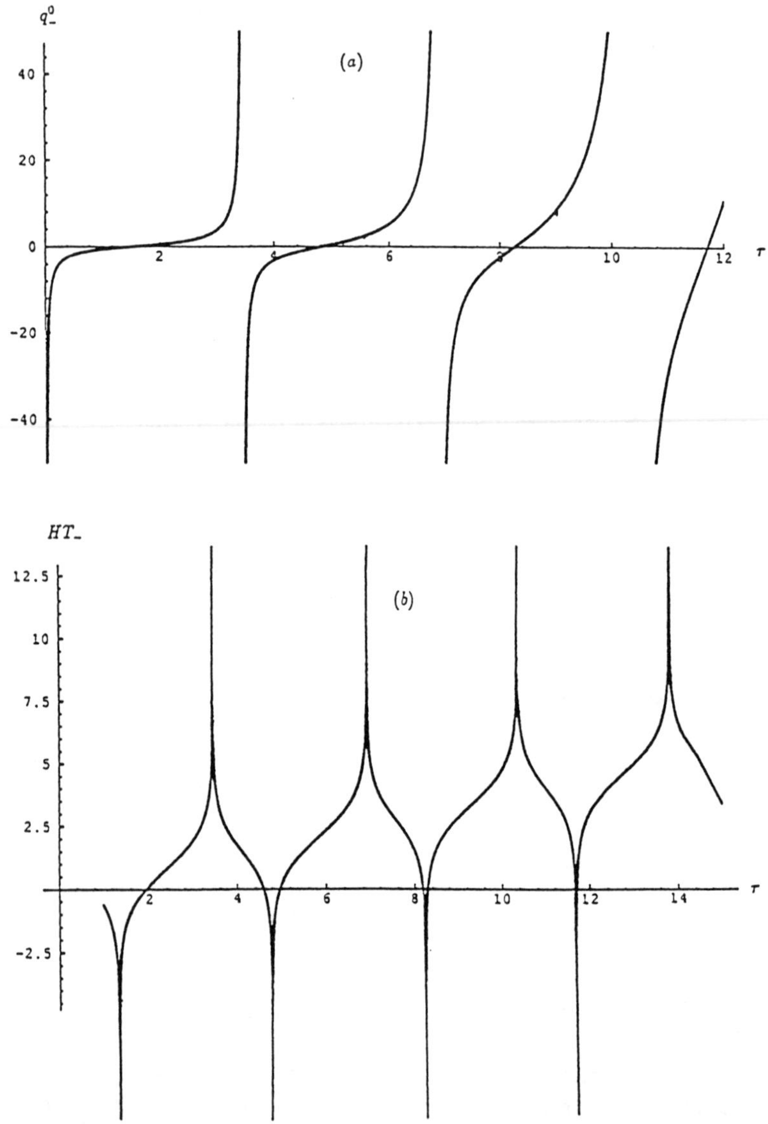

Figure 3. (a) The hyperboloid time q_-^0 as a function of τ in the elliptic case $bH^2\alpha'^2 < 1/4$. Each of the infinitely many branches corresponds to one string. (b) The comoving time T_- as a function of τ in the elliptic case $bH^2\alpha'^2 < 1/4$. Notice that T_- is not periodic.

branch $\tau \in [0, 2\omega]$, the string starts at $\tau = 0$ ($q_-^0 = -\infty$) with infinite size, then contracts to the minimal size $HS_- = \sqrt{(1+\sqrt{1-4bH^2\alpha'^2})/2}$ and eventually expands towards infinite size at $\tau = 2\omega$ ($q_-^0 = \infty$). These solutions never collapse to $r = 0$.

For the solution r_+ of the $bH^2\alpha'^2 < 1/4$ case, the string dynamics takes place inside the horizon. r_+, being a regularly oscillating function of τ, is then also a regularly oscillating function of the physical times q_+^0, T_+ and t_+. The static coordinate time t_+, from which one easily deduces q_+^0 and T_+, is expressed in terms of theta-functions [8]. The solution r_+ describes one stable string oscillating between its minimal size $S_+ = 0$ (collapse) and its maximal size $HS_+ = \sqrt{(1-\sqrt{1-4bH^2\alpha'^2})/2}$. It must be noticed that the string oscillations here do not follow a pure harmonic motion as in flat Minkowski spacetime, but they are precise superpositions of all frequencies $(2n-1)\Omega$, ($\Omega = \pi\mu/(2K)$, $n = 1, 2, ..., \infty$) [8]; here the non-linearity of the string equations of motion fixes the relation between the mode coefficients, and the basic frequency Ω depends on the string energy via b.

For $bH^2\alpha'^2 > 1/4$ two real independent solutions are obtained for $\tau_o = 0$ and $\tau_o = \omega_2'$, where ω_2' is the imaginary semi-period of the Weierstrass function:

$$H^2 r_-^2(\tau) = \wp(\tau) + 1/3, \tag{2.14}$$

$$H^2 r_+^2(\tau) = \wp(\tau + \omega_2') + 1/3. \tag{2.15}$$

In this case r_- again describes infinitely many strings, all of them are unstable. The difference with the $bH^2\alpha'^2 < 1/4$ case, is that here the strings have a collapse during their evolution. For instance, in the branch $\tau \in [0, 2\omega_2]$, where ω_2 is the real semi-period of the Weierstrass function, the string starts with infinite size at $\tau = 0$ ($q_-^0 = -\infty$), it then contracts until it collapses to a point and then it expands towards infinite size again at $\tau = 2\omega_2$ ($q_-^0 = \infty$). In contrast to the $bH^2\alpha'^2 < 1/4$ case, the solution r_+ is here just a time translated version of r_-:

$$r_+^2(\tau) = r_-^2(\tau + \omega_2), \tag{2.16}$$

and describes therefore essentially the same features as the solution r_-.

A summary of the main features and conclusions of the results of this section is presented in Table 1. Further details can be found in Ref.[8].

Table 1. Circular string evolution in de Sitter spacetime. For each value of $bH^2\alpha'^2$, there exists two independent solutions r_- and r_+:

$bH^2\alpha'^2$	r_-	r_+
$< 1/4$	Infinitely many different strings. All are unstable ($Hr_-^{max} = \infty$), and never collapse. ($Hr_-^{min} = \sqrt{(1+\sqrt{1-4bH^2\alpha'^2})/2} > 0$).	One stable oscillating string ($0 \leq r_+ \leq r_+^{max}$; $Hr_+^{max} = \sqrt{(1-\sqrt{1-4bH^2\alpha'^2})/2}$).
$> 1/4$	Infinitely many strings. All of them are unstable ($Hr_-^{max} = \infty$) and they collapse to a point ($r_-^{min} = 0$).	Infinitely many strings similar to r_-. In this case r_+ is just a time-translation of r_-: $r_+(\tau) = r_-(\tau + \omega_2)$.
$= 1/4$	Two different and non-oscillating strings $r_-^{(I)}$ and $r_-^{(II)}$. $r_-^{(I)}$ is unstable and $r_-^{(II)}$ is stable for large de Sitter radius.	One stable and non-oscillating string (it makes only one oscillation). $r_+^{min} = 0$, $Hr_+^{max} = 1/\sqrt{2}$.

3 Physical Interpretation, Energy, Pressure

In this section we discuss the physical properties (energy, pressure) of a gas of circular strings in de Sitter and anti de Sitter spacetimes. In static coordinates, we thus consider the spacetimes:

$$ds^2 = -a(r)dt^2 + \frac{dr^2}{a(r)} + r^2 d\phi^2. \tag{3.1}$$

For simplicity we consider the string dynamics in a 2+1 dimensional spacetime. All our solutions can however be embedded in a higher dimensional spacetime, where they will describe plane circular strings. We are interested in the three cases:

$$\begin{aligned}
a(r) &= 1 \quad \text{for Minkowski spacetime,} \\
a(r) &= 1 - H^2r^2 \quad \text{for de Sitter spacetime,} \\
a(r) &= 1 + H^2r^2 \quad \text{for anti de Sitter spacetime.}
\end{aligned}$$

The equations describing the evolution of circular strings are:

$$\dot{t} = \frac{\sqrt{b}\alpha'}{a(r)} \tag{3.2}$$

$$\dot{r}^2 + V(r) = b\alpha'^2; \quad V(r) = r^2 a(r) \tag{3.3}$$

and generalize Eqs.(2.2)-(2.3).

Properties like energy and pressure of the strings are more conveniently discussed in comoving (cosmological) coordinates:

$$ds^2 = -(dT)^2 + a^2(T)\,\frac{dR^2 + R^2\,d\phi^2}{(1+\frac{k}{4}R^2)^2}, \tag{3.4}$$

including as special cases Minkowski, de Sitter and anti de Sitter spacetimes:

$$\begin{aligned} a(T) &= 1, \;\; k=0 \quad \text{for Minkowski spacetime,} \\ a(T) &= e^{HT}, \;\; k=0 \quad \text{for de Sitter spacetime,} \\ a(T) &= \cos HT, \;\; k=-H^2 \quad \text{for anti de Sitter spacetime.} \end{aligned}$$

The spacetime energy-momentum tensor is given in $2+1$ dimensions by:

$$\sqrt{-g}\,T^{\mu\nu} = \frac{1}{2\pi\alpha'}\int d\tau d\sigma\,(\dot{X}^\mu\dot{X}^\nu - X'^\mu X'^\nu)\,\delta^{(3)}(X - X(\tau,\sigma)). \tag{3.5}$$

After integration over a spatial volume that completely encloses the string [13], the energy-momentum tensor for a circular string takes the form of a fluid:

$$T^\mu_\nu = \text{diag.}(-E,\ P,\ P), \tag{3.6}$$

where, in the comoving coordinates (3.4):

$$E(X) = \frac{1}{\alpha'}\dot{T}, \tag{3.7}$$

$$P(X) = \frac{1}{2\alpha'}\,\frac{a^2(T)}{(1+\frac{k}{4}R^2)^2}\,\frac{\dot{R}^2 - R^2}{\dot{T}}, \tag{3.8}$$

represent the string energy and pressure, respectively.

We have already described all the circular string solutions in de Sitter spacetime, Section 2. The circular string solutions in anti de Sitter spacetime were obtained in Ref.[14]. The general solution is:

$$Hr(\tau) = \frac{\bar{k}}{\sqrt{1-2\bar{k}^2}}\text{cn}[\frac{\tau}{\sqrt{1-2\bar{k}^2}},\ \bar{k}], \tag{3.9}$$

where:

$$\bar{k}^2 = \frac{-1+\sqrt{1+4bH^2\alpha'^2}}{\sqrt{1+4bH^2\alpha'^2}}. \tag{3.10}$$

For arbitrary values of the elliptic modulus $\bar{k} \in [0, \sqrt{1/2}\ [$, the solution describes one oscillating string.

Let us now return to the energy-momentum tensor for circular strings. The circular string solutions in de Sitter and anti de Sitter spacetimes depend on the elliptic modulus $k \equiv \sqrt{1-\mu^2}\ /\mu$ and $\bar{k}$, respectively, as we have seen. From these exact solutions we find the corresponding energy-momentum tensors. They turn out to have the perfect fluid form with an equation of state:

$$P = (\gamma - 1)E, \tag{3.11}$$

where γ in general is time-dependent and depends on the elliptic modulus (k or $\bar{k}$) as well. We have analyzed the equation of state, Eq.(3.11), for all circular string solutions in de Sitter and anti de Sitter spacetimes, Ref.[9].

In de Sitter spacetime, for strings expanding from zero radius towards infinity, the equation of state changes continuosly from the ultra-relativistic matter-type $P = +E/2$, (in $2+1$ dimensions) when $r \approx 0$, to the unstable string-type [4, 5], $P = -E/2$, when $r \to \infty$, see Fig.4.

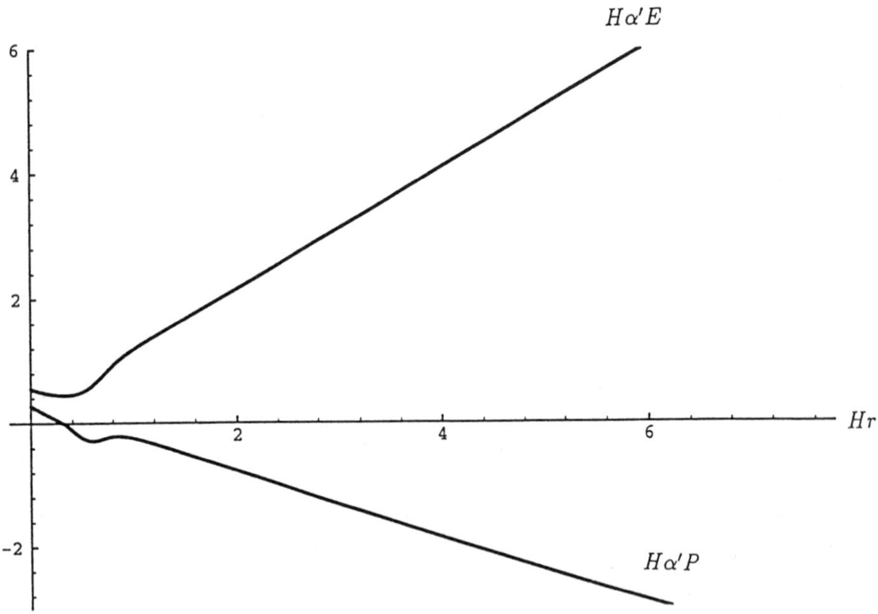

Figure 4. The energy and pressure for a string expanding from $r = 0$ towards infinity (unstable string) in de Sitter spacetime. The curves are drawn for the case $bH^2\alpha'^2 = 0.3$.

On the other hand, for an oscillating stable string in de Sitter spacetime, γ oscillates between $\gamma(r = 0) = 3/2$ and $\gamma(r = r_{\max}) = 1/2 + k^2/(1 + k^2)$, where $k \in [0, 1]$. Averaging over one oscillation period, the pressure *vanishes* [9]. That is, these stable string solutions actually describe (in average) *cold matter*. See also Fig.5.

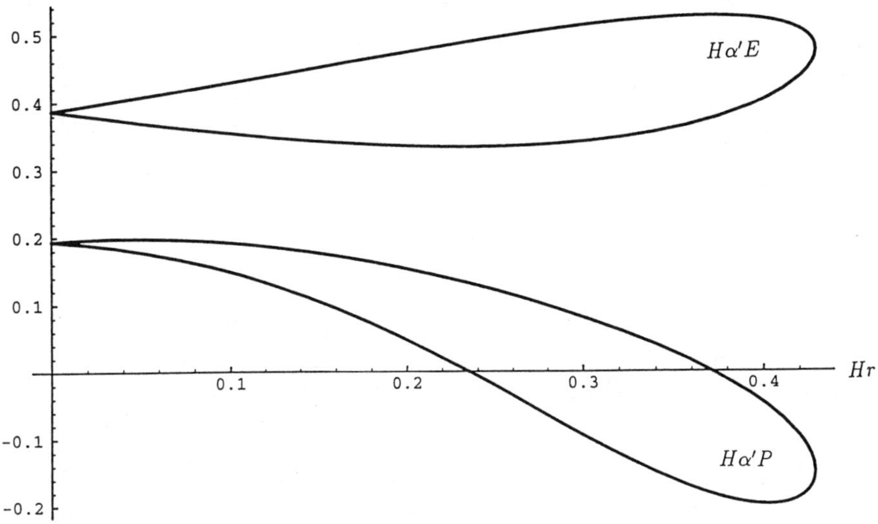

Figure 5. The energy and pressure for an oscillating (stable) string in de Sitter spacetime. The curves describe one period of oscillation. The curves are drawn for the case $bH^2\alpha'^2 = 0.15$.

In anti de Sitter spacetime, only the oscillating (stable) circular string solutions, Eq.(3.9), exist. We find that γ oscillates between $\gamma(r = 0) = 3/2$ and $\gamma(r = r_{\max}) = 1/2$, i.e. the equation of state "oscillates" between $P = +E/2$ and $P = -E/2$. This is similar to the situation in flat Minkowski spacetime. When averaging over an oscillation period in anti de Sitter spacetime, we find that γ takes values from 1 to $1 + 1/\pi^2$ for the allowed range of the elliptic modulus. That is, the average pressure over one oscillation period is always *positive* in anti de Sitter spacetime.

In conclusion, positive pressure characterizes the regime in which the string radius is small relative to the string maximal size, while negative pressure is characteristic for the regime in which the string radius is large. In Minkowski spacetime, the two regimes are of equal "size", in the sense that the average pressure is identically zero. The influence of the spacetime curvature is among other effects, to modify the relative "size" of the two regimes.

Table 2. Circular string energy and pressure in Minkowski, de Sitter and anti de Sitter spacetimes.

Circular String	Energy	Pressure	Equation of State
MINKOWSKI:			
$r(\tau) = \sqrt{b}\alpha' \mid \cos\tau \mid$			$P(r \to 0) = +E/2$, i.e.
Oscillating between	$E = \sqrt{b} = \text{const.}$	$P = \frac{-\sqrt{b}}{2}\cos 2\tau$	ultrarel. in 2+1 dim.
0 and $r_{\text{max}} = \sqrt{b}\alpha'$			$P(r \to r_{\text{max}}) = -E/2$
Average values:	$< E >= E = \sqrt{b}$	$< P >= 0$	Cold matter
DE SITTER:			
(i) $H^2\alpha'^2 b > 1/4$:	$E(r \to 0) = \sqrt{b}$	$P(r \to 0) = \sqrt{b}/2$	$P(r \to 0) = +E/2$
Strings expanding	$E(r \to \infty) = r/\alpha'$	$P(r \to \infty) = -r/2\alpha'$	$P(r \to \infty) = -E/2$
from 0 towards	E increases and P decreases as strings expand,		
$r = \infty$	P becoming negative.		
(ii) $H^2\alpha'^2 b \leq 1/4$:	$E(r \to 0) = \sqrt{b}$	$P(r \to 0) = \sqrt{b}/2$	$P(r \to 0) = +E/2$
Strings oscillating	$E(r_{\text{max}}) = \frac{\nu}{H\alpha'\sqrt{1-\nu^2}}$	$P(r_{\text{max}}) = \frac{\nu(\nu^2-1/2)}{H\alpha'\sqrt{1-\nu^2}}$	$P(r_{\text{max}}) = (\gamma - 1)E$
between 0 and			$\gamma = \nu^2 + 1/2$
$Hr_{\text{max}} \equiv \nu =$	$\nu << 1$: (as in Minkowski space)		$P(r_{\text{max}}) = -E/2$
$\sqrt{\frac{1-\sqrt{1-4H^2\alpha'^2 b}}{2}}$	$\nu \to 1/\sqrt{2}$: (as cold matter)		$P(r_{\text{max}}) = 0$
Average values:	$< E >= \frac{2\nu}{H\alpha' T_r}\Pi(\nu^2, \frac{\nu}{\sqrt{1-\nu^2}})$	$< P >= 0$	Cold matter
ANTI DE SITTER:			
Strings oscillating			
between 0 and	$E(r \to 0) = \frac{\bar{\nu}\sqrt{1+\bar{\nu}^2}}{H\alpha'}$	$P(r \to 0) = \frac{\bar{\nu}\sqrt{1+\bar{\nu}^2}}{2H\alpha'}$	$P(r \to 0) = +E/2$
$Hr_{\text{max}} \equiv \bar{\nu} =$	$E(r_{\text{max}}) = \frac{\bar{\nu}}{H\alpha'}$	$P(r_{\text{max}}) = -\frac{\bar{\nu}}{2H\alpha'}$	$P(r_{\text{max}}) = -E/2$
$\sqrt{\frac{-1+\sqrt{1+4H^2\alpha'^2 b}}{2}}$			
Average values:	$< E >= \frac{2\bar{k}}{H\alpha' T_r}\sqrt{\frac{1-2\bar{k}^2}{1-\bar{k}^2}}\Pi\left(\frac{\bar{k}^2}{1-\bar{k}^2}, \bar{k}\right)$	$< P >= (\gamma - 1) < E >\neq 0$	
$(\bar{k} \equiv \bar{\nu}/\sqrt{1+2\bar{\nu}^2}$	$< E >_{\bar{k}<<1} \approx \frac{\bar{k}}{H\alpha'}$	$< P >_{\bar{k}<<1} \approx \frac{\bar{k}^3}{32H\alpha'}$	$\gamma_{\bar{k}<<1} \approx 1 + \frac{\bar{k}^2}{32}$
$\in [0, 1/\sqrt{2}\,[\,)$	$< E >_{\bar{k}\to\frac{1}{\sqrt{2}}} \approx \frac{\pi}{2H\alpha' K(1/\sqrt{2})\sqrt{1-2\bar{k}^2}}$	$< P >_{\bar{k}\to\frac{1}{\sqrt{2}}} \approx \frac{1}{\pi^2} < E >_{\bar{k}\to\frac{1}{\sqrt{2}}}$	

4 Semi-Classical Quantization

In this section we perform a semi-classical quantization of the circular string configurations discussed in the previous sections. We use an approach developed in field theory by Dashen et. al. [15, 16], based on the stationary phase approximation of the partition function. The method can be only used for time-periodic solutions of the classical equations of motion. In our string problem, these solutions however, include all the circular string solutions in Minkowski and in anti de Sitter spacetimes, as well as the oscillating circular strings ($bH^2\alpha'^2 \leq 1/4$, c.f. Section 2) in de Sitter spacetime.

The result of the stationary phase integration is expressed in terms of the function:

$$W(m) \equiv S_{\text{cl}}(T(m)) + m\,T(m), \tag{4.1}$$

where S_{cl} is the action of the classical solution, m is the mass and the period $T(m)$ is implicitly given by:

$$\frac{dS_{\text{cl}}}{dT} = -m. \tag{4.2}$$

In string theory we must choose T to be the period in a physical time variable. For example, when a light cone gauge exists, T is the period in $X^0 = \alpha' p\tau$. The bound state quantization condition then becomes [15, 16]:

$$W(m) = 2\pi n, \quad n \in N_0 \tag{4.3}$$

for n 'large'. The method has been successfully used in many cases from quantum mechanics to quantum field theory. For integrable field theories the semi-classical quantization happens in fact, to be exact. It must be noticed that string theory in de Sitter spacetime is exactly integrable [10, 11].

The actual computation of $W(m)$ is straightforward but tedious; the details are given in Ref.[9].

For completeness we first considered flat Minkowski spacetime. In that case, Eq.(4.3) reproduces the exact mass spectrum, except for the intercept. That is, the mass spectrum becomes $\alpha' m^2 = 4\,n, \quad n \in N_0$.

We find for de Sitter (anti de Sitter) spacetime that the mass is exactly proportional to the square-root of the Casimir operator $C = -L_{\mu\nu}L^{\mu\nu}$ of the $O(3,1)$-de Sitter [$O(2,2)$-anti de Sitter] group:

$$m = \frac{\sqrt{C}}{\pi H \alpha'}. \tag{4.4}$$

In Fig.6 we give parametric plots of $H^2\alpha' W$ as a function of $H^2 m^2 \alpha'^2$ for $k \in [0,\ 1]$ for de Sitter spacetime and for $\bar{k} \in [0,\ 1/\sqrt{2}\,[$ for anti de Sitter spacetime, respectively. We find for de Sitter spacetime:

$$\alpha' m^2_{\text{dS}} \approx 4\,n - 5H^2\alpha'\,n^2, \quad n \in N_0 \tag{4.5}$$

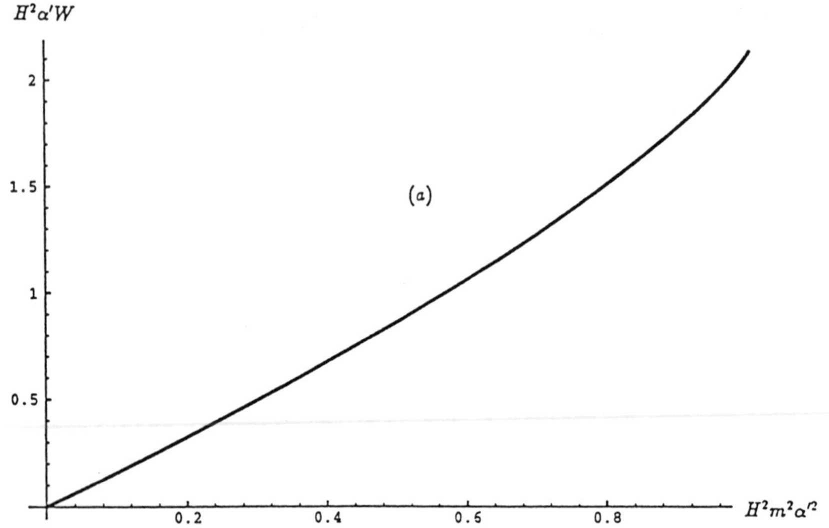

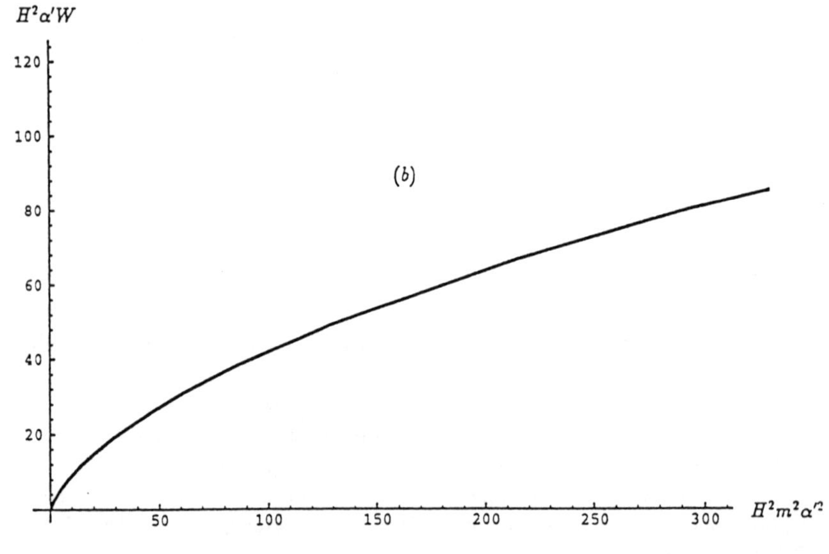

Figure 6. (a) Parametric plot of $H^2\alpha' W$ as a function of $H^2 m^2 \alpha'^2$ for $k \in [0,\ 1]$ in de Sitter spacetime. There is only a finite number of states. (b) Parametric plot of $H^2\alpha' W$ as a function of $H^2 m^2 \alpha'^2$ in anti de Sitter spacetime. There are infinitely many states.

This is different from the mass spectrum in Minkowski spacetime. The level spacing is however still approximately constant, but the levels are less separated than in Minkowski spacetime. Notice in particular that the level spacing slightly decreases for larger and larger n. In de Sitter spacetime there is only a *finite* number of levels as can be seen from Fig.1. The number of quantized circular string states can be estimated to be:

$$N_q \approx \frac{0.34}{H^2\alpha'}. \tag{4.6}$$

It is interesting to compare this result with the number of particle states obtained using canonical quantization [1]. One finds in this way a maximum number of states:

$$N_{\text{max}} \approx \frac{0.15}{H^2\alpha'}, \tag{4.7}$$

which is of the same order as the semi-classical value Eq.(4.6). It must be noticed that in de Sitter spacetime, these states can *decay* quantum mechanically due to the possibility of quantum mechanical tunneling through the potential barrier, see Fig.1. Semi-classically, the decay probability is however highly suppressed for $H^2\alpha' << 1$ and for any value of the elliptic modulus k, except near $k = 1$ where the barrier disappears, and for which the tunneling probability is close to unity, see Ref.[9].

In anti de Sitter spacetime arbitrary high mass states exist. The quantization of the high mass states yields (see Ref.[9]):

$$\alpha' m^2_{\text{AdS}} \approx H^2\alpha'\, n^2. \tag{4.8}$$

Thus the (mass)^2 grows like n^2 and the level spacing grows proportionally to n. This is a completely different behaviour as compared to Minkowski spacetime where the level spacing is constant. A similar result was recently found, using canonical quantization of generic strings in anti de Sitter spacetime [14, 17]. The physical consequences, especially the non-existence of a critical string temperature (Hagedorn temperature), of this kind of behaviour is discussed in detail in Ref.[17].

For both de Sitter and anti de Sitter spacetimes we find thus a strong qualitative agreement between the results obtained using canonical quantization, based on generic string solutions (string perturbation approach), and the results obtained using the semi-classical approach, based on oscillating circular string configurations.

A summary of these features are presented in Table 3. More details are given in Ref.[9].

Table 3. Semi-classical quantization of oscillating circular strings in Minkowski, de Sitter and anti de Sitter spacetimes.

SEMICLASSICAL QUANTIZATION OF PERIODIC SOLUTIONS

Stationary phase: $W(M) = S_{cl}(T(M)) + M\,T(M)$

Stationary phase point: $\frac{\partial S_{cl}}{\partial T} = -M,$ determines the period T

Quantization condition: $W(M) = 2\pi n, \quad n \in N_0$

MINKOWSKI:

Circular string: $r(\tau) = \frac{AT}{2\pi}\cos\frac{2\pi\tau}{T}$

Classical action: $S_{cl} = -\frac{A^2T^2}{2\pi\alpha'}$

Mass: $m = \frac{AT}{\pi\alpha'}$

Quant. condition: $\alpha' m^2 = 4n, \quad n \in N_0$

DE SITTER:

Circular string: $Hr(\tau) = \frac{k}{\sqrt{1+k^2}}\mathrm{sn}[\frac{\tau}{\sqrt{1+k^2}},\ k]$

Classical action: $S_{cl} = \frac{8}{H^2\alpha'}\frac{E(k)-K(k)}{\sqrt{1+k^2}}$

Mass: $m = \frac{2}{H\alpha'}\frac{k}{1+k^2}$

Quant. condition: $\alpha' m^2 \approx 4n - 5H^2\alpha' n^2, \quad n \in N_0$

Level spacing is still approximately constant, but levels are less separated than in Minkowski spacetime. Furthermore, it turns out that $H^2\alpha'^2 m^2 \in [0,\ 1]$ and $H^2\alpha' W \in [0,\ 4\sqrt{2} - 4\log(1+\sqrt{2})]$, so that there is only a *finite* number of states:

$$N_q \approx 1 + \mathrm{Int}\left(\frac{2\sqrt{2} - 2\log(1+\sqrt{2})}{\pi H^2\alpha'}\right) \approx \frac{0.34}{H^2\alpha'} \qquad \text{for} \quad H^2\alpha' << 1$$

The states decay quantum mechanically with probability:

$$\mathcal{T} \propto \mathrm{Exp}\left[\frac{-4}{H^2\alpha'}\ \frac{E(k') - \Pi(-(k'/k)^2, k')}{\sqrt{1+k^2}}\right]$$

ANTI DE SITTER:

Circular string: $Hr(\tau) = \frac{\bar{k}}{\sqrt{1-2\bar{k}^2}}\mathrm{cn}[\frac{\tau}{\sqrt{1-2\bar{k}^2}},\ \bar{k}]$

Classical action: $S_{cl} = \frac{8}{H^2\alpha'}\frac{(1-\bar{k}^2)K(\bar{k}) - E(\bar{k})}{\sqrt{1-2\bar{k}^2}}$

Mass: $m = \frac{2\bar{k}}{H\alpha'}\frac{\sqrt{1-\bar{k}^2}}{1-2\bar{k}^2}$

Quant. condition: $\alpha' m^2 \approx H^2\alpha' n^2, \quad n \in N_0$ (for high states, $\bar{k} \to \sqrt{\frac{1}{2}}$)

Level spacing grows as n. Notice that $H^2\alpha'^2 m^2 \in [0,\ \infty[$ and $H^2\alpha' W \in [0,\ \infty[$, so that there are infinitely many states.

Agreement with canonical quantization in the string perturbation series approach.

5 Conclusion

We have found the exact general evolution of circular strings in $2+1$ dimensional de Sitter spacetime. We have expressed it closely and completely in terms of elliptic functions, see Ref.[8] for the details. The solution generically describes infinitely many (different and independent) strings, and depends on one constant parameter b (for fixed H, α') related to the string energy. We have computed *exactly* the equation of state of the circular string solutions in de Sitter and anti de Sitter spacetimes. The string equation of state has the perfect fluid form $P = (\gamma - 1)E$, with P and E expressed closely and completely in terms of elliptic functions (see Ref.[9] for the details) and the instantaneous parameter γ depending on an elliptic modulus. We have quantized the time-periodic (oscillating) string solutions within the semi-classical (stationary phase approximation) approach and found the mass formulas in de Sitter and anti de Sitter spacetimes. The semi-classical quantization of the *exact* circular string solutions and the canonical quantization of generic strings (string perturbation series approach), provide the same qualitative results.

Acknowledgements:
A.L. Larsen is supported by the Danish Natural Science Research Council under grant No. 11-1231-1SE

References

[1] H.J. de Vega and N. Sánchez, Phys. Lett. B197 (1987) 320.

[2] N. Sánchez in "String Quantum Gravity and Physics at the Planck Energy Scale", Proceedings of the Erice Workshop held in June 1992. Edited by N. Sánchez, World Scientific, 1993. Pages 73-139, and references given therein.

[3] H.J. de Vega in "String Quantum Gravity and Physics at the Planck Energy Scale", Proceedings of the Erice Workshop held in June 1992. Edited by N. Sánchez, World Scientific, 1993. Pages 140-185, and references given therein.

[4] N. Sánchez and G. Veneziano, Nucl. Phys. B333 (1990) 253.

[5] M. Gasperini, N. Sánchez and G. Veneziano, IJMP A6 (1991) 3853, Nucl. Phys. B364 (1991) 365.

[6] F. Combes, H.J. de Vega, A.V. Mikhailov and N. Sánchez, Phys. Rev. D50 (1994) 2754.

[7] H.J. de Vega, A.V. Mikhailov and N. Sánchez, Teor. Mat. Fiz. 94 (1993) 232, Mod. Phys. Lett. A29 (1994) 2745.

[8] H.J. de Vega, A.L. Larsen and N. Sánchez, Nucl. Phys. B427 (1994) 643.

[9] H.J. de Vega, A.L. Larsen and N. Sánchez, "Semi-Classical Quantization of Circular Strings in de Sitter and anti de Sitter Spacetimes", DEMIRM-Paris-94049, submitted to Phys. Rev. D.

[10] V.E. Zakharov and A.V. Mikhailov, JETP 47 (1979) 1017.
H. Eichenherr, in "Integrable Quantum Field Theories", ed. J. Hietarinta and C. Montonen (Springer, Berlin, 1982).

[11] H.J. de Vega and N. Sánchez, Phys. Rev. D47 (1993) 3394.

[12] M. Abramowitz and I.A. Stegun, Handbook of Mathematical Functions (Dover Publications Inc, New York, 1964) page 599.

[13] H.J. de Vega and N. Sánchez, Int. J. Mod. Phys. A7 (1992) 3043.

[14] A.L. Larsen and N. Sánchez, Phys. Rev. D50, 7493 (1994).

[15] R. Dashen, B. Hasslacher and A. Neveu, Phys. Rev. D11, 3424 (1975).

[16] H.J. de Vega and J.M. Maillet, Phys. Rev. D28, 1441 (1983).

[17] A.L. Larsen and N. Sánchez, "Mass spectrum of strings in anti de Sitter spacetime", DEMIRM 94-48 preprint, to appear in Phys. Rev. D.

THE TWO-DIMENSIONAL STRINGY BLACK-HOLE: A NEW APPROACH AND A NEW EFFECT

H.J. DE VEGA

Laboratoire de Physique Théorique et Hautes Energies, Paris[⋆]

J. RAMÍREZ MITTELBRUN, M. RAMÓN MEDRANO

Departamento de Física Teórica, Madrid[†]

N. SÁNCHEZ

Observatoire de Paris, Demirm[*]

⋆ *Laboratoire Associé au CNRS UA 280*
Postal address:
L.P.T.H.E., Tour 16, 1er étage, Université Paris VI,
4, Place Jussieu, 75252, Paris cedex 05, FRANCE

† *Postal address:*
Facultad de Ciencias Físicas, Universidad Complutense, Ciudad Universtaria,
E-28040, Madrid, ESPAÑA.

* *Laboratoire Associé au CNRS UA 336, Observatoire de Paris et École Normale Supérieure.*
Postal address:
DEMIRM, Observatoire de Paris. 61, Av. de l'Observatoire, 75014 Paris, FRANCE.

ABSTRACT

The string propagation in the two-dimensional stringy black-hole is investigated from a new approach . We completely solve the classical and quantum string dynamics in the lorentzian and euclidean regimes. In the lorentzian case all the physics reduces to a massless scalar particle described by a Klein-Gordon type equation with a singular effective potential. The scattering matrix is found and it reproduces the results obtained by coset CFT techniques. It factorizes into two pieces : an elastic coulombian amplitude and an absorption part. In both parts, an infinite sequence of imaginary poles in the energy appear. The generic features of string propagation in curved D-dimensional backgrounds (string stretching, fall into spacetime singularities) are analyzed in the present case. A **new** physical phenomenon specific to the present black-hole is found : the quantum renormalization of the speed of light. We find $c_{quantum} = \sqrt{\frac{k}{k-2}}\, c_{classical}$, where k is the integer in front of the WZW action. Only for $k \to \infty$ this new effect disappears (although the conformal anomaly is present). We analyze all the classical euclidean string solutions and exactly compute the quantum partition function. No critical Hagedorn temperature appears here.

1. Contextual Background

The space-time metric associated to the $SL(2,R)/U(1)$ coset model (and its different versions), and interpreted as a two dimensional black hole [1,2,3] , arose enormous attention recently [4,5,6].

In this paper, we look at this problem from our string gravity point of view, that we started and developed (see refs.[7-21]) for instance), independently and before the interest on two dimensional black holes flourished.

The two dimensional black hole model is interesting in the sense that it is an exact solution of the renormalization group equations of string theory. As it is known, these equations define the backgrounds in which strings propagate consistently.

Although two dimensional models have many attractive tractable aspects and can be used to test and get insights on particular features, $D = 2$ is not for string theory neither for gravity the most physically appealing dimension.

Strings in two dimensional black hole backgrounds have been extensively treated in the literature in the context of the conformal Field Theory (CFT) techniques. (See ref. [4] for a complete description). In this paper, we present a different view and a different approach to this problem which yields in a simple and physical way the full exact quantum result; giving new insights with respect the already known CFT descriptions.

2. Introduction

We study string propagation in the two-dimensional black hole background of refs.[1-3]

$$dS^2 = k\left[dr^2 - \tanh^2 r\, dt^2\right]\ ,\ \Phi = \log(\cosh^2 r)\ ,\ \ 0 \leq r < \infty,\ -\infty < t < \infty. \qquad (2.1)$$

where k is a positive integer and Φ is the dilaton field. (Classically, the string propagates in this metric, and Φ is decoupled from them). In Kruskal null type coordinates (u,v) the metric eq.(2.1) reads

$$dS^2 = k\,\frac{dudv}{1-uv}\ ,\ \ 0 \leq u,v < \infty. \qquad (2.2)$$

We completely solve the string equations of motion and constraints in the Lorentzian and on the Euclidean (i.e. $t = i\Theta$) regimes. The solutions fall into four types. In the Lorentzian regime in which there are no compact dimensions in the space-time, the σ-dependence can be completely gauged away, and all the solutions describe just the geodesics of a massless point particle. The only physical degree of freedom is the one associated to the center of mass of the string ; we are left more with a point particle field theory rather than a string theory.

In the Euclidean regime, the Schwarzchild imaginary time is periodic , $(0 \leq \Theta \leq 2\pi)$, and then the σ-dependence of the string solutions remains. The solution must be also euclidean in the world sheet (σ becames purely imaginary and identified to the imaginary time). There are two types of euclidean solutions : (i) a string winding n times along the Θ - direction in the "cigar" manifold $dS^2 = k\left[dr^2 + \tanh^2 r\, d\Theta^2\right]$ from $r = 0$ to $r = \infty$, and (ii) a string winding n times along Θ in the "trumpet" manifold $dS^2 = k\left[ds^2 + \coth^2 s\ d\Theta^2\right]$, $0 < s < \infty$. It must be noticed that these are *rigid strings* staying around the manifolds without any oscillation. In $D = 2$, $e^{in\sigma}$ does not describes the string fluctuations, thus the name modes ("winding modes" or "momentum modes", as commonly referred (see for example ref.[4]), are misleading.

For these solutions, the metric on the world-sheet is

$$
\begin{aligned}
dS^2_{cigar} &= k\,\frac{n^2}{1+e^{2n\tau}}\,(d\tau^2 + d\sigma^2) \\
dS^2_{trumpet} &= k\,\frac{n^2}{1-e^{-2n\tau}}\,(d\tau^2 + d\sigma^2)
\end{aligned}
\tag{2.3}
$$

The proper string length (between two points σ and $\sigma + d\sigma$ at fixed τ) is

$$
\Delta S_{cigar} = \sqrt{k}\,\frac{n}{\sqrt{1+e^{2n\tau}}}\,d\sigma \quad , \quad \Delta S_{trumpet} = \sqrt{k}\,\frac{n}{\sqrt{1-e^{2n\tau}}}\,d\sigma \tag{2.4}
$$

In $D = 2$ the string world sheet and the physical space-time can be identified ; the world sheet covers completely (generically, n times) the physical space. The trumpet manifold has a curvature singularity at $s = 0$. Near $s = 0$, i.e. $\tau \to 0$, $\Delta S_{trumpet}$ blows up, the string stretches infinitely. This typical feature of string instability [13-16,21] , just corresponds here to the location of the string in the open extremity of the trumpet. The euclidean black hole manifold is non singular, and the string length (ΔS_{cigar}), is always finite in it, since the cigar radius is finite. We also discuss the cosmological version of the black hole string solutions (in the euclidean regime) ; in this case the string solutions are the same as the trumpet solution.

We recall that in $D = 2$ de Sitter space-time with Lorentzian signature, the general solution is a string wound n times around de Sitter space (the circle S^1) and evolving with it. For $\tau \to 0$, $\Delta S_{deSitter} \to \infty$; the string expands (or contracts, $\Delta S_{deSitter} \to 0$) with the universe itself.

It is interesting to point out the relation between τ and the physical time : τ interpolates between the Kruskal time ($\tau \to 0$) and the Schwarzschild time ($\tau \to \infty$). This is like strings in cosmological backgrounds in which τ interpolates between the conformal time ($\tau \to 0$) and the cosmic time ($\tau \to \infty$). (And like strings in gravitational wave backgrounds). The logarithmic relation between the Schwarzchild time and τ for $\tau \to 0$ is exactly like the cosmic time and τ in de Sitter space.

In this two dimensional black hole, the string takes an infinite time τ to approach the curvature singularity $uv = 1$, and then, never reaches it. This is to be contrasted with the higher dimensional black hole [8, 21]. For all $D > 2$, the string falls down into the physical singularity $uv = 1$ in a finite proper time τ, and classically as well as quantum mechanically, the string is trapped by the singularity [21].

We study the string quantization in this background (in the lorentzian and in the euclidean regimes) and follow the spirit of ref.[7], in which one starts from the c.m. motion, and then takes the σ-dependence as fluctuations around. Here, in the lorentzian case, there are no σ-fluctuations, and the quantization of the c.m. Hamiltonian

$$\hat{\mathcal{H}} = -\frac{1}{2k}\left(\frac{\partial^2}{\partial r^2} + (\coth r + \tanh r)\frac{\partial}{\partial r} - \coth^2 r\,\frac{\partial^2}{\partial t^2}\right) \tag{2.5}$$

yields the full and exact description of the system.

Eq.(2.5) is the same as the L_0 operator of ref.[4] (hereafter referred as DVV). They only differ by additive and multiplicative constants, but as operators are identically the same). We solve the eigenfunction problem

$$\hat{\mathcal{H}}\Psi = \frac{\lambda}{2k}\Psi$$

describing the scattering of the (massless) tachyon field by the black hole, which yields a Schrödinger type equation with an energy-dependent potential

$$\begin{aligned} V_{eff} &= -\frac{E^2 + 1/4}{\sinh^2 r} + \frac{1}{4\cosh^2 r} - E(E-1) - \lambda + 1 \\ V_{eff} &\overset{r\to 0}{=} -\frac{\gamma}{r^2} + O(1)\ , \quad \gamma \equiv E^2 + 1/4\ . \end{aligned} \tag{2.6}$$

Since $\gamma \geq 1/4$, there exists absorption (fall into the event horizon $r = 0$) for all energy E.

This is like the $D > 2$ black holes.

The constant λ is determined from the $r \to \infty$ behaviour, by requiring the tachyon to be massless

$$\lambda - 1 = (1 - c^2)p^2$$

p being the momentum, $p = E/c$.

The eigenfunctions Ψ are given in terms of hypergeometric functions

$$\Psi^{\pm}_{E,c}(r,t) = e^{-iE_t} \ (\sinh r)^{\mp iE} \ F\left(\frac{1}{2} \mp \frac{iE}{2}(1+\frac{1}{c}), \frac{1}{2} \mp \frac{iE}{2}(1-\frac{1}{c}); 1 \mp iE\,; -\sinh^2 r\right) \tag{2.7}$$

They exhibit the typical behaviour of the wave functions near the black hole horizon

$$\Psi^{\pm}_{E,c}(r,t) \stackrel{r \to 0}{=} e^{-iE(t \pm \log r)} \ , \tag{2.8}$$

Ψ^+ describing purely incoming particles at the future event horizon. (Ψ^- describes outcoming particles from the past horizon). The tachyon field incident from spatial infinity is partially absorbed and reflected by the black hole

$$\Psi^{+}_{E,c}(r,t) \stackrel{r \to \infty}{=} e^{-ipct} \left[e^{-ipr} + S(p,c)\, e^{ipr}\right] \tag{2.9}$$

We find

$$S(p,c) = S_{coul}(p)\ \tilde{S}(p,c)$$

where

$$\begin{aligned} S_{coul}(p) &= 2^{-2ip} \frac{\Gamma(ip)}{\Gamma(-ip)} = 2^{-2ip} e^{2i\arg\Gamma(ip)} \\ \tilde{S}(p,c) &= \left[\frac{\Gamma\left(\frac{1}{2} - \frac{ip}{2}(c+1)\right)}{\Gamma\left(\frac{1}{2} - \frac{ip}{2}(c-1)\right)}\right]^2 \end{aligned} \tag{2.10}$$

S_{coul} takes into account the large r interaction and the purely elastic scattering; $\tilde{S}$ describes the genuine black hole features :

$$\begin{aligned} |S_{coul}(p)| &= 1 \\ |\tilde{S}(p,c)| &= \left[\frac{\cosh \frac{\pi p}{2}(c-1)}{\cosh \frac{\pi p}{2}(c+1)}\right]^2 \ < \ 1 \ , \end{aligned} \tag{2.11}$$

which describes the absorption by the black hole.

$S(p,c)$ exhibits an infinity sequence of imaginary poles at the values $ip = n, (n = 0, 1, ...)$, which are like the Coulombian-type poles, and also an infinite sequence of imaginary poles at $ip(c+1) = 2n+1,\ n = 0, 1...$ (in the $\tilde{S}$ part).

It must be stressed that $S(E,c)$ depends on two physical parameters : E and c . For each energy E, we have a monoparametric family of solutions depending on c, each c yields a different S-matrix. c is a **purely quantum mechanical parameter**, which is not fixed by any special requirement, and can take any value ; c accounts for a renormalization of the speed of light. Classically, we choose our units such that $c = 1$ (see eq.(2.1) for $r \to \infty$). But, quantum mechanically, c is no more unit in this problem. We find that c is related to the parameter k of the WZW model :

$$c = \sqrt{\frac{k}{k-2}} \tag{2.12}$$

and we reproduce the results of ref.[4]. In ref.[4], the effect of the renormalization of the speed of light is also present (although it has not been noticed). Classically, before quantization, DVV have $c = 1$, but after quantization, since they choose $k = 9/4$, they have $c = 3$. This can be seen from the asymptotic behaviour of the metric for $r \to \infty$. Classically, $dS^2(r \to \infty) = dr^2 + d\Theta^2$, and after CFT quantization of the $SL(2,R)/U(1)$ model :

$$dS^2_{DVV} \overset{r\to\infty}{=} 2(k-2)\left[dr^2 + \frac{k}{k-2}d\theta^2\right] = \frac{4}{c^2-1}[dr^2 + c^2 d\theta^2]$$

Notice that in the classical limit $k \to \infty$, c takes its classical value. The S-matrix of ref.[4] is a particular case of our results for $c = 3$:

$$S(p, c=3) = S_{DVV}(p) = 2^{-2ip}\frac{\Gamma(ip)}{\Gamma(-ip)}\left[\frac{\Gamma\left(\frac{1}{2} - 2ip\right)}{\Gamma\left(\frac{1}{2} - ip\right)}\right]^2 \tag{2.13}$$

[The factor 2^{-2ip} is missing in ref. [4]]. Notice that nothing special happens, however, at the conformal invariant point $k = 9/4$. The physical relevance of the $k = 9/4$ point in the two dimensional black hole scattering matrix is not clear. Also notice that the presence or absence of conformal anomalies is totally irrelevant for the black-hole singularity $uv = 1$. [The conformal anomaly vanishes for $k = \frac{9}{4}$].

The computation of the Hawking radiation follows directly from the by now well known treatment of QFT in curved space-time and does not present any particular feature here. The vacuum spectrum is a Planckian distribution at the temperature $1/2\pi$.

We also quantize the system in the euclidean regime and exactly compute the partition function of the two dimensional stringy black hole at temperature β^{-1}. It is given by

$$Z(\beta) = \frac{1}{\pi} \sum_{m,n=-\infty}^{+\infty} \sqrt{m^2 + (kn)^2 + 1}\; K_1\left(\frac{\beta}{2k}\sqrt{m^2 + (kn)^2 + 1}\right) \tag{2.14}$$

where $K_1(z)$ is a modified Bessel function. $Z(\beta)$ is analytic for all positive temperatures and hence no critical Hagedorn temperature appears here.

In conclusion, the quantization of the center of mass of the string, i.e. of the classical solution, with the requirement that $m^2 = 0$, yields all the physics of the problem.We have obtained the full exact quantum result without introducing any correction $1/k$ to the space time metric, neither to the dilaton.

The two dimensional CFT constructions can be avoided in a problem like the two dimensional stringy black hole. They introduce a lot of technicality in a problem in which all the physics can be described by the straightforward quantization of a two-dimensional massless scalar particle. Perhaps the CFT tools would be really necessary for the problem of interacting (higher genus) strings in the two dimensional black hole background, problem, which unfortunately, has not been treated until now.

Finally, let us comment about the renormalization of the speed of light. Normally, c is never affected by quantum corrections. Once one chooses units such that $c = 1$ classically, this value remains true in the quantum theory (relativistic quantum mechanics, QFT, string theory, etc). In the two dimensional stringy black-hole, it turns out that $c = \sqrt{\frac{k}{k-2}}$ for the quantum massless particle described by the string, whereas the wave equation in the metric (2.2) gives unit speed of propagation. Only for $k = \infty$ both speeds coincide (but the conformal anomaly is present).

We think that this effect is specific of the two dimensional string dynamics in such curved background. More precisely, this effect is to be traced back to the dilaton background which couples with the string only at the quantum level and as a surface effect (infinite distance).

A pathological situation arises when one has a quantum string model like the present one and a field theory. Assuming we choose units such that $c = 1$ at the classical level in both models. Then, one would find $c = \sqrt{\frac{k}{k-2}}$ for the massles particles in the string model while $c = 1$ for the classical and quantum particles in the field theory.

3. Strings in two dimensional geometries and dilaton backgrounds

The action for a string in a two dimensional space-time can be written as

$$\mathcal{S} = -\frac{1}{2\pi}\int d\sigma d\tau\sqrt{h}\left[\frac{h_{\mu\nu}(\sigma,\tau)}{2}G_{AB}(X)\,\partial^{\mu}X^{A}(\sigma,\tau)\,\partial^{\nu}X^{B}(\sigma,\tau) - \frac{R^{(2)}}{4}\Phi(X)\right] \tag{3.1}$$

$$+\ \text{surface terms}\ . \qquad\qquad A,B = 1,2\ ;\ \mu,\nu = 1,2.$$

where $h_{\mu\nu}$ is the metric on the world sheet, $G_{AB}(X)$ is the space time metric, $R^{(2)}$ is the world sheet curvature and $\Phi(X)$ is the dilaton field. $\Phi(X)$ and $G_{AB}(X)$ are constrained by the vanishing of the beta functions. To one loop level this implies

$$R_{AB} = D_A D_B \Phi \tag{3.2}$$

Classically, one has the string propagating in the background G_{AB}, and the dilaton is decoupled from them.

We can always choose the conformal gauge on the world sheet, such that

$$h_{\mu\nu}(\sigma,\tau) = \exp[\,\varphi(\sigma,\tau)\,]\ \mathrm{diag}(-1,+1)$$

Then,

$$\sqrt{h}\,R^{(2)} = (\partial_\tau^2 - \partial_\sigma^2)\varphi(\sigma,\tau) \tag{3.3}$$

and

$$\mathcal{S} = -\frac{1}{2\pi}\int d\sigma d\tau\left[\frac{1}{2}G_{AB}(X)\,\partial_\mu X^{A}(\sigma,\tau)\,\partial^{\mu}X^{B}(\sigma,\tau) - \frac{1}{4}\,\partial^{\mu}\Phi(X)\partial_\mu\varphi\right]. \tag{3.4}$$

Here $\partial^\mu\alpha\partial_\mu\beta = -\dot{\alpha}\dot{\beta}+\alpha'\beta'$ and we made a precise choice of the surface terms. Recall that under a Weyl transformation

$$h_{\mu\nu}(\sigma,\tau) \to e^{\lambda(\sigma,\tau)}\,h_{\mu\nu}(\sigma,\tau) \tag{3.5}$$

the first term in eq.(3.1) is (classically) invariant.

This is not the case of the dilaton term that transforms as

$$\sqrt{h}\,R^{(2)} \to \sqrt{h}\,R^{(2)} + \partial^2\lambda$$

Since the beta function vanishes here [eq.(3.2)], there are no quantum anomalies to one-loop level. The Weyl invariance holds for example, if the dilaton term is a total divergence in eq.(3.4). This happens when $\partial_\mu\varphi$ = constant. [This term has actually a factor α' in front].

For a genus zero euclidean world-sheet (dominant quantum level) we can start from a flat metric on the complex plane

$$dS^2 = dz\, d\bar{z} \tag{3.6}$$

and, map it into the euclidean world sheet (σ, τ') as

$$z = e^{\tau' + i\sigma} \ , \quad 0 \le \sigma < 2\pi \ , \ -\infty < \tau' < +\infty \tag{3.7}$$

The Lorentzian world sheet follows by Wick rotation, $\tau' = i\tau$ where τ is the (real) world sheet time. It follows from eqs.(3.6) and (3.7) that

$$dS^2 = e^{2\tau'}\left[d\tau'^2 + d\sigma^2\right] = e^{2i\tau}\left[-d\tau^2 + d\sigma^2\right] \tag{3.8}$$

That is,

$$\varphi(\sigma, \tau) = 2i\tau \tag{3.9}$$

Therefore,

$$\frac{1}{4}\,\partial^\mu\Phi(X)\partial_\mu\varphi = -\frac{i}{2}\frac{\partial\Phi(X)}{\partial\tau}$$

and the last term in eq.(3.4) becomes a total derivative. It is therefore irrelevant for the classical solutions and obviously respects Weyl invariance. Of course, we could use a real Lorentzian metric on the world-sheet. Actually, the choice of the metric is completely irrelevant at the classical level since it decouples from the dynamics of the string. The action eq.(3.4) is then φ -independent and hence, real, except for the surface term.

The reason why we prefer the complex form (3.8)-(3.9) for the world-sheet metric will become apparent for the quantization of the string in sec. 6.

As is known, since the space time is two dimensional, we can always find conformal coordinates (U, V) where

$$G_{AB}(X) = e^{2\omega(U,V)} \begin{pmatrix} 0 & 1 \\ 1 & 0 \end{pmatrix} . \tag{3.10}$$

Then, the action (3.4) takes the form

$$\mathcal{S} = -\frac{1}{2\pi} \int d\sigma d\tau \left[e^{2\omega(U,V)} \partial_\mu U(\sigma,\tau)\, \partial^\mu V(\sigma,\tau) + \frac{i}{2} \frac{\partial \Phi(X)}{\partial \tau} \right] . \tag{3.11}$$

The equations of motion result given by

$$\partial_+ \partial_- V + 2 \frac{\partial \omega}{\partial V} \partial_+ V\, \partial_- V = 0 , \tag{3.12}$$

$$\partial_+ \partial_- U + 2 \frac{\partial \omega}{\partial U} \partial_+ U\, \partial_- U = 0 , \tag{3.13}$$

and the constraints are given by

$$T_{\pm\pm} = e^{2\omega(U,V)}\, \partial_\pm U\, \partial_\pm V = 0 \tag{3.14}$$

where

$$\partial_\pm = \frac{1}{2}(\partial_\sigma \pm \partial_\tau) \quad , \quad x_\pm = \sigma \pm \tau .$$

Eqs.(3.12) - (3.14) posses the following solutions

(*i*)

$$U = f(\sigma - \tau) \quad , \quad V = g(\sigma + \tau) \tag{3.15}$$

(*ii*)

$$U = f(\sigma + \tau) \quad , \quad V = g(\sigma - \tau) \tag{3.16}$$

where f and g are arbitrary functions of the indicated variables.

(*iii*)

$$U = u_0 = \text{constant and } V = V(\sigma, \tau) \text{ is a solution of eq.(3.12)} \tag{3.17}$$

(*iv*)

$$V = v_0 = \text{constant and } U = U(\sigma, \tau) \text{ is a solution of eq.(3.13)} \tag{3.18}$$

In the case (iii), in order to solve eq.(3.12) with $U = v_0$, we set $W = F(V)$. Then,

$$\partial_+\partial_- W = F'(V)\left[\partial_+\partial_- V + \frac{F''(V)}{F'(V)}\,\partial_+ V\,\partial_- V\right] \tag{3.19}$$

By choosing

$$\frac{F''(V)}{F'(V)} = 2\,\frac{\partial\omega(u_0, V)}{\partial V} \quad \text{and} \quad \partial_+\partial_- W = 0\ , \tag{3.20}$$

eq.(3.12) is fulfilled with $U = u_0$. Now, by integrating eq.(3.20) we find

$$\begin{gathered} F(V) = A\int_0^V dv e^{2\omega(u_0,v)} + B\ , \\ W = f(\sigma - \tau) + g(\sigma + \tau)\ , \text{ and } V = F^{-1}(W) \end{gathered} \tag{3.21}$$

where A and B are arbitrary constants.

Similarly, for the solutions of type (iv) eq.(3.18) we find

$$\begin{gathered} U = G^{-1}(W) \text{ where } \partial_+\partial_- W = 0 \text{ and} \\ G(U) = C\int_0^U du e^{2\omega(u,v_0)} + D\ , \end{gathered} \tag{3.22}$$

C and D being arbitrary constants.

Let us analyze the physical content of the solutions. The solutions of type (iii) and (iv) describe geodesics propagating on the characteristics $U = u_0$ and $V = v_0$ respectively. We can make on these solutions a conformal transformation $\sigma \pm \tau \to \phi_\pm(\sigma \pm \tau)$ making $W = \tau$. Then, we obtain σ-independent solutions

$$U = u_0 \quad , \quad V = F^{-1}(\tau) \tag{3.23}$$

Or,

$$U = G^{-1}(\tau) \quad , \quad V = v_0 \tag{3.24}$$

where $F(V)$ and $G(U)$ are defined by eqs. (3.21) - (3.22) .

In other words, the solutions of type (iii) and (iv) reduce in an appropriate gauge to the **geodesics** for massless particles.

Let us discuss now the type (i) - solutions. In general, we can write

$$U(\sigma-\tau)=q_U+(\sigma-\tau)\,p_U+\sum_{n\neq 0}\alpha_n\,e^{in(\sigma-\tau)}$$
$$V(\sigma+\tau)=q_V+(\sigma+\tau)\,p_V+\sum_{n\neq 0}\tilde{\alpha}_n\,e^{-in(\sigma+\tau)} \tag{3.25}$$

Therefore,

$$U(\sigma+2\pi-\tau)-U(\sigma-\tau)=2\pi\,p_U \quad , \quad V(\sigma+2\pi+\tau)-U(\sigma+\tau)=2\pi\,p_V$$

If the physical space-time *does not* enjoy periodic directions, we must set $p_U=p_V=0$. Then, we can locally gauge away completely the α_n and $\tilde{\alpha}_n$-dependence leaving no oscillator degrees of freedom. This is exactly like a string in two dimensional Minkowski space-time.

When *periodic directions do exist*, the situation is different. Let us call a and b the periods in the coordinates $X=U-V$ and $T=U+V$ respectively. Then, periodicity requires

$$p_U=\frac{Mb+Na}{4\pi} \quad , \quad p_V=\frac{Mb-Na}{4\pi} \tag{3.26}$$

where M and N are integers. We can call them winding numbers. If there is periodicity only on X, we should set $Mb=0$ (and viceversa). That is, we can locally write the solutions as

$$T(\sigma,\tau)=q_X+\frac{Mb}{2\pi}\sigma-\frac{Na}{2\pi}\tau \quad , \quad X(\sigma,\tau)=q_T+\frac{Na}{2\pi}\sigma-\frac{Mb}{2\pi}\tau \tag{3.27}$$

In conclusion, in the absence of periodical directions we are left with solutions of type (iii) and (iv). The σ-components can be completely gauged away and we are left merely with (massless) point particle geodesics. The only physical degree of freedom is the one associated to the center of mass. That is, in the absence of transverse dimensions in two dimensions, the strings has no excitations and hence, it only describes one massless scalar particle. This is more a point particle field theory than a string theory.

4. String propagation in two dimensional black hole geometries

Let us now discuss the string propagation in the two-dimensional black hole background of refs.[1-3]. In this case

$$dS^2 = k\left[dr^2 - \tanh^2 r\, dt^2\right] \ , \quad 0 < r < \infty, \ -\infty < t < +\infty. \tag{4.1}$$

where k is a positive integer and the dilaton field takes the form

$$\phi = \log(\cosh^2 r) \tag{4.2}$$

k must be an integer in order to have e^{iS} univalued since it is introduced as the coefficient of a multivalued WZW term [3].

Here, (r,t) are Schwarzschild type coordinates related to Kruskal coordinates (u,v) by

$$u = \sinh r\ e^{t} \ , \ v = -\sinh r\ e^{-t} \ , \quad 0 < r < \infty, \ -\infty < t < +\infty. \tag{4.3}$$

In terms of (u,v), the length element eq.(4.1) takes the form

$$dS^2 = k\ \frac{dudv}{1-uv} \ , \quad 0 < u, v < +\infty. \tag{4.4}$$

The hyperbola $uv = 1$ is a singularity of the space-time. (The curvature is $R = \frac{4}{\cosh^2 r} = \frac{4}{1-uv}$). The null lines $uv = 0$ (at which $r = 0$ and $t = \pm\infty$) correspond to the event horizon.

It is convenient to define the variable $r^\star$ by

$$\sinh r = e^{r^\star} \qquad -\infty < r^\star < +\infty. \tag{4.5}$$

in terms of which the metric is written as

$$dS^2 = k\ \frac{e^{2r^\star}}{1+e^{2r^\star}}\ [(dr^\star)^2 - dt^2] \tag{4.6}$$

and we have simply

$$u = e^{r^\star + t} \ , \quad v = -e^{r^\star - t} \ . \tag{4.7}$$

In the black hole background with *Lorentzian* signature, there are no periodic dimensions. Then, the solutions of the string equations of motion in the *Lorentzian* black hole background reduce to the type (iii) and type (iv) solutions discussed in the previous section.

That is, they are just the geodesics of **massless point** particles. Namely,

$$u = u_0 = \text{const.} \quad v = \frac{1}{u_0}\left[1 - e^{u_0(a\tau+b)}\right] \tag{4.8}$$

Or,

$$v = v_0 = \text{const.} \quad u = \frac{1}{v_0}\left[1 - e^{v_0(a\tau+b)}\right] \tag{4.9}$$

That is,

$$r^\star = \frac{1}{2}\log\left[e^{u_0(a\tau+b)} - 1\right] \quad , \quad t = -\frac{1}{2}\log\left[\frac{e^{u_0(a\tau+b)} - 1}{u_0^2}\right] \tag{4.10}$$

It is interesting to discuss the relation between the world sheet time (τ), and the physical-Schwarzschild (t) or Kruskal ($T = u + v$) time. (Let us take the $u =$ const. case). We see that :

$$v(\tau \to 0) \to A + B\tau \quad , \quad t(\tau \to 0) \to -\frac{1}{2}\log\tau \tag{4.11}$$

and

$$v(\tau \to \infty) \to e^{u_0 a\tau} \quad , \quad t(\tau \to \infty) \to -\frac{1}{2}(a\tau + b) \tag{4.12}$$

We see that the world sheet time τ **interpolates** between the Kruskal time ($\tau \to 0$) and the Schwarzschild time ($\tau \to \infty$). This is like strings in cosmological backgrounds in which the world sheet time interpolates between conformal time ($\tau \to 0$) and cosmic time ($\tau \to \infty$). A similar relation holds also in gravitational wave backgrounds [13]. The logarithmic relation between Schwarzschild time and τ for $\tau \to 0$ in eq.(4.11) is exactly like in de Sitter space between cosmic time and τ. More generally, $\tau \to \infty$ characterizes the time with respect to which strings in D-dimensional curved spacetime oscillate, then ingoing and outgoing scattering states can be defined and the usual particle interpretation can be applied. On the other hand, the $\tau \to 0$ regime characterizes the time governing the string evolution in the strong gravitational field regime where the string does not oscillate in time.

It must be noticed that inside the two dimensional black hole, the string approaches the hyperbola $uv = 1$ for $u_0 a\tau \to -\infty$. The affine parameter λ of the null geodesic eq.(4.9) can be taken proportional to the coordinate v. Therefore, for $u_0 a\tau \to -\infty$, the string takes a finite time λ to reach the singularity at $uv = 1$.

In the higher dimensional case, for all $D > 2$, the string reaches the singularity in a finite proper time τ. Depending on the singularity strength, the string crosses smoothly the space-time singularity, or, alternatively, it is trapped by it [13]. In the D-dimensional black hole the string falls down into the space-time $uv = 1$ singularity (in a *finite* time τ), and classically, as well as quantum mechanically , the string is trapped by the singularity [21] .

A generic feature of string propagating in D-dimensional curved backgrounds which *is lost* in the $D = 2$ case with Lorentzian signature is the *string stretching.* As it is known [13-16,21], when strings propagate in inflationary cosmological backgrounds and near space-time singularities, the string proper length grows indefinitely. This is accompanied by a non oscillatory behaviour in time, typical of this string instability. This is not present in the $D = 2$ case with Lorentzian signature, since the transverse oscillators are absent, and the σ-dependence is gauged away, the string in $D = 2$ behaving like a point particle rather than an extended object.

5. Two dimensional Euclidean black hole. Cigar solutions and Trumpet solutions

In the $D = 2$ black hole space-time with Lorentzian signature, in which compact directions are absent, the only string physical degree of freedom is the center of mass (point particle). In the Euclidean signature case, however, compact dimensions are present and the σ-dependence remains. Let us consider now the solutions in the euclidean regime. The Schwarzschild time is purely imaginary, namely

$$t = i\Theta \quad \text{i.e.} \quad dS^2 = k\left[dr^2 + \tanh^2 r\, d\Theta^2\right] \ , \tag{5.1}$$

and the Kruskal coordinates are

$$u = \sinh r\ e^{i\Theta} \ , \ v = -\sinh r\ e^{-i\Theta} \ , \quad 0 < r < \infty,\ 0 < \Theta < 2\pi. \tag{5.2}$$

The "euclidean time" Θ is periodic with period 2π.

For solutions of the type (i) or (ii) [eqs. (3.15) or (3.16)] in the euclidean regime, periodicity in σ requires

$$\sigma = i\hat{\sigma} = i\frac{\Theta}{n}$$

that is, the world sheet signature must also be euclidean with $\hat{\sigma}$ playing the rôle of the euclidean time - (i.e., $t = n\sigma$).

The solution is given by

$$u = e^{n(\tau + i\hat{\sigma})} \quad , \quad v = -e^{n(\tau - i\hat{\sigma})} \tag{5.3}$$

that is,

$$\sinh r = e^{n\tau} \quad , \quad i.e. \;\; r^{\star} = n\tau \;\; ; \;\; \Theta = n\hat{\sigma} \;\; , \tag{5.4}$$

with the dilaton given by

$$\Phi = \log[1 + e^{2n\tau}] \; . \tag{5.5}$$

This is a non singular real manifold with metric

$$dS^2 = k \, \frac{n^2}{1 + e^{-2n\tau}} \, (d\tau^2 + d\hat{\sigma}^2) \tag{5.6}$$

For $\tau \to \infty$ the metric is flat.

$\Theta = n\hat{\sigma}$ are winding configurations. They correspond to the so-called "momentum modes", in the terminology of the literature (see ref.[4]), since here, $\hat{\sigma}$ is an imaginary time. The physical interpretation of these modes is a string winding n times along the angular Θ direction, lying in the semi-infinite cigar from $r = 0$ to $r = \infty$. At $\tau = -\infty$, the string is at $r = 0$ (the horizon is shrank to a point in the euclidean space). Notice, that this is a *rigid string* configuration staying around the cigar without *any oscillations*. In $D = 2$, $e^{in\hat{\sigma}}$ does not describes string fluctuations as in the $D > 2$ cases, thus the name **modes** ("winding modes" or "momentum modes") referring to strings in $D = 2$ are misleading.

Also notice that in the $D = 2$ euclidean regime, the string world sheet *covers* completely the target space ; world sheet and physical space became the same ; (for $n = 1$, both are identified). This is characteristic of strings in $D = 2$ compact spaces. Recall that in $D = 2$ de Sitter space [14] , the general solution of the string equations of motion describes a string winding n times around de Sitter universe (the circle S^1) and evolving with it. In that case, the winding string solution naturally exists in the *Lorentzian* signature regime in both target and world sheet space-times ; (it can be also extended to the Euclidean regime). The winding string de Sitter solution is given by [14]

$$dS^2_{deSitter} = \frac{1}{H^2} \left[-du^2 + \cosh^2 u \; dv^2 \right] \tag{5.7}$$

with $\quad v = n\sigma, \qquad u = \log(\tan \frac{n\tau}{2}) \;, \qquad 0 < \sigma \leq 2\pi \;, \qquad 0 < \tau \leq \frac{2\pi}{n}.$

That is,

$$dS^2_{deSitter} = \frac{n^2}{H^2 \sin^2 n\tau}(d\sigma^2 \; - \; d\tau^2) \tag{5.8}$$

[here (u, v) are the coordinates on the de Sitter hyperboloid].

In the black hole background, winding string solutions are only possible in the Euclidean signature regime.

It must be noticed that the transformation

$$r = s + \frac{i\pi}{2} \quad , \quad \Theta = \Theta \tag{5.9}$$

maps the manifold eq. (5.1) into the dual or "trumpet" manifold

$$dS^2 = k\left[ds^2 + \coth^2 s \; d\Theta^2\right] \tag{5.10}$$

This is a real manifold with curvature singularity at $s = 0$, $(uv = 1)$. The Kruskal coordinates are complex conjugate one to another, i.e.

$$u = \cosh s \; e^{i(\Theta + \frac{\pi}{2})} \quad , \quad v = \cosh s \; e^{-i(\Theta + \frac{\pi}{2})} \tag{5.11}$$

The string solution eqs.(5.3) - (5.4) transforms under eq.(5.9) into the different real solution

$$\cosh s = e^{n\tau} \quad , \quad \Theta = n\hat{\sigma} \tag{5.12}$$

(the dilaton here is $\Phi = \log[1 - e^{2n\tau}]$). That is,

$$u = e^{n(\tau + i\hat{\sigma}) + i\pi/2} \quad , \quad v = e^{n(\tau - i\hat{\sigma}) - i\pi/2} \tag{5.13}$$

For this solution, the metric is

$$dS^2 = k\; \frac{n^2}{1 - e^{-2n\tau}}\; (d\tau^2 + d\hat{\sigma}^2) \tag{5.14}$$

The proper string length (between two points $\hat{\sigma}$ and $\hat{\sigma} + d\hat{\sigma}$ at fixed τ) is given by

$$\Delta S_{trumpet} = \sqrt{k}\; \frac{n}{\sqrt{1 - e^{2n\tau}}}\; d\hat{\sigma} \tag{5.15}$$

Near the curvature singularity $s \to 0$ (*i.e.* $\tau \to 0$), ΔS blows up, the string stretches indefinitely, corresponding to the location of the string in the open extremity of the trumpet.

5.1 Cosmological solutions

The cosmological version of the black hole solution eq.(4.1) is just obtained by setting

$$r = iT \quad , \quad t = x \tag{5.16}$$

in the original metric. The cosmological background is described by

$$\begin{aligned} dS^2 &= k\left[-dT^2 + \tan^2 T\, dx^2\right] , \\ \Phi &= -\log \sin T \ , \quad 0 < T < \pi/2, \ -\infty < x < +\infty. \end{aligned} \tag{5.17}$$

which is singular in $uv = 1$, i. e. $T = \pi/2$. Here

$$u = i \sin T \ e^{x} \quad , \quad v = -i \sin T \ e^{-x} \tag{5.18}$$

It is conveniently written in the variable $T^\star$

$$\sin T = e^{T^\star} \quad , \quad -\infty < T^\star < 0$$

i.e.

$$dS^2 = \frac{1}{1 - e^{-2T^\star}} \ \left[(-dT^\star)^2 + dx^2\right] \tag{5.19}$$

The analysis of the string solutions done in section IV for the black hole, holds here with the corresponding mapping eq.(5.16). There are no solutions describing a string winding n times in the Lorentzian regime. In the Euclidean regime

$$\begin{aligned} x &= i\Theta \ , \quad 0 \le \Theta < 2\pi, \\ -dS^2 &= \frac{1}{1 - e^{-2T^\star}} \ \left[(dT^\star)^2 + d\Theta^2\right] . \end{aligned} \tag{5.20}$$

We have the solution

$$T^\star = n\tau \ , \quad \Theta = n\hat{\sigma} \ , \quad -\infty < \tau \le 0 \ , \quad 0 \le \hat{\sigma} < \frac{2\pi}{n}$$

That is

$$\begin{aligned} u &= i\ e^{T^\star + i\Theta} &= i\ e^{n(\tau + i\hat{\sigma})}, \\ v &= -\,i\ e^{T^\star - i\Theta} &= -i\ e^{n(\tau - i\hat{\sigma})}, \end{aligned} \tag{5.21}$$

describes a string winding n times in the euclidean cosmology. This solution yields

$$dS^2 = k\ \frac{n^2}{1 - e^{-2n\tau}} \ (d\tau^2 + d\hat{\sigma}^2) . \tag{5.22}$$

The string length blows up at $\tau \to 0$.

The cosmological version of the trumpet manifold is obtained by setting

$$\begin{aligned} T &= \; - iU + \pi/2 \;\; , \;\; \text{i.e.} \\ dS^2 &= k\,[ds^2 + \coth^2 U \; d\Theta^2] = k\,\frac{1}{1 - e^{-2U^\star}}\;[(dU^\star)^2 + d\Theta^2] \end{aligned} \tag{5.23}$$

with

$$u = \cosh U \; e^{i\theta} = e^{U^\star + i\theta} \quad , \quad v = \cosh U \; e^{-i\theta} = e^{U^\star - i\theta}$$

The winding string solution here is given by

$$U^\star = n\tau \quad , \quad \Theta = n\hat{\sigma} \quad , \quad -\infty < \tau < 0 \quad , \quad 0 < \hat{\sigma} < \frac{2\pi}{n} \tag{5.24}$$

The solutions found in sections (IV) and (V) exhaust **all** the real string solutions in the lorentzian and euclidean regimes. The σ-dependent solutions exist only in the euclidean regime. These solutions are of the type (i) and (ii) [eqs.(3.15) and (3.16)]. There are no real σ-dependent euclidean solutions of the type (iii) and (iv) [eqs.(3.17) and (3.18)].

There are two types of euclidean string solutions : cigar solutions and trumpet solutions. These correspond to a string winding n times in the cigar and in the trumpet manifolds respectively. In the cosmological version of the black hole manifold (in the euclidean regime) the string solutions are the same as the trumpet solutions.

6. String quantization in the two dimensional black-hole

In the Lorentzian two dimensional black hole background eqs.(4.1) - (4.2) the action eq.(3.4) takes the form

$$\mathcal{S} = -\frac{1}{2\pi}\int d\sigma d\tau \left\{ \frac{k}{2}\,[\dot{r}^2 - r'^2 - \tanh^2 r\;(\dot{t}^2 - t'^2)] - i\,\dot{r}\tanh r \right\} \tag{6.1}$$

where we also used eq.(3.8) for the world sheet metric. Notice that the dilaton field leaves only a surface term analogous to a background charge in conformal field theory [22].

The Hamiltonian associated to the action (6.1) has the form

$$\mathcal{H} = \frac{1}{2k}\,[(p_r - i\tanh r)^2 - p_t^2\coth^2 r] + \frac{k}{2}\,(r'^2 - \tanh^2 r\;t'^2) \tag{6.2}$$

As previously discussed, in absence of periodic directions in the space-time, the parameter σ can be classically gauged away. Since this is the case for black holes with Lorentzian signature, we can drop the second term in eq.(6.2) and take

$$\mathcal{H} = \frac{1}{2k}\left[(p_r - i\tanh r)^2 - p_t^2\coth^2 r\right] \tag{6.3}$$

Let us now consider the quantum theory. We will follow the spirit of ref.[7] where one starts from the center of mass motion and then, takes into account the σ-dependence as fluctuations around. Here, in $D = 2$, there are no physical fluctuations, so we can expect that the quantum version of the center of mass hamiltonian eq.(6.3) should give the full description of the model.

In order to quantize this system, we use the canonical prescription

$$p_r \to \hat{p}_r = -i\frac{\partial}{\partial r} \quad , \quad p_t \to \hat{p}_t = -i\frac{\partial}{\partial t} \tag{6.4}$$

Now, eq.(6.3) suffers from ordering ambiguities at the quantum level. We choose the following ordering prescription. The classical piece $p_r^2 - p_t^2\coth^2 r$ in eq. (6.3) is replaced by the D'Alambertian in this space-time :

$$p_r^2 - p_t^2\coth^2 r \to \partial^2 = \frac{1}{\tanh r}\frac{\partial}{\partial r}(\tanh r\frac{\partial}{\partial r}) - \coth^2 r\,\frac{\partial^2}{\partial t^2} \tag{6.5}$$

The operator ∂^2 is self adjoint with respect to the integration measure $\sqrt{G} = k\tanh r$.

The double product $-2ip_r\tanh r$ in eq.(6.3) is ordered symmetrically

$$-2ip_r\tanh r \to -i(\tanh r\;\hat{p}_r + \hat{p}_r\tanh r) \tag{6.6}$$

The last term $-\tanh^2 r$ does not present ordering problem. We find in this way

$$\mathcal{H} \to \hat{\mathcal{H}} = -\frac{1}{2k}\left(\frac{\partial^2}{\partial r^2} + (\coth r + \tanh r)\frac{\partial}{\partial r} - \coth^2 r\,\frac{\partial^2}{\partial t^2}\right) \tag{6.7}$$

Let us write

$$\mathcal{H} \to \hat{\mathcal{H}} = \frac{1}{2k}h \quad , \quad \partial_t = -iE$$

then,

$$h = -\left[\frac{d^2}{dr^2} + 2\coth(2r)\frac{d}{dr} + E^2\coth^2 r\right] \; , \tag{6.8}$$

and the eigenvalue problem

$$h\Psi = \lambda\Psi \quad , \tag{6.9}$$

yields

$$\Psi'' + 2\coth(2r)\Psi' + (E^2\coth^2 r + \lambda)\Psi = 0 \tag{6.10}$$

The constant λ describes the zero point quantum fluctuations and it will be determined below ; (the prime ' stands for $\frac{d}{dr}$). Eq.(6.10) describes the interaction of the tachyon with the black hole geometry. It is convenient to set here

$$\Psi = \frac{\chi}{\sqrt{\sinh 2r}} \tag{6.11}$$

then, eq.(6.10) reads

$$\chi'' + \left[\frac{E^2 + \frac{1}{4}}{\sinh^2 r} - \frac{1}{4\cosh^2 r} + E^2 + \lambda - 1\right]\chi = 0 \tag{6.12}$$

which is a Schrödinger type equation with the energy-dependent effective potential

$$\begin{aligned} V_{eff} &= -\frac{E^2 + \frac{1}{4}}{\sinh^2 r} + \frac{1}{4\cosh^2 r} - E^2 - \lambda + 1 + E \\ V_{eff} &\overset{r\to 0}{=} -\frac{E^2 + 1/4}{r^2} + O(1) \,, \end{aligned} \tag{6.13}$$

For $r \to 0$, this is a strongly attractive singular potential $-\frac{\gamma}{r^2}$. Since $\gamma > \frac{1}{4}$ there is absorption (fall into the event horizon $r = 0$) for all energy E, even $E = 0$. This is exactly like the horizon behaviour of the effective potential describing the interaction of the black hole with a free scalar field in $D = 4$, where $\gamma = E^2 + 1/4$ too.

The Hamiltonian eq.(6.8) , is the same Hamiltonian L_0 as in ref.[4]. In order to compare with ref.[4], (hereafter referred as DVV), please notice that

$$r_{DVV} = 2r \quad , \quad E = 2\omega_{DVV} \tag{6.14}$$

Then, eq.(4.11) in DVV reads

$$L_0^{DVV} = -\frac{1}{k-2}\left[\frac{\partial^2}{\partial r^2} + (\coth r + \tanh r)\frac{\partial}{\partial r} - \left(\coth^2 r - \frac{2}{k}\right)\frac{\partial^2}{\partial t^2}\right] \tag{6.15}$$

and the condition $(L_0 - 1)\chi$ yields

$$\left[\frac{d^2}{dr^2} + \frac{E^2 + \frac{1}{4}}{\sinh^2 r} - \frac{1}{4\cosh^2 r} + E^2 + 1 + 4(k-2)\right] \chi = 0 \tag{6.16}$$

$\hat{\mathcal{H}}$ in eq. (6.7) and the L_0^{DVV} only differ by additive and multiplicative constants. As operators they are identically the same.

Let us now determine the constant λ. It can be determined from the behaviour at $r \to \infty$ of the solution. For $r \to \infty$, from eq.(6.10) we have

$$\Psi'' + 2\Psi' + (E^2 + \lambda)\Psi = 0 \tag{6.17}$$

And then, (including the time dependence of the total solution) :

$$\Psi(r \to \infty, t) = e^{i(pr - Et)} \tag{6.18}$$

with the momentum $p = \sqrt{E^2 + \lambda - 1}$. The parameter λ becomes determined by requiring the tachyon to be massless. In order to describe a relativistic and massless particle, (the string ground state particle is massless in $D = 2$ due to the absence of transverse dimensions), it must be $E = cp$, that is $\lambda - 1 = (1 - c^2)p^2$. For comparison with DVV, notice that for $r \to \infty$, DVV have

$$\Psi'' + 2\Psi' + E^2(1 - \frac{2}{k})\Psi = 0 \tag{6.19}$$

which corresponds to have $\lambda = 1 - 2E^2/k$. That is, eqs.(6.17) and (6.19) yield the relation

$$c = \sqrt{\frac{k}{k-2}} \tag{6.20}$$

We will comment about the interpretation of λ and this relation below.

In order to describe the solutions of eq.(6.10), it is useful to define the variable

$$z \equiv \cosh^2 r$$

and to express Ψ , as

$$\Psi_E(z) = (z-1)^{iE/2}\ \psi_E(z) \tag{6.21}$$

Then, eq.(6.10) reads

$$z(1-z)\psi_E(z)'' + [1 - z(iE+2)]\psi_E(z)' - \left(\frac{iE}{2} + \frac{\lambda}{4}\right)\psi_E(z) = 0 \tag{6.22}$$

whose solutions are the standard hypergeometric functions.

The general solution is given by

$$\Psi(r,t) = \left[\, A\, U_{E,c}(r) + B\, U_{E,c}(r)^* \,\right] e^{-iEt} \tag{6.23}$$

where A and B are arbitrary constants and

$$U_{E,c}(r) = (\sinh r)^{-iE} F\left(\frac{1}{2} - \frac{iE}{2}(1+\frac{1}{c}), \frac{1}{2} - \frac{iE}{2}(1-\frac{1}{c}); 1-iE\,; -\sinh^2 r\right) \tag{6.24}$$

These solutions describe the scattering of the (massless) tachyon by the black hole geometry. The solutions $U_{E,c}$ and $U^*_{E,c}$ exhibit the typical behaviour of the wave functions near the black hole horizon, namely

$$U_{E,c}(r)\, e^{-iEt} \overset{r\to 0}{=} e^{-iE(t+\log r)} \quad , \quad U_{E,c}(r)^*\, e^{-iEt} \overset{r\to 0}{=} e^{-iE(t-\log r)} \tag{6.25}$$

$U_{E,c}(r,t)$ corresponds to have purely incoming particles at the (future) horizon. $U^*_{E,c}$ describes purely outcoming particles from the (past) horizon. The solution $U_{E,c}$ describes the physical process in which the tachyon field coming in from spatial infinity is partially reflected and partially absorbed by the black hole. The asymptotic behaviour of $U_{E,c}(r,t)$ for $r \to \infty$ is given by

$$U_{E,c}(r)\; e^{-iEt} \overset{r\to\infty}{=} e^{-iEt}\; \left[e^{-iEr/c} + S(E,c)e^{iEr/c}\right]. \tag{6.26}$$

From eq.(6.24) we obtain for the scattering matrix $S(E,c)$:

$$S(E,c) = 2^{-2iE/c}\frac{\Gamma(iE/c)}{\Gamma(-iE/c)}\left[\frac{\Gamma\left(\frac{1}{2} - \frac{iE}{2}(1+\frac{1}{c})\right)}{\Gamma\left(\frac{1}{2} - \frac{iE}{2}(1-\frac{1}{c})\right)}\right]^2 \tag{6.27}$$

or, in terms of p eqs.(2.10) . Notice that the S-matrix factorizes into two parts, one which is a pure phase, equal to the coulombian S-matrix, and another one which describes both absorption and reflection, namely

$$S(p,c) = S_{coul}(p)\; \tilde{S}(p,c)$$

where

$$\begin{aligned} S_{coul}(p) &= 2^{-2ip}\frac{\Gamma(ip)}{\Gamma(-ip)} = 2^{-2ip}e^{2i\arg\Gamma(ip)} \\ \tilde{S}(p,c) &= \left[\frac{\Gamma\left(\frac{1}{2} - \frac{ip}{2}(c+1)\right)}{\Gamma\left(\frac{1}{2} - \frac{ip}{2}(c-1)\right)}\right]^2 \end{aligned} \tag{6.28}$$

S_{coul} takes into account the large r interaction while $\tilde{S}$ describes the genuine black hole features. Notice that

$$|S_{coul}(p)| = 1$$

$$|\tilde{S}(p,c)| = \left[\frac{\cosh\frac{\pi p}{2}(c-1)}{\cosh\frac{\pi p}{2}(c+1)}\right]^2 \; < \; 1\,, \tag{6.29}$$

which describes the absorption by the black hole. Also notice that $S(E,c)$ exhibits imaginary poles. An infinite sequence of purely imaginary poles occur for the values $ip = n$ $(n = 0, 1, ...)$ which are like the Coulombian-type poles (describing the relativistic hydrogen bound-state spectrum). In addition, the part $\tilde{S}(E,c)$ exhibits an infinity sequence of purely imaginary poles at $ip(c+1) = 2n+1, n = 0, 1, 2....$

It must be stressed that the scattering matrix $S(E,c)$ depends on two parameters E and c . That is, for each energy we have a monoparametric family of well defined solutions depending on the parameter c. For each value of c, we have a different S-matrix. It must be noticed that c is a purely quantum mechanical parameter which is not fixed at this level by any special requirement, c is a physical parameter which accounts for a renormalization of the speed of light. At the classical level, c can take any value, (we have taken $c = 1$), the physics does not depend on it. But, quantum mechanically, c is not more equal to 1 in this problem.

The parameter c is related to the parameter k of the conformal field theory construction, $c = \sqrt{\frac{k}{k-2}}$ [see eq.(6.20)]. In particular, for $k = 9/4$, *i.e.* $c = 3$ we reproduce the results of ref.[4] . In this respect, notice that in ref.[4] , the effect of the renormalization of the speed of light is also present, (although it has not been noticed there). Classically, before quantization, DVV have $c = 1$, and after quantization (since they choose $k = 9/4$), they have $c = 3$. This can be seen from the asymptotic behaviour of the metric for $r \to \infty$.

Classically,

$$dS^2 = k\left[\, dr^2 + \tanh^2 r \, d\Theta^2 \,\right] \overset{r\to\infty}{=} dr^2 + d\Theta^2 \;\; \text{i.e. ,} \;\; c = 1.$$

After quantization of the euclidean black hole CFT coset $SL(2,R)/U(1)$ model :

$$dS^2_{DVV} = \frac{1}{2}(k-2)\left[\, dr^2_{DVV} \; + \; \beta^2(r_{DVV})\, d\Theta^2 \,\right] ,$$

$$\text{where} \quad \beta(r) = 2\left(\coth^2\frac{r}{2} - \frac{2}{k}\right)^{-1/2} \quad \text{i.e.} \;, \;\; \beta(\infty) = \frac{2}{\sqrt{1-2/k}} \,, \tag{6.30}$$

$$dS^2_{DVV} \overset{r\to\infty}{=} \frac{1}{2}(k-2)\left[\, dr^2_{DVV} \; + \; \frac{4k}{k-2}\, d\Theta^2 \,\right] ,$$

Recall that $r_{DVV} = 2r$, i. e.

$$dS^2_{DVV} \stackrel{r\to\infty}{=} 2(k-2)\left[\, dr^2 \,+\, \frac{k}{k-2}\, d\Theta^2 \,\right] = \frac{4}{c^2-1}\left[\, dr^2 \,+\, c^2\, d\Theta^2 \,\right] \tag{6.31}$$

We see that in ref. [4] , it also appears that $c^2 = \frac{k}{k-2}$ in particular there, $k_{DVV} = 9/4$. Notice that nothing special happens at the conformal invariant point $k = 9/4$. The physical relevance of the $k = 9/4$ point in the two-dimensional black hole scattering matrix is not clear. The S-matrix of ref.[4] is a particular case of our results for $c = 3$. The S-matrix, eq.(5.22) of ref.[4] is given by

$$S_{DVV}(p) = 2^{-2ip}\frac{\Gamma(ip)}{\Gamma(-ip)}\left[\frac{\Gamma\left(\frac{1}{2}-2ip\right)}{\Gamma\left(\frac{1}{2}-ip\right)}\right]^2 \tag{6.32}$$

[The factor 2^{-2ip} is missing in ref. [4]]. From eq.(6.28) , we see that

$$S(p, c=3) = S_{DVV}(p) \tag{6.33}$$

In conclusion, in this problem, the quantization of the *center of mass of the string*, that is the quantization of the classical solution, together with the requirement that $m^2 = 0$, yields *all* the physics of the problem. [In this sense, this is like the light cone gauge for strings in flat spacetime for D=26]. We have obtained the complete quantum result without introducing any correction $1/k$ to the space-time metric, neither to the dilaton field.

The 2-dimensional conformal field theory construction can be avoided for such a simple problem like the two-dimensional stringy black hole. They introduce a lot of technicality in a problem in which all the physics is described by the standard quantization of a two dimensional massless scalar particle. Indeed, CFT techniques are necessary to compute the value of the central charge (see for example ref.[24]). Actually, the central charge value can be changed at will by adding extra conformal fields ('compact dimensions'). In addition, the CFT tools will be probably needed for the problem of interacting (higher genus) strings in the two dimensional black hole background, a problem which, unfortunately, has not been treated until now.

Finally, let us comment about the only non trivial physical feature, perhaps, appearing in this two dimensional stringy black hole problem, namely the renormalization of the speed of light. Normally, the speed of light is never affected by quantum corrections. That is, once one chooses units such that $c = 1$ classically, this value remains true in the quantum theory (relativistic quantum mechanics, or quantum field theory or string theory, etc...). In the present case, $c = 1$, for the classical propagation in the background eq.(2.1)), while $c = \sqrt{\frac{k}{k-2}}$ for the quantum propagation of the massless particle described by the string. Only for $k \to \infty$, this effect dissapears (but the conformal anomaly is present). In that case:

$$\begin{aligned} S(p, c = 1) &= S_{coul}(p)\, \frac{1}{\pi} \left[\Gamma\left(\frac{1}{2} - ip\right)\right]^2 \\ |S(p, c = 1)| &= \frac{1}{\cosh^2 \pi p} \end{aligned} \quad . \tag{6.34}$$

6.1 Quantum Partition Function

Let us now consider the quantum statistical mechanics for this system by computing its partition function $Z = Tr\left[e^{-\beta \hat{\mathcal{H}}_E}\right]$, where $\hat{\mathcal{H}}_E$ is the euclidean version of the Lagrangian in eq.(6.1) . In the euclidean regime ($t = i\Theta$) the variable Θ is periodic with period 2π and in an appropriate conformal gauge, generic configurations take the form

$$\Theta(\sigma, \tau) = n\hat{\sigma} + \hat{\Theta}(\tau) \tag{6.35}$$

The whole σ-dependence can be gauged away from r . Thus, using eq.(6.35) , we find an extra term in eq.(6.1) :

$$\frac{1}{2} k n^2 \tanh^2 r$$

This term plus the Wick rotated version of $\hat{\mathcal{H}}$ [eq.(6.7)] yields the euclidean hamiltonian

$$\hat{\mathcal{H}}_E = \frac{1}{2k}\, h_E$$

with

$$h_E = -\frac{\partial^2}{\partial r^2} - 2\coth(2r)\, \frac{\partial}{\partial r} - \coth^2 r\, \frac{\partial^2}{\partial \Theta^2} + (kn)^2 \tanh^2 r \tag{6.36}$$

The eigenfunctions of h_E take the form

$$W(r, \Theta) = e^{im\Theta}\; w_{m,n}(r)$$

where $m \epsilon \mathcal{Z}$ in order to have 2π periodicity and $w_{m,n}(r)$ is a solution of the eigenvalue equation

$$\left[-\frac{d^2}{dr^2} - 2\coth(2r)\frac{d}{dr} + m^2\coth^2 r + (kn)^2\tanh^2 r\right] w_{m,n}(r) \; = \; \Lambda \, w_{m,n}(r) \qquad (6.37)$$

Regular solutions behave as

$$w_{m,n}(r) \overset{r\to 0}{=} r^{|m|} \quad , \quad w_{m,n}(r) \overset{r\to\infty}{=} e^{-r}\left[e^{ipr} + S_{m,n}(p)\right] \qquad (6.38)$$

where $p \in \mathcal{R}$ is the euclidean momentum. The eigenvalues Λ of h_E on such eigenfunctions are

$$\Lambda_{p,m,n} = \sqrt{p^2 + m^2 + (kn)^2 + 1} \qquad n, m \in \mathcal{Z} \ , \ p \in \mathcal{R}.$$

Then, the partition function results:

$$Z(\beta) = Tr\left[e^{-\beta\hat{\mathcal{H}}_E}\right] = \sum_{m,n=-\infty}^{+\infty} \int_{-\infty}^{+\infty} \frac{dp}{2\pi} \; e^{-\frac{\beta}{2k}\Lambda_{p,m,n}}$$

The integral over p can be computed with the result

$$Z(\beta) = \frac{1}{\pi} \sum_{m,n=-\infty}^{+\infty} \sqrt{m^2 + (kn)^2 + 1} \; K_1\left(\frac{\beta}{2k}\sqrt{m^2 + (kn)^2 + 1}\right) \qquad (6.39)$$

There is a manifest duality invariance $m \leftrightarrow kn$.

For large β (low temperatures), $Z(\beta)$ decreases exponentially as

$$Z(\beta) \overset{\beta\to\infty}{=} \sqrt{\frac{k}{\pi\beta}} \; e^{-\frac{\beta}{2k}}$$

due to the presence of a gap in the eigenvalues :$\Lambda_{p,m,n} \geq 1$. For high temperatures, $Z(\beta)$ grows as β^{-3} :

$$Z(\beta) \overset{\beta\to 0}{=} \frac{16k^2}{\beta^3}$$

Notice that $Z(\beta)$ is analytic for all $\infty \geq \beta > 0$ and hence no phase transition shows up here for finite temperatures. The absence of a Hagedorn critical temperature is due to the lack of transverse string modes in two dimensions, as it is well known.

7. Hawking radiation

The computation of the Hawking radiation follows directly from the standard treatment of the by now well known QFT in curved space time and does not present any particular problem here. For the sake of completeness we point out the essential points to be considered and which allow to avoid unnecessary complications. The formulation of quantum (point particle) field theory, as well as quantum string theory in non trivial spacetime exhibits new fundamental features with respect to the usual formulation in flat space time: (i) the possibility for a given field or string theory to have different alternative well defined ground states, and thus different Fock spaces or "sectors" of the theory. (ii) the presence of intrinsic statistical features (temperature, entropy) arising from the non trivial structure (geometry, topology) of the space time and not from a superimposed statistical description of the quantum matter fields or strings.

An important step in field as well as string quantization is the definition of positive frequency states and its associated ground state. Different possible choices of the physical time lead to different alternative positive frequency basis.

The modes $U_{E,c}(r,t)$ (see eq.(6.24)) define positive frequencies with respect to the Schwarzschild time and are the analogous of the Rindler modes for an uniformely accelerated observer $r =$ const. Creation and annihilation operators a_E, a^*_E are associated to these modes which define the ground state $|0_S>$ or "Schwarzschild" ground state.

The modes $U_{E,c}$ are however, a linear combination of positive and negative frequency modes with respect to the Kruskal time T. For a locally inertial observer near the horizon or Kruskal observer, the positive frequency basis is

$$\varphi_k \overset{r\to 0}{=} \{\frac{1}{2\sqrt{\pi k}}\, e^{-iku} \,, \frac{1}{2\sqrt{\pi k}}\, e^{-ikv}\} \,, \quad k > 0 \,,$$

with respect to which the global or Kruskal ground state is defined $\alpha_k \, |0_K >= 0$. The Bogoliubov coefficients relating such two basis, namely,

$$\varphi_k = \mathcal{A}_{kE} \; U_{E,c}(r,t) + \mathcal{B}_{kE} \; U^*_{E,c}(r,t) \quad [\text{ or } \alpha_k = \mathcal{A}_{kE} \;\; a_E + \mathcal{B}_{kE} \;\; a^*_E \;] \,,$$

are defined by the scalar products

$$\mathcal{A}_{kE} \;\; = < U_E, \varphi^*_k > \;\; , \quad \mathcal{B}_{kE} \;\; = < U_E, \varphi_k > \,,$$

and explicitly given by

$$\mathcal{A}_{kE} = e^{\pi E}\, \mathcal{B}_{kE} \quad , \quad \mathcal{B}_{kE} \overset{k\to\infty}{=} -\frac{\sqrt{E}\, k^{iE}}{2\pi\sqrt{k}}\, \Gamma(-iE)\, e^{-\pi E/2}$$

The behaviour of the basis $\{U_E\}$ and $\{\varphi_k\}$ near the horizon (and thus, the $k \to \infty$ behaviour of the Bogoliubov coefficients) is enough to determine the relevant quantities of interest. The expectation value of the Schwarzschild number operator in the Kruskal vacuum is determined by the function

$$N(E, E') \equiv < 0_K|\alpha^*_E\, \alpha_{E'}|0_K > = \int_{-\infty}^{\infty} dk\, \mathcal{B}^*_{kE}\, \mathcal{B}_{kE'} \tag{7.1}$$

The $E \to E'$ limit is determined by the $k \to \infty$ behavior of eq.(7.1), namely

$$N(E, E') = N_V(E)\, \delta(E - E') \quad , \quad N_V(E) = \frac{1}{e^{2\pi E} - 1}$$

$N_V(E)$ is the number of created modes per unit frequency and per unit volume of the Schwarzschild manifold, at temperature $T = 1/(2\pi)$.

This can be seen directly from the mapping eq.(4.7) $u = f(u')$, i.e.,

$$u = e^{r^\star + t} \quad , \quad v = -e^{r^\star - t} \; .$$

which defines the maximal extension of the manifold [23]. This mapping is holomorphic in the complex variable $r^\star - i\Theta$ and has a critical point defining the event horizon, i.e. $f'(-\infty) = 0$. The inverse of the periodicity in the imaginary time Θ is the temperature characterizing the vacuum spectrum $N_V(\lambda)$.

A complete classification of the vacuum spectra and thermal properties in a large class of curved and accelerated manifolds can be done directly in terms of the properties of the holomorphic mappings $u = f(u')$ defining the maximal analytic extension of the manifold, of which the exponential mapping eq.(4.7) is just a particular case, (u being global or Kruskal-like coordinates, u' being accelerated or Schwarzschild-like ones) [23].

ACKNOWLEDGMENTS

The authors acknowledge the french-spanish scientific cooperation for partial support sponsored by Ministerio de Educación y Ciencia (DGICYT) (Spain), Ministère des Affaires Etrangères and Ministère de L'Education Nationale (France).

REFERENCES

1. I. Bars and D. Nemeschansky, *Nucl. Phys.* **B348**, 89 (1991).
   G. Mandal and A. M. Sengupta and S. R. Wadia, *Mod. Phys. Lett.* **A6**, 1685(1991).
2. K. Bardacki, A. Forge and E. Rabinovici, *Nucl. Phys.* **B344**,344 (1990).
   S. Elitzur, A. Forge and E. Rabinovici, *Nucl. Phys.* **B359**, 581 (1991).
3. E. Witten, *Phys. Rev.* **D 44** , 314 (1991).
4. R. Dijkgraaf, H. Verlinde and E. Verlinde, *Nucl. Phys.* **B371**, 269 (1992).
5. I. Bars and K. Sfetsos, *Mod. Phys. Lett.* **A7**, 1091 (1992).
   I. Bars in "String Quantum Gravity and Physics at the Planck Energy Scale"
   Erice School, N. Sánchez Editor, World Sci. Publ. 1993.
   M. Crescimanno, *Mod. Phys. Lett.* **A7**, 489 (1992).
   E. S. Fradkin and V. Ya. Linetsky, *Phys. Lett.* **277B**, 73 (1992).
   J H Horne and G T Horowitz, *Nucl. Phys.* **B368** , 444 (1992).
   C. J. Ahn et al., *Phys. Rev.* **D 47**, 1699 (1993).
6. A. A. Tseytlin, Imperial College preprint/TP/92-93/10
7. H. J. de Vega and N. Sánchez, *Phys. Lett.* **197B**, 320 (1987).
8. H. J. de Vega and N. Sánchez, *Nucl. Phys.* **B309**, 552 and 577(1988).
9. H. J. de Vega, in "String Quantum Gravity and Physics at the Planck Energy Scale"
   Erice School, N. Sánchez Editor, World Sci. Publ. 1993,
   N. Sánchez, ibidem.
10. H. J. de Vega and N. Sánchez, *Nucl. Phys.* **B317**, 706 and 731 (1989).
11. H. J. de Vega and N. Sánchez, *Nucl. Phys.* **B299**, 818(1988).
12. M. Costa and H. J. de Vega, *Ann. Phys. (N.Y.)* **211**, 223 and 235 (1991).
13. H. J. de Vega and N. Sánchez, *Phys. Rev.* **D 45** , 2783 (1992).
    H. J. de Vega, M. Ramón Medrano and N. Sánchez, *Class. and Quantum Gravity* **10**, 2007 (1993).
14. H. J. de Vega and N. Sánchez, *Phys. Rev.* **D 47**, 3394 (1993),
    H. J. de Vega, A. V. Mikhailov and N. Sánchez, *Theor. Math. Phys.* **94**, 232 (1993)
15. N. Sánchez and G. Veneziano, *Nucl. Phys.* **B333**, 253 (1990),
    M. Gasperini, N.Sánchez and G. Veneziano,
    Int. J. Mod. Phys. **A6**, 3853 (1991) and *Nucl. Phys.* **B364**, 365 (1991).

16. V. P. Frolov and N. Sánchez, *Nucl. Phys.* **B349**, 815 (1991).

17. H. J. de Vega and N. Sánchez, *Phys. Rev.* **D 42** , 3969 (1990) and
H. J. de Vega, M. Ramón Medrano and N. Sánchez, *Nucl. Phys.* **B374**, 405 (1992).

18. H. J. de Vega and N. Sánchez, *Phys. Lett.* **244B**, 215(1990),
Phys. Rev. Lett. **65** (C), 1517(1990) and *Int. J. Mod. Phys.* **A7**, 3043 (1992).

19. H. J. de Vega, M. Ramón Medrano and N. Sánchez, *Nucl. Phys.* **B351**, 277 (1991),
Nucl. Phys. **B374**, 425 (1992) and *Phys. Lett.* **285B**,206(1992) .

20. C.O.Loustó and N. Sánchez, *Nucl. Phys.* **B355**, 231 (1991), *Nucl. Phys.* **B383**, 377 (1992) and *Phys. Rev.* **D 46**, 4520 (1992).

21. C.O.Loustó and N. Sánchez, to appear in *Phys. Rev.* **D**

22. S L Lukyanov and V A Fateev,
Soviet Scientific Reviews, sect A, vol. 15, part 2, p. 117 .

23. N. Sánchez, *Phys. Rev.* **D 24**, 2100 (1981),
N. Sánchez and B. F. Whiting, , *Phys. Rev.* **D 34**, 1056 (1986).

24. L J Dixon, M E Peskin and J Lykken, *Nucl. Phys.* **B325**, 329 (1989).

RELIC GRAVITONS FROM THE PRE-BIG BANG: WHAT WE KNOW AND WHAT WE DO NOT KNOW*

MAURIZIO GASPERINI

Theory Division, CERN, CH-1211 Geneva 23, Switzerland
and *Dipartimento di Fisica Teorica, Università di Torino, Turin, Italy*

I discuss the status of present knowledge about a possible background of relic gravitons left by an early, pre-big bang cosmological epoch, whose existence in the past of our Universe is suggested by the duality symmetries of string theory.

Table of Contents

*Based on talks presented at the Meeting on *"Detection of high-frequency gravitational waves"* (CERN, January 1996), at the INTAS Meeting on *"Fundamental problems in classical, quantum and string gravity"* (Turin, May 1996), at the *"XLth Conference of the Italian Astronomical Society"* (Osservatorio di Roma, Rome, May 1996), and at the Network Meeting on *"String Gravity"* (Observatoire de Paris, June 1996).

1. Introduction

The cosmological solutions of the low energy string effective action, and their symmetry properties, have recently motivated the study of a very early, "pre-big bang" cosmological phase[1,2,3], whose kinematical properties are the "dual" counterpart of the present, standard cosmological phase. The purpose of this paper is to give a short introduction to the pre-big bang scenario, to explain why the amplification of perturbations is more efficient in such a context than in the standard inflationary scenario, and to discuss the general properties of a possible cosmic graviton background associated to the transition from the pre- to the post-big bang epoch.

The main ideas and results reported here are the fruit of a research programme in string cosmology that started from earlier work done with Norma Sanchez and Gabriele Veneziano on the motion of strings in accelerated backgrounds[4,5], and that was subsequently developed in collaboration with Gabriele Veneziano and with a number of other researchers: Ramy Brustein, Massimo Giovannini, Jnan Maharana, Krzysz Meissner and Slava Mukhanov (listed here in alphabetical order). The interested reader can find an updated list of references and papers about the pre-big bang scenario on the WWW "home page" devoted to string cosmology, available at the address:

`http://www.to.infn.it/teorici/gasperini/`

2. The Pre-Big Bang Scenario

I will call, for short, pre-big bang scenario, a cosmological scenario that is based on the equations obtained from the string effective action, and which includes an initial phase of accelerated evolution and growing curvature[1,2,3]. The most revolutionary aspect of the class of models describing this scenario is probably the fact that the initial state of the Universe, instead of being hot and dense as in the standard cosmological context, is the string perturbative vacuum, namely a state with flat metric, vanishing gauge coupling ($g = e^{\phi/2} = 0$, $\phi = -\infty$), and no matter present (with the possible exception of an incoherent, highly diluted gas of non-interacting strings[3]). As a consequence, in this class of models the initial evolution of the Universe can be consistently described in terms of the lowest order string effective action,

$$S = -\int d^{d+1}x\sqrt{|g|}\left[e^{-\phi}\left(R+(\nabla\phi)^2+\alpha'\text{corrections}\right)+V(\phi,\text{nonperturbative})\right] + \text{loops}\left(e^{\phi}\right), \tag{2.1}$$

neglecting both finite-size effects (α' corrections) and higher (field theory) loops in the coupling constant $g = e^{\phi/2}$. The initial evolution is simply driven by the kinetic energy of the dilaton field, with negligible contributions also from a possible non-perturbative potential $V(\phi)$, which is known, at small coupling, to approach zero very rapidly[6] as $V \sim \exp[-\exp(-\phi)]$. Here I will neglect, for simplicity, also a possible anisotropic contribution of the string antisymmetric tensor to the cosmological equations; see Ref. [7] for the inclusion of such a contribution.

Irrespective of the possible initial presence of matter sources[2,3], the (generally anisotropic) solution of the string cosmology equations, after inserting the perturbative vacuum as initial condition, can be asymptotically parametrized as

$$a_i = (-t)^{\beta_i}, \quad \phi = \left(\sum_i \beta_i - 1\right) \ln(-t), \quad \sum_i \beta_i^2 = 1, \quad t < 0, \quad t \to 0 \qquad (2.2)$$

($i = 1, .., d$). This solution describes a phase in which the curvature scale and the dilaton are both growing, since $|H_i| \sim (-t)^{-1} \sim |\dot{\phi}|$ for $t \to 0_-$, where $H = \dot{a}/a$, and the dot denotes derivatives with respect to cosmic time t. A typical example of background is illustrated in Fig. 1.

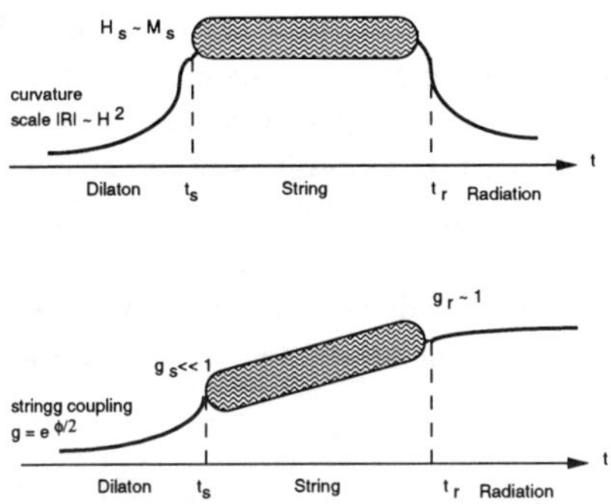

Figure 1: Qualitative evolution of the curvature scale and of the string coupling as a function of time, for a generic pre-big bang type of background.

Of course, when the curvature reaches the string scale, determined by the string mass parameter $M_s \equiv \lambda_s^{-1}$, the background enters a truly "stringy" regime, in which all higher derivative terms in the α' expansion become important (our present imperfect knowledge of the detailed evolution, in that region, is represented by the wavy lines of Fig. 1). At the end of the string phase the dilaton eventually approaches the strong coupling regime, $g_r \equiv g(t_r) \sim 1$, where both the quantum loop effects and the semiclassical back-reaction of the produced radiation become important, and the transition to the standard decelerated evolution (with frozen dilaton) is expected to occur. Some concrete examples of how the inclusion of one-loop corrections can stop the growth of the curvature[8] have been presented recently and discussed in the context of four-[9] and two-dimensional[10] cosmology.

At the beginning of the string phase, on the contrary, the value of the string coupling $g_s \equiv g(t_s)$ can be arbitrarily small (see Fig. 1); it represents one of the main parameters of this class of backgrounds. The other important parameter is the duration of the string phase, measured by the ratio $|t_s/t_r|$, or by the ratio of the associated scale factors, $a(t_s)/a(t_r)$. If g_s is already of order 1 at the beginning of the string phase, the duration of such a phase presumably shrinks to a number of order unity (in string units). The presence of such a phase, however, cannot be avoided; otherwise, to the tree-level in g_s, and to zeroth order in α', the growth of the curvature and of the dilaton is unbounded[11]. Even in that case, however, one can compute quantum-mechanically, via the Wheeler-De Witt equation, the transition probability from the initial to the final background configuration. Such a transition can be represented as a wave reflection in minisuperspace[12,13], and the probability turns out to be finite and non-vanishing even in the presence of a singularity that disconnects, classically, the two asymptotic regimes.

2.1. *Motivations*

There are at least two motivations, in a string theory context, that lead to introduce an early, pre-big bang cosmological phase in the past history of our Universe, as a natural complement of the present, post-big bang cosmological era. The first motivation is provided by the solutions of the classical string equations of motion embedded in inflationary cosmological backgrounds[4,5]. In the presence of shrinking event horizons, a gas of classical string develops in fact an effective negative pressure, so that the background evolution can be consistently self-sustained by the string gas itself as a source[3,14].

The second motivation is provided by the duality symmetries of the lowest-order string effective action, which can be written, including the antisymmetric tensor $B_{\mu\nu}$,

$$S = -\int d^{d+1}x\sqrt{|g|}e^{-\phi}\left[R + (\nabla_\mu\phi)^2 - \frac{1}{2}\left(\partial_{[\mu}B_{\nu\alpha]}\right)^2\right], \quad \mu,\nu = 1,2,...,d+1. \quad (2.3)$$

If $a(t)$ and $\phi(t)$ represent an exact, homogeneous and isotropic solution of this action (with $B = 0$ and zero spatial curvature), then

$$\tilde{a} = a^{-1}, \qquad \tilde{\phi} = \phi - 2d\ln a \qquad (2.4)$$

represent another exact solution, thanks to the scale factor duality symmetry[15,16]. This symmetry is only a particular case of a more general covariance property of the action (2.3) under global $O(d,d)$ transformations[17,18], which mixes non-trivially the spatial components of the metric and of the antisymmetric tensor,

$$M \to \tilde{M} = \Lambda^T M \Lambda, \qquad M = \begin{pmatrix} G^{-1} & -G^{-1}B \\ BG^{-1} & G - BG^{-1}B \end{pmatrix} \qquad (2.5)$$

($\Lambda \in O(d,d)$ and $G = g_{ij}$, $B = B_{ij}$). Such an $O(d,d)$ covariance is also preserved by the addition of source terms to the cosmological equations[19], provided they represent "bulk" string matter, whose components satisfy the string equation of motion in the given background. In the presence of sources, and in the perfect fluid approximation, a scale factor duality transformation is simply associated to a "reflection" of the equation of state[15], $p/\rho \to -p/\rho$.

The importance of these symmetries[1], in our cosmological context, follows from the fact that by operating simultaneously a duality transformation (for instance according to eq. (2.4)) and a time-reversal transformation on a given decelerated, post-big bang solution,

$$\dot{a} > 0, \qquad \ddot{a} < 0, \qquad \dot{H} < 0, \tag{2.6}$$

it is always possible to obtain a new accelerated solution, with growing curvature, of the pre-big bang type:

$$\dot{a} > 0, \qquad \ddot{a} > 0, \qquad \dot{H} > 0. \tag{2.7}$$

In $d = 3$, for instance, the standard radiation-dominated solution

$$a = t^{1/2}, \qquad \phi = \text{const}, \qquad p = \rho/3, \qquad t > 0 \tag{2.8}$$

(which is still an exact solution for the action (2.3) supplemented by perfect fluid sources) is mapped to the pre-big bang solution

$$a = (-t)^{-1/2}, \qquad \phi = -3\ln(-t), \qquad p = -\rho/3, \qquad t < 0 \tag{2.9}$$

(see also Refs. [18,19] for other, less trivial examples).

This natural association of pre- and post-big bang solutions, which always come in pair from the lowest-order string cosmology equations, is impossible in the standard cosmological context based on the Einstein equations: in that case there is in fact no field playing the role of the dilaton, and the above duality symmetries cannot be implemented. The challenge is, of course, to formulate a complete and consistent string cosmology scenario in which the two duality-related solutions are smoothly joined, in the context of a background satisfying (probably only asymptotically) the self-dual condition[1] $a(t) = a^{-1}(-t)$. We shall assume in the rest of the paper that such a smooth transition (impossible to lowest order) can be successfully implemented when higher-order corrections are included, and we shall discuss some phenomenological aspects of the emerging cosmological scenario.

2.2. *Kinematical aspects*

From a kinematical point of view, the epoch of pre-big bang evolution can be invariantly characterized as a phase of shrinking event horizons[1]. This means that the scale factor $a(t)$, for an isotropic phase of pre-big bang evolution, can be parametrized in cosmic time as

$$a(t) = (-t)^{\beta}, \qquad \beta < 1, \qquad t < 0, \tag{2.10}$$

where the condition $\beta < 1$ is required for the convergence of the integral defining the comoving size of the event horizon:

$$\int_t^0 \frac{dt'}{a(t')} < \infty \quad \Rightarrow \quad \beta < 1. \tag{2.11}$$

This condition can be satisfied in two ways:

- $\beta < 0 \Rightarrow \dot{a} > 0, \ddot{a} > 0, \dot{H} > 0$,
  which corresponds to a metric describing a phase of accelerated expansion and growing curvature, usually called superinflation[21] (or pole inflation).

- $0 < \beta < 1 \Rightarrow \dot{a} < 0, \ddot{a} < 0, \dot{H} < 0$,
  which corresponds to a metric describing accelerated contraction and, again, growing curvature scale.

The first type of metric provides a representation of the pre-big bang scenario in the String frame (also called Brans-Dicke frame), in which test strings move along geodesic surfaces, the second in the more conventional Einstein frame, in which the gravi-dilaton action appears diagonalized in the standard, canonical way. The String frame was explicitly used above for presenting the effective action (2.1) and the duality transformations, the Einstein frame will be used below for discussing the amplification of perturbations. The choice of the frame is indeed a matter of convenience, as both frames give consistent and physically equivalent descriptions of the pre-big bang scenario (see Refs. [2,3] for the explicit relation between the two frames and for a discussion of their physical equivalence).

In the context of this paper, what is important is the fact that both classes of backgrounds, irrespective of the sign of β, are accelerated, and in both classes the amplification of perturbations is more efficient, at high frequency, than in a standard inflationary background. This occurs for two reasons:

- First because the curvature scale is growing and, as a consequence, the perturbation spectrum[22,23] grows with frequency[21,24];

- Second because the comoving amplitude of perturbations may grow in time even outside the horizon[2,25], instead of being frozen as in the standard scenario.

These two effects are crucial for obtaining a strong graviton background at high frequency, and will thus be discussed in some detail in the next section.

3. Growth of perturbations

For an accelerated background, the scale factor can be conveniently parametrized in conformal time (η, such that $dt = ad\eta$), and in the negative time range, as

$$a(\eta) = (-\eta)^{\alpha}, \qquad \eta < 0. \tag{3.1}$$

Positive α corresponds to accelerated contraction, negative α to accelerated expansion, $\alpha = -1$ corresponds in particular to the standard de Sitter-like inflation (see Fig. 2).

Consider for instance the canonical variable $h_{\mu\nu}$ representing tensor perturbations, in the Einstein frame, normalized to an initial vacuum fluctuation spectrum,

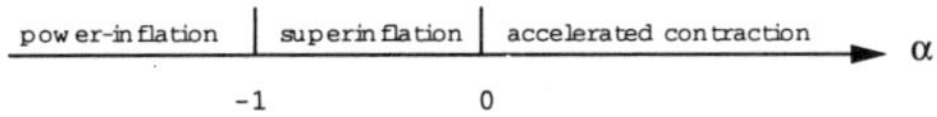

Figure 2: Different classes of accelerated backgrounds, parametrized in conformal time by the power α.

i.e. $h_k \sim \left(e^{-ik\eta}/aM_p\sqrt{k}\right)$, for $\eta \to -\infty$ (M_p is the Planck mass). The Fourier component of each polarization mode h_k satisfies the well-known equation[22,23]

$$(ah_k)'' + \left(k^2 - \frac{a''}{a}\right) ah_k = 0. \tag{3.2}$$

In a background of the pre-big bang type the horizon is shrinking, so that all modes are "pushed out" of the horizon, for $\eta \to 0_-$. For a mode whose wavelength is larger than the horizon size ($|k\eta| << 1$), the asymptotic solution of the above perturbation equation is given by

$$h_k = A_k + B_k\,|\eta|^{1-2\alpha}, \qquad \eta \to 0_- \tag{3.3}$$

where A_k and B_k are integration constants. We have in general two possibilities.

- If $\alpha < 1/2$ the first term is dominant in the asymptotic solution (3.3), the comoving amplitude h_k tends to stay constant outside the horizon, and the typical (dimensionless) perturbation amplitude $\delta_h(k)$ (see Appendix A) can be expressed as usual in terms of the Hubble factor at horizon crossing:

$$|\delta_h(k)| \equiv k^{3/2}\,|h_k| \simeq \left(\frac{H}{M_p}\right)_{h.c.}, \qquad \text{horizon crossing} \Leftrightarrow |k\eta| \sim 1. \tag{3.4}$$

  This amplitude is constant in time. According to the horizon crossing condition, however, higher-frequency modes cross the horizon later in time, and then at higher value of the Hubble factor if the curvature scale is growing, as in our class of backgrounds. As a consequence, the amplitude of higher-frequency modes is enhanced with respect to lower frequency modes, in contrast with the usual scale-invariant (Harrison-Zeldovich) spectrum, where the amplitude is the same for all modes.

- If $\alpha > 1/2$ the solution (3.3) is dominated by the second term, and h_k tends to grow outside the horizon. The typical perturbation amplitude becomes time-dependent,

$$|\delta_h(k,\eta)| \simeq \left(\frac{H}{M_p}\right)_{h.c.} |k\eta|^{1-2\alpha}, \qquad \eta \to 0_-, \tag{3.5}$$

  and grows not only with frequency, but also with time. This second effect is not automatic, however, in any background of the pre-big bang type, but requires a fast enough contraction (in the Einstein frame) of the scale factor.

3.1. *Formal problems*

The analysis of the perturbation spectrum, in the context of the pre-big bang scenario, is complicated in general by two types of formal problems.

The first is that, just because of this "anomalous" growth in time of the amplitude, the linear approximation tends to break down for scalar perturbations, in the standard longitudinal gauge. This problem may be solved, at least for a class of dilaton-driven backgrounds, by moving to an off-diagonal gauge[26] (also called "uniform-curvature" gauge[27]), in which the scalar part of the perturbed metric is parametrized by two scalar potentials, Φ and B, as follows

$$ds^2 = a^2 \left[d\eta^2(1+2\Phi) - dx_i^2 - 2\partial_i B dx^i d\eta\right]. \tag{3.6}$$

In this way the dangerous growing mode is "gauged down", enough to restore the validity of the linear approximation. For other backgrounds, however, the growth of perturbations outside the horizon may be a true physical effect, which cannot be eliminated in a convenient gauge; we have to restrict ourselves to a reduced portion of parameter space for the linear approximation to be valid.

A second problem is related to the normalization of the perturbations to the initial spectrum of the quantum fluctuations of the vacuum, in a background that is in general higher-dimensional and anisotropic. Such a normalization requires the knowledge of the so-called "normal modes" of oscillation of an anisotropic background, i.e. of the variables that diagonalize the action perturbed to second order, and that satisfy canonical commutation relations. This problem has been solved, up to now, only for spatially flat higher-dimensional backgrounds, in which the translations along internal dimensions are isometries of the full perturbed metric[28].

In any case, even if we have to restrict ourselves to a limited class of backgrounds in which the linear approximation is valid, and the canonical normalization is known, we find that the enhanced amplification of perturbation, in the context of the pre-big bang scenario, is in general associated to new and interesting phenomenological consequences.

3.2. *Phenomenological consequences*

There are two effects, in particular, worth mentioning. The first is the amplification of the vacuum fluctuations of the electromagnetic field (and of the gauge fields, in general), due to their coupling to a dynamical dilaton, according to the effective interaction

$$e^{-\phi}\sqrt{-g}F_{\mu\nu}F^{\mu\nu}. \tag{3.7}$$

The electromagnetic field is also coupled to the metric, but in four dimensions such a coupling is conformally invariant, and there is no parametric amplification of perturbations in a conformally flat metric background[29], like that of a typical inflationary model. In a string cosmology context, the amplification of the electromagnetic perturbations is a consequence of the dilaton running during the phase of pre-big bang evolution. Such an amplification may be large enough to produce the "seed fields" required for the origin of the galactic magnetic field[30], and possibly also to produce the observed anisotropy[31] of the Cosmic Microwave Background (CMB) radiation[32].

A second important effect is the production of a relic graviton background, much stronger, at high frequency, than predicted by the standard inflationary scenario[33].

The present spectral energy density of such a background, in units of critical energy density ρ_c, is defined as

$$\Omega_G(\omega, t_0) = \frac{\omega}{\rho_c}\frac{d\rho_G(\omega, t_0)}{d\omega}, \qquad \rho_c(t_0) = \frac{3M_p^2 H_0^2}{8\pi}, \tag{3.8}$$

where

$$\frac{d\rho_G(\omega, t_0)}{d\omega} = 2\omega\overline{n}(\omega)\left[\frac{4\pi\omega^2}{(2\pi)^3}\right], \tag{3.9}$$

and where $\overline{n}(\omega)$ is the number density of produced particles with proper energy $\omega(t) = k/a(t)$ (the factor 2 is because of the two graviton polarization modes; the factor in square brackets refers to integration in momentum space). The peak value of Ω_G, for any model of cosmological evolution, can be conveniently referred to the present CMB energy density Ω_γ (see next section), which is a number of order[34] 10^{-4}–10^{-5} (the uncertainty depends on the present uncertainty about the value of ρ_c). In the standard inflationary scenario the energy density of the graviton background is constrained by the high degree of isotropy of the CMB radiation[31], $(\Delta T/T)_\gamma \lesssim 10^{-5}$, which imposes, at all scales $\omega > 10^{-16}$Hz, the bound[35,36]

$$\Omega_G \lesssim \Omega_\gamma \left(\frac{\Delta T}{T}\right)_\gamma^2 \lesssim 10^{-15}. \tag{3.10}$$

In the context of the pre-big bang scenario, on the contrary, the graviton background is in general too low, on the COBE scale $\omega_0 \sim 10^{-18}$Hz, to be constrained by the CMB isotropy[3,26,33]. The peak value of the spectrum is simply controlled by the present value of the fundamental ratio of the string and Planck mass, which is expected[37] to be a number in the range 0.3–0.03:

$$\Omega_G \lesssim \Omega_\gamma \left(\frac{M_s}{M_p}\right)^2 \lesssim 10^{-6}. \tag{3.11}$$

This implies a possible signal, in the frequency range of present and near-future gravity wave detectors, up to *nine* orders of magnitude higher than predicted by the standard inflationary scenario. The next section is devoted to a detailed discussion of this possibility.

4. The Expected Graviton Background

4.1. *Minimal model of pre-big bang*

In order to discuss the expected graviton background we shall first consider a "minimal" model of pre-big bang scenario (the same as used in Refs. [29, 32]), in which there are only three phases: an initial phase of dilaton-driven inflation for $t < t_s$, a high curvature string phase for $t_s < t < t_r$, and the standard radiation-dominated evolution for $t > t_r$ (see also Fig. 1).

At low frequency, namely for modes with $\omega << \omega_s \simeq (a_s\eta_s)^{-1}$, which cross the horizon before the beginning of the string phase, the slope of the spectrum can be computed exactly from the perturbation equation obtained from the low-energy action, and it is found to be cubic[3,26,33] (modulo small corrections due to a logarithmic growth in time of the comoving amplitude outside the horizon, during the dilaton-driven phase):

$$\Omega_G(\omega, t_0) \propto \left(\frac{\omega}{\omega_s}\right)^3 \ln^2\left(\frac{\omega}{\omega_s}\right), \qquad \omega < \omega_s. \tag{4.1}$$

In this range of frequencies, the spectrum can easily be computed by noting that, for the given model of background, the tensor perturbation equation (3.2) reduces to a Schrödinger-like equation in conformal time[22,23]:

$$\psi_k'' + \left[k^2 - V(\eta)\right]\psi_k = 0, \tag{4.2}$$

with an effective potential ($V = a''/a$ in the Einstein frame) whose modulus grows during the dilaton phase, keeps growing in the string phase where it reaches a maximum around the transition scale η_r, and rapidly approaches zero at the beginning of the radiation era (see Fig. 3).

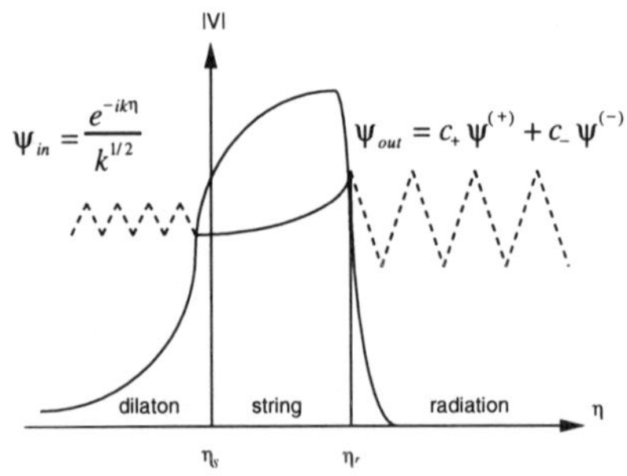

Figure 3: Parametric amplification of the perturbation mode ψ_k due to the effective potential barrier $V = a''/a$ ("anti-tunnelling" effect).

Normalizing the solution at $\eta \to -\infty$ to an initial vacuum fluctuation spectrum, $\psi_k \sim e^{-ik\eta}/\sqrt{k}$, the asymptotic solution at late times in the radiation era, for $\eta \to +\infty$, is in general a linear combination of modes of positive and negative frequency with respect to the initial vacuum state:

$$\psi_k = c_+\psi_k^{(+)} + c_-\psi_k^{(-)}, \qquad \eta \to +\infty, \tag{4.3}$$

where the mixing coefficients $c_\pm$ parametrize the unitary Bogoliubov transformation connecting the set of |in⟩ and |out⟩ annihilation and creation operators,

$$\{\psi_{in}, b_k, b_k^\dagger\} \Longrightarrow \{\psi_{out}, a_k, a_k^\dagger\}, \tag{4.4}$$

$$a_k = c_+ b_k + c_-^* b_{-k}^\dagger, \qquad a_{-k}^\dagger = c_- b_k + c_+^* b_{-k}^\dagger. \tag{4.5}$$

So, even starting from the vacuum, one ends up with a final expectation number of gravitons, which is, in general, non-vanishing[38,39,40]

$$\overline{n}_i = \langle 0|b^\dagger b|0\rangle, \qquad \overline{n}_f = \langle 0|a^\dagger a|0\rangle = |c_-|^2 \neq 0. \tag{4.6}$$

The Bogoliubov coefficients $c_\pm(k)$ can be computed by solving the perturbation equation (4.2) in the three different regions of Fig. 3, and by matching $\psi(\eta)$, $\psi'(\eta)$ at $\eta = \eta_s$ and $\eta = \eta_r$. The expectation number $\overline{n}(k) = |c_-(k)|^2$ then gives the spectral energy distribution $\Omega_G(\omega)$ according to eqs. (3.8), (3.9).

This process of parametric amplification[22] can be seen as an "anti-tunnelling" effect for the "wave function" ψ_k. Indeed, if we look at Fig. 3 from the right to the left, and imagine to replace the time variable η with a spacelike coordinate, we obtain the description of a scattering process in which $\psi^{(+)} \sim e^{-ik\eta}$ plays the role of a wave incident on the barrier from $+\infty$, $\psi^{(-)} \sim e^{+ik\eta}$ is the reflected wave, and ψ_{in} is the transmitted wave,

$$\begin{aligned} \eta \to +\infty\ , \qquad & \psi \sim A_{in} e^{-ik\eta} + A_{ref} e^{ik\eta} \\ \eta \to -\infty\ , \qquad & \psi \sim A_{tr} e^{-ik\eta}. \end{aligned} \tag{4.7}$$

In the semiclassical approximation, which corresponds in our case to the limit of a very large number of produced particles, $\overline{n}_f >> 1$, the reflection coefficient $R = |A_{ref}|^2/|A_{in}|^2$ is approximately 1, and the Bogoliubov coefficient $|c_-|^2$ becomes the inverse of the tunnelling coefficient $T = |A_{tr}|^2/|A_{in}|^2$:

$$|c_-|^2 = \frac{|A_{ref}|^2}{|A_{tr}|^2} = \frac{R}{T} \simeq \frac{1}{T}. \tag{4.8}$$

Hence the name anti-tunnelling.

At high frequencies, namely for modes with $\omega >> \omega_s$, crossing the horizon (or "hitting" the effective potential barrier) during the string phase, the exact slope of the spectrum is at present unknown. In fact, even assuming that the background curvature stays constant, $H \simeq M_s$ (in string units) during the string phase[30,33], we do not know the corresponding evolution of the dilaton background. Also, and most important, we do not know the exact form of the perturbation equation, because of possible higher derivatives corrections due to higher orders in the α' expansion (eventually, also higher-loop contributions cannot be excluded, a priori).

It is important to stress, however, that the latter problem can at most affect the amplitude, but not the slope (4.1) of the "dilatonic" branch of the spectrum. In fact, a transverse and traceless mode of constant amplitude is always a solution of a linearized and covariant perturbation equation, to all orders in the background curvature[41]. This preserves the spectral distribution of modes whose amplitude is frozen already during the dilaton-driven phase.

Concerning the former problem, we can parametrize our ignorance of the background evolution with a power-like behaviour of the dilaton background[30,33] (see also next section). By using the low-energy perturbation equation, we can thus predict for the graviton spectrum a slope that is at most cubic, but in general flatter than at lower frequencies. Interestingly enough, this property is valid for the *exact* solution of the low energy perturbation equation[41], to all orders in the k expansion (3.3), and for any given time-dependent background, not only of the power-law type. Since this result is a consequence of the duality symmetry characterizing the low-energy evolution of perturbations[42], one might argue that a flatter slope at higher frequency is in general to be expected, provided the duality symmetry is not spoiled by higher derivatives and higher-loop corrections.

What we can do, also, irrespective of a detailed knowledge of the background kinematics, is to provide an accurate estimate of the present value of the end point of the spectrum, namely of the maximal frequency undergoing parametric amplification[43]. This frequency is of the same order as the frequency corresponding to the production of one graviton per polarization and per unit phase-space volume (it is known that, for higher frequencies, graviton production is exponentially suppressed[29,39]). To this aim we rescale such a frequency, which we call $\omega_1(t)$, to the beginning of the radiation era, i.e. $\omega_1(t_0) = \omega_1(t_r)a_r/a_0$, where by definition

$$H_r^2 = \frac{8\pi}{3M_p^2}\frac{\pi^2 N_r}{30}T_r^4 \tag{4.9}$$

(N_r is the total effective number of massless degrees of freedom in thermal equilibrium at $t = t_r$). Also, we rescale T_r to the present CMB temperature $T_0 = 2.7$K, by taking into account a possible production of radiation due to some reheating process occurring well below the string scale. The effect of such processes can be parametrized by the associate fraction of present thermal entropy density, δs, as

$$s_0 \equiv \frac{2\pi^2}{45}n_0(a_0T_0)^3 = \frac{2\pi^2}{45}n_r(a_rT_r)^3 + s_0\delta s, \tag{4.10}$$

where n_0 and n_r are the number of species contributing (with their own statistical weight) to the thermal entropy at t_0 and t_r, respectively ($n_0 = 3.91$, see for instance Ref. [33]).

In a string cosmology context, the amplification of tensor perturbations is always accompanied by a copious production of gauge bosons (according to eq. (3.7)) and other ultrarelativistic particles, which are expected to thermalize rapidly, unlike gravitons and dilatons, which interact only gravitationally[44,45]. For the formulation of a consistent scenario in which the energy density of such particles, produced from the vacuum, becomes critical and start dominating the background evolution at $t = t_r$, we shall assume that both H_r and $\omega_1(t_r)$ are of the same order as the string scale at t_r, i.e. $H_r \simeq M_s(t_r) \simeq \omega_1(t_r)$. In that case we obtain, from eqs. (4.9) and (4.10),

$$\omega_1(t_0) \simeq 1.1\, T_0 \left[\frac{M_s(t_r)}{M_p}\right]^{1/2} \left(\frac{10^3}{n_r}\right)^{1/12} (1-\delta s)^{1/3}, \qquad (4.11)$$

where we have used the fact that $n_r \simeq N_r$, and that N_r is expected to be a number of order 10^2–10^3.

On the other hand, the fact that the radiation temperature cannot exceed the string mass scale in a string cosmology context, $T_r \lesssim M_s(t_r)$, implies that the total energy density of the quantum fluctuations (equal to ρ_r, at $t = t_r$) is dominated by the energy density ρ_1 corresponding to the end-point frequency ω_1, i.e. $\rho_1 = \omega_1^4/\pi^2$. In fact,

$$\frac{\rho_r}{N_r\omega_1^4(t_r)} \simeq \frac{\rho_r}{N_r M_s^4(t_r)} = \frac{\pi^2}{30}\frac{T_r^4}{M_s^4(t_r)} \lesssim 1. \qquad (4.12)$$

In addition, the fact that $\rho_r \simeq N_r M_s^4$ is critical at $t = t_r$ implies that the string scale is already quite close to its present value, as $M_p/M_s(t_r) \simeq N_r^{1/2} \sim 10$–$30$. Consequently, the peak of the graviton background must be of the same order as the end-point value $\Omega_G(\omega_1, t_0) = \omega_1(t_0)/\pi^2\rho_c(t_0)$, with M_s fixed by a dilaton expectation value already in its present range.

So, the spectral graviton distribution has to be on the average non-decreasing from ω_s to ω_1, with an upper bound that can be written in critical units as[43]

$$\begin{aligned}\Omega_G(\omega_1, t_0) &\simeq 2.6\, \Omega_\gamma(t_0) \left(\frac{M_s}{M_p}\right)^2 \left(\frac{10^3}{n_r}\right)^{1/3} (1-\delta s)^{4/3} \\ &\simeq 7\times 10^{-5} h_{100}^{-2} \left(\frac{M_s}{M_p}\right)^2 \left(\frac{10^3}{n_r}\right)^{1/3} (1-\delta s)^{4/3}, \end{aligned} \qquad (4.13)$$

where $h_{100} = H_0/(100 \text{ km sec}^{-1}\text{Mpc}^{-1})$. Note that this peak value can be integrated from ω_1 down to the Hertz scale without contradicting the bound following from the standard nucleosynthesis analysis[31],

$$h_{100}^2 \int \Omega_G(\omega, t_0) d\ln\omega \lesssim 0.5\times 10^{-5}. \qquad (4.14)$$

The above situation is illustrated in Fig. 4, which shows the maximum allowed graviton energy density for the case $n_r = 10^3$, in the Hz to GHz range. For any given δs, the uncertainty of the end point (4.11) and of the peak (4.13) is mainly due to the present theoretical uncertainty of the fundamental ratio M_s/M_p, which has been chosen, for the illustrative purpose of Fig. 4, to range from 0.1 to 0.01. This uncertainty is represented by the shaded boxes of Fig. 4. To the left of the end point the spectrum can be at most flat, so that in the absence of significant reheating at scales much lower than t_r, the expected maximal strength of the graviton background should lie within the dashed lines in the band labelled by $\delta s = 0$.

A non-zero δs would instead imply that the radiation that becomes dominant at the end of the string phase is only a fraction of the thermal radiation that we observe today, with a consequent dilution of the graviton background (produced at the string scale) with respect to the total present radiation energy density. However, we can see from the picture that even if 99% of the present entropy would be produced during the latest stage of evolution, the intensity of the relic graviton background would stay well above the full line labelled "de Sitter" in Fig. 4, which represents the most optimistic prediction of the standard inflationary scenario[36], in this range of frequency.

Also shown in the picture are three lines of constant sensitivity of a gravity wave detector, $S_h^{1/2} = 10^{-23}, 10^{-25}, 10^{-27}\mathrm{Hz}^{-1/2}$, given in terms of the spectral amplitude $S_h^{1/2}$, related to the spectral energy density Ω_G by

$$S_h(\nu) = \frac{3H_0^2}{4\pi^2\nu^3}\,\Omega_G(\nu), \qquad \nu = \omega/2\pi, \tag{4.15}$$

(see Appendix B). As clearly shown in Fig. 4, a sensitivity of 10^{-25} in the kHz range (which is a typical range for present, Earth-based detectors) would already be enough to enter the region where we expect a signal, thus constraining (even with a negative result) the parameters of our class of string cosmology models[43].

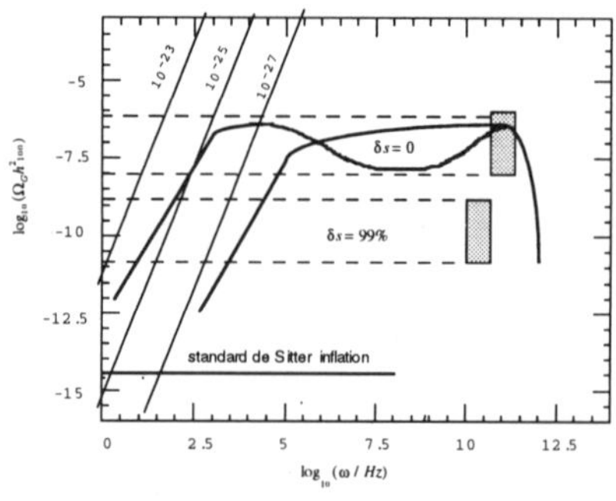

Figure 4: Position of the peak and of the end point of the relic graviton spectrum for a typical pre-big bang scenario.

In fact, any spectrum with a shape that reaches the end point, growing not faster than ω^3, is in principle allowed in this context, as the two examples shown by the thick curves in Fig. 4. The slope has to be cubic at low enough frequencies, for modes crossing the horizon in the dilaton-driven phase. At higher frequencies the spectrum has to be on the average non-decreasing. However, it could also be non-monotonic, if $\dot{\phi}$ does not follow exactly, during the string phase, the constant behaviour of the spacetime curvature, but decreases or undergoes small oscillations around the string scale (see the next section). Interestingly enough,, an oscillating spectrum at high frequency seems to be possible even for a monotonic evolution of the dilaton background[46].

Finally, it may be interesting to point out that for a long enough string phase it seems possible, in principle, to have a spectrum that is normalized to the end point value at high frequency, and which is nevertheless strong enough at the COBE scale to be significantly constrained by the present data on the CMB anisotropy.

Suppose in fact that all scales inside our present horizon ($\gtrsim \omega_0 \sim 10^{-18}$ Hz) crossed the horizon during the high curvature string phase, and let us parametrize the present (unknown) distribution of tensor perturbations with a spectral index n, such that $n = 1$ corresponds to the flat, scale-invariant spectrum. For all scales reentering the horizon after the time of matter–radiation equilibrium, $\omega < \omega_{eq} \sim 10^{-16}$Hz, there is an additional effect of parametric amplification due to the matter-dominated phase, which imposes an extra ω^{-2}-dependence on the spectrum[39,47]. By choosing, from eqs. (4.11) and (4.13), the typical values $\omega_1 \sim 10^{11}$Hz and $\Omega_G(\omega_1) \sim 10^{-6}$ for illustrative purpose, we then have the possible normalized distribution

$$\begin{aligned} \Omega_G(\omega_1, t_0) &= 10^{-6} \left(\frac{\omega}{\omega_1}\right)^{n-1}, & \omega_{eq} < \omega < \omega_1 \\ &= 10^{-6} \left(\frac{\omega}{\omega_1}\right)^{n-1} \left(\frac{\omega_{eq}}{\omega}\right)^2, & \omega_0 < \omega < \omega_{eq}. \end{aligned} \tag{4.16}$$

For the dilatonic branch of the spectrum the spectral index is $n = 4$, and Ω_G is consequently negligible at the present horizon scale ω_0. For the string branch of the spectrum, however, n is unknown, and can be bounded from below by imposing the COBE normalization, which implies[35] $\Omega_G(\omega_0, t_0) \sim 10^{-10}$. This condition is perfectly compatible with the end-point normalization, and gives $n = 37/29 \simeq 1.27$. The corresponding spectral behaviour in the range $\omega_0 < \omega < \omega_{eq}$, namely $\Omega_G \sim \omega^{-1.73}$, is not flat enough to match the observed CMB anisotropy (a gravity wave interpretation of such an isotropy is problematic also in the standard inflationary context[49]). However, for $\omega > \omega_{eq}$, the resulting, slightly growing, "blue" spectrum is well consistent with the bound obtained from pulsar timing data[48], $\Omega_G(\omega_p) \lesssim 10^{-8}$ at $\omega_p \sim 10^{-8}$Hz, as illustrated in Fig. 5.

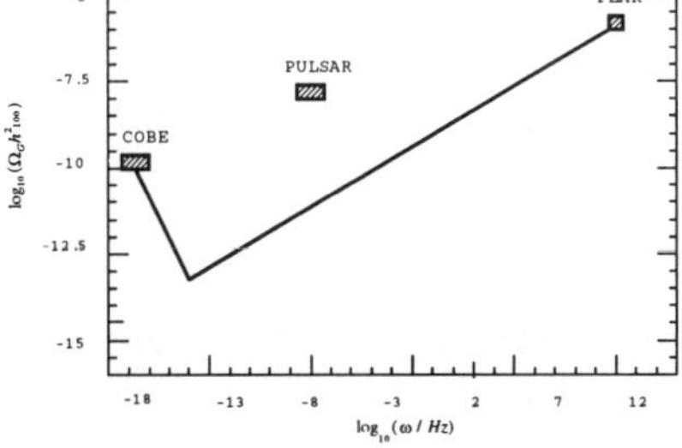

Figure 5: Example of possible graviton spectrum which has the correct peak normalization at high frequency, and that satisfies also the COBE normalization at the frequency scale of our present horizon.

Note that, by assuming the normalization at the COBE scale as input, since the end-point normalization is fixed, we may *predict* a spectrum with $\Omega_G \sim 10^{-8}$ in the kHz range, which seems to be accessible to the sensitivity of gravity-wave detectors of the second generation (see Section 4.5).

4.2. *A diagrammatic approach to the perturbation spectrum*

At this point, two remarks are in order. The first concerns the possibility of constraining the graviton spectrum through the associated amplification of electromagnetic fluctuations, which in our context is always expected to accompany graviton production. The second concerns the possible qualitative modification of the spectrum, in models of background more complicated than the minimal one previously considered.

A complete discussion of these points would require a detailed knowledge of the kinematics and of the higher-order corrections during the string phase. In the absence of such knowledge, I will approximate the background with a finite number of phases, each of them characterized by a different type of power-law evolution. Also, I will use the lowest-order perturbation equation, working however only in the asymptotic regimes $|k\eta| << 1$ and $|k\eta| >> 1$ (also called, sometimes improperly, outside and inside the horizon), assuming that the qualitative behaviour of the asymptotic solution is not spoiled by higher-order corrections (see the previous section for arguments supporting this assumption).

Under the above assumptions, the graviton spectrum can be conveniently obtained through a sort of "diagrammatic" technique, in which one has only to specify the power (in conformal time) of the scale factor of the various phases, and the time of horizon crossing of the frequency band considered.

This technique is based on the fact that the exact solution of the perturbation equation (4.2), for a generic power-law evolution of the Einstein frame scale factor, $a \sim |\eta|^\alpha$, can be written in terms of the first-kind ($H_\nu^{(1)}$) and second-kind ($H_\nu^{(2)}$) Hankel functions, of index $\nu = |\alpha - 1/2|$, as follows:

$$a = |\eta|^\alpha \Rightarrow \psi_k = |\eta_k|^{1/2}\left[A_+(k)H_\nu^{(2)}(|k\eta|) + A_-(k)H_\nu^{(1)}(|k\eta|)\right], \qquad \nu = |\alpha - 1/2| \tag{4.17}$$

($A_\pm$ are integration constants). When matching the solution from a phase with power α_2 to a phase with power α_1, at the time $\eta = \eta_1$,

$$\psi_k = \psi_k^2(A_\pm^2, \nu_2), \qquad \eta < \eta_1; \qquad \psi_k = \psi_k^1(A_\pm^1, \nu_1), \qquad \eta > \eta_1; \tag{4.18}$$

we may thus easily distinguish the two asymptotic regimes $|k\eta_1| << 1$ and $|k\eta_1| >> 1$, corresponding respectively to the comoving frequency of a mode that hits and one that does not hit the potential barrier of eq. (4.2), whose height at η_1 is $|V_1|^{1/2} \sim |\eta_1|^{-1}$. In the first case, by using the small argument limit of the Hankel functions[50], the continuity of ψ and ψ' gives

$$\begin{aligned}\left(A_+^1 + A_-^1\right) &= b_1\left(A_+^2 + A_-^2\right)x_1^{\nu_2-\nu_1} + b_2\left(A_+^2 - A_-^2\right)x_1^{-\nu_2-\nu_1}\\ \left(A_+^1 - A_-^1\right) &= b_3\left(A_+^2 + A_-^2\right)x_1^{\nu_2+\nu_1} + b_4\left(A_+^2 - A_-^2\right)x_1^{-\nu_2+\nu_1},\end{aligned} \tag{4.19}$$

where $x_1 = |k\eta_1|$ and b_1, b_2, b_3, b_4 are complex numbers with modulus of order 1 (for $\nu = 0$, x^ν has to be replaced by $\ln x$). In the second case, $|k\eta| >> 1$, the perturbation is approximately unaffected by the background transition, and the large argument limit of the Hankel functions gives

$$A^1_+ = A^2_+, \qquad\qquad A^1_- = A^2_-. \tag{4.20}$$

Suppose now to approximate the background evolution with $n+1$ different phases, with powers $\alpha_1, \alpha_2, ..., \alpha_{n+1}$, and solutions $\psi_1, \psi_2, ..., \psi_{n+1}$ of the perturbation equations, with the corresponding Bessel index $\nu_1, \nu_2, ..., \nu_{n+1}$. The transitions occur, respectively, at the times $\eta_n < \eta_{n-1} < ... < \eta_2 < \eta_1$, and the height of the effective potential at the transition time η_i is $|V_i|^{1/2} \sim |\eta_i|^{-1} \equiv k_i$ (see Fig. 6). Normalizing the initial solution to a vacuum fluctuation spectrum,

$$A^{n+1}_+ = 1, \quad A^{n+1}_- = 0, \qquad \psi_{n+1} = |\eta|^{1/2} H^{(2)}_{\nu_{n+1}}(|k\eta|), \tag{4.21}$$

the Bogoliubov coefficient determining the expectation number of particles, appearing in the final spectral distribution Ω_G (see eq.(3.9)), is then given by $c_-(k) = A^1_-(k)$.

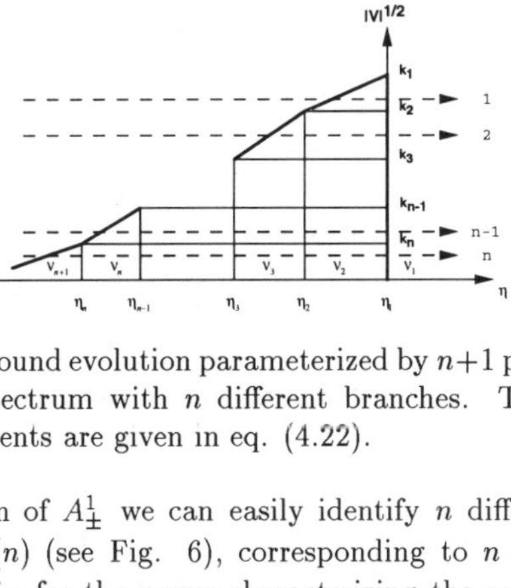

Figure 6: A background evolution parameterized by $n+1$ phases is in general associated to a spectrum with n different branches. The corresponding Bogoliubov coefficients are given in eq. (4.22).

In the computation of $A^1_\pm$ we can easily identify n different branches of the spectrum, $(1), (2), ..., (n)$ (see Fig. 6), corresponding to n asymptotic frequency bands, which only differ for the power characterizing the scale factor at the time of horizon crossing. By applying, step by step, the matching conditions (4.19) and (4.20), one can easily compute the Bogoliubov coefficients $c_1, c_2, ..., c_n$, for the frequency bands that are affected, respectively, by $1, 2, ..., n$ background transitions. Neglecting numerical coefficients of the order of unity, we have

$$|c_1| \sim x_1^{-\nu_1} x_1^{-\nu_2}, \qquad k_2 < k < k_1$$

$$|c_2| \sim x_1^{-\nu_1} \left(\frac{x_1}{x_2}\right)^{-\epsilon_2\nu_2} x_2^{-\nu_3}, \qquad k_3 < k < k_2$$

$$|c_3| \sim x_1^{-\nu_1} \left(\frac{x_1}{x_2}\right)^{-\epsilon_2\nu_2} \left(\frac{x_2}{x_3}\right)^{-\epsilon_3\nu_3} x_3^{-\nu_4}, \qquad k_4 < k < k_3$$

$$\cdots\cdots \sim \cdots\cdots\cdots\cdots\cdots\cdots\cdots\cdots\cdots\cdots\cdots\cdots$$

$$|c_n| \sim x_1^{-\nu_1}\left(\frac{x_1}{x_2}\right)^{-\epsilon_2\nu_2}\left(\frac{x_2}{x_3}\right)^{-\epsilon_3\nu_3}\cdots\cdots\left(\frac{x_{n-1}}{x_x}\right)^{-\epsilon_n\nu_n} x_n^{-\nu_{n+1}}, \; k < k_n \tag{4.22}$$

where

$$x_i = |k\eta_i|, \qquad \epsilon_i = \text{sign}\ln\left(\frac{x_i}{x_{i-1}}\right), \qquad i = 1, 2, ..., n. \tag{4.23}$$

The ϵ_i coefficients have been inserted to take into account the possibility of a non-monotonic potential (unlike that of Fig. 6), since for $x_i < x_{i-1}$ the ratio x_{i-1}/x_i enters the Bogoliubov coefficients with the opposite power. Note also that a phase with vanishing effective potential, like radiation-dominated evolution, $\alpha_i = 0$, has a Bessel index $\nu_i = 1/2$.

As a simple application of the general prescription (4.22), let us write down the Bogoliubov coefficients for a background characterized by four different phases and by a non-monotonic effective potential, which has a first break at η_s, a peak at η_1, and which becomes radiation-dominated at η_r, with $x_s > x_r > x_1$ (see Fig. 7). The Bessel indices are, respectively, ν_0, ν_s, ν_1 and $\nu_r = 1/2$. The graviton spectrum is characterized by three branches, differing by the number of background transitions by which a given mode is affected. The corresponding Bogoliubov coefficients can be immediately deduced from the diagram of Fig. 7, by applying the general prescription (4.22):

$$\begin{aligned}
|c_1| &\sim x_1^{-\nu_1} x_1^{-\nu_s}, && k_r < k < k_1 \\
|c_2| &\sim x_r^{-1/2}\left(\frac{x_r}{x_1}\right)^{\nu_1} x_1^{-\nu_s}, && k_s < k < k_r \\
|c_3| &\sim x_r^{-1/2}\left(\frac{x_r}{x_1}\right)^{\nu_1}\left(\frac{x_1}{x_s}\right)^{-\nu_s} x_s^{-\nu_0}, && k < k_s
\end{aligned} \tag{4.24}$$

A similar type of background will be discussed in Section 4.4.

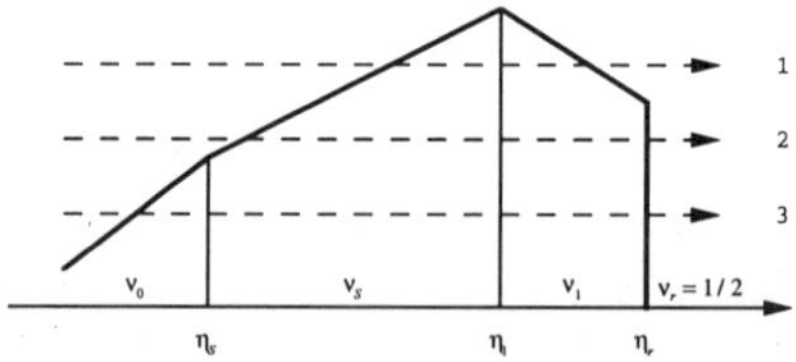

Figure 7: Example of background with non-monotonic potential. .

As discussed in the previous section, the square of the Bogoliubov coefficient determines the expectation number of the produced gravitons and the spectral distribution Ω_G, according to eqs. (3.8) and (3.9). In critical units,

$$\Omega_G(\omega, t) = \frac{\omega}{\rho_c}\frac{d\rho}{d\omega} = \frac{8\omega^4}{3\pi M_p^2 H^2(t)}|c_-|^2. \tag{4.25}$$

By expressing all ratios in terms of proper frequencies, $x_i/x_j \equiv |k\eta_i|/|k\eta_j| \equiv k_j/k_i = \omega_j/\omega_i$, and referring the spectrum to the proper frequency of the final transition, $\omega_1(t) = k_1/a(t) \sim H_1 a_1/a$, we easily obtain, from eqs. (4.22) and (4.25), the spectral distribution $\Omega(\omega)$ corresponding to the n branches of Fig. 6:

$$\Omega_1 \sim \left(\frac{H_1}{M_p}\right)^2 \left(\frac{H_1}{H}\right)^2 \left(\frac{a_1}{a}\right)^4 \left(\frac{\omega}{\omega_1}\right)^{4-2\nu_1-2\nu_2}, \qquad \omega_2 < \omega < \omega_1$$

$$\Omega_2 \sim \left(\frac{H_1}{M_p}\right)^2 \left(\frac{H_1}{H}\right)^2 \left(\frac{a_1}{a}\right)^4 \left(\frac{\omega}{\omega_1}\right)^{4-2\nu_1-2\nu_3} \left(\frac{\omega_2}{\omega_1}\right)^{-2\epsilon_2\nu_2+2\nu_3}, \qquad \omega_3 < \omega < \omega_2$$

$$\Omega_3 \sim \left(\frac{H_1}{M_p}\right)^2 \left(\frac{H_1}{H}\right)^2 \left(\frac{a_1}{a}\right)^4 \left(\frac{\omega}{\omega_1}\right)^{4-2\nu_1-2\nu_4} \left(\frac{\omega_2}{\omega_1}\right)^{-2\epsilon_2\nu_2} \left(\frac{\omega_3}{\omega_2}\right)^{-2\epsilon_3\nu_3} \left(\frac{\omega_1}{\omega_3}\right)^{-2\nu_4},$$
$$\omega_4 < \omega < \omega_3$$

$$\ldots \sim \ldots\ldots\ldots\ldots\ldots\ldots\ldots\ldots\ldots\ldots\ldots\ldots$$

$$\Omega_n \sim \left(\frac{H_1}{M_p}\right)^2 \left(\frac{H_1}{H}\right)^2 \left(\frac{a_1}{a}\right)^4 \left(\frac{\omega}{\omega_1}\right)^{4-2\nu_1-2\nu_{n+1}} \left(\frac{\omega_2}{\omega_1}\right)^{-2\epsilon_2\nu_2} \left(\frac{\omega_3}{\omega_2}\right)^{-2\epsilon_3\nu_3} \ldots.$$
$$\ldots. \left(\frac{\omega_n}{\omega_{n-1}}\right)^{-2\epsilon_n\nu_n} \left(\frac{\omega_1}{\omega_n}\right)^{-2\nu_{n+1}}, \qquad \omega < \omega_n. \tag{4.26}$$

It is important to point out that the slope of the spectrum, for a given branch Ω_i, is sensitive to the particular time evolution of the background only when hitting and when leaving the barrier: in fact, $\Omega_i(\omega) \sim \omega^{4-2\nu_1-2\nu_{i+1}}$. Note also that, if we wish to refer the spectrum to a frequency ω_k other than ω_1 (for instance to the peak frequency, which may be different from ω_1 if the potential is non-monotonic), we have to simply multiply the i-th branch of the spectrum by $(\omega_k/\omega_1)^{4-2\nu_1-2\nu_{i+1}}$.

Let us now apply the general prescription (4.26) to a typical pre-big bang scenario, in which the background evolves from an initial dilaton-driven phase with $\nu_{n+1} = 0$. Eventually, the background becomes radiation-dominated, so that $\nu_1 = 1/2$, and the ratio

$$\left(\frac{H_1}{H}\right)^2 \left(\frac{a_1}{a}\right)^4 = \Omega_r(t) \tag{4.27}$$

gives the radiation energy density, in critical units, rescaled to a time $t > t_1$. The various phases from η_n to η_1 correspond to possible types of background during the string era, and the ratio

$$g_1 = H_1/M_p \tag{4.28}$$

is taken to be of the same order as the present value of the fundamental ratio between string and Planck mass, $g_1 \sim M_s/M_p \sim 0.3 - 0.03$. Let us assume, also, that the evolution of the metric background, in the Einstein frame, is always accelerated from η_n to η_1, so that the potential $|V(\eta)|$ is non-decreasing, ω_1 is also the maximal amplified frequency, and $\epsilon_i = 1$ for all the transitions. The spectral distribution (4.26) for the i-th band of frequency, crossing the horizon during the string phase

$(i < n)$, and affected by i background transitions, can thus be written

$$\Omega_i(\omega) \sim g_1^2\, \Omega_r(t) \left(\frac{\omega}{\omega_1}\right)^{3-2\nu_{i+1}} \left(\frac{\omega_2}{\omega_1}\right)^{-2\nu_2} \left(\frac{\omega_3}{\omega_2}\right)^{-2\nu_3} \cdots \\ \cdots\cdots \left(\frac{\omega_i}{\omega_{i-1}}\right)^{-2\nu_i} \left(\frac{\omega_1}{\omega_i}\right)^{-2\nu_{i+1}}, \qquad \omega_{i+1} < \omega < \omega_i. \tag{4.29}$$

As anticipated in the previous section, since $\nu_{i+1} \geq 0$, this spectrum is characterized by a maximal slope $\Omega \sim \omega^3$ (which is the same slope as that of the frequency band crossing the horizon in the dilaton-driven phase, modulo log corrections). Also, since $\omega_n < \omega_{n-1} < ... < \omega_2 < \omega_1$, the peak of the spectrum, $g_1^2\Omega_r$, is reached at $\omega = \omega_1$, so that the spectral distribution has to be on the average non-decreasing from ω_n to ω_1, i.e. $\Omega_i \leq \Omega_1$. However, the spectrum could be non-monotonic during the string phase (in spite of the monotonicity of the potential). Consider in fact the ratio

$$\frac{\Omega_i(\omega_i)}{\Omega_{i-1}(\omega_{i-1})} = \left(\frac{\omega_i}{\omega_{i-1}}\right)^{3-2\nu_i}. \tag{4.30}$$

As $\omega_i < \omega_{i-1}$, the spectrum may be decreasing in the frequency band $\omega_i < \omega < \omega_{i-1}$, provided $3 < 2\nu_i = |2\alpha_i - 1|$. If we assume a constant curvature evolution in the String frame, the decreasing slope may be due to a phase of decreasing dilaton.

Suppose in fact that, during the string phase, there is a period in which the scale factor and the dilaton evolve in time (in the String frame) as

$$\tilde{a}_i = (-\eta)^{-1}, \qquad \tilde{\phi}_i = \gamma \ln \tilde{a}_i, \qquad \gamma < 0 \tag{4.31}$$

(corresponding to $\tilde{H}_i$ =const and decreasing dilaton, $\dot{\tilde{\phi}}_i = \gamma \tilde{H}_i < 0$). Tranforming to the Einstein frame[2,3] we have

$$a_i = \tilde{a}_i e^{-\phi_i/2} = (-\eta)^{-1+\gamma/2}, \qquad \phi_i = \tilde{\phi}_i = -\gamma \ln(-\eta) \tag{4.32}$$

(conformal time is the same in the two frames). The slope of the corresponding frequency band is then decreasing, since

$$3 - 2\nu_i = 3 - |2\alpha_i - 1| = 3 + 2\left(\frac{\gamma}{2} - 1\right) - 1 = \gamma < 0. \tag{4.33}$$

4.3. *Constraints from photon production*

In the context of the pre-big bang scenario, the time variation of the background fields that amplifies the perturbations of the metric tensor also amplifies the quantum fluctuations of the electromagnetic field[30,51]. The electromagnetic perturbation equation has the same form as eq. (4.2), with the only difference that the effective potential is generated by a time-dependent dilaton field,

$$V(\eta) = \frac{\phi'^2}{4} - \frac{\phi''}{2}. \tag{4.34}$$

For a cosmological evolution that can be approximated by various phases, characterized by a logarithmic variation of the dilaton, $\phi_i = -\beta_i \ln|\eta|$, the electromagnetic perturbation spectrum can thus be computed with the diagrammatic technique illustrated in the previous section. The result is a spectrum with the same general structure as that of the graviton spectrum, but with different (in general) values of the Bessel indices ν_i. The phenomenological bounds imposed on electromagnetic perturbations constrain the parameters of the cosmological background and then, indirectly, also the unknown slope of the graviton spectrum during the string phase.

In order to discuss this effect we shall work in the context of the minimal model of Section 4.1, characterized by three phases (dilaton → string → radiation), and we shall parametrize the unknown background evolution during the string phase (in the Einstein frame) with two powers α and β,

$$a \sim |\eta|^{\alpha}, \qquad \phi \sim -2\beta \ln|\eta|, \qquad \eta_s < \eta < \eta_r. \tag{4.35}$$

For the dilaton-driven phase ($\eta < \eta_s$) and for the radiation-dominated phase ($\eta > \eta_r$) the background evolution is known[2,3], and is given respectively by

$$\begin{aligned} a \sim |\eta|^{1/2}, \qquad & \phi \sim -\sqrt{3}\ln|\eta|, \qquad && \eta < \eta_s, \\ a \sim |\eta|, \qquad & \phi \sim \text{const}, \qquad && \eta > \eta_r. \end{aligned} \tag{4.36}$$

The effective potential $|V(\eta)|$ grows monotonically for $\eta < \eta_r$ and goes rapidly to zero for $\eta > \eta_r$, both for gravitons and photons. The spectra have two branches, which can easily be deduced from the diagram of Fig. 8.

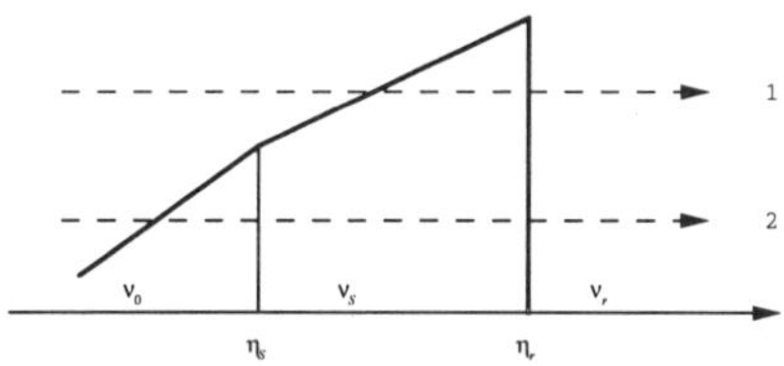

	ν_0	ν_s	ν_r
gravitons	0	$\vert\alpha-1/2\vert$	1/2
photons	$(\sqrt{3}-1)/2$	$\vert\beta-1/2\vert$	1/2

Figure 8: Diagram and Bessel indices for the graviton and photon spectrum in a minimal, two-parameter model of background.

The Bessel indices ν_i are fixed by the time evolution of the background, according to eqs. (4.35) and (4.36). The two parameters α and β are not independent, however, if we require[30,33] (as seems natural) that, during the string phase, the curvature remains controlled by the string scale $M_s = \lambda_s^{-1}$. This implies, in the Einstein frame[45],

$$\left|\frac{H_s}{H_1}\right| \simeq \left|\frac{\eta_r}{\eta_s}\right|^{1+\alpha} \simeq \frac{g_s}{g_r} \simeq \left|\frac{\eta_r}{\eta_s}\right|^{\beta}, \quad g_s = e^{\phi_s/2}, \; g_r = e^{\phi_r/2}, \tag{4.37}$$

from which

$$1+\alpha \simeq \beta \simeq -\frac{\log(g_s/g_r)}{\log|\eta_s/\eta_r|}. \tag{4.38}$$

Defining the convenient parameters

$$\begin{aligned} x &= \log_{10}|\eta_s/\eta_r| = \log_{10} z_s, \\ y &= \log_{10}(g_s/g_r) = \log_{10}\frac{H_s/M_p}{H_r/M_p}, \end{aligned} \tag{4.39}$$

related respectively to the duration of the string phase, $z_s = |\eta_s/\eta_r|$, and to the corresponding shift of the dilaton coupling, g_s/g_r, we obtain

$$\alpha - \frac{1}{2} = \frac{3}{2} + \frac{y}{x}, \quad \beta - \frac{1}{2} = \frac{1}{2} + \frac{y}{x}, \quad \frac{\omega_r}{\omega_s} = \frac{H_r a_r}{H_s a_s} \simeq \left|\frac{\eta_s}{\eta_r}\right| = 10^x. \tag{4.40}$$

By applying the rules of the previous section we can now parametrize the spectra in terms of x and y. For the two branches of the graviton spectrum we obtain, from Fig. 8,

$$\begin{aligned} \Omega_G &\sim g_1^2\, \Omega_r(t) \left(\frac{\omega}{\omega_1}\right)^{3-|3+2y/x|}, && 10^{-x} < \frac{\omega}{\omega_1} < 1 \\ &\sim g_1^2\, \Omega_r(t) \left(\frac{\omega}{\omega_1}\right)^{3} 10^{|3x+2y|}, && \frac{\omega}{\omega_1} < 10^{-x} \end{aligned} \tag{4.41}$$

(we have chosen to refer all frequencies to the maximal amplified one, ω_1). With the same rules we obtain the two branches of the electromagnetic spectrum,

$$\begin{aligned} \Omega_{em} &\sim g_1^2\, \Omega_r(t) \left(\frac{\omega}{\omega_1}\right)^{3-|1+2y/x|}, && 10^{-x} < \frac{\omega}{\omega_1} < 1 \\ &\sim g_1^2\, \Omega_r(t) \left(\frac{\omega}{\omega_1}\right)^{4-\sqrt{3}} 10^{x(1-\sqrt{3})+|x+2y|}, && \frac{\omega}{\omega_1} < 10^{-x}. \end{aligned} \tag{4.42}$$

The back reaction of the electromagnetic perturbations on the geometry turns out to be negligible, consistently with the use of a homogeneous and isotropic background, provided we impose on the two-dimensional parameter space of our model the condition[30] $y \gtrsim -2x$. Also, the amplified perturbations are large enough to act as "seeds" for the galactic magnetic field, provided[52] $\Omega_G(10^{-14}\text{Hz}) \gtrsim 10^{-34}\Omega_r$. These two conditions define an allowed region in the $\{x, y\}$ plane, and constrain the possible slope of the string branch of the graviton spectrum.

The first condition, by itself, does not provide significant constraints. When combined with the second, however, we obtain the constraint

$$2 < 3 - \left|3 + 2\frac{y}{x}\right| < 4, \tag{4.43}$$

which for the graviton spectrum (4.41) defines the allowed region shown in Fig. 9 (where I have used $g_1 = 0.1$ and $\omega_1 = 10^{11}$Hz). The bold dashed line in Fig. 9

represents the prediction for the graviton spectrum for the case that the amplified electromagnetic perturbations are strong enough, on a large scale, to contribute in a significant way to the observed CMB anisotropy[32]. This possibility would require indeed a flat enough string branch of the electromagnetic spectrum, $\Omega_{em} \sim \omega^{\epsilon}$, $0 < \epsilon < 1$, and could be implemented consistently with other phenomenological constraints only for a background in which the comoving amplitude of perturbations grows in time also outside the horizon[32]. In terms of our parameters this gives the conditions $y > -x/2$ and $y/x = 1 + \epsilon$ and implies, for the string branch of the spectrum, a slope $\Omega_G \sim \omega^{2+\epsilon}$.

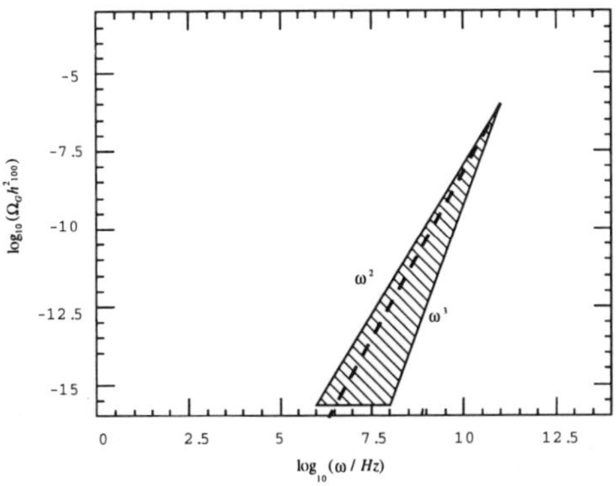

Figure 9: The shaded area shows the allowed region for the graviton spectrum, in a model in which the electromagnetic fluctuations are amplified enough to act as "seeds" for the galactic magnetic field.

It is important to note that the allowed region of Fig. 9 corresponds to a very small energy density of the relic graviton background in the kHz range, $\Omega_G \lesssim 10^{-21}$; this is certainly outside the sensitivity of gravity wave detectors now existing or planned (see Section 4.5). A possible future detection of a relic graviton background in the kHz range could thus be used to exclude, in our cosmological scenario, a primordial origin of the seed fields and a possible electromagnetic origin of the CMB anisotropy (at least in the context of a minimal model, and in the range of validity of the low-energy perturbation equation). Conversely, if we believe that the "dilatonic" amplification of the electromagnetic fluctuations is efficient enough to be responsible for the above effects, we should expect a graviton background outside the detector sensitivities, in the kHz range. This may provide strong motivations for developing detectors with improved sensitivities in the very high frequency range ($\sim$ GHz), near the end point of the spectrum, where the background is expected to reach the peak intensity. At lower frequencies, the absence of a signal at the sensitivity level of 10^{-6} (or better) in Ω_G would nevertheless be important, since it would support, indirectly, a primordial dilatonic origin of the cosmic magnetic fields.

4.4. *Non-minimal models*

In the minimal model of pre-big bang assumed in the previous Sections, the end of the high curvature string phase was nearly coincident with the beginning of the radiation-dominated, constant dilaton phase. It is not impossible, however, to imagine more complex models of backgrounds, in which the dilaton is still growing while the curvature starts decreasing and where the radiation era with frozen dilaton starts at much later times, when the curvature scale is much below the string scale. An example of such a background is illustrated in Fig. 10 (the minimal model corresponds to the particular case $\eta_r \sim \eta_1$, and $g_1 \sim 1$).

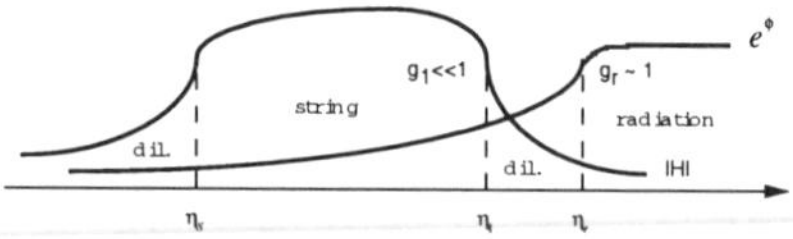

Figure 10: Non-minimal model of pre-big bang background.

It is conceivable, in fact, that the combined action of loops and α' corrections may stop the growth of the curvature and may induce a change of branch of the solution, while the string coupling remains growing[10]. Only later could the dilaton be attracted to a fixed point of the non-perturbative potential. In that context, however, the amplified vacuum fluctuations are not expected to become naturally dominant, and the radiation should arise from other sources (for instance, coherent dilaton oscillations).

If we accept a model like that of Fig. 10, the non-vanishing part of the effective potential appearing in the perturbation equation (4.2) is no longer monotonic. Consequently, high-frequency modes may reenter the horizon (or leave the potential barrier) *before* the beginning of the radiation era, i.e. for $\eta_1 < \eta < \eta_r$. This modifies the slope of the spectrum in the high frequency sector. For the background of Fig. 10, in particular, the spectrum has three branches, whose slope depends on the relative duration of the string phase and of the new phase inserted before the radiation era. If $|\eta_r/\eta_1| < |\eta_s/\eta_1|$, i.e. $\omega_r > \omega_s$, the spectrum corresponds to the diagram of Fig. 7. If $\omega_r < \omega_s$ the three branches of the spectrum are instead obtained from the diagram of Fig. 11, where I have supposed that the decelerated phase from η_1 to $\eta_r > 0$ is again a dilaton-driven solution of the lowest order effective action, such that

$$a \sim \eta^{1/2}, \qquad e^{\phi} \sim \eta^{\sqrt{3}}, \qquad \left(\frac{H_1}{H_r}\right) \simeq \left(\frac{g_r}{g_1}\right)^{\sqrt{3}} \simeq \left(\frac{a_r}{a_1}\right)^3, \qquad \eta > 0. \tag{4.44}$$

By referring the spectrum to the maximal amplified frequency ω_1, and applying the rules of Sect. 4.2, we obtain from Fig. 11

$$\Omega_1 \sim g_1^2 \left(\frac{g_r}{g_1}\right)^{2/\sqrt{3}} \Omega_r(t) \left(\frac{\omega}{\omega_1}\right)^{4-2\nu_s}, \qquad \omega_s < \omega < \omega_1$$

$$\sim g_1^2 \left(\frac{g_r}{g_1}\right)^{2/\sqrt{3}} \Omega_r(t) \left(\frac{\omega}{\omega_1}\right)^4 \left(\frac{\omega_s}{\omega_1}\right)^{-2\nu_s}, \qquad \omega_r < \omega < \omega_s$$

$$\sim g_1^2 \left(\frac{g_r}{g_1}\right)^{2/\sqrt{3}} \Omega_r(t) \left(\frac{\omega}{\omega_1}\right)^3 \left(\frac{\omega_s}{\omega_1}\right)^{-2\nu_s} \left(\frac{\omega_r}{\omega_1}\right), \quad \omega < \omega_s \tag{4.45}$$

(neglecting logarithmic corrections, and using the relation $\omega_1/\omega_r = (H_1/H_r)^{2/3}$, valid for the background (4.44)).

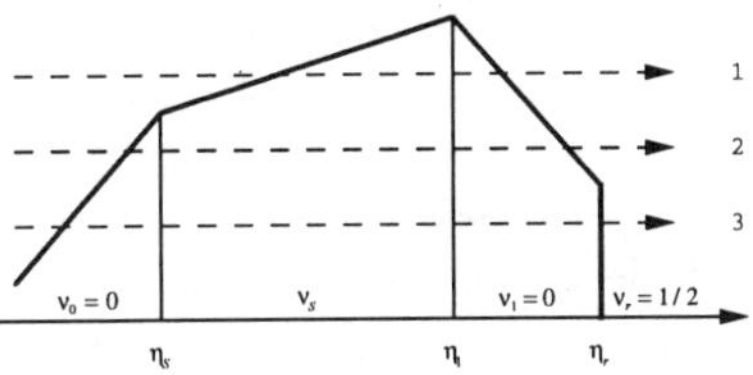

Figure 11: The three branches of the graviton spectrum for the background of Fig. 10, and for the case $\omega_r < \omega_s$.

For the case $\omega_s < \omega_r$ the spectrum is similarly obtained from the diagram of Fig. 7. This spectrum differs from that of the minimal model in three important aspects:

- The end point value of the spectrum, $\Omega_1(\omega_1)$, is shifted to arbitrarily low values $\sim \Omega_r g_1^{2(1-1/\sqrt{3})} << \Omega_r$, since $g_1 << g_r \sim 1$. The maximal amplified frequency is correspondingly shifted,

$$\omega_1(t_0) = \frac{\omega_1}{\omega_r}\frac{H_r a_r}{a(t_0)} \sim g_1^{1/2} T_0 \left(\frac{H_1}{H_r}\right)^{1/6} \sim g_1^{1/2} \left(\frac{g_r}{g_1}\right)^{1/2\sqrt{3}} \times 10^{11}\text{Hz} << 10^{11}\text{Hz} \tag{4.46}$$

  (T_0 is the present CMB temperature).

- The maximal allowed slope is steeper, $\Omega \sim \omega^4$ instead of $\Omega \sim \omega^3$.

- The spectrum may be truly non-monotonic, with a peak value at ω_s higher than the end point value, $\Omega(\omega_s) > \Omega(\omega_1)$, provided $4 < 2\nu_s$.

So, in spite of the fact that the end point is shifted to lower values, the relic background corresponding to a non-minimal model could be detected even more easily than in the minimal case, because of a possible peak at lower frequencies, more accessible to present detectors. It is true that the upper limit, $\Omega_g \sim 10^{-2}\Omega_r$, cannot be obtained naturally in a non-minimal scenario. However, the analysis of photon production in this scenario shows that such a maximal peak signal would not be excluded by the bounds on the electromagnetic spectrum, provided $\omega_r < \omega_s$, i.e. provided the post-big bang dilaton-driven phase is long enough with respect to the string phase. It is thus important to point out that, in the context of a non-minimal model, the possibility is not excluded to see a graviton background with a peak $\Omega_G \sim 10^{-6}$ at an arbitrarily low frequency ω_s ($> 10^{-8}$Hz, however, because of the pulsar bound[48]), without implying that the spectrum is flat from ω_1 to ω_s, thus evading the otherwise constraining bound on the total integrated energy following from nucleosynthesis (see eq. (4.14)).

4.5. *Experimental sensitivities*

It may be interesting, at this point, to compare the relic graviton background expected in the context of the pre-big bang scenario with the experimental sensitivity of gravity wave detectors. A precise and complete discussion of the present experimental possibilities of detecting a stochastic gravity wave background is, of course, outside the scope of this paper (see for instance the reviews of Refs. [52,53]). Here I will simply report, at the level of an order of magnitude estimate, the sensitivities accessible to some detectors that are already existing and operating, and some other that are now under construction and will be operating in the near future.

The "theory versus experiments" situation is qualitatively sketched in Fig. 12, where the bold upper border line defines the maximal allowed region for the relic graviton spectrum in a string cosmology context, obtained from Fig. 4; the bold lower line gives the corresponding border in the standard inflationary context.

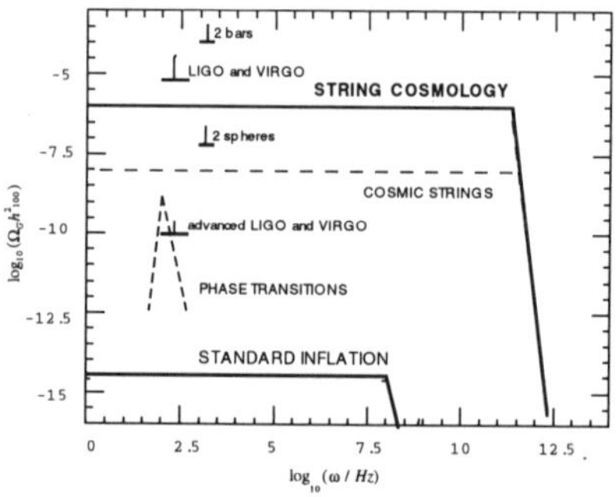

Figure 12: Present and future experimental sensitivities, compared with the allowed region for the graviton spectrum.

Available data, at present, come from the cryogenic, resonant-mass detector EXPLORER, operating at CERN, and provide the upper bound[55] $\Omega_G \lesssim 500$ at a frequency $\nu = 923$ Hz, which is too high to be significant for a relic cosmological spectrum. However, by cross-correlating the data obtained from two bar detectors with the same characteristics as those already existing (EXPLORER, NAUTILUS, AURIGA), it will be possible to reach, in very few years, the level of sensitivity[55] $\Omega_G \sim 8 \times 10^{-4}$ around the kHz range. This is still outside, but not so far off, the border of the interesting region, as shown in Fig. 12. Even better sensitivities, in the same frequency range, can be reached through the cross-correlation of a bar and an interferometric detector[56,57]. And in fact the first generation of interferometric detectors, LIGO[58] and VIRGO[59], when operating, will soon reach a sensitivity[53] $\Omega_G \sim 10^{-5}$ centred around the 100 Hz frequency band. The possible improvement in the sensitivity of microwave cavities available at present, operating as gravity wave detectors, is also under active study[60].

Also shown in Fig. 12 are the sensitivities expected from the cross-correlation of two resonant spherical detectors[55,57], $\Omega_G \sim 10^{-7}$ at $\nu \sim 1$ kHz, and the planned sensitivity of the advanced interferometric detectors[53,61], $\Omega_G \sim 10^{-10}$ at $\nu \sim 10^2$kHz. Both are well inside the allowed region, and are thus able to constrain significantly, in case of no detection, the parameters of our class of string cosmology models. To the best of my knowledge, in this range of frequency there are only two other possible primordial backgrounds, competing in intensity with the relics of a pre-big bang phase. The first is the background (with very flat spectrum) generated by cosmic strings[62] or by other topological defects[63], characterized by $\Omega_G \sim 10^{-8}$. The second is a background sharply peaked around $10^2 - 10^3$Hz, generated by first order phase transitions[62,64] in the context of extended inflation models, with peak value $\Omega_G \sim 10^{-8}$–10^{-9}. Both these two possibilities are illustrated by the dashed curves of Fig. 12.

Finally, it should be recalled that for non-minimal models of the pre-big bang era, the allowed string cosmology region of Fig. 12 can be extended to the left, in principle without further suppression down to frequencies of 10^{-7}Hz. The space interferometer LISA[65], operating around the mHz range, will thus provide significant bounds on non-minimal models and, at the maximal expected sensitivity level of $\Omega_G \sim 10^{-10}$, also on minimal models whose spectrum decreases enough to evade the nucleosynthesis bound.

5. Conclusion

The cosmological scenario in which I have discussed the production of a relic graviton background was based on two main assumptions:

- the initial state of the Universe is the string perturbative vacuum;
- the pre-big bang phase evolves smoothly into a final decelerated state, dominated by radiation and with constant dilaton.

While the first assumption represents an arbitrary choice of the initial conditions, the second one reflects our present ignorance about string cosmology in the high-curvature and non-perturbative regime, where the transition from the pre- to the post-big bang epoch is expected to occur. Our present lack of knowledge about the kinematics of such a transition forbids, in particular, a precise prediction of the detailed shape of the graviton spectrum.

If we accept the above two assumptions, however, we can already conclude that from a phenomenological point of view the pre-big bang scenario is significantly different from the standard inflationary scenario, and that these differences lead to effects that can be tested (at least in principle) even more easily than the corresponding effects of the standard scenario.

In particular, there is no compelling reason in this context to exclude a relic graviton background as high as

$$\Omega_g(\omega) \sim 10^{-6} h_{100}^{-2}, \tag{5.1}$$

in the whole range from 1 Hz to 100 GHz. Thus, experimental measurements performed at this level of sensitivity would be true tests of Planck-scale physics. The detection of such a cosmic graviton background would represent an event of an importance comparable to the detection of the cosmic black-body radiation[66]. Even more important than this, in some sense, because the CMB photons are relic radiation from the hot, "big bang" phase that occurred in the past history of our Universe. Those gravitons would, instead, be relics of a much earlier "pre-big bang" epoch.

Acknowledgements: This work is supported in part by EC contract ERBCHRX-CT94-0488. I am grateful to Ramy Brustein, Massimo Giovannini and Gabriele Veneziano for useful discussions and for fruitful collaboration on the relic graviton background in string cosmology.

6. Appendix A
The dimensionless amplitude $|\delta_h|$

Consider tensor perturbations, in a conformally flat metric background, and in the transverse traceless gauge,

$$g_{\mu\nu} + \delta g_{\mu\nu}, \qquad \delta g_{\mu\nu} = a^2(\eta) h_{\mu\nu}, \qquad h_\mu^\mu = 0 = \nabla_\nu h_\mu^\nu. \tag{A.1}$$

The gravitational action, perturbed up to second order in each of the two physical polarization modes $h(x)$,

$$\frac{M_p^2}{16\pi} \int d^3x d\eta \; a^2 \left[h'^2 - (\nabla_i h)^2\right], \tag{A.2}$$

defines the canonical variable[24,26,28,67] u, which diagonalizes the perturbed action:

$$u = ahM_p, \tag{A.3}$$

and whose Fourier components u_k are required to satisfy canonical commutation relations:

$$[u_k, \dot{u}_{k'}] = i\delta_{kk'}. \tag{A.4}$$

This fixes the dimensionality of the mode h_k, $[h_k] = [u_k/M_p] = [k^{-3/2}]$, and gives the correct normalization for the Fourier transform in a finite volume V,

$$h(\vec{x},t) = \frac{1}{\sqrt{V}} \sum_{\vec{k}} h_{\vec{k}}(t) e^{i\vec{k}\cdot\vec{x}}. \tag{A.5}$$

In the continuum limit

$$h(\vec{x},t) = \frac{\sqrt{V}}{(2\pi)^3} \int d^3k \; h(\vec{k},t) e^{i\vec{k}\cdot\vec{x}}. \tag{A.6}$$

Consider now the two-point correlation function

$$\xi(\vec{r}) = \langle h(\vec{x}) h(\vec{x}+\vec{r}) \rangle, \tag{A.7}$$

where the brackets denote the quantum expectation value on a given state, if we work in the second-quantization formalism in which the perturbation field is expanded into annihilation and creation operators. We recall, however, that the perturbation background is a stochastic background as a consequence of its quantum origin, since it is obtained by amplifying the quantum fluctuations of the vacuum[68], for which

$$\langle a_k^\dagger a_{k'} \rangle = |f|^2 \delta^3(k-k'), \tag{A.8}$$

where f is an appropriate normalization coefficient. This means that, if we perform the classical limit by replacing quantum expectation values with ensemble averages, we are led to the stochastic condition

$$\langle h(\vec{k}) h(-\vec{k}') \rangle = |h(\vec{k})|^2 \frac{(2\pi)^3}{V} \delta^3(k-k'). \tag{A.9}$$

For a stochastic background the correlation function (A.7) thus becomes

$$\xi(\vec{r}) = \int \frac{d^3k}{(2\pi)^3} |h(k)|^2 e^{-i\vec{k}\cdot\vec{r}} = \frac{1}{2\pi^2} \int \frac{dk}{k} \frac{\sin kr}{kr} |\delta_h(k)|^2 , \tag{A.10}$$

where we have defined

$$|\delta_h(k)| = k^{3/2} |h(k)|. \tag{A.11}$$

Note that we have used the reality condition, $h^*(k) = h(-k)$, and the so-called isotropy condition, i.e. the assumption that $|h(k)|$ is a function only of $k = |\vec{k}|$. The brackets of eq. (A.7) can also be formally defined as a macroscopic average over space:

$$\xi(\vec{r}) = \frac{1}{V} \int d^3x \; h(\vec{x}) h(\vec{x}+\vec{r}); \tag{A.12}$$

the result for $\xi(\vec{r})$ is exactly the same as in eq. (A.10). By evaluating the correlation function for a distance $r = k^{-1}$ we finally obtain from (A.10)

$$\left[\xi^{1/2}(r)\right]_{r=k^{-1}} \sim |\delta_h(k)| . \tag{A.13}$$

The variable $|\delta_h(k)|$ can thus be interpreted as the typical, dimensionless amplitude of tensor perturbations over a comoving length scale $r = k^{-1}$.

The spectral density of perturbations, $\Omega_G(\omega)$, can be conveniently expressed in terms of $|\delta_h(k)|$ by recalling that the average energy density, summing over polarization, is defined by[34]

$$\rho_g = \frac{dE_g}{a^3 d^3x} = \frac{M_p^2}{8\pi}\langle \dot{h}^2\rangle = -\frac{M_p^2}{a^3 8\pi}\frac{V}{(2\pi)^6}\int d^3k d^3k' \omega\omega' e^{i\vec{x}\cdot(\vec{k}+\vec{k}')}\langle h(k)h(k')\rangle, \quad (A.14)$$

where $\omega = |\vec{k}|/a$. By using the stochastic condition (A.9), in the hypothesis $|\delta_h(k)| = |\delta_h(-k)|$,

$$\frac{dE_g}{a^3 d^3x\, d\ln k} = \frac{M_p^2}{(2\pi a)^3}|\delta_h(k)|^2\omega^2. \quad (A.15)$$

The spectral distribution of the proper energy density, in critical units of $\rho_c = 3M_p^2H^2/8\pi$, can thus be expressed as

$$\frac{d(\rho_g/\rho_c)}{d\ln\omega} = \frac{dE_g}{\rho_c a^3 d^3x\, d\ln\omega} = \Omega_G(\omega,t) = \frac{1}{3\pi^2}|\delta_h(\omega)|^2\left(\frac{\omega}{H}\right)^2. \quad (A.16)$$

7. Appendix B
The spectral amplitude S_h

Consider the Fourier transform of the polarized gravity wave amplitude h,

$$h(\nu) = \int dt\; h(t)e^{-2\pi i\nu t}, \qquad \nu = \omega/2\pi. \quad (B.1)$$

The average proper energy density, summing over polarizations, can be written

$$\begin{aligned}\rho_g &= \frac{dE_g}{a^3d^3x} = \frac{M_p^2}{8\pi}\langle\dot{h}^2\rangle = \\ &= -4\pi^2\frac{M_p^2}{8\pi}\int d\nu d\nu' \nu\nu'\langle h(\nu)h(\nu')\rangle e^{2\pi i(\nu+\nu')t},\end{aligned} \quad (B.2)$$

where the brackets denote time or ensemble average. By defining the spectral amplitude $S_h(\nu)$, such that[69]

$$\langle h(\nu)h^*(-\nu')\rangle = \frac{1}{2}\delta(\nu+\nu')S_h(\nu), \quad (B.3)$$

the energy distribution Ω_g, in critical units, can be expressed as

$$\Omega_G(\nu,t) = \frac{1}{\rho_c}\frac{d\rho_G}{d\ln\nu} = \frac{4\pi^2}{3H^2(t)}\nu^3 S_h(\nu) \quad (B.4)$$

(I have assumed $S_h(\nu) = S_h(-\nu)$).

A particular value of $S_h(\nu)$, representing the experimental sensitivity at a given frequency ν, corresponds to a present graviton energy density[43]

$$\Omega_G(\nu, t_0) h_{100}^2 \simeq 1,25 \times 10^{36} \nu^3 S_h(\nu) \ \mathrm{Hz}^{-2}. \tag{B.5}$$

Reaching the level of expected maximal signal, $\Omega_G h_{100}^2 \lesssim 10^{-6}$ (see Sect. 4. 1), would thus require a sensitivity[43]

$$S_h^{1/2}(\nu) \lesssim 3 \times 10^{-26} \left(\frac{\mathrm{kHz}}{\nu}\right)^{3/2} \mathrm{Hz}^{-1/2}. \tag{B.6}$$

Comparing eqs. (B.4) and (A.16), we can finally obtain a useful relation between the spectral amplitude S_h and the dimensionless amplitude δ_h,

$$S_h\left(\frac{\omega}{2\pi}\right) = \frac{2}{\pi\omega} \left|\delta_h(\omega)\right|^2. \tag{B.7}$$

1. M. Gasperini and G. Veneziano, Astropart. Phys. 1, 317 (1993).
2. M. Gasperini and G. Veneziano, Mod. Phys. Lett. A8, 3701 (1993).
3. M. Gasperini and G. Veneziano, Phys. Rev. D50, 2519 (1994).
4. M. Gasperini, N. Sanchez and G. Veneziano, Nucl. Phys. B364, 365 (1991).
5. N. Sanchez, *"String theory in curved spacetime"*, this book.
6. G. Veneziano, *"Classical and quantum string cosmology"*, this book.
7. E. J. Copeland, A. Lahiri and D. Wands, Phys. Rev. D53, 4247 (1994).
8. I. Antoniadis, J. Rizos and K. Tamvakis, Nucl. Phys. B415, 497 (1994).
9. R. Easther and K. Maeda, *"One-loop superstring cosmology and the non-singular Universe"*, hep-th/9605173.
10. S. J. Rey, *"Back reaction and graceful exit in string inflationary cosmology"*, hep-th/9605176; M. Gasperini and G. Veneziano, *"Singularity and exit problems in two-dimensional string cosmology"*, CERN-TH/96-165, hep-th/9607126.
11. R. Brustein and G. Veneziano, Phys. Lett. B329, 429 (1994); N. Kaloper, R. Madden and K. Olive, Nucl. Phys. B452, 677 (1995); Phys. Lett. B371, 34 (1996); R. Easther, K. Maeda and D. Wands, Phys. Rev. 53, 4247 (1996).
12. M. Gasperini, J. Maharana and G. Veneziano, hep-th/9602087, Nucl. Phys. B (1996), in press; M. Gasperini and G. Veneziano, hep-th/9602096, Gen. Rel Grav. 28 (1996), in press.
13. J. E. Lidsey, *"Inflationary and deflationary branches in extended pre-big bang cosmology"*, gr-qc/9605017.
14. M. Gasperini, M. Giovannini, K. Meissner and G. Veneziano, *"Evolution of a string network in background with rolling horizons "*, this book (hep-th/9502130).
15. G. Veneziano, Phys. Lett. B265, 287 (1991).
16. A. A. Tseytlin, Mod. Phys. Lett. A6, 1721 (1991).
17. K. A. Meissner and G. Veneziano, Phys. Lett. B267, 33 (1991); Mod. Phys. Lett. A6, 3397 (1991).
18. A. Sen, Phys. Lett. B 271, 295 (1991).
19. M. Gasperini and G. Veneziano, Phys. Lett. B277, 256 (1992).

20. M. Gasperini, J. Maharana and G. Veneziano, Phys. Lett. B272, 277 (1991); B296, 51 (1992).
21. F. Lucchin and S. Matarrese, Phys. Lett. B164, 282 (1985).
22. L. P. Grishchuk, Sov. Phys. JEPT 40 (1975) 409.
23. A. A. Starobinski, JEPT Lett. 30 (1979) 682.
24. M. Gasperini and M. Giovannini, Phys. Rev. D47, 1519 (1993).
25. R. B. Abbott, B. Bednarz and S. D. Ellis, Phys. Rev. D33, 2147 (1986).
26. R. Brustein, M. Gasperini, M. Giovannini, V. Mukhanov and G. Veneziano, Phys. Rev. D51, 6744 (1995).
27. J. Hwang, Ap. J. 375, 443 (1991).
28. M. Giovannini, PhD Thesis (Turin University, 1996); M. Gasperini and M. Giovannini, *"Normal modes for metric fluctuations in a class of anisotropic background"*, gr-qc/9604002.
29. N. D. Birrel and P. C. W. Davies, *"Quantum fields in curved space"* (Cambridge Univ. Press, Cambridge, 1982).
30. M. Gasperini, M. Giovannini and G. Veneziano, Phys. Rev. Lett. 75, 3796 (1995).
31. G. F. Smoot et al., Ap. J. 396, L1 (1992); C. L. Bennet et al., Ap. J. 430, 423 (1994).
32. M. Gasperini, M. Giovannini and G. Veneziano, Phys. Rev. D52, 6651 (1995); R. Durrer et al., in preparation.
33. R. Brustein, M. Gasperini, M. Giovannini and G. Veneziano, Phys. Lett. B361, 45 (1995).
34. See for instance E. W. Kolb and M. S. Turner, *The Early Universe* (Addison-Wesley Publ. Co., New York, 1990).
35. L. M. Krauss and M. White, Phys. Rev. Lett. 69, 869 (1992).
36. D. Polarski and A. A. Starobinski, Class. Quantum Grav. 13, 377 (1996).
37. V. Kaplunovsky, Phys. Rev. Lett. 55, 1036 (1985).
38. L. F. Abbott and M. B. Wise, Nucl. Phys. B244, 54 (1984).
39. B. Allen, Phys. Rev. D37, 2078 (1988).
40. L. P. Grishchuk, Class. Quantum Grav. 10, 2449 (1993).
41. G. Veneziano, private communication.
42. R. Brustein, M. Gasperini and G. Veneziano, in preparation.
43. R. Brustein, M. Gasperini and G. Veneziano, *" Peak and end point of the relic graviton background in string cosmology"*, hep-th/9604804.
44. G. Veneziano, in *"String gravity and physics at the Planck energy scale"*, ed. by N. Sanchez and A. Zichichi (Kluwer Acad. Publ., Dordrecht, 1996), p. 285.
45. M. Gasperini, *ibid*, p. 305.
46. A. Buonanno, M. Maggiore and C. Ungarelli, *"Spectrum of relic gravitational waves in string cosmology"*, gr-qc/9605072.
47. L. P. Grishchuk, Sov. Phys. Usp. 31, 940 (1988); V. Shani, Phys. Rev. D42, 453 (1990).
48. V. Kaspi, J. Taylor and M. Ryba, Ap. J. 428, 713 (1994).
49. D. H. Lyth, *"What would we learn by detecting a gravitational wave signal in the CMB anisotropy?"*, hep-ph/9606387.
50. M. Abramowitz and I. A. Stegun, *"Handbook of mathematical functions"* (Dover, New, York, 1972).
51. D. Lemoine and M. Lemoine, Phys. Rev D52, 1955 (1995).
52. M. S. Turner and L. M. Widrow, Phys. Rev. D37, 2743 (1988).
53. B. Allen, *"Stochastic gravity-wave background: sources and detectors"*, gr-qc/9604033.
54. E. Coccia, *"Resonant mass gravitational wave detectors"*, to appear in Proc. of the 14th Int. Conf. on General Relativity, Florence, August 1995.
55. P. Astone et al., *" Upper limit for a gravitational wave stochastic background*

measured with the EXPLORER and NAUTILUS gravitational wave resonant detectors, INFN Preprint, Rome, February 1996, to appear.

56. P. Astone, J. A. Lobo and B. F. Schutz, Class. Quantum Grav. 11, 2093 (1994).
57. S. Vitale, M. Cerdonio, E. Coccia and A. Ortolan, *"Gravitational-wave stochastic background detection with resonant-mass detectors"*, in preparation.
58. A. Abramovici et al., Science 256, 325 (1992).
59. C. Bradaschia et al., Nucl. Instrum. Meth. A289, 518 (1990).
60. E. Picasso, private communication.
61. K. S. Thorne, in *" 300 years of gravitation"*, ed. by S. W. Hawking and W. Israel (Cambridge Univ. Press, Cambridge, 1987), p. 330.
62. R. A. Battye and E. P. S. Shellard, *"Primordial gravitational waves: a probe of the very early Universe"*, astro-ph/9604059.
63. X. Martin and A. Vilenkin, *"Gravitational wave background from hybrid topological defects"*, astro-ph/9606022.
64. M. S. Turner and F. Wilczek, Phys. Rev. Lett. 65, 3080 (1990).
65. J. R. Jafry, J. Cornelisse and R. Reinhard, ESA Journal 8, 219 (1994).
66. E. Penzias and R. W. Wilson, Ap. J. 142, 419 (1965).
67. V. F. Mukhanov, H. A. Feldman and R. H. Brandenberger, Phys. Rep. 215, 203 (1992).
68. L. P. Grishchuk, Phys. Rev. Lett. 70, 2371 (1993).
69. E. Flanagan, Phys. Rev D48, 2389 (1993).

CLASSICAL SPLITTING OF FUNDAMENTAL STRINGS.

H.J. de Vega[(a)],

J. Ramírez Mittelbrun[(b)], M. Ramón Medrano[(b)],

N. Sánchez[(c)]

[(a)] Laboratoire de Physique Théorique et Hautes Energies, Paris.

Laboratoire Associé au CNRS UA 280, Université Paris VI (P. et M. Curie) et Université Paris VII (D. Diderot).[1]

[(b)] Departamento de Física Teórica, Madrid.[2]

[(c)] Observatoire de Paris, DEMIRM.

Laboratoire Associé au CNRS UA 336, Observatoire de Paris et École Normale Supérieure.[3]

---

[1]Postal address: LPTHE, Tour 16, 1er étage, Université Paris VI,

4, Place Jussieu, 75252, Paris cedex 05, FRANCE.

[2]Postal address: Facultad de Ciencias Físicas, Universidad Complutense, Ciudad Universtaria, E-28040, Madrid, ESPAÑA.

[3]Postal address: Observatoire de Paris, DEMIRM, 61, Avenue de l'Observatoire, 75014 Paris, FRANCE.

Abstract

We find exact solutions of the string equations of motion and constraints describing the *classical* splitting of a string into two. We show that for the same Cauchy data, the strings that split have **smaller** action than the string without splitting. This phenomenon is already present in flat space-time. The mass, energy and momentum carried out by the strings are computed. We show that the splitting solution describes a natural decay process of one string of mass M into two strings with a smaller total mass and some kinetic energy. The standard non-splitting solution is contained as a particular case. We also describe the splitting of a closed string in the background of a singular gravitational plane wave, and show how the presence of the strong gravitational field increases (and amplifies by an overall factor) the negative difference between the action of the splitting and non-splitting solutions.

11.25.-w, 98.80.Cq, 11.27+d

I. INTRODUCTION

A great amount of work has been devoted to string theory in the lasts years. However, very little attention has been paid to the theory of fundamental strings as a classical theory. The interaction of strings through joining and splitting has been mostly considered at the *quantum* level, although self-intersections of classical cosmic strings have been studied in [1], and the classical string splitting for hadronic strings has been discussed in [2]. At the quantum level, the usual procedure to compute the quantum string scattering amplitudes is based on the evaluation of the correlation functions of vertex operators or functionals, which are constructed out of a particular type of solution of the classical string equations of motion and constraints: one in which the string propagates without splitting and sweeping a world sheet that has the topology of a cylinder or a strip [3]. In this paper, we will show that besides these classical solutions which are used to build vertex operators, there also exist *classical* solutions in which the string *splits*. More precisely, let us consider as our classical action the Polyakov action [4], (we set here $2\alpha' = 1$)

$$S = \frac{1}{2\pi} \int_{W.S.} d\sigma d\tau \sqrt{-g} g^{ab} G_{AB}(X) \partial_a X^A \partial_b X^B \tag{1.1}$$

where g_{ab} is the world sheet metric and G_{AB} is the background space-time metric. Since we are going to consider classical solutions, both g_{ab} and G_{AB} have Lorentzian signature. We will show that there exists stationary points of S which correspond to a classical splitting of one string into two. Moreover, we will show that for some fixed Cauchy data $X^A(\sigma, \tau_0)$ and $\dot{X}^A(\sigma, \tau_0)$, the solutions corresponding to a string that splits into two, have *smaller* action than the one in which the string does not split, i.e. the sum of the areas swept by the two pieces in which the string splits, is smaller than the area swept out by the string that does not split. Of course, the existence of different kinds of solutions for the same Cauchy data is due to the fact that the world sheet metric is not fixed by any dynamical equation, and so we are free to choose it at will. Indeed, if we enforce the conformal gauge *globally* on our world sheet, we are left with a world sheet that has the topology of a cylinder (for closed strings), and therefore with a string that does not split. However, nothing prevents us to consider a different world sheet *topology*, and the interesting point is that the solutions so obtained describe strings that split and have a *smaller* action than the string that does not split. In order to show explicit solutions of this type, we give here some interesting examples, saving a more general discussion for future work. We will consider first an example in flat space-time, to show how the phenomenon of smaller action for the string that splits is already present in this case. Secondly, we will consider a closed string moving in the background given by a singular plane wave, and show how the presence of a strong gravitational field *increases* and *amplifies* the negative difference between the action for the string that splits and the action for the string that does not split. In fact, such difference is amplified by an overall factor and becomes infinitely negative at the space-time singularity. The classical splitting string solution contains the standard (non-splitting) solution as a particular case. We also show that the splitting solution describes a natural desintegration process of one string of mass M decaying into two pieces of a smaller total mass with some kinetic energy. In this sense, we want to stress that the splitting string solutions which we obtain are exact solutions of the D'Alambert equation (2.4) and constraints (2.5) (it must be noticed that $\partial_\sigma^2 X$ and $\partial_\tau^2 X$ exhibit equal Dirac delta singular terms for each of the two pieces in which the string splits).

It is also possible to construct string solutions where one string splits into more than two pieces.

II. STRING SPLITTING IN FLAT SPACE-TIME

Let us consider a closed string $X(\sigma, \tau)$ moving in a D dimensional flat Minkowski space-time. In order to describe a string that splits, we choose a world sheet $\mathcal{M}$ with the topology of a pant, and call (σ_0, τ_0) the point at which the string breaks into two pieces. To construct a solution for the string with this world sheet topology, we consider in $\mathcal{M}$ the three regions I, II, and III given by

$$
\begin{aligned}
I \;\; &\equiv \{(\sigma,\tau):\; 0 \le \sigma < 2\pi,\; \tau_i \le \tau < \tau_0\} \\
II \;\; &\equiv \{(\sigma,\tau):\; 0 \le \sigma < \sigma_0,\; \tau_0 < \tau \le \tau_f\} \qquad (2.1) \\
III &\equiv \{(\sigma,\tau):\; \sigma_0 \le \sigma < 2\pi,\; \tau_0 < \tau \le \tau_f\}
\end{aligned}
$$

and impose the continuity of the dynamical variables $X(\sigma,\tau)$ and $\dot{X}(\sigma,\tau)$ at the splitting world sheet time $\tau = \tau_0$

$$
X^{(I)}(\sigma,\tau_0) = \begin{cases} X^{(II)}(\sigma,\tau_0) & \text{if } 0 \le \sigma < \sigma_0 \\ X^{(III)}(\sigma,\tau_0) & \text{if } \sigma_0 \le \sigma < 2\pi \end{cases} \tag{2.2}
$$

$$
\dot{X}^{(I)}(\sigma,\tau_0) = \begin{cases} \dot{X}^{(II)}(\sigma,\tau_0) & \text{if } 0 \le \sigma < \sigma_0 \\ \dot{X}^{(III)}(\sigma,\tau_0) & \text{if } \sigma_0 \le \sigma < 2\pi \end{cases} \tag{2.3}
$$

In each of the regions we are in the conformal gauge. Thus, the equation of motion and the string constraints read the same for the three functions $X^{(\mathcal{J})}(\sigma,\tau)$, with $\mathcal{J} = I, II, III$

$$
(\partial_\tau^2 - \partial_\sigma^2) X^{(\mathcal{J})} = 0 \tag{2.4}
$$

$$
(\dot{X}^{(\mathcal{J})} \pm X'^{(\mathcal{J})})^2 = 0 \tag{2.5}
$$

[We shall use in this paper the symbol $\mathcal{J}$ to label the world-sheet regions I, II and III].

However, we must impose different boundary conditions for each of the functions $X^{(\mathcal{J})}$. Since we want to describe a closed string splitting into two closed strings, the appropriate boundary conditions are the periodicity conditions

$$
X^{(\mathcal{J})}(\sigma + \lambda_{\mathcal{J}}, \tau) = X^{(\mathcal{J})}(\sigma,\tau), \quad \mathcal{J} = I, II, III, \tag{2.6}
$$

where

$$
\lambda_I = 2\pi \; , \; \lambda_{II} = \sigma_0 \; , \; \lambda_{III} = 2\pi - \sigma_0 \tag{2.7}
$$

Of course, for the splitting to be possible, the string configuration $X^{(I)}(\sigma,\tau)$ with which we start must satisfy the consistency condition

$$X^{(I)}(0,\tau_0) = X^{(I)}(\sigma_0,\tau_0) = X^{(I)}(2\pi,\tau_0) \tag{2.8}$$

The general solution of equations (2.4) with the periodic boundary conditions (2.6) is

$$X^{(\mathcal{J})}(\sigma,\tau) = \sum_{n=-\infty}^{\infty} X_n^{(\mathcal{J})}(\tau) e^{i\frac{2\pi}{\lambda_\mathcal{J}}n\sigma} \tag{2.9}$$

where

$$X_0^{(\mathcal{J})} = q^{(\mathcal{J})} + p^{(\mathcal{J})}\tau \tag{2.10}$$

and

$$X_n^{(\mathcal{J})}(\tau) = A_n^{(\mathcal{J})}\, e^{-i\frac{2\pi}{\lambda_\mathcal{J}}n\tau} + B_n^{(\mathcal{J})}\, e^{i\frac{2\pi}{\lambda_\mathcal{J}}n\tau} \quad \text{for } n \neq 0 \tag{2.11}$$

where $\mathcal{J} = I, II, III$ as defined above.

Now, in order to construct a solution of the equations of motion and constraints corresponding to a string that splits, we begin with a function $X^{(I)}(\sigma,\tau)$ that satisfies equations (2.4) and the condition (2.8) at $\tau = \tau_0$, and then we construct $X^{(II)}$ and $X^{(III)}$ by determining their Fourier coefficients $X_n^{(II)}(\tau)$ and $X_n^{(III)}(\tau)$ through the matching conditions (2.2) and (2.3). Thus, from equations (2.9), (2.2) and (2.3) we obtain

$$\begin{aligned} X_n^{(II)}(\tau_0) &= i \sum_{m=-\infty}^{\infty} \frac{1-e^{im\sigma_0}}{m\sigma_0 - 2\pi n} X_m^{(I)}(\tau_0) \\ \dot{X}_n^{(II)}(\tau_0) &= i \sum_{m=-\infty}^{\infty} \frac{1-e^{im\sigma_0}}{m\sigma_0 - 2\pi n} \dot{X}_m^{(I)}(\tau_0) \end{aligned} \tag{2.12}$$

and

$$\begin{aligned} X_n^{(III)}(\tau_0) &= -ie^{-i\frac{2\pi}{2\pi-\sigma_0}n\sigma_0} \sum_{m=-\infty}^{\infty} \frac{1-e^{im\sigma_0}}{m(2\pi-\sigma_0) - 2\pi n} X_m^{(I)}(\tau_0) \\ \dot{X}_n^{(III)}(\tau_0) &= -ie^{-i\frac{2\pi}{2\pi-\sigma_0}n\sigma_0} \sum_{m=-\infty}^{\infty} \frac{1-e^{im\sigma_0}}{m(2\pi-\sigma_0) - 2\pi n} \dot{X}_m^{(I)}(\tau_0) \end{aligned} \tag{2.13}$$

so that equations (2.10), (2.11) and their derivatives at $\tau = \tau_0$ determine the constant coefficients $(q, p, A_n, B_n)^{(II)}$ and $(q, p, A_n, B_n)^{(III)}$ in terms of the initial data $(q, p, A_n, B_n)^{(I)}$. In this way we obtain the solutions for the string pieces II and III from a general string solution I. This last must satisfy the consistency condition (2.8).

It is important also to notice that the constraints (2.5) are also fulfilled by the strings $X^{(II)}$ and $X^{(III)}$. The matching conditions (2.2) and (2.3) imply that both $\dot{X}(\sigma, \tau)$ and $X'(\sigma, \tau)$ are continuous at $\tau = \tau_0$. Therefore

$$\begin{aligned} (\dot{X}^{(II)} \pm X'^{(II)})^2(\sigma, \tau_0) &= (\dot{X}^{(I)} \pm X'^{(I)})^2(\sigma, \tau_0) = 0 \\ (\dot{X}^{(III)} \pm X'^{(III)})^2(\sigma, \tau_0) &= (\dot{X}^{(I)} \pm X'^{(I)})^2(\sigma, \tau_0) = 0 \end{aligned} \tag{2.14}$$

Then, taking into account the equation of motion (2.4), it follows that the constraints hold for all $\tau \geq \tau_0$.

Let us now consider a particular example given by a circular string that winds r times upon itself:

$$\begin{aligned} T^{(I)} &= M\tau \\ X^{(I)} &= \frac{M}{r} \sin r\tau \, \cos r\sigma \\ X^{(I)} &= \frac{M}{r} \sin r\tau \, \sin r\sigma \\ X^{i(I)} &= 0 \text{ for } i = 3, \cdots, D-1 \end{aligned} \tag{2.15}$$

which satisfy

$$X^{(I)}(\sigma, 0) = 0 = Y^{(I)}(\sigma, 0) = T^{(I)}(0) \tag{2.16}$$

and

$$\dot{X}^{(I)}(\sigma, 0) = M \cos r\sigma \ , \ \dot{Y}^{(I)}(\sigma, 0) = M \sin r\sigma \ , \ \dot{T}^{(I)}(0) = M \tag{2.17}$$

It is inmediate to check that equations (2.15) are a solution of equation (2.4) and satisfy the constraints (2.5).

In order to obtain a *splitting* string solution, let us choose without loss of generality $\tau_0 = 0$. (One can always replace τ by $\tau - \tau_0$ in equations (2.15)). Then the splitting consistency condition (2.8) is trivially fulfilled because at $\tau = 0$ the string (2.15) collapses to a point.

For the string coordinate T, the matching conditions (2.2)-(2.3) yield

$$T^{(II)} = M\tau \quad , \quad T^{(III)} = M\tau \tag{2.18}$$

Next, to obtain $X^{(II)}(\sigma,\tau)$, $Y^{(II)}(\sigma,\tau)$, $X^{(III)}(\sigma,\tau)$ and $Y^{(III)}(\sigma,\tau)$, we observe that from equations (2.15) and (2.9)

$$\begin{aligned} T^{(I)}(0) &= T_0^{(I)}(0) = 0 \\ \dot{T}^{(I)}(0) &= \dot{T}_0^{(I)}(0) = M \end{aligned} \tag{2.19}$$

and

$$\begin{aligned} X_n^{(I)}(0) &= Y_n^{(I)}(0) &&= 0 \\ \dot{X}_r^{(I)}(0) &= \dot{X}_{-r}^{(I)}(0) &&= \frac{1}{2}M \\ \dot{Y}_r^{(I)}(0) &= -\dot{Y}_{-r}^{(I)}(0) &&= -\frac{i}{2}M \\ \dot{X}_n^{(I)}(0) &= \dot{Y}_n^{(I)}(0) &&= 0 \ \text{ for } n \neq \pm r \end{aligned} \tag{2.20}$$

Therefore, the matching equations (2.12) and (2.13) read in this case

$$\begin{aligned} X_n^{(II)}(0) &= 0 = Y_n^{(II)}(0) \\ X_n^{(III)}(0) &= 0 = Y_n^{(III)}(0) \end{aligned} \tag{2.21}$$

and

$$
\begin{aligned}
\dot{X}_n^{(II)}(0) &= M\phi_n(\sigma_0) \\
\dot{Y}_n^{(II)}(0) &= M\psi_n(\sigma_0) \\
\dot{X}_n^{(III)}(0) &= Me^{-i\frac{2\pi}{2\pi-\sigma_0}n\sigma_0}\phi_n^*(2\pi-\sigma_0) \\
\dot{Y}_n^{(III)}(0) &= -Me^{-i\frac{2\pi}{2\pi-\sigma_0}n\sigma_0}\psi_n^*(2\pi-\sigma_0)
\end{aligned}
\tag{2.22}
$$

where

$$
\begin{aligned}
\phi_n(\sigma_0) &= -\frac{r\sigma_0\sin r\sigma_0 + i\,2\pi n\,(1-\cos r\sigma_0)}{4\pi^2n^2 - r^2{\sigma_0}^2} \\
\psi_n(\sigma_0) &= -\frac{r\sigma_0(1-\cos r\sigma_0) - i\,2\pi n\,\sin r\sigma_0}{4\pi^2n^2 - r^2{\sigma_0}^2}
\end{aligned}
\tag{2.23}
$$

Computing now the two sets of constants $(q,p,A_n,B_n)^{(II)}$ and $(q,p,A_n,B_n)^{(III)}$ from expressions (2.10) and (2.11), we obtain the time dependent Fourier coefficients $X_n^{(II)}(\tau)$, $Y_n^{(II)}(\tau)$, $X_n^{(III)}(\tau)$ and $Y_n^{(III)}(\tau)$. Finally the two pieces in which the string splits read

$$
\begin{aligned}
X^{(II)}(\sigma,\tau) &= M\tau\,\frac{\sin r\sigma_0}{r\sigma_0} + M\,\frac{\sigma_0}{2\pi}\sum_{n\neq 0}\frac{\phi_n(\sigma_0)}{n}\sin\left[\frac{2\pi}{\sigma_0}n\tau\right]\,e^{i\frac{2\pi}{\sigma_0}n\sigma} \\
Y^{(II)}(\sigma,\tau) &= M\tau\,\frac{1-\cos r\sigma_0}{r\sigma_0} + M\,\frac{\sigma_0}{2\pi}\sum_{n\neq 0}\frac{\psi_n(\sigma_0)}{n}\sin\left[\frac{2\pi}{\sigma_0}n\tau\right]\,e^{i\frac{2\pi}{\sigma_0}n\sigma}
\end{aligned}
\tag{2.24}
$$

$$
\begin{aligned}
X^{(III)}(\sigma,\tau) &= -M\tau\,\frac{\sin r\sigma_0}{r(2\pi-\sigma_0)} + M\,\frac{2\pi-\sigma_0}{2\pi}\sum_{n\neq 0}\left\{\frac{1}{n}\sin\left[\frac{2\pi}{2\pi-\sigma_0}n\tau\right]\right. \\
&\qquad\left.\phi_n^*(2\pi-\sigma_0)\,e^{i\frac{2\pi}{2\pi-\sigma_0}n(\sigma-\sigma_0)}\right\} \\
Y^{(III)}(\sigma,\tau) &= -M\tau\,\frac{1-\cos r\sigma_0}{r(2\pi-\sigma_0)} - M\,\frac{2\pi-\sigma_0}{2\pi}\sum_{n\neq 0}\left\{\frac{1}{n}\sin\left[\frac{2\pi}{2\pi-\sigma_0}n\tau\right]\right. \\
&\qquad\left.\psi_n^*(2\pi-\sigma_0)\,e^{i\frac{2\pi}{2\pi-\sigma_0}n(\sigma-\sigma_0)}\right\}
\end{aligned}
\tag{2.25}
$$

Let us now discuss the properties of the splitting string solution given by equations (2.15), (2.24) and (2.25). First, we notice that for $n \neq 0$ and $\sigma_0 \to 2\pi$,

$$\phi_n(\sigma_0) \underset{\sigma_0 \to 2\pi}{\longrightarrow} \frac{1}{2}(\delta_{nr} + \delta_{n(-r)})$$

$$\psi_n(\sigma_0) \underset{\sigma_0 \to 2\pi}{\longrightarrow} -\frac{i}{2}(\delta_{nr} - \delta_{n(-r)}) \tag{2.26}$$

Therefore,

$$X^{(II)}(\sigma, \tau) \underset{\sigma_0 \to 2\pi}{\longrightarrow} \frac{M}{r} \sin r\tau \ \cos r\sigma \ = \ X^{(I)}(\sigma, \tau)$$

$$Y^{(II)}(\sigma, \tau) \underset{\sigma_0 \to 2\pi}{\longrightarrow} \frac{M}{r} \sin r\tau \ \sin r\sigma \ = \ Y^{(I)}(\sigma, \tau) \tag{2.27}$$

Similarly for $\sigma_0 \to 0$,

$$X^{(III)}(\sigma, \tau) \underset{\sigma_0 \to 0}{\longrightarrow} X^{(I)}(\sigma, \tau)$$

$$Y^{(III)}(\sigma, \tau) \underset{\sigma_0 \to 0}{\longrightarrow} Y^{(I)}(\sigma, \tau) \tag{2.28}$$

Thus, the splitting string solution given by equations (2.24) and (2.25) gives the solution $X^{(I)}(\sigma, \tau)$ in the limits $\sigma_0 \to 2\pi$ and $\sigma_0 \to 0$ as it should be. In this sense, the splitting solution generalizes the standard string solution without splitting. Moreover, it can be checked from equations (2.24)and (2.25) that the two string pieces II and III satisfy the (homogeneous) D'Alambert equation (2.4) and the constraints (2.5). It should be noticed that $X^{(\mathcal{J})}$ and $Y^{(\mathcal{J})}$ are continuous periodic functions with periods σ_0 and $2\pi - \sigma_0$; their first derivatives with respect to σ and τ have step discontinuities, and their second derivatives exhibit Dirac delta singularities. These singularities sit on the world sheet characteristics through the splitting point (σ_0, τ_0). However the Dirac delta terms in the expressions for the second derivatives with respect to σ and τ are equal and cancel in the d'Alambertian.

The energy and momentum carried out by each of the strings $\mathcal{J} = I, II$ and III is given by

$$E^{(\mathcal{J})} = \frac{1}{2\pi}\int_0^{\lambda_{\mathcal{J}}} d\sigma\ \dot{T}^{(\mathcal{J})}$$

$$P_X^{(\mathcal{J})} = \frac{1}{2\pi}\int_0^{\lambda_{\mathcal{J}}} d\sigma\ \dot{X}^{(\mathcal{J})} \tag{2.29}$$

$$P_Y^{(\mathcal{J})} = \frac{1}{2\pi}\int_0^{\lambda_{\mathcal{J}}} d\sigma\ \dot{Y}^{(\mathcal{J})}$$

Then, using the Fourier series expansions (2.15), (2.24) and (2.25) we obtain

$$(E, P_X, P_Y)^{(I)} = M\,(1, 0, 0)$$

$$(E, P_X, P_Y)^{(II)} = M\,(\frac{\sigma_0}{2\pi}, \frac{\sin r\sigma_0}{2\pi r}, \frac{1-\cos r\sigma_0}{2\pi r}) \tag{2.30}$$

$$(E, P_X, P_Y)^{(III)} = M\,(\frac{2\pi-\sigma_0}{2\pi}, -\frac{\sin r\sigma_0}{2\pi r}, -\frac{1-\cos r\sigma_0}{2\pi r})$$

From equations (2.30) we see that the energy-momentum of the string before and after the splitting is conserved, as it should be. The masses of the three strings are given by

$$M_I = M$$

$$M_{II} = \frac{M}{2\pi}\sqrt{\sigma_0^2 - \frac{4}{r^2}\sin^2\frac{r\sigma_0}{2}} \tag{2.31}$$

$$M_{III} = \frac{M}{2\pi}\sqrt{(2\pi-\sigma_0)^2 - \frac{4}{r^2}\sin^2\frac{r\sigma_0}{2}}$$

Again, we see that

$$M_{II} \underset{\sigma_0\to 2\pi}{\longrightarrow} M_I$$

$$M_{III} \underset{\sigma_0\to 0}{\longrightarrow} M_I \tag{2.32}$$

From equations (2.31) it also follows that

$$M_I \geq M_{II} + M_{III} \tag{2.33}$$

This tell us that the classical splitting string solution describes a natural desintegration process, in which a string of mass M decays into two pieces with a smaller total mass and some kinetic energy, which depends on the point where the splitting takes place. Furthermore, one can see in equations (2.30), (2.24) and (2.25) that the two outgoing pieces II and III go away one from each other with opposite momenta.

From equations (2.32) it follows that the kinetic energy

$$K(\sigma_0) = M_I - (M_{II} + M_{III}) \tag{2.34}$$

vanishes when

$$r\sigma_0 = 2l\pi \ ; \ l = 0, 1, \cdots, r \tag{2.35}$$

This corresponds to cutting the initial string $X^{(I)}$ into two pieces, $X^{(II)}$ and $X^{(III)}$, which contain an integer number of turns: l and $r-l$ respectively. In this case, the equations (2.30) tell us that the momentum of the two pieces vanishes, and moreover the series (2.24) and (2.25) sum up to $X^{(I)}(\sigma,\tau)$ and $Y^{(I)}(\sigma,\tau)$ respectively. That is, cutting the string (2.15) into two pieces which contain an integer number of circumferences, is equivalent to not cutting it at all.(In other words, a circular string $X^{(I)}$ wound r times is equivalent to two concentric strings $X^{(II)}$ and $X^{(III)}$ wound l and $r-l$ times respectively). And more generally, a circular string with $r = n$ turns is equivalent to r strings with $n = 1$ turn each. Of course, this is due to the fact that, in this case, the periodicity conditions (2.6) that we are enforcing for $\tau > 0$ are already present in $X^{(I)}(\sigma,\tau)$ for $\tau < 0$.

From equations (2.33), (2.34) and (2.35) it follows that in each of the r equal windings in σ described by the string I, there is a splitting point σ_0^i, $i = 1, 2, \cdots r$ that maximizes the kinetic energy of the pieces II and III. That is, for each of the turns of the string upon itself, there is an intermediate point where the splitting of the string is most energetically favorable.

Also from eq. (2.31) we see that the kinetic energy $K(\sigma_0)$ of the pieces II and III decreases with the growing of r. In fact, $K(\sigma_0) \to 0$ for $r \to \infty$ and reaches its maximum value for $r = 1$ and $\sigma_0 = \pi$. The value of this maximum kinetic energy is

$$K_{\max} = M\left(1-\sqrt{1-\frac{4}{\pi^2}}\right) \simeq 0.229M \tag{2.36}$$

So, the most energetically favorable case correspond to a string with $r = 1$, which cuts into to equally long pieces. This also indicates that for a string with $r > 1$ the most energetically favorable process is not the breaking of the string into two pieces, but the breaking into r pieces at the midpoints of each winding. Thus, the fundamental case to be considered is that of a string with $r = 1$ that cuts into two pieces.

Let us now discuss the string action. Let S_I be the area swept by the classical solution $X^{(I)}$ that does not split, when it evolves in τ from 0 to τ_f; and S_{II}, S_{III} the areas swept by the two pieces of the splitting string evolving for the same τ interval. We are interested in comparing S_I with $S_{II} + S_{III}$, for long enough evolution time, i.e. $\tau_f \gg 2\pi$.

In flat space-time the string action (1.1) takes the form

$$S = \frac{1}{2\pi}\int_0^{\tau_f} d\tau \int_0^{2\pi} d\sigma\, (-\dot{X}^A\dot{X}_A + X'^A X'_A) \tag{2.37}$$

that using the constraints (2.5), can be rewritten as

$$S = \frac{1}{\pi}\int_0^{\tau_f} d\tau \int_0^{2\pi} d\sigma\, (\partial_\sigma X)^2 \tag{2.38}$$

Now for the string I, using the equation (2.15) we have

$$S_I = \frac{1}{\pi}\int_0^{\tau_f} d\tau \int_0^{2\pi} d\sigma\, M^2 \sin^2 r\tau \;\sim\; M^2\tau_f \;\;,\;\; (\tau_f \to \infty) \tag{2.39}$$

For the string II, using equations (2.18) and (2.24) we obtain

$$\begin{aligned} S_{II} &= \frac{1}{\pi}\int_0^{\tau_f} d\tau \int_0^{\sigma_0} d\sigma\, \{(\partial_\sigma X^{(II)})^2 + (\partial_\sigma Y^{(II)})^2\} \\ &= M^2\,\frac{\sigma_0}{\pi}\sum_{n\neq 0}\left(|\phi_n(\sigma_0)|^2 + |\psi_n(\sigma_0)|^2\right)\int_0^{\tau_f} d\tau\, \sin^2\left(\frac{2\pi}{\sigma_0}n\tau\right) \end{aligned} \tag{2.40}$$

For large τ_f and $n \neq 0$

$$\int_0^{\tau_f} d\tau\, \sin^2\left(\frac{2\pi}{\sigma_0}n\tau\right) \;\sim\; \frac{1}{2}\tau_f \tag{2.41}$$

On the other hand

$$\sum_{n\neq 0}\left(|\phi_n(\sigma_0)|^2+|\psi_n(\sigma_0)|^2\right)=1-\frac{4}{r^2\sigma_0^2}\sin^2\left(\frac{r\sigma_0}{2}\right) \tag{2.42}$$

Thus, for large τ_f

$$S_{II}=M^2\tau_f\,\frac{\sigma_0}{2\pi}\left[1-\frac{4}{r^2\sigma_0^2}\sin^2\left(\frac{r\sigma_0}{2}\right)\right] \tag{2.43}$$

Notice that when $\sigma_0\to 2\pi$, $S_{II}\to S_I$. In addition, the use of the large τ_f approximation deserves the following comment: one has to wait long enough time to appreciate the difference between the areas swept by the string I, and the strings II and III. In fact, as it can be easily seen from equations (2.24) and (2.25), at first order in the Taylor expansion in τ around $\tau=0$, $X^{(I)}$, $X^{(II)}$ and $X^{(III)}$ coincide, and therefore when $\tau_f\to 0$:

$$\frac{1}{\tau_f}(S_I-S_{II}-S_{III})\underset{\tau_f\to 0}{\longrightarrow}0 \tag{2.44}$$

Let us come back to the comparison between S_I and $S_{II}+S_{III}$, for large τ_f. First, the behaviour of S_{III} is obtained from equation (2.43) through the replacement $\sigma_0\to 2\pi-\sigma_0$

$$S_{III}\;\sim\;M^2\tau_f\,\frac{2\pi-\sigma_0}{2\pi}\left[1-\frac{4}{r^2(2\pi-\sigma_0)^2}\sin^2\left(\frac{r\sigma_0}{2}\right)\right] \tag{2.45}$$

Then

$$S_{II}+S_{III}\;\sim\;M^2\tau_f\left\{1-\frac{2}{\pi r^2}\left(\frac{1}{\sigma_0}+\frac{1}{2\pi-\sigma_0}\right)\sin^2\left(\frac{r\sigma_0}{2}\right)\right\} \tag{2.46}$$

and from equations (2.39) and (2.46) we obtain

$$\Delta S=(S_{II}+S_{III})-S_I=-\,M^2\tau_f\,\frac{2}{\pi r^2}\left(\frac{1}{\sigma_0}+\frac{1}{2\pi-\sigma_0}\right)\sin^2\frac{r\sigma_0}{2} \tag{2.47}$$

which is a negative quantity. Therefore, the string that splits sweeps a smaller area than the string that does not split. From eq. (2.47) it also follows that the decrease in area $|\Delta S|$ vanishes for $r\sigma_0=2l\pi$. This corresponds to the splitting into two pieces containing an integer number of circumferences, which is equivalent to non-splitting.

For the fundamental case $r=1$, the relative decrease in area is

$$\eta_1(\sigma_0)=\frac{|\Delta S_1|}{S_{1,I}}=\frac{2}{\pi}\left(\frac{1}{\sigma_0}+\frac{1}{2\pi-\sigma_0}\right)\sin^2\left(\frac{\sigma_0}{2}\right) \tag{2.48}$$

which reaches its maximum value for $\sigma_0 = \pi$

$$\eta_{1,\max} = \frac{4}{\pi^2} \simeq 0.405 \tag{2.49}$$

Thus, for the string configuration that we have chosen (eq. (2.15) the area swept by the strings that splits into two equally long pieces reduce to a 60% of the area swept by the string that does not split.

III. STRING SPLITTING IN A SINGULAR PLANE WAVE BACKGROUND

We discuss now the splitting solution for a closed string in a strong gravitational field. We consider a D dimensional singular gravitational plane wave described by the metric

$$ds^2 = \frac{\alpha}{U^2}\,(X^2 - Y^2)\,dU^2 - dU dV + \sum_{j=2}^{D-1}(dX^j)^2 \tag{3.1}$$

where $U = X^0 - X^1$, and $V = X^0 + X^1$ are light cone coordinates, $X^2 \equiv X$, $X^3 \equiv Y$, and α is a constant. This space-time is sourceless and has curvature on the null plane $U = 0$. In this space-time, the classical string equations of motion and constraints have been solved for the ordinary non-splitting string in [5]. In this metric, the equation for U is simply $\partial^2 U = 0$. This allows us to take the light cone gauge exactly for all τ

$$U = p\tau \tag{3.2}$$

In this gauge the string equations for X and Y reduce to the linear equations

$$\begin{aligned} \left(-\partial_\tau^2 + \partial_\sigma^2 + \frac{\alpha}{\tau^2}\right) X &= 0 \\ \left(-\partial_\tau^2 + \partial_\sigma^2 - \frac{\alpha}{\tau^2}\right) Y &= 0 \end{aligned} \tag{3.3}$$

which can be solved by Fourier expanding $X(\sigma, \tau)$ and $Y(\sigma, \tau)$ in σ. Then, the τ dependent Fourier coefficients $X_n(\tau)$ and $Y_n(\tau)$ express in terms of Bessel functions.

The remaining transverse coordinates $j = 4, \cdots, D-1$ satisfy the flat space-time equations

$$\left(-\partial_\tau^2 + \partial_\sigma^2\right) X^j = 0 \;\; ; \;\; j = 4, \cdots, D-1 \tag{3.4}$$

Finally, the longitudinal coordinate V is determined through the constraints

$$G_{AB}\, \partial_\pm X^A\, \partial_\pm X^B = 0 \tag{3.5}$$

which yield

$$\begin{aligned} p\,\partial_\sigma V &= 2\sum_{j=2}^{D-1} \partial_\tau X^j \partial_\sigma X^j \\ p\,\partial_\tau V &= \frac{\alpha}{\tau^2}\,(X^2 - Y^2) + \sum_{j=2}^{D-1} \left\{(\partial_\tau X^j)^2 + (\partial_\sigma X^j)^2\right\} \end{aligned} \tag{3.6}$$

Let us describe now the splitting solution. We consider a generic string configuration evolving in the region of negative τ; i.e. before the string reaches the singularity at $U = 0$, and splitting at a certain point (σ_0, τ_0) with $\tau_0 < 0$. We choose for the three strings $\mathcal{J} = I, II, III$, a solution of the form

$$\begin{aligned} U^{(\mathcal{J})} &= p\tau \\ X^{(\mathcal{J})} &= \sum_{n=-\infty}^{\infty} X_n^{(\mathcal{J})}(\tau)\, e^{i\frac{2\pi}{\lambda_{\mathcal{J}}}n\sigma} \\ Y^{(\mathcal{J})} &= 0 \\ X^{j(\mathcal{J})} &= 0 \;\; , \;\; j = 4, \cdots, D-1 \end{aligned} \tag{3.7}$$

with the string coordinate V determined through the constraints

$$\begin{aligned} p\,\partial_\sigma V^{(\mathcal{J})} &= 2\,\partial_\tau X^{(\mathcal{J})}\,\partial_\sigma X^{(\mathcal{J})} \\ p\,\partial_\tau V^{(\mathcal{J})} &= \frac{\alpha}{\tau^2}\,(X^{(\mathcal{J})})^2 + (\partial_\tau X^{(\mathcal{J})})^2 + (\partial_\sigma X^{(\mathcal{J})})^2 \end{aligned} \tag{3.8}$$

The Fourier coefficients $X_n^{(\mathcal{J})}(\tau)$ are solutions of the equations

$$\ddot{X}_n^{(\mathcal{J})} + \left[\left(\frac{2\pi n}{\lambda_{\mathcal{J}}}\right)^2 - \frac{\alpha}{\tau^2}\right] X_n^{(\mathcal{J})} = 0 \tag{3.9}$$

which can be written in the form

$$\begin{aligned} X_n^{(\mathcal{J})}(\tau) &= C_n^{(\mathcal{J})}\sqrt{-\tau}\; J_{-\nu}\left(-\frac{2\pi}{\lambda_{\mathcal{J}}}|n|\tau\right) + D_n^{(\mathcal{J})}\sqrt{-\tau}\; J_\nu\left(-\frac{2\pi}{\lambda_{\mathcal{J}}}|n|\tau\right), \; (n \neq 0) \\ X_0^{(\mathcal{J})}(\tau) &= C_0^{(\mathcal{J})}(-\tau)^{\frac{1}{2}-\nu} + D_0^{(\mathcal{J})}(-\tau)^{\frac{1}{2}+\nu} \end{aligned} \tag{3.10}$$

where $J_{-\nu}$ and J_ν are Bessel functions with index

$$\nu = \sqrt{\frac{1}{4} + \alpha} \tag{3.11}$$

and $\mathcal{J} = I, II, III$ as defined above.

As in the flat space-time case, the functions $(U, V, X)^{(II)}$ and $(U, V, X)^{(III)}$ which describe the evolution of strings II and III, are fixed by the initial string $(U, V, X)^{(I)}$, and the matching conditions (2.2) and (2.3). However, since we are working now in the light cone gauge, the following two remarks are in order. First, the choice of the light cone gauge for string I, and the matching conditions (2.2) and (2.3) for U, imply that the light cone gauge holds for the pieces II and III as well, as stated in eqs. (3.7). Second, the string coordinates $V^{(II)}$ and $V^{(III)}$ are determined through the constraints (3.8), instead through the matching conditions (2.2)-(2.3). However, this is consistent because the matching conditions for the string coordinate X, together with the constraints (2.2)-(2.3) imply that $V^{(\mathcal{J})}$ $(J = I, II, III)$ also satisfy the matching conditions (2.2)-(2.3).

Let us choose now a specific initial configuration for the string coordinate $X^{(I)}$ given by

$$X^{(I)} = \frac{k}{r} f_\nu(\tau) \cos r\sigma \tag{3.12}$$

where

$$f_\nu(\tau) = r\sqrt{\tau\tau_0}\,[J_\nu(-r\tau_0)J_{-\nu}(-r\tau) - J_{-\nu}(-r\tau_0)J_\nu(r\tau)] \tag{3.13}$$

This describes a straight string along the X-axis. According to eq. (3.10) we have chosen the constants $\dot{C}_n^{(I)}$ and $D_n^{(I)}$ in the form

$$
\begin{aligned}
C_r^{(I)} &= C_{-r}^{(I)} = \frac{k}{2}\,\sqrt{-\tau_0}\,J_\nu(-r\tau_0) \\
D_r^{(I)} &= D_{-r}^{(I)} = \frac{k}{2}\,\sqrt{-\tau_0}\,J_{-\nu}(-r\tau_0) \\
C_n^{(I)} &= D_n^{(I)} = 0 \quad \text{for } n \neq \pm r
\end{aligned}
\tag{3.14}
$$

This initial string configuration $(U, V, X)^{(I)}$ that we have chosen must satisfy the splitting consistency condition (2.8). For U (eq.(3.2)) this condition is trivially satisfied. On the other hand, eqs. (3.12) and (3.13) yield

$$
f_\nu(\tau_0) = 0 \tag{3.15}
$$

Thus

$$
X^{(I)}(\sigma, \tau_0) = 0 \tag{3.16}
$$

and so the condition (2.8) holds for the X string coordinate. Finally, using eqs. (3.12) and (3.15) in the constraints (3.8), we obtain

$$
\begin{aligned}
p\,\partial_\sigma V^{(I)}(\sigma, \tau) &= -k\dot{f}_\nu(\tau) f_\nu(\tau)\,\sin(2r\sigma) \\
p\,\partial_\sigma V^{(I)}(\sigma, \tau_0) &= 0
\end{aligned}
\tag{3.17}
$$

i.e. $V^{(I)}(\sigma, \tau_0)$ is independent of σ and also satisfies the consistency condition (2.8).

We can determine now the constants $(C_n^{(II)}, D_n^{(II)})$ and $(C_n^{(III)}, D_n^{(III)})$ from the initial data $(C_n^{(I)}, D_n^{(I)})$, by using the matching conditions (2.2) and (2.3). These are

$$
\begin{aligned}
C_0^{(II)} &= K\,\frac{\sin \nu\pi}{\nu\pi}\,(-\tau_0)^{\frac{1}{2}+\nu}\,\frac{\sin r\sigma_0}{r\sigma_0} \\
C_n^{(II)} &= K\,\sqrt{-\tau_0}\,J_\nu\left(-\frac{2\pi}{\sigma_0}|n|\tau_0\right)\phi_n(\sigma_0)\,, \quad \text{for } n \neq 0 \\
D_0^{(II)} &= -K\,\frac{\sin \nu\pi}{\nu\pi}\,(-\tau_0)^{\frac{1}{2}-\nu}\,\frac{\sin r\sigma_0}{r\sigma_0} \\
D_n^{(II)} &= -K\,\sqrt{-\tau_0}\,J_{-\nu}\left(-\frac{2\pi}{\sigma_0}|n|\tau_0\right)\phi_n(\sigma_0)\,, \quad \text{for } n \neq 0
\end{aligned}
\tag{3.18}
$$

and

$$
\begin{aligned}
C_0^{(III)} &= -K\ \frac{\sin\nu\pi}{\nu\pi}\ (-\tau_0)^{\frac{1}{2}+\nu}\ \frac{\sin r\sigma_0}{r\sigma_0} \\
C_n^{(III)} &= K\ e^{-i\frac{2\pi}{2\pi-\sigma_0}n\sigma_0}\ \sqrt{-\tau_0}\ J_\nu\left(-\frac{2\pi}{2\pi-\sigma_0}|n|\tau_0\right)\phi_n{}^*(2\pi-\sigma_0)\ ,\ \text{for } n\neq 0 \\
D_0^{(III)} &= K\ \frac{\sin\nu\pi}{\nu\pi}\ (-\tau_0)^{\frac{1}{2}-\nu}\ \frac{\sin r\sigma_0}{r\sigma_0} \\
D_n^{(III)} &= -K\ e^{-i\frac{2\pi}{2\pi-\sigma_0}n\sigma_0}\ \sqrt{-\tau_0}\ J_{-\nu}\left(-\frac{2\pi}{2\pi-\sigma_0}|n|\tau_0\right)\phi_n{}^*(2\pi-\sigma_0)\ ,\ \text{for } n\neq 0
\end{aligned}
\tag{3.19}
$$

Hence, the Fourier expansions (3.7) for the X coordinates of strings II and III read

$$
\begin{aligned}
X^{(II)}(\sigma,\tau) &= X_0^{(II)}(\tau)+ \\
&\frac{\sigma_0}{2\pi}\ k\ \sum_{n\neq 0}\left\{\frac{1}{|n|}\ F_\nu\left(\frac{2\pi}{\sigma_0}|n|\tau,\frac{2\pi}{\sigma_0}|n|\tau_0\right)\phi_n(\sigma_0)\ e^{i\frac{2\pi}{\sigma_0}n\sigma}\right\}
\end{aligned}
\tag{3.20}
$$

and

$$
\begin{aligned}
&X^{(III)}(\sigma,\tau) = X_0^{(III)}(\tau) \\
&+\frac{2\pi-\sigma_0}{2\pi}\ k\ \sum_{n\neq 0}\left\{\frac{1}{|n|}\ F_\nu\left(\frac{2\pi}{2\pi-\sigma_0}|n|\tau,\frac{2\pi}{2\pi-\sigma_0}|n|\tau_0\right)\phi_n{}^*(2\pi-\sigma_0)\ e^{i\frac{2\pi}{2\pi-\sigma_0}n(\sigma-\sigma_0)}\right\}
\end{aligned}
\tag{3.21}
$$

where

$$
F_\nu(u,v) = \sqrt{uv}\,[J_\nu(-v)J_{-\nu}(-u) - J_{-\nu}(-v)J_\nu(-u)] \tag{3.22}
$$

and

$$
X_0^{(II)}(\tau) = -X_0^{(III)}(\tau) = k\ \frac{\sin\nu\pi}{\nu\pi}\ \frac{\sin r\sigma_0}{r\sigma_0}\ \sqrt{\tau_0\tau}\left[\left(\frac{\tau_0}{\tau}\right)^\nu - \left(\frac{\tau}{\tau_0}\right)^\nu\right] \tag{3.23}
$$

Let us discuss now the properties of the splitting string solution given by (3.20) and (3.21). First, we notice that from eq. (2.26)

$$
\begin{aligned}
X^{(II)}(\sigma,\tau) &\underset{\sigma_0\to 2\pi}{\longrightarrow} X^{(I)}(\sigma,\tau) \\
X^{(III)}(\sigma,\tau) &\underset{\sigma_0\to 0}{\longrightarrow} X^{(I)}(\sigma,\tau)
\end{aligned}
\tag{3.24}
$$

That is, the splitting solution with strings II and III contains the one string non-splitting solution as a particular case. In addition, for

$$r\sigma_0 = 2l\pi \quad ; \quad l = 1, \cdots, r-1 \tag{3.25}$$

the Fourier series (3.20) and (3.21) sum up to $X^{(I)}(\sigma,\tau)$. Thus again, cutting the string by an integer number of windings is equivalent to not cutting it at all.

Let us study now the string action. We want to compare the area S_I swept by the string without splitting, with the areas S_{II} and S_{III} swept by the strings II and III. We consider the evolution of the three strings for the same τ interval

$$\tau_0 \leq \tau \leq \tau_f < 0 \tag{3.26}$$

in the ingoing region, where the string has not yet reached the singularity at $\tau = 0$. We shall compute the action for a long τ interval, i.e. $\tau_f - \tau_0 \gg 2\pi$, as we did in flat space-time. However, in this case we shall implement this approximation by letting $\tau_0 \to -\infty$, and allowing $\tau_f \to 0^-$ in order to incorporate the effect of the space-time singularity at $\tau = 0$.

In the space-time (3.1) the string action is

$$S = \int\int_{W.S.} d\tau d\sigma \left\{ \frac{\alpha}{U^2}(X^2 - Y^2)\, \partial_a U \partial^a U - \partial_a U \partial^a V + \sum_{j=2}^{D-1} \partial_a X^j \partial^a X^j \right\} \tag{3.27}$$

In the light cone gauge, using the constraints (3.8) and for the particular string configuration (3.7), the action for the three strings takes the form

$$S_{\mathcal{J}} = \frac{1}{\pi} \int_{\tau_0}^{\tau_f} d\tau \int_0^{\lambda_{\mathcal{J}}} d\sigma \; (\partial_\sigma X^{(\mathcal{J})})^2 \tag{3.28}$$

where $\mathcal{J} = I, II, III$. Then, using eqs. (3.12),(3.20) and (3.22), we have

$$S_I = k^2 \int_{\tau_0}^{\tau_f} d\tau \; F_\nu^2(r\tau, r\tau_0) \tag{3.29}$$

and

$$S_{II} = k^2 \, \frac{\sigma_0}{\pi} \int_{\tau_0}^{\tau_f} d\tau \sum_{n\neq 0} |F_\nu\left(\frac{2\pi}{\sigma_0}|n|\tau, \frac{2\pi}{\sigma_0}|n|\tau_0\right)|^2 \; |\phi_n(\sigma_0)|^2 \tag{3.30}$$

From eq. (2.27) we see that $S_{II} \to S_I$ when $\sigma_0 \to 2\pi$ as it should be.

We shall do the comparison of the two areas S_I and $S_{II} + S_{III}$ in two regimes: first for $\alpha = 0$ which corresponds to flat space-time, and then for $\alpha \geq 3/4$ which correspond to a strong gravitational wave.

For $\alpha = 0$, the index ν is 1/2 and the Bessel functions reduce to circular functions:

$$F_{1/2}\left(\frac{2\pi}{\sigma_0}|n|\tau, \frac{2\pi}{\sigma_0}|n|\tau_0\right) = \frac{2}{\pi}\sin\left(\frac{2\pi}{\sigma_0}|n|(\tau-\tau_0)\right) \tag{3.31}$$

Then, for $\tau_0 \to -\infty$, eqs. (3.29) and (3.30) yield

$$S_I^{(\alpha=0)} = \frac{4k^2}{\pi^2}\int_{\tau_0}^{\tau_f} d\tau \, \sin^2 r(\tau-\tau_0) \sim \frac{2k^2}{\pi^2}(\tau_f - \tau_0) \tag{3.32}$$

and

$$\begin{aligned} S_{II}^{(\alpha=0)} &= \frac{4k^2}{\pi^3}\,\sigma_0 \int_{\tau_0}^{\tau_f} d\tau \sum_{n\neq 0} \sin^2\left(\frac{2\pi}{\sigma_0}|n|(\tau-\tau_0)\right)\, |\phi_n(\sigma_0)|^2 \sim \\ &\sim \frac{2k^2}{\pi^2}\,(\tau_f - \tau_0)\,\frac{\sigma_0}{\pi}\sum_{n\neq 0}|\phi_n(\sigma_0)|^2 = \\ &= \frac{2k^2}{\pi^2}\,(\tau_f-\tau_0)\,\frac{\sigma_0}{2\pi}\left(1 + \frac{\sin 2r\sigma_0}{2r\sigma_0} - \frac{2\sin^2 r\sigma_0}{r^2\sigma_0^2}\right) \end{aligned} \tag{3.33}$$

In addition, replacing $\sigma_0 \to 2\pi - \sigma_0$, we have

$$S_{III}^{(\alpha=0)} \sim \frac{2k^2}{\pi^2}\,(\tau_f-\tau_0)\,\frac{2\pi-\sigma_0}{2\pi}\left[1 - \frac{\sin 2r\sigma_0}{2r(2\pi-\sigma_0)} - \frac{2\sin^2 r\sigma_0}{r^2(2\pi-\sigma_0)^2}\right] \tag{3.34}$$

Thus

$$\Delta S^{(\alpha=0)} = (S_{II}+S_{III}) - S_I = -\frac{1}{\pi r^2}\left(\frac{1}{\sigma_0} + \frac{1}{2\pi-\sigma_0}\right)\sin^2\! r\sigma_0 \; S_I^{(\alpha=0)} \tag{3.35}$$

which is a negative quantity. Again, the string that splits has smaller action than the non-splitting one. However the result (3.35) is different from (2.47), because we have here a different string configuration. In particular, the action difference (3.35) vanishes in the present case for

$$r\sigma_0 = (2s+1)\pi \ , \quad s = 0, 1, \cdots, \left[\frac{r-1}{2}\right] \tag{3.36}$$

i.e. it vanishes not only for an integer number of windings $r\sigma_0 = 2l\pi$, $l = 0, 1, \cdots, r$, but also for a half integer number of windings. This is so because the barycentric term (eq.(3.23)) vanish for an integer or half integer number of windings. However, the Fourier expansions (3.20) and (3.21) sum up to $X^{(I)}(\sigma,\tau)$ for an integer number of windings, but do not for a half integer number of windings. This happens here because when σ_0 corresponds to a half integer number of windings, we have a straight string configuration with $X'(0,\tau_0) = X'(\sigma_0,\tau_0) = 0$. Hence the initial closed string may split into two open strings. Thus, in this case the strings II and III are open strings that stay together and change their respective shapes compared with string I.

For the fundamental case $r = 1$, the relative decrease in area is

$$\eta_1^{(\alpha=0)}(\sigma_0) = \frac{|\Delta S_1|}{S_{1,I}} = \frac{1}{\pi}\left(\frac{1}{\sigma_0} + \frac{1}{2\pi - \sigma_0}\right)\sin^2\sigma_0 \tag{3.37}$$

which has two symmetric maxima at the points $\sigma_0 = 1.291$ and $\sigma_0 = 2\pi - 1.291$ with heigth

$$\eta_{1,\max} = 0.287 \tag{3.38}$$

Notice that in this case $\sigma_0 = \pi$ gives a minimum with $\eta_1(\pi) = 0$, corresponding to the splitting of the closed string into two open strings.

Let us turn now to the discussion of the regime $\alpha \geq 3/4$, $(\nu \geq 1)$. In this case, the integral (3.29) diverges in the limit $\tau_f \to 0^-$. Thus, for $\nu \geq 1$ the behaviour of S_I for $\tau_f \to 0^-$ is dominated by the upper limit in the integral (3.29). That is, the most important contribution to S_I comes from the region near the singularity. It is in this sense that we talk of a strong enough gravitational wave for $\alpha \geq 3/4$. For $\tau_f \to 0^-$ we have

$$S_I \sim k^2 \frac{2^{2\nu}\,\tau_0}{\Gamma^2(1-\nu)\,(2-2\nu)\,r^{2\nu-2}}\,J_\nu^2(-r\tau_0)\left(\frac{\tau_f}{\tau_1}\right)^{2-2\nu} \tag{3.39}$$

where τ_1 is an intermediate point in the interval $\tau_0 < \tau_1 < \tau_f < 0$, and we assume that τ_0 is such that

$$J_\nu^2(-r\tau_0) \neq 0 \tag{3.40}$$

Let us consider the series in the integrand of eq. (3.30). The terms with $|n\tau| \ll 1$ behave as

$$|F_\nu\left(\frac{2\pi}{\sigma_0}|n|\tau, \frac{2\pi}{\sigma_0}|n|\tau_0\right)|^2 \sim$$
$$\sim \frac{2^{2\nu}}{\Gamma^2(1-\nu)}\left(\frac{2\pi|n|}{\sigma_0}\right)^{2-2\nu}(-\tau_0)\; J_\nu^2\left(-\frac{2\pi}{\sigma_0}|n|\tau_0\right)(-\tau)^{1-2\nu} \tag{3.41}$$

For $|n\tau| \geq 1$, and small enough $|-\tau|$ (i.e. $|n| \gg 1$), the terms in the series of eq. (3.30) are very much suppressed because for $n \to \infty$, $|\phi_n(\sigma_0)|^2 \sim 1/n^2$. Therefore, the behaviour of the series for $\tau \to 0-$ is

$$\sum_{n\neq 0}|F_\nu\left(\frac{2\pi}{\sigma_0}|n|\tau, \frac{2\pi}{\sigma_0}|n|\tau_0\right)|^2\; |\phi_n(\sigma_0)|^2 \sim$$
$$\sim \frac{2^{2\nu}}{\Gamma^2(1-\nu)}\left(\frac{2\pi}{\sigma_0}\right)^{2-2\nu}(-\tau_0)\; -\tau^{1-2\nu}\sum_{n\neq 0}\frac{J_\nu^2\left(-\frac{2\pi}{\sigma_0}|n|\tau_0\right)}{|n|^{2\nu-2}}\;|\phi_n(\sigma_0)|^2 \tag{3.42}$$

Then, for $\nu \geq 1$ and $\tau_0 \to 0-$ the behaviour of S_{II} is dominated by the upper limit of the integral. So, inserting eq (3.42) into eq. (3.30) and taking into account eq. (3.39) we get

$$S_{II} \sim \frac{\sigma_0}{\pi}\left(\frac{2\pi}{\sigma_0}\right)^{2-2\nu}\frac{r^{2\nu-2}}{J_\nu^2(-r\tau_0)}\sum_{n\neq 0}\frac{J_\nu^2\left(-\frac{2\pi}{\sigma_0}|n|\tau_0\right)}{|n|^{2\nu-2}}\;|\phi_n(\sigma_0)|^2\; S_I \tag{3.43}$$

Now we take the long τ interval approximation by doing $\tau_0 \to -\infty$ in (3.43). Thus

$$S_{II} \sim \frac{\sigma_0}{\pi}\left(\frac{\sigma_0}{2\pi}\right)^{2\nu-1}\sum_{n\neq 0}\left(\frac{r}{|n|}\right)^{2\nu-1}\frac{\cos^2\left(-\frac{2\pi}{\sigma_0}|n|\tau_0 - \frac{\pi}{2}\nu - \frac{\pi}{4}\right)}{\cos^2\left(-r\tau_0 - \frac{\pi}{2}\nu - \frac{\pi}{4}\right)}\;|\phi_n(\sigma_0)|^2\; S_I \tag{3.44}$$

The factor

$$\frac{1}{\cos^2\left(-r\tau_0 - \frac{\pi}{2}\nu - \frac{\pi}{4}\right)} \tag{3.45}$$

comes from the reciprocal of the Bessel functions $J_\nu^2(-r\tau_0)$ for $\tau_0 \to -\infty$, entering in (3.43), and which was assumed not to vanish. In particular we can choose τ_0 in such a way that the factor (3.45) is 1. This yields an upper bound estimate of the action S_{II} of the form

$$S_{II} \leq \left(\frac{\sigma_0}{2\pi}\right)^{2\nu-1}\frac{\sigma_0}{\pi}\sum_{n\neq 0}\left(\frac{r}{|n|}\right)^{2\nu-1}\;|\phi_n(\sigma_0)|^2\; S_I \tag{3.46}$$

and for string III we have

$$S_{III} \leq \left(\frac{2\pi - \sigma_0}{2\pi}\right)^{2\nu-1} \frac{2\pi - \sigma_0}{\pi} \sum_{n\neq 0} \left(\frac{r}{|n|}\right)^{2\nu-1} |\phi_n(2\pi - \sigma_0)|^2 \; S_I \tag{3.47}$$

Notice that this upper bound becomes exact for $\sigma_0 = 0$ and $\sigma_0 = 2\pi$ (the non-splitting solution).

In order to get a better insight on the behaviour of ΔS, we choose $\alpha = 2$ $(\nu = 3/2)$, in which case, the series (3.46) and (3.47) can be sumed in closed form. For $\alpha = 2$ we have

$$\begin{aligned} S_{II}^{(\alpha=2)} \;\leq\; & \frac{\sigma_0}{2\pi}\left(1 + \frac{\sin 2r\sigma_0}{2r\sigma_0} - \frac{2\sin^2 r\sigma_0}{r^2\sigma_0^2} + \right. \\ & \left. + \frac{\sin^2 r\sigma_0}{6} + \frac{2\,\sin r\sigma_0}{r\sigma_0} - \frac{8\sin^2\frac{1}{2}r\sigma_0}{r^2\sigma_0^2}\right) S_I \end{aligned} \tag{3.48}$$

Thus

$$\begin{aligned} & \Delta S^{(\alpha=2)} = (S_{II}^{(\alpha=2)} + S_{III}^{(\alpha=2)}) - S_I^{(\alpha=2)} \leq \\ & \leq \left\{ -\frac{1}{\pi r^2}\left(\frac{1}{\sigma_0} + \frac{1}{2\pi - \sigma_0}\right)\sin^2 r\sigma_0 + \frac{\sin^2 r\sigma_0}{6} - \frac{4}{\pi r^2}\left(\frac{1}{\sigma_0} + \frac{1}{2\pi - \sigma_0}\right)\sin^2\frac{r\sigma_0}{2}\right\} S_I \end{aligned} \tag{3.49}$$

For the fundamental case $r = 1$, the action difference takes the form

$$\Delta S_1^{(\alpha=2)} = \left\{\left[\frac{1}{6} - \frac{1}{\pi}\left(\frac{1}{\sigma_0} + \frac{1}{2\pi - \sigma_0}\right)\right]\sin^2\sigma_0 - \frac{4}{\pi}\left(\frac{1}{\sigma_0} + \frac{1}{2\pi - \sigma_0}\right)\sin^2\frac{\sigma_0}{2}\right\} S_{1,I} \tag{3.50}$$

which is a negative quantity for all values of σ_0 in the interval $[0, 2\pi]$. Thus, in the background of the singular gravitational plane wave $(\alpha = 2)$, the action of the splitting solution is smaller than the action of the non-splitting one. Moreover, this effect is magnified by an overall divergent τ_f dependent factor when we approach the singularity at $\tau = 0$, because the action $S_I^{(\alpha=2)}$ is multiplied by such a factor in this limit (eq.(3.39). In addition, the effect of smaller action for the splitting solution is also increased in relative terms, as a consequence of the new terms appearing in (3.50). This is easily seen in terms of the lower bound that we have for the relative decrease in area. According to (3.50)

$$\eta_1^{(\alpha=2)}(\sigma_0) = \frac{|\Delta S_1^{(\alpha=2)}|}{S_{1,I}^{(\alpha=2)}} \geq h(\sigma_0) \tag{3.51}$$

where

$$h(\sigma_0) = \left[\frac{1}{\pi}\left(\frac{1}{\sigma_0} + \frac{1}{2\pi - \sigma_0}\right) - \frac{1}{6}\right] \sin^2\sigma_0 + \frac{4}{\pi}\left(\frac{1}{\sigma_0} + \frac{1}{2\pi - \sigma_0}\right) \sin^2\frac{\sigma_0}{2} \tag{3.52}$$

The lower bound $h(\sigma_0)$ has its maximum at $\sigma_0 = \pi$ with a heigth $h_{\max} = 0.811$. Therefore

$$\eta_{1,\max}^{(\alpha=2)} > 0.811 \tag{3.53}$$

which is much larger than the relative decrease in area for the same string configuration in flat space-time, given in (3.38).

IV. CONCLUDING REMARKS

In this paper we have explored a new type of solutions of the string equations of motion and constraints. We are mainly thinking in the theory of fundamental strings. We show that the solutions in which the string splits into two are perfectly natural within the classical theory of strings based on the Polyakov action. There is no need of extra interactions, i.e: extra terms in the action to produce the splitting. The only difference with the non-splitting solutions are the boundary conditions. In this sense the choice of a splitting or a non-splitting solution is somehow similar to the choice between an open or a closed string. Of course, open and closed strings can be used to describe different physics and this can also be true for the splitting and non splitting string solutions. In particular, these solutions could be relevant for the description of cosmic strings breaking.

The solutions that we have constructed describe the splitting of strings as a natural decay process that takes place in real (lorentzian signature) space-time. This process occurs at the classical level and this is natural because the string is an extended object. The splitting solutions are already present in flat space-time, and they correspond to stationary points of the action (area) with lower value than the non splitting strings. In order to explore the effect of a gravitational field on the splitting solutions, we have considered a gravitational singular plane wave background. In the case that we analyze, the gravitational field produces an enhacement of the effect of smaller area for the splitting solution. It would be interesting to settle to what extent these results are universal, and work in this direction is now in progress by the present authors.

On the other hand, string splitting is usually considered and discussed within a quantum formulation, namely the euclidean path integral functional approach to the quantum string scattering amplitudes in the light-cone gauge [7]. In this context, the stationary points of the euclidean action correspond to solutions of Dirichlet or Neumann boundary value problems for elliptic operators (laplacians) on bordered Riemann surfaces [8], i.e. the classical string equations of motion and constraints lead to solve boundary value problems for elliptic operators with Dirichlet or Neumann boundary conditions. Of course, these are different from the solutions considered in this paper, in which we solve the hyperbolic (lorentzian) evolution equations for the Cauchy data $X^A(\sigma, \tau_0)$ and $\dot{X}^A(\sigma, \tau_0)$ with some fixed topology. This topology should be viewed as enforced by the world-sheet metric used to construct the D'Alambertian operator. Notice that quantum mechanically the initial data $X^A(\sigma, \tau_0)$ and $\dot{X}^A(\sigma, \tau_0)$ can not be given simultaneously. Instead, one gives the initial and final string shapes to compute a transition amplitude between them.

Although our splitting solutions are purely classical, string splitting for massive strings is also present at the quantum level. The relevant magnitude to be computed in that case is the probability amplitude for such a process. In fact, such probability has been computed in [6] for flat space-time. It would be very interesting to explore the relationships between the classical splitting solutions and the quantum probability for string desintegration, and also the effect of a gravitational field on such probability. The quantum probability amplitude for string splitting in a singular plane wave will be discussed by the present authors in a forthcoming paper.

We conclude with a final remark concerning the classical solutions of the string equations of motion in curved spaces-times. It has been stablished (see for instance [9] and references therein) that these solutions present a phenomenon of indefinitely string stretching near space-time singularities, due to the absorption by the string of energy from the background gravitational field. Of course, there must be a mechanism that avoids this indefinite string growing, and indeed the strings can radiate away energy by emiting gravitons or other particle like excitations. However, another natural mechanism to avoid string growing is string splitting, and it would be interesting to elucidate its quantitative relevance to avoid the indefinite stretching of strings in strong gravitational fields.

Acknowledgements

One the authors (J.R.M.) acknowledges the hospitality at LPTHE (Université de Paris VI-VII) and at DEMIRM (Observatoire de Paris) where part of this work was carried out. Also J.R M. acknowledges the Dirección General de Investigación Científica y Técnica (DG-ICYT) for financial support.

REFERENCES

[1] T. Kibble and N. Turok, *Phys. Lett.* **116 B**, 141, (1982).

[2] X. Artru, *Phys. Rep.* **97**, 147, (1983)

[3] See for example:

M. Green, J. Schwarz, E. Witten, *Superstring Theory* . Cambridge University Press. 1987.

[4] A. M. Polyakov, *Phys. Lett.* **103B**, 207 (1981).

[5] H.J. de Vega and N. Sánchez, *Phys. Rev.* **D45**, 2783 (1992).

H.J. de Vega, M. Ramón Medrano and N. Sánchez,

Class. and Quantum Gravity **10**, 2007 (1993).

[6] J. Dai and J. Polchinski, *Phys. Lett.* **220B**, 387 (1989).

R.B. Wilkinson, N. Turok and D. Mitchell, *Nuclear Physics* **B332**, 131 (1990).

[7] S. Mandelstam in *Superstrings: the first 15 years of superstring theory* Vol 2, Ed: J. Schwarz. World Scientific. 1985; and related articles therein.

[8] J. Ramírez Mittelbrunn and M.A. Martín Delgado,

Int. Journal of Mod. Phys., **A 6**, 1719 (1991).

[9] H.J. de Vega and N. Sánchez in *String Quantum Gravity and Physics at the Planck Energy Scale* Erice, 1992. World Scientific.

Cosmic Strings and Black Holes

A.L. Larsen*

Theoretical Physics Institute, Department of Physics, University of Alberta, Edmonton, Canada T6G 2J1

Abstract

In the first part of this talk, I consider some exact string solutions in curved spacetimes. In curved spacetimes with a Killing vector (timelike or spacelike), the string equations of motion and constraints are reduced to the Hamilton equations of a relativistic point-particle in a scalar potential, by imposing a particular ansatz. As special examples I consider circular strings in axially symmetric spacetimes, as well as stationary strings in stationary spacetimes, and I review some recent work done in collaboration with N. Sànchez and H.J. de Vega.

In the second part of the talk, I then consider in more detail the stationary strings in the Kerr-Newman geometry. It is shown that the world-sheet of a stationary string, that passes the static limit of the 4-D Kerr-Newman black hole, describes a 2-D black hole. Mathematical results for 2-D black holes can therefore be applied to physical objects; (say) cosmic strings in the vicinity of Kerr black holes. As an immediate general result, it follows that the string modes are thermally excited. This second part of the talk is a review of some recent work done in collaboration with V. Frolov and S. Hendy.

*Electronic address: alarsen@phys.ualberta.ca

1 Introduction

In a generic curved spacetime, the classical equations of motion of a bosonic string take the simplest form in the orthonormal gauge:

$$\ddot{x}^{\mu} - x''^{\mu} + \Gamma^{\mu}_{\rho\sigma}(\dot{x}^{\rho}\dot{x}^{\sigma} - x'^{\rho}x'^{\sigma}) = 0, \tag{1.1}$$

supplemented by the constraints:

$$g_{\mu\nu}\dot{x}^{\mu}x'^{\nu} = g_{\mu\nu}(\dot{x}^{\mu}\dot{x}^{\nu} + x'^{\mu}x'^{\nu}) = 0. \tag{1.2}$$

Here dot and prime denote differentiation with respect to the world-sheet coordinates tau and sigma, respectively. $\Gamma^{\mu}_{\rho\sigma}$ is the Christoffel symbol corresponding to the spacetime metric $g_{\mu\nu}$.

The main complication, as compared to the case of flat Minkowski spacetime, is related to the non-linearity of the equations of motion (1.1). It makes it possible to obtain the complete analytic solution only in a very few special cases like conical spacetime [1] and plane-wave/shock-wave backgrounds [2]. There are however also very general results concerning integrability and solvability for maximally symmetric spacetimes [3, 4] and gauged WZW models [5, 6]. These are the exceptional cases; generally the string equations of motion in curved spacetimes are not integrable and even if they are, it is usually an extremely difficult task to actually separate the equations, integrate them and finally write down the complete solution in closed form. Fortunately, there are several different ways to "attack" a system of coupled non-linear partial differential equations, using either numerical, approximative or exact methods (for some recent reviews on strings in curved spacetimes, see for instance [7, 8]).

In this paper, we consider some exact string solutions in curved spacetimes, obtained by imposing a particular ansatz. In Section 2, it is shown that in curved spacetimes with a Killing vector (timelike or spacelike), the string equations of motion and constraints (1.1)-(1.2) reduce to the Hamilton equations of a relativistic point-particle in a very simple scalar potential. As special examples we consider circular strings in axially symmetric spacetimes as well as stationary strings in stationary spacetimes. In Section 3, we consider in more detail the stationary strings in the Kerr-Newman geometry. It is shown that the world-sheet of a stationary string, that passes the static limit of the 4-D Kerr-Newman black hole in the equatorial plane, describes a 2-D Reissner-Nordström black hole. Mathematical results for 2-D black holes can therefore be applied to physical objects; (say) cosmic strings in the vicinity of Kerr black holes. An immediate general result is that the string modes are thermally excited.

2 Exact String Solutions

In this section we shall consider some exact string solutions in curved spacetimes. We first consider, as special examples, circular strings and stationary strings. Eventually we then give a more general description valid in curved spacetimes with an arbitrary Killing vector.

A. Circular Strings

The condition for a curved spacetime to allow for circular strings, is that the spacetime is axially symmetric. We can thus use a coordinate system $x^\mu = (t, r, \theta, \phi)$ where the metric is explicitly independent of the azimuthal angle ϕ :

$$ds^2 = g_{\mu\nu}(t, r, \theta) dx^\mu dx^\nu. \tag{2.1}$$

We consider for simplicity a 4-D spacetime, since we have mainly cosmic strings in mind. The ansatz which describes a dynamical circular string is provided by:

$$t = t(\tau), \qquad r = r(\tau), \qquad \theta = \theta(\tau), \qquad \phi = \sigma + f(\tau). \tag{2.2}$$

That is, we identify the spatial world-sheet coordinate σ with the azimuthal angle ϕ up to the addition of a function of the world-sheet coordinate τ, and let the other coordinates depend only on τ; this guarantees that the string is circular. The four functions $t(\tau)$, $r(\tau)$, $\theta(\tau)$ and $f(\tau)$ are now to be determined by the equations of motion and constraints (1.1)-(1.2).

The invariant definition of a circular string in an axially symmetric spacetime is that the Killing vector corresponding to the axial symmetry is tangent to the string world-sheet. It is then straightforward to show that the ansatz (2.2) is in fact the most general ansatz describing circular strings, at least up to residual gauge transformations and constant rescalings of σ.

By inserting the ansatz (2.2) into eqs.(1.1), it can be shown that the equations of motion reduce to the Hamilton equations of the point-particle Hamiltonian:

$$\mathcal{H} = \frac{1}{2} g^{\mu\nu} \mathcal{P}_\mu \mathcal{P}_\nu + \frac{1}{2} g_{\phi\phi}, \tag{2.3}$$

while the two constraints eqs.(1.2) become:

$$\mathcal{H} = 0, \qquad \mathcal{P}_\phi = 0. \tag{2.4}$$

The dynamics of a circular string in an axially symmetric curved spacetime is thus mathematically equivalent to the dynamics of a *massless* zero-angular-momentum point-particle, moving in the same curved spacetime as well as in a scalar potential given simply by the $(\phi\phi)$-component of the metric. By a conformal transformation, this system is also equivalent to the system of a *massive* point-particle moving in some unphysical spacetime (see [9] for a discussion of this point in the more general case of charge-current carrying circular strings, as well as the original references [10, 11] for the similar discussion for stationary strings. See also Section 3); however, the latter approach will not be considered in this section. We find it more convenient to work directly with the physical metric $g_{\mu\nu}$.

It must be stresset that the reduction of the system (1.1)-(1.2) to the system (2.3)-(2.4) of course does not mean that we have solved anything. In fact, Hamiltonian systems of the form (2.3), with or without the scalar potential $g_{\phi\phi}$, are generally not separable, and even if the "standard" point-particle part $g^{\mu\nu}\mathcal{P}_\mu\mathcal{P}_\nu$ of the Hamiltonian is separable, the separability is generally destroyed by the scalar potential $g_{\phi\phi}$. This is precisely what happens in the case of the Kerr-Newman black hole: the $g^{\mu\nu}\mathcal{P}_\mu\mathcal{P}_\nu$-part of the Hamiltonian (2.3) is separable, as is well known, but the $g_{\phi\phi}$-part is not of separable type [9, 12].

In most cases it is therefore necessary to make further reductions. Consider for instance 4-D spacetimes of the form:

$$ds^2 = g_{tt}(r,\theta)dt^2 + g_{rr}(r,\theta)dr^2 + g_{\theta\theta}(r,\theta)d\theta^2 + g_{\phi\phi}(r,\theta)d\phi^2 + 2g_{t\phi}(r,\theta)dtd\phi, \tag{2.5}$$

which covers the black hole spacetimes, some cosmological spacetimes (in static coordinates) the $2+1$ black hole and black string spacetimes (taking $\theta = const.$) etc. If we further consider circular strings in the equatorial plane $\theta = \pi/2$, then it can be shown that the equation determining the radius of the string-loop, as obtained from the Hamiltonian (2.3), takes the form:

$$\dot{r}^2 + V(r) = 0, \tag{2.6}$$

where:

$$V(r) = g^{rr}(E^2 g^{tt} + g_{\phi\phi}); \qquad E = const. \tag{2.7}$$

That is, everything is solved in quadratures, and the dynamics of the circular string is exactly known in closed form.

Using this approach, it is now a relatively easy task to describe the circular string dynamics in any curved spacetime of the form (2.5). Simply by looking at the line-element, one can immediately read off the potential $V(r)$, which determines the dynamics of the circular string. Next, one can then solve the equations exactly and the physical properties (energy-density, pressure,..) can be extracted.

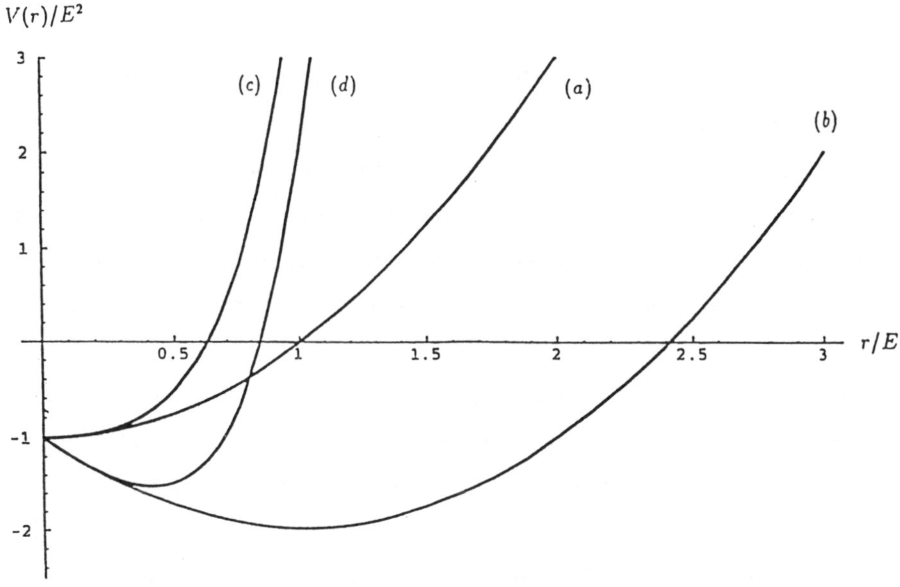

Figure 1. The potential $V(r)$, eq.(2.7), for a circular string in the equatorial plane of the four $3+1$ dimensional spacetimes: (a) Minkowski, (b) Schwarzschild black hole, (c) anti de Sitter and (d) Schwarzschild anti de Sitter. The details are explained in the text.

Circular strings were first discussed in Minkowski spacetime in [13]. More recently the dynamics of circular strings has been investigated in detail in both cosmological and black hole spacetimes [14-26] (charged circular strings in curved spacetimes were discussed in [12] and references given therein). We refer the interested reader to these publications and only give a couple of examples here. In general the dynamics of a circular string is governed by the tension, which always tries to contract the string, and the local gravitational field which can be either attractive or repulsive.

In the absence of a repulsive gravitational field, the circular string will always collapse from some maximal size. This is illustrated in Fig.1 in the cases of (a) Minkowski, (b) Schwarzschild, (c) anti de Sitter and (d) Schwarzschild-anti de Sitter. In all these cases the string collapses to zero size. There are however differences due to the causal structure of the spacetimes. In Minkowski and anti de Sitter spacetimes, the string dynamics is truely oscillatory (we neglect the gravitational backreaction), while in the black hole spacetimes the dynamics stops when the string falls into the spacetime singularity.

In the presence of a repulsive gravitational field, the situation is more complicated. A circular string can be oscillating between a maximal size and zero size, oscillating between a maximal size and a non-zero minimal size, it can be expanding forever or it can even be of constant size. This is illustrated in Fig.2 in the cases of (a) de Sitter and (b) Kerr black hole. The potential shown for de Sitter spacetime corresponds to the special case where $H^2E^2 < 1/4$, where H is the Hubble constant and E is the integration constant introduced in (2.7); the more general analysis for arbitrary values of H^2E^2 is presented in [22]. From the potential (a) follows that small strings will be oscillating between a maximal size and zero size, while large strings will bounce on the barrier and re-expand forever. The other types of circular string dynamics in de Sitter spacetime are described in detail in [22].

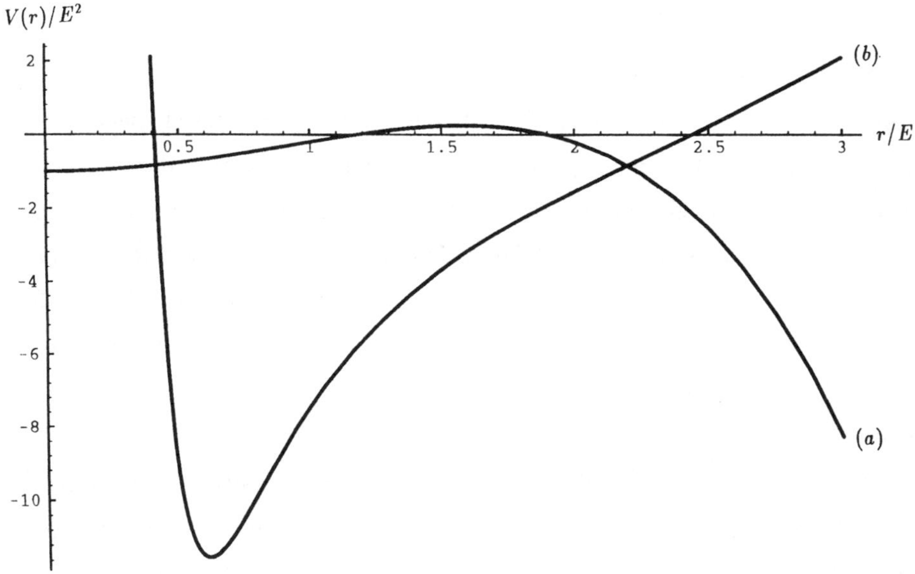

Figure 2. The potential $V(r)$, eq.(2.7), for a circular string in the equatorial plane of the two $3+1$ dimensional spacetimes: (a) de Sitter and (b) Kerr black hole. The details are explained in the text.

The potential shown for the Kerr Black hole corresponds to the special case where $a^2 > E^2$, where a is the angular momentum of the black hole; the more general potential for arbitrary values of a is given in [23]. From the potential (b) follows that a circular string contracts from its maximal size to a non-zero minimal size. When the string reaches its minimal size, it is actually inside the event horizon and thus cannot re-expand because of the causal structure. However, such solution can be interpreted as a string contracting in one spacetime, expanding in another etc [20, 23].

B. Stationary Strings

As another example of a family of exact string solutions in curved spacetimes, we now consider stationary strings. It is well known that in flat Minkowski spacetime there is only one stationary string configuration, namely the straight string; a string with any other shape will start vibrating. However, we shall now show that the situation in curved spacetimes is completely different. In fact, in some curved spacetimes stationary strings can have infinitely many completely different shapes, and there can be very interesting physics associated with them.

The condition for a curved spacetime to allow for stationary strings, is that the spacetime itself is stationary. We can thus use a coordinate system $x^\mu = (t, r, \theta, \phi)$ where the metric is explicitly independent of the time-coordinate t :

$$ds^2 = g_{\mu\nu}(r, \theta, \phi) dx^\mu dx^\nu. \tag{2.8}$$

The ansatz which describes a stationary string is:

$$t = \tau + f(\sigma), \qquad r = r(\sigma), \qquad \theta = \theta(\sigma), \qquad \phi = \phi(\sigma). \tag{2.9}$$

That is, we identify the temporal world-sheet coordinate τ with the time-coordinate t up to the addition of a function of the world-sheet coordinate σ, and let the other coordinates depend only on σ; this guarantees that the string is stationary. The four functions $f(\sigma)$, $r(\sigma)$, $\theta(\sigma)$ and $\phi(\sigma)$ are now to be determined by the equations of motion and constraints (1.1)-(1.2).

C.f. the previous discussion of circular strings in axially symmetric spacetimes, the invariant definition of a stationary string in a stationary spacetime is that the timelike Killing vector ∂_t is tangent to the string world-sheet. It is then straightforward to show that the ansatz (2.9) is the most general ansatz describing stationary strings up to residual gauge transformations and constant rescalings of τ.

By inserting the ansatz (2.9) into eqs.(1.1), it can be shown that the equations of motion reduce to the Hamilton equations of the point-particle Hamiltonian:

$$\mathcal{H} = \frac{1}{2}g^{\mu\nu}\mathcal{P}_\mu\mathcal{P}_\nu + \frac{1}{2}g_{tt}, \tag{2.10}$$

while the two constraints eqs.(1.2) become:

$$\mathcal{H} = 0, \qquad \mathcal{P}_t = 0, \tag{2.11}$$

compare with eqs.(2.3)-(2.4). Thus mathematically the problem of finding the stationary string configurations looks very similar to the problem of describing the dynamical circular strings. It is however a very interesting result that in the case of the Kerr-Newman black hole, the Hamiltonian system (2.10) is actually separable [10]. Thus contrary to the case of the circular strings in the Kerr-Newman background, the g_{tt}-part of the Hamiltonian (2.10) does not destroy the separability of the system, and therefore the stationary string configurations can be described completely and explicitly [10].

In most other cases it is however still necessary to make further reductions to solve the system. If we consider again the 4-D spacetimes of the form (2.5), and restrict ourselves to describing only stationary strings in the equatorial plane $\theta = \pi/2$, then it can be shown that the equation determining the radial coordinate of the stationary string, as obtained from the Hamiltonian (2.10), takes the form:

$$r'^2 + U(r) = 0, \tag{2.12}$$

where:

$$U(r) = g^{rr}(L^2 g^{\phi\phi} + g_{tt}); \qquad L = const. \tag{2.13}$$

That is, everything is solved in quadratures, and the stationary string configuration is exactly known in closed form.

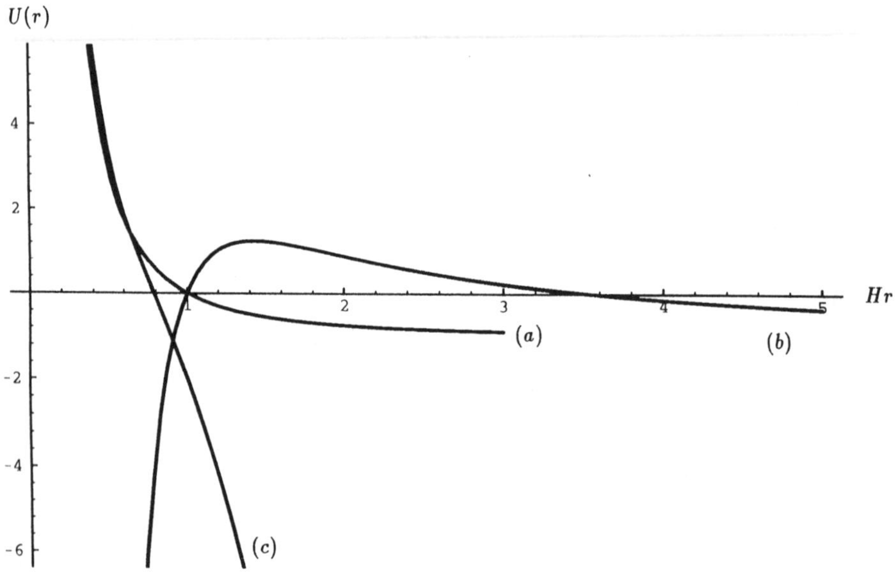

Figure 3. The potential $U(r)$, eq.(2.13), for a stationary string in the equatorial plane of the three $3+1$ dimensional spacetimes: (a) Minkowski, (b) anti de Sitter and (c) Schwarzschild black hole. The details are explained in the text.

Using this approach, it is now a relatively easy task to describe the stationary strings in any curved spacetime of the form (2.5). Everything goes exactly as in the previous case of the circular strings: simply by looking at the line-element, one can immediately read off the potential $U(r)$, which determines the location of the stationary string. Next, one can then solve the equations exactly and the physical properties (energy-density, pressure,..) can be extracted.

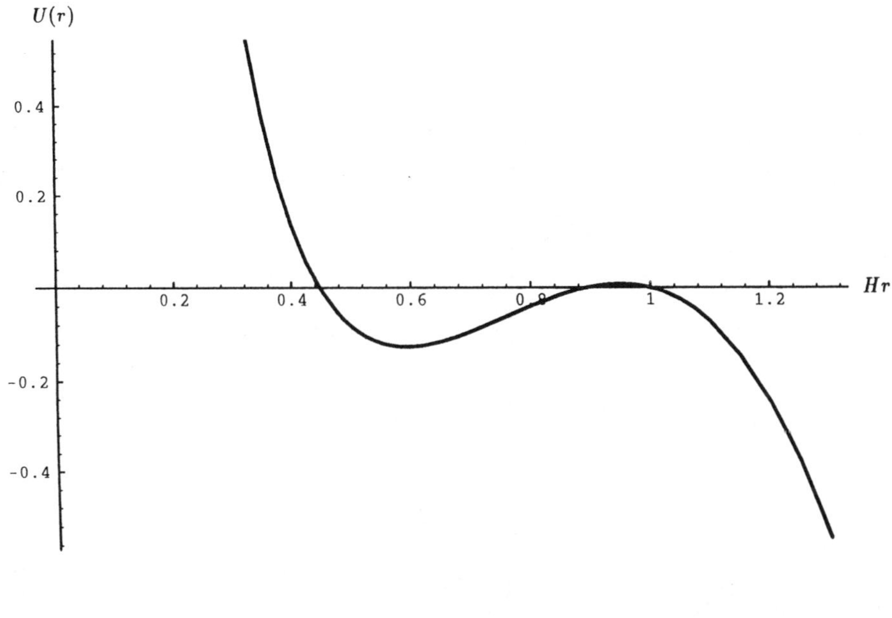

Figure 4. The potential $U(r)$, eq.(2.13), for a stationary string in the equatorial plane of de Sitter spacetime. The details are explained in the text.

As already mentioned, the only stationary string in Minkowski spacetime is the straight string. In stationary spacetimes with an attractive gravitational field, the stationary strings typically extend out to spatial infinity. This is illustrated in Fig.3 in the cases of (a) Minkowski, (b) Schwarzschild and (c) anti de Sitter spacetimes. For the potentials shown, the two ends of the stationary string are at spatial infinity, and the string passes $r = 0$ with some impact parameter related to the integration constant L. For a more complete discussion of these stationary strings we refer the reader to [10, 27].

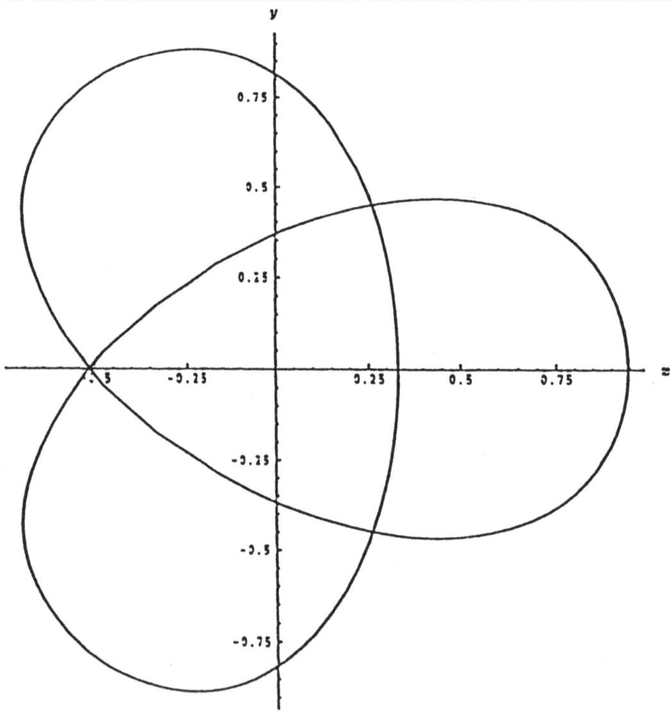

Figure 5. The so-called $(N, M) = (3, 2)$ *stationary string solution inside the horizon of de Sitter spacetime.*

If the spacetime has a repulsive gravitational field it is also possible to have stationary strings in a small compact region. This is illustrated in Fig.4 in the case of de Sitter spacetime. In this case the stationary string is located somewhere between two finite values of the radial coordinate (both being smaller than the horizon size). It turns out that there are infinitely many completely different stationary string configurations in de Sitter spacetime classified, using a simple underlying mathematical structure, in terms of two integers [27]. An example of a stationary string in de Sitter spacetime is presented in Fig.5. The complete discussion is given in [27].

C. General Formalism

It is appearent from the previous discussions that the mathematics of the circular strings is very similar to the mathematics of the stationary strings, although the physics is of course completely different. By comparing the formulas (2.1)-(2.7) with (2.8)-(2.13) we notice that there is actually a kind of "duality" between circular strings in axially symmetric spacetimes and stationary strings in stationary spacetimes [9, 27]; the "duality-transformation" being simply:

$$t \longleftrightarrow \phi\,, \qquad \tau \longleftrightarrow \sigma \tag{2.14}$$

It must be stressed, however, that there is absolutely nothing mysterious about this "duality". In fact, the previous discussions of circular strings and stationary strings are just special examples of a more general approach for obtaining special exact solutions of the string equations of motion and constraints in curved spacetimes with a Killing vector:

Consider an arbitrary curved spacetime with an arbitrary Killing vector ξ. Then take a coordinate system such that the Killing vector has the standard form:

$$\xi^\mu = \delta^\mu_i\,, \qquad \text{for some } i. \tag{2.15}$$

Now make the following ansatz for the string:

$$x^i = \sigma^1 + f(\sigma^2), \qquad x^j = x^j(\sigma^2); \quad j \neq i \tag{2.16}$$

where (σ^1, σ^2) are the two world-sheet coordinates. That is, the targetspace coordinate corresponding to the Killing vector is identified with one of the world-sheet coordinates plus a function of the other, while the other targetspace coordinates are only functions of the other world-sheet coordinate. In that case it is easy to show that the string equations of motion and constraints (1.1)-(1.2) reduce to the Hamiltonian system:

$$\mathcal{H} = \frac{1}{2} g^{\mu\nu} \mathcal{P}_\mu \mathcal{P}_\nu + \frac{1}{2} g_{ii}, \tag{2.17}$$

$$\mathcal{H} = 0, \qquad \mathcal{P}_i = 0. \tag{2.18}$$

This is just the Hamiltonian of a massless point-particle moving in the curved spacetime as well as in a scalar potential given simply by the (ii)-component of the metric. If the Killing vector is timelike, this system describes stationary strings, while if it is spacelike, it describes dynamical strings of some non-trivial shape.

3 Stationary Strings and 2-D Black Holes

(This section is essentially a reproduction of [28])
The aim of this section is to attract the attention to the remarkable fact, that the study of propagation of perturbations along a stationary cosmic string located in the gravitational field of a stationary 4-dimensional black hole, is directly related with physics of 2-dimensional black holes. If one neglects the gravitational effects of a cosmic string and assumes that its thickness is zero, then a string configuration in a given gravitational field is a minimal surface, and its equations of motion can be obtained by variation of the

Goto-Nambu action; these equations are of course equivalent to eqs.(1.1)-(1.2). An important case is when both the gravitational field and a string configuration are stationary. This case corresponds to a physical situation of a string which is in equilibrium in the gravitational field. The problem of finding the equilibrium string configurations in a stationary spacetime can be reduced to the problem of solving geodesic equations in a 3-dimensional unphysical space [10]. The remarkable property of the gravitational field of a stationary black hole is that these geodesic equations can be solved in quadratures [10, 11]. Consider a 2-dimensional time-like world-sheet Σ of the string defined by the equations $x^\mu = x^\mu(\sigma^A)$ (x^μ ($\mu = 0, ..., 4$) denote spacetime coordinates and σ^A ($A = 0, 1$) are the coordinates on the world-sheet). The metric G_{AB} induced on Σ reads:

$$G_{AB} = g_{\mu\nu} x^\mu_{,A} x^\nu_{,B}. \tag{3.1}$$

where $g_{\mu\nu}$ is the spacetime metric. Consider a stationary spacetime and let ξ^μ be the corresponding 4-dimensional Killing vector. We call a string stationary if ξ^μ is tangent to its world sheet Σ. Denote by η^A a 2-dimensional vector $\eta^A = G^{AB} x^\mu_{,B} \xi_\mu$. It is easy to show that:

$$\xi^\mu = \eta^A x^\mu_{,A}, \tag{3.2}$$

$$\eta_{A|B} = x^\mu_{,A} x^\nu_{,B} \xi_{\mu;\nu}, \tag{3.3}$$

where semicolon and vertical line denote covariant derivatives with respect to 4 and 2-dimensional metrics, respectively. The last relation implies that η is a Killing vector for the induced metric G_{AB}. Denote:

$$F = -g_{\mu\nu} \xi^\mu \xi^\nu = -G_{AB} \eta^A \eta^B. \tag{3.4}$$

Then the induced line-element $dl^2 = G_{AB} d\sigma^A d\sigma^B$ on Σ can be written as:

$$dl^2 = -F d\hat{\tau}^2 + F^{-1} d\hat{\sigma}^2, \tag{3.5}$$

where $\hat{\tau} = \hat{\tau}(\sigma^A)$, $\hat{\sigma} = \hat{\sigma}(\sigma^A)$. This representation is valid in the regions where $F \neq 0$, and hence ξ is either time-like or space-like. We assume that the 4-dimensional spacetime is asymptotically flat and contains a black hole. We assume also that $\hat{\sigma} = \infty$ corresponds to the points of the string located in the asymptotically flat region of the physical spacetime, so that $F(\hat{\sigma} = \infty) = 1$. The metric, eq.(3.5), describes a 2-dimensional black hole if $F = 0$ at finite value of $\hat{\sigma}$. At this point the Killing 2-vector η is null. It happens at the points where the string world-sheet crosses the infinite-red-shift surface (the static limit), i.e. the surface where $\xi^2 = 0$. For a static black hole this surface coincides with the event horizon. For stationary (rotating) black holes it is lying outside the horizon. The region located between infinite-red-shift surface and the event horizon of a stationary black hole is known as the ergosphere. Points of the string located inside the ergosphere thus correspond to the interior of the 2-dimensional black hole.

The main observation we would like to make now is that perturbations propagating along the string can be described by a coupled system of a pair of scalar field equations in the 2-dimensional metric G_{AB} [29]. For this reason, if a stationary string is passing through the ergosphere of a 4-dimensional black hole, the physics of string excitations is effectively reduced to the physics of 2-dimensional black holes.

We shall first derive the differential equations describing the perturbations propagating along a stationary string configuration embedded in an arbitrary stationary 4-dimensional spacetime. We then consider, as a special example, stationary strings in the background of a Kerr-Newman black hole.

To be more specific, we write the metric of a generic 4-dimensional stationary spacetime in the form:

$$g_{\mu\nu} = \begin{pmatrix} -F & -FA_i \\ -FA_i & -FA_iA_j + H_{ij}/F \end{pmatrix}, \tag{3.6}$$

where $\partial_t F = 0,\ \partial_t A_i = 0,\ \partial_t H_{ij} = 0$. That is to say, the Killing vector is given explicitly by:

$$\xi^\mu = (1,\ 0,\ 0,\ 0), \qquad \xi_\mu = (-F,\ -FA_i), \tag{3.7}$$

consistent with the notation of eq.(3.4). A stationary string configuration is parametrized in the following way:

$$t = x^0 = \tau, \qquad x^i = x^i(\sigma)\ ; \qquad (\tau \equiv \sigma^0,\ \sigma \equiv \sigma^1). \tag{3.8}$$

Then the equations of motion corresponding to the Goto-Nambu action reduce to [10]:

$$x^{i\prime\prime} + \tilde{\Gamma}^i_{jk} x^{j\prime} x^{k\prime} = 0, \qquad H_{ij} x^{i\prime} x^{j\prime} = 1, \tag{3.9}$$

where $\tilde{\Gamma}^i_{jk}$ is the Christoffel connection for the metric H_{ij} and a prime denotes differentiation with respect to σ. The induced metric on the world-sheet now takes the form:

$$G_{AB} = \begin{pmatrix} -F & -FA \\ -FA & -FA^2 + 1/F \end{pmatrix}; \qquad A \equiv A_i x^{i\prime}, \tag{3.10}$$

so that det $G = -1$. The following coordinate transformation on the world-sheet:

$$\hat{\tau} = \tau + \int^\sigma A\, d\sigma, \qquad \hat{\sigma} = \sigma, \tag{3.11}$$

brings the induced line element into the form of eq.(3.5), that is:

$$\hat{G}_{AB} = \begin{pmatrix} -F & 0 \\ 0 & 1/F \end{pmatrix}. \tag{3.12}$$

A covariant approach describing the propagation of perturbations along an arbitrary string configuration embedded in an arbitrary spacetime, was developed in Ref.[29] (see also [30, 31]). The general transverse (physical) perturbation around a background Goto-Nambu string configuration is written as:

$$\delta x^{\mu} = \Phi^{R} n^{\mu}_{R}, \qquad (R = 2, 3), \tag{3.13}$$

where the two vectors n^{μ}_{R}, normal to the string world-sheet, fulfill:

$$g_{\mu\nu} n^{\mu}_{R} n^{\nu}_{S} = \delta_{RS}, \qquad g_{\mu\nu} x^{\mu}_{,A} n^{\nu}_{R} = 0, \tag{3.14}$$

as well as the completeness relation:

$$g^{\mu\nu} = G^{AB} x^{\mu}_{,A} x^{\nu}_{,B} + \delta^{RS} n^{\mu}_{R} n^{\nu}_{S}. \tag{3.15}$$

It can be shown [29] that the perturbations Φ^{R} are determined by the following effective action:

$$\mathcal{S}_{\text{eff.}} = \int d^{2}\zeta \sqrt{-G}\, \Phi^{R} \left\{ G^{AB} (\delta^{T}_{R} \nabla_{A} + \mu_{R}{}^{T}{}_{A})(\delta_{TS} \nabla_{B} + \mu_{TSB}) + \mathcal{V}_{RS} \right\} \Phi^{S}, \tag{3.16}$$

where $\mathcal{V}_{RS} = \mathcal{V}_{(RS)}$ are scalar potentials and $\mu_{RS}{}^{A} = \mu_{[RS]}{}^{A}$ are vector potentials on Σ, determined by its embedding into the 4-dimensional spacetime, and ∇_{A} is the covariant derivative with respect to the metric G_{AB}. The vector potentials μ_{RSA} coincide with the normal fundamental form:

$$\mu_{RSA} = g_{\mu\nu} n^{\mu}_{R} x^{\rho}_{,A} \nabla_{\rho} n^{\nu}_{S}, \tag{3.17}$$

where ∇_{ρ} is the covariant derivative with respect to the metric $g_{\mu\nu}$. The scalar potentials $\mathcal{V}_{RS}$ are defined as:

$$\mathcal{V}_{RS} \equiv \Omega_{RAB} \Omega_{S}{}^{AB} - G^{AB} x^{\mu}_{,A} x^{\nu}_{,B} R_{\mu\rho\sigma\nu} n^{\rho}_{R} n^{\sigma}_{S}, \tag{3.18}$$

where Ω_{RAB} is the second fundamental form:

$$\Omega_{RAB} = g_{\mu\nu} n^{\mu}_{R} x^{\rho}_{,A} \nabla_{\rho} x^{\nu}_{,B}, \tag{3.19}$$

and $R_{\mu\rho\sigma\nu}$ is the Riemann tensor corresponding to the metric $g_{\mu\nu}$. The equation describing the propagation of perturbations along the string world-sheet is:

$$\left\{G^{AB}(\delta_R^T\nabla_A + \mu_R{}^T{}_A)(\delta_{TS}\nabla_B + \mu_{TSB}) + \mathcal{V}_{RS}\right\}\Phi^S = 0. \tag{3.20}$$

We obtain now explicit expressions for vector $\mu_{RS}{}^A$ and scalar $\mathcal{V}_{RS}$ potentials which enter the propagation equations for a stationary string obeying eqs.(3.9), in the background (3.6). Using eqs.(3.14), the normal vectors n^μ_R take the form:

$$n^\mu_R = (-A_i n^i_R,\ n^i_R), \qquad n_{\mu R} = (0,\ F^{-1}H_{ij}n^j_R) \equiv (0,\ n_{iR}). \tag{3.21}$$

It is convenient to introduce also 3-dimensional vectors

$$\tilde{n}^i_R = |F|^{-1/2} n^i_R, \tag{3.22}$$

which together with $x^{i\prime}$ form an orthonormal system $(x^{i\prime}, \tilde{n}^i_2, \tilde{n}^i_3)$ in the 3-dimensional space with metric H_{ij}. We note that there is an ambiguity $\tilde{n}^i_R \to \tilde{n}^i_R + \delta(\tilde{n}^i_R)$ in the choice of pair of normal vectors $\tilde{n}^i_R$:

$$\delta(\tilde{n}^i_R) = \Lambda_R{}^S \tilde{n}^i_S\ ; \qquad \Lambda_{RS}(\zeta) = -\Lambda_{SR}(\zeta). \tag{3.23}$$

Under these 'gauge' transformations, which leave invariant the perturbations (3.13) and the effective action (3.16), the quantities Φ^S, μ_{RSA}, and $\mathcal{V}_{RS}$ are transformed as follows:

$$\begin{aligned} \delta(\Phi^R) &= \Lambda^R{}_S\Phi^S, \\ \delta(\mu_{RSA}) &= \Lambda_R{}^T\mu_{TSA} - \Lambda_S{}^T\mu_{TRA} - \Lambda_{RS,A}, \\ \delta(\mathcal{V}_{RS}) &= \Lambda_R{}^T\mathcal{V}_{TS} + \Lambda_S{}^T\mathcal{V}_{RT}. \end{aligned} \tag{3.24}$$

We fix the freedom of 'gauge' by choosing the vectors $\tilde{n}^i_R$ to be covariantly constant in the 3-dimensional space [29]. After straightforward but tedious calculations we get:

$$\mu_{RS}{}^{A} = \eta^A \frac{A_{ij}}{2} n^i_R n^j_S = \eta^A \frac{1}{2F^2} \xi_{[0,i}\xi_{j]} n^i_R n^j_S, \tag{3.25}$$

$$\begin{aligned} \mathcal{V}_{RS} &= -x^{i'} x^{j'} \tilde{R}_{iklj} n^k_R n^l_S + \frac{\delta_{RS}}{2} x^{i'} x^{j'} (\tilde{\nabla}_i \tilde{\nabla}_j F) \\ &- \frac{\delta_{RS}}{4F} x^{i'} x^{j'} (\tilde{\nabla}_i F)(\tilde{\nabla}_j F) + \frac{F}{4} \delta^{TU} A_{il} A_{kj} n^i_U n^j_T n^k_R n^l_S. \end{aligned} \tag{3.26}$$

Here the "field-strength" A_{ij} is defined by:

$$A_{ij} = A_{i,j} - A_{j,i} = \tilde{\nabla}_j A_i - \tilde{\nabla}_i A_j, \tag{3.27}$$

where $\tilde{\nabla}_i$ is the covariant derivative with respect to the metric H_{ij} and $\tilde{R}_{iklj}$ is the Riemann tensor corresponding to the metric H_{ij}. It is easy to verify that for our choice of 'gauge' the vector potentials $\mu_{RS}{}^{A}$ obey the analog of the Lorentz gauge conditions $\nabla_A \mu_{RS}{}^{A} = 0$. The anti-symmetric products of normal vectors, appearing in the scalar and vector potentials, eqs.(3.25)-(3.26), can be eliminated using the identity:

$$\epsilon^{RS} \tilde{n}^i_R \tilde{n}^j_S = \tilde{e}^{ijk} H_{kl} x^{l'}; \qquad \tilde{e}^{ijk} = (H)^{-1/2} \epsilon^{ijk}. \tag{3.28}$$

In particular we have:

$$x^{i'} x^{j'} \tilde{R}_{iklj} n^k_R n^l_S = F \delta_{RS} (\frac{1}{2} \tilde{R} - \tilde{R}_{ij} x^{i'} x^{j'}) - \tilde{R}_{ij} n^i_R n^j_S, \tag{3.29}$$

$$\frac{F}{4} \delta^{TU} A_{il} A_{kj} n^i_U n^j_T n^k_R n^l_S = \frac{F^3}{4} \delta_{RS} (A_{ik} A_j{}^{k} x^{i'} x^{j'} - \frac{1}{2} A_{ij} A^{ij}). \tag{3.30}$$

By using the equations (3.29)-(3.30) we get:

$$\mu_{RS}{}^{A} = \mu \epsilon_{RS} \eta^A \ ; \qquad \mu = \frac{F}{4} A_{ij} \tilde{e}^{ijk} H_{kl} x^{l'}, \tag{3.31}$$

$$\begin{aligned} \mathcal{V}_{RS} &= \delta_{RS} [\frac{x^{i'} x^{j'}}{2} (\tilde{\nabla}_i \tilde{\nabla}_j F) - \frac{1}{4F} x^{i'} x^{j'} (\tilde{\nabla}_i F)(\tilde{\nabla}_j F) - F (\frac{1}{2} \tilde{R} - \tilde{R}_{ij} x^{i'} x^{j'}) \\ &+ \frac{F^3}{4} (A_{ik} A_j{}^{k} x^{i'} x^{j'} - \frac{1}{2} A_{ij} A^{ij})] + \tilde{R}_{ij} n^i_R n^j_S. \end{aligned} \tag{3.32}$$

As a special application of the above formalism, we now consider perturbations propagating along a stationary string in the Kerr-Newman black hole background. The Kerr-Newman metric in Boyer-Lindquist coordinates [32] reads:

$$ds^2 = -\frac{\Delta}{\rho^2}[dt - a\sin^2\theta d\phi]^2 + \frac{\sin^2\theta}{\rho^2}[(r^2+a^2)d\phi - adt]^2 + \frac{\rho^2}{\Delta}dr^2 + \rho^2 d\theta^2, \quad (3.33)$$

where $\Delta = r^2 - 2Mr + Q^2 + a^2$ and $\rho^2 = r^2 + a^2\cos^2\theta$. This metric is of the form (3.6) with:

$$H_{rr} = \frac{\Delta - a^2\sin^2\theta}{\Delta}, \qquad H_{\theta\theta} = \Delta - a^2\sin^2\theta, \qquad H_{\phi\phi} = \Delta\sin^2\theta, \quad (3.34)$$

$$F = \frac{\Delta - a^2\sin^2\theta}{r^2 + a^2\cos^2\theta}, \qquad A_\phi = a\sin^2\theta\,\frac{2Mr - Q^2}{\Delta - a^2\sin^2\theta}. \quad (3.35)$$

The unperturbed stationary string configurations are obtained by solving eqs.(3.9), [10]:

$$\begin{aligned}(H_{rr}r')^2 &= \frac{a^2b^2}{\Delta^2} - \frac{q^2}{\Delta} + 1,\\ (H_{\theta\theta}\theta')^2 &= q^2 - \frac{b^2}{\sin^2\theta} - a^2\sin^2\theta \equiv q^2 - U(\theta),\\ (H_{\phi\phi}\phi')^2 &= b^2,\end{aligned} \quad (3.36)$$

where b and q are integration constants. Equations (3.36) are evidently invariant with respect to the reflection $\phi \to -\phi$. It means that if $(r(\sigma), \theta(\sigma), \phi(\sigma))$ is a solution of eqs.(3.36) then $(r(\sigma), \theta(\sigma), -\phi(\sigma))$ is also a solution.

We consider at first the case when a stationary string is located in the equatorial plane $\theta = \pi/2$, corresponding to $|b| \geq |a|$ and $q^2 = a^2 + b^2$. In that case, the two covariantly constant normal vectors $\tilde{n}^i_R$, introduced in eq.(3.22), are given by:

$$\tilde{n}^i_2 \equiv \tilde{n}^i_\perp = \frac{1}{\sqrt{|r^2 - 2Mr + Q^2|}}(0,\ 1,\ 0), \quad (3.37)$$

$$\tilde{n}^i_3 \equiv \tilde{n}^i_{\|} = \frac{1}{\sqrt{|r^2 - 2Mr + Q^2|}}(-b,\ 0,\ H_{rr}r'). \tag{3.38}$$

It is now straightforward to compute the vector and scalar potentials given by eqs.(3.31),(3.32):

$$\mu = 0, \tag{3.39}$$

$$\begin{aligned}
\mathcal{V}_{\perp\perp} &= \mathcal{V}_{\|\,\|} + \frac{2(M^2 - Q^2)(a^2 - b^2)}{r^2(\Delta - a^2)^2} \\
&= \frac{M^2 - Q^2}{r^2(\Delta - a^2)}[1 + \frac{2(a^2 - b^2)}{\Delta - a^2}] + \frac{(r - M)(b^2 - a^2)}{(\Delta - a^2)^2}[\frac{M}{r^2} - \frac{Q^2}{r^3}] \\
&+ \frac{\Delta - b^2}{(\Delta - a^2)^2}[-\frac{2M}{r} + \frac{3Q^2}{r^2} + \frac{3M^2}{r^2} - \frac{6MQ^2}{r^3} + \frac{2Q^4}{r^4}].
\end{aligned} \tag{3.40}$$

In the generic case, $|b| > |a|$, the stationary configuration in the equatorial plane describes an infinitely long open string with two "arms" in the asymptotically flat regions and a turning point *outside* the ergosphere [10]. The perturbation equations (3.20) with the potentials (3.39)-(3.40) then determine the reflection and transmission of waves ('phonons') travelling along the string between the asymptotically flat regions [29, 33]. However, in the special case $|b| = |a|$, the string passes the static limit and spirals inside the ergosphere towards the horizon. This is the case we are interested in, c.f. the discussion after eq.(3.5). Notice that the potential (3.40), in general being divergent at the static limit ($\Delta = a^2$), is finite for $|b| = |a|$, i.e. the divergences precisely cancel out in the particular case where the string actually crosses the static limit.

Let us now consider the case $|b| = |a|$ in more detail. We use the notations $\Sigma_\pm$ for a pair of string configurations connected by the reflection transformation, $\phi \to -\phi$, discussed after eq.(3.36). In order to make this prescription unique we choose $dr/d\phi < 0$ for Σ_+ and $dr/d\phi > 0$ for Σ_-. Then we find from eq.(3.11):

$$\hat{\sigma} = r, \qquad \hat{\tau} = t \pm a^2 \int^r \frac{2Mr - Q^2}{\Delta(\Delta - a^2)}\, dr, \tag{3.41}$$

where the signs $\pm$ correspond to the string configuration $\Sigma_\pm$. For both configurations $\Sigma_\pm$:

$$F = 1 - \frac{2M}{r} + \frac{Q^2}{r^2}, \tag{3.42}$$

thus the world-sheet line element (3.5) takes the form of a 2-dimensional Reissner-Nordström black hole. The parameters of mass M and charge Q of this 2-dimensional black hole are the same as the corresponding parameters of the original 4-dimensional Kerr-Newman metric. The world-sheet spatial coordinate $\hat{\sigma}$ equals the Boyer-Lindquist radial coordinate r, while the world-sheet time $\hat{\tau}$ approaches the Boyer-Lindquist time t in the asymptotically flat region, $r \to \infty$. The coordinates $(\hat{\tau}, r)$ cover only the exterior region of the string hole. One can easily obtain the string metric valid in a wider region. For this purpose it is convenient to introduce the Eddington-Finkelstein coordinates $(u_\pm, \varphi_\pm)$:

$$dt = du_\pm \mp \Delta^{-1}(r^2 + a^2)dr, \qquad d\phi = d\varphi_\pm \mp \Delta^{-1} a dr, \tag{3.43}$$

and to rewrite the Boyer-Lindquist metric (3.33) as [34]:

$$\begin{aligned} ds^2 &= -\frac{\Delta}{\rho^2}[du_\pm - a\sin^2\theta d\varphi_\pm]^2 + \frac{\sin^2\theta}{\rho^2}[(r^2 + a^2)d\varphi_\pm - a du_\pm]^2 \\ &+ \rho^2 d\theta^2 \pm 2dr[du_\pm - a\sin^2\theta d\varphi_\pm]. \end{aligned} \tag{3.44}$$

In these coordinates, the strings $\Sigma_\pm$ are described by equations $\theta = \pi/2$, $\varphi_\pm = \text{const.}$, so that the induced metric on $\Sigma_\pm$ is:

$$ds^2 = -F du_\pm^2 \pm 2dr du_\pm. \tag{3.45}$$

This metric for Σ_+ describes a black hole, while for Σ_- it describes a white hole. In both cases the perturbation equations (3.20) on $\Sigma_\pm$ reduce to:

$$\Box\Phi^R = \frac{\Box r}{r}\Phi^R; \qquad R = \perp, \parallel \tag{3.46}$$

where $\Box$ is the d'Alambertian on the world-sheet, $\Box = G^{AB}\nabla_A\nabla_B$. We also have $\Box r = F_{,r}$.

Several interesting remarks are now in order. First notice that the perturbation equations for the two transverse polarizations Φ^R are decoupled and *identical.* Secondly, eq.(3.46) is precisely the s-wave scalar field equation in the 4-dimensional Reissner-Nordström black hole background:

$$\Box^{(4)}\phi = 0, \qquad g^{(4)}_{\mu\nu} = \text{diag}(-F,\ 1/F,\ r^2,\ r^2\sin^2\theta), \tag{3.47}$$

with F given by eq.(3.42). The decomposition $\phi = \sum r^{-1}\Phi_l(r,t)\ Y_{lm}(\theta,\phi)$ yields:

$$-\frac{1}{F}\partial_t^2\Phi_l + \partial_r(F\partial_r\Phi_l) = \frac{F_{,r}}{r}\Phi_l + \frac{l(l+1)}{r^2}\Phi_l, \tag{3.48}$$

which is identical to equation (3.46) for $l = 0$, as is most easily seen by writing eq.(3.46) in the $(\hat{\tau}, r)$ world-sheet coordinate system. (Equation (3.48) has been studied in detail in the literature, see for instance [35] and references given therein.) The string exitations for a stationary string passing through the ergosphere in the equatorial plane of a Kerr-Newman black hole are therefore described mathematically by the $l = 0$ scalar waves in the background of a 4-dimensional Reissner-Nordström black hole.

The above results allow generalization to the case when a stationary string is located not in the equatorial plane but on the cone $\theta = \theta_0 \neq \pi/2$. For this case the parameters which enter the equations (3.36) are related as $q^2 = 2|ab|$. The corresponding θ_0 is determined as the minimum of the potential $U(\theta)$ and is, $\sin^2\theta_0 = |b/a|$. This relation implies that $|b| < |a|$. The remarkable fact is that such a string allows a simple geometrical description. The Kerr-Newman metric possesses two principal null geodesic congruences, one of them is formed by incoming and the other by outgoing principal null rays [34]. Take one of these null geodesics $\gamma_{\pm}$ ($-$ stands for an outgoing and $+$ stands for an incoming ray). Consider two-dimensional surfaces $\Sigma_{\pm}$ formed by Killing trajectories passing through $\gamma_{\pm}$. It is possible to prove that $\Sigma_{\pm}$ is a minimal surface and it describes an "equilibrium" string configuration with the parameter $q^2 = 2|ab|$. Two string configurations $\Sigma_{\pm}$ differ by signs of r' in (3.36). The metric induced on both surfaces $\Sigma_{\pm}$ possesses the Killing horizon. In case of Σ_{+} it describes a black hole, while in case of Σ_{-} it describes a white hole (for a white hole a future directed timelike curve crossing the Killing horizon enters the black hole exterior). The above considered case of a stationary string located on the equatorial plane and crossing the infinite-red-shift surface is a special example of stationary cone strings. The perturbation equations for cone strings have been considered in [36].

To summarize, we have shown that the metric induced on a stationary string crossing the infinite-red-shift surface describes a two-dimensional geometry of a black or white hole. It opens remarkable possibilities to apply results of mathematical study of two-dimensional black holes to physical objects (cosmic strings in the vicinity of a black hole), which at least in principle allow experimental observations. In particular, in the presence of the horizon on Σ_{+} (for the two-dimensional string black hole) the conditions that the quantum state is regular near the horizon implies that the string perturbations Φ^R are to be thermally excited. It means that there will be a thermal flux of the string excitations ('phonons') propagating to infinity which forms the corresponding Hawking radiation. For a radial string crossing the event horizon of the Schwarzschild black hole this effect was considered in [37]. We would like to stress that the analogous radiation will also be present when a stationary string crosses the infinite-red-shift surface (which for a rotating black hole is located outside the horizon). String perturbations ('phonons') generated in the region lying inside the 2-horizon and propagating along the string Σ_+ cannot escape to infinity. But it is well known that the causal signals emitted in the ergosphere and propagating in the 4-dimensional spacetime can reach a distant observer. One can use such signals ('photons') in order to get information from the interior of the two-dimensional string black hole. This situation is similar to one which happens in a 'dumb' hole considered by Unruh [38]. In order to define a 'dumb' hole one uses the causal structure connected with the propagation of phonons. The photons propagating with supersonic velocity can escape a 'dumb' hole interior. The nice feature which differs our model is that it is constructed in the framework of completely relativistic theory.

Possibility of getting information from a black hole interior was also discussed in [39], where a gedanken experiment was proposed in which a traversable wormhole is used. The mechanism discusses in the present paper which makes it possible to extract information from the interior of string black holes is different. It is connected with the presence of extra-dimensions and does not require non-trivial topology. The possibility of the information extraction from the interior of a string black hole might give new insight to the problem of information loss in black holes. In particular the arguments of Ref.[40] being applied to string black holes indicate that black hole complementarity may be inconsistent, at least for these black hole models.

Acknowledgements

I would like to thank N. Sánchez and H.J. de Vega for organizing this very interesting and stimulating meeting. I also thank N. Sánchez and H.J. de Vega as well as V. Frolov and S. Hendy for collaboration on different parts of the material presented in this talk. I finally acknowledge the financial support from NSERC and from NATO Collaboration Grant CRG 941287.

References

[1] H.J. de Vega and N. Sánchez, Phys. Rev. D42 (1990) 3969.
H.J. de Vega, M. Ramón-Medrano and N. Sánchez, Nucl. Phys. B374 (1992) 405.

[2] H.J. de Vega and N. Sánchez, Phys. Lett. B244 (1990) 215; Phys. Rev. Lett. 65C (1990) 1517; IJMP A7 (1992) 3043; Nucl. Phys. B317 (1989) 706 and ibid 731.
D. Amati and C. Klimcik, Phys. Lett. B210 (1988) 92.
M. Costa and H.J. de Vega, Ann. Phys. 211 (1991) 223 and ibid 235.
C. Loustó and N. Sánchez, Phys. Rev. D46 (1992) 4520.

[3] V.E. Zakharov and A.V. Mikhailov, JETP 47 (1979) 1017.
H. Eichenherr, in "Integrable Quantum Field Theories", ed. J. Hietarinta and C. Montonen (Springer, Berlin, 1982).

[4] H.J. de Vega and N. Sánchez, Phys. Rev. D47 (1993) 3394.

[5] I. Bars and K. Sfetsos, Mod. Phys. Lett. A7 (1992) 1091.

[6] H.J. de Vega, J.R. Mittelbrunn, M.R. Medrano and N. Sánchez, Phys. Lett. B232 (1994) 133 and Int. J. Mod. Phys. A 11 (1996) 1463.

[7] H.J. de Vega and N. Sánchez, in "String Quantum Gravity and Physics at the Planck Energy Scale". Proceedings of the Erice Conference held in September 1995, ed. N. Sánchez and A. Zichichi (Kluwer Academic Publishers, 1996) p. 11.

[8] A.L. Larsen and N. Sánchez, in "String Quantum Gravity and Physics at the Planck Energy Scale". Proceedings of the Erice Conference held in September 1995, ed. N. Sánchez and A. Zichichi (Kluwer Academic Publishers, 1996) p. 65.

[9] A.L. Larsen, Class. Quantum Grav. 10 (1993) 1541.

[10] V.P. Frolov, V.D. Skarzhinsky, A.I. Zelnikov and O. Heinrich, Phys. Lett. B224 (1989) 255.

[11] B. Carter and V.P. Frolov, Class. Quantum Grav. 6 (1989) 569.
B. Carter, V.P. Frolov and O. Heinrich, Class. Quantum Grav. 8 (1991) 135.

[12] A.L. Larsen, Class. Quantum Grav. 11 (1994) 1201.

[13] A. Vilenkin, Phys. Rev. D24 (1981) 2082.

[14] K.G. Akdeniz, S. Erbil, H. Mutus and E. Rizaoglu, Phys. Lett. B237 (1990) 192.

[15] A. Vilenkin, Phys. Rev. D43 (1991) 1060.

[16] R. Basu, A.H. Guth and A. Vilenkin, Phys. Rev. D44 (1991) 340.

[17] J. Garriga and A. Vilenkin, Phys. Rev. D44 (1991) 1007; D45 (1992) 3469; D47 (1993) 3256.

[18] X. Li and J. Zhang, Phys. Lett. B312 (1993) 62.

[19] H.J. de Vega, A.V. Mikhailov and N. Sánchez, Teor. Mat. Fiz. 94 (1993) 232; Mod. Phys. Lett. A29 (1994) 2745.

[20] H.J. de Vega and I.L. Egusquiza, Phys. Rev. D50 (1994) 763.

[21] A.L. Larsen, Phys. Rev. D50 (1994) 2623.

[22] H.J. de Vega, A.L. Larsen and N. Sánchez, Nucl. Phys. B427 (1994) 643.

[23] A.L. Larsen and N. Sánchez, Phys. Rev. D50 (1994) 7493.

[24] A.L. Larsen, Phys. Rev. D51 (1991) 4330.

[25] H.J. de Vega, A.L. Larsen and N. Sánchez, Phys. Rev. D51 (1995) 6917.

[26] A.L. Larsen and N. Sánchez, "The Effect of Spatial Curvature on the Classical and Quantum Strings", to appear in IJMPA.

[27] A.L. Larsen and N. Sánchez, Phys. Rev. D51 (1995) 6929.

[28] V.P. Frolov and A.L. Larsen, Nucl. Phys. B449 (1995) 149.

[29] A.L. Larsen and V.P. Frolov, Nucl. Phys. B414 (1994) 129.

[30] B. Carter, Phys. Rev. D48 (1993) 4835.

[31] J. Guven, Phys. Rev. D48 (1993) 5562.

[32] R.H. Boyer and R.W. Lindquist, J. Math. Phys. 8 (1967) 265.

[33] A.L. Larsen, Nucl. Phys. B412 (1994) 372.

[34] C. W. Misner, K. S. Thorne and J. A. Wheeler, Gravitation, (W. H. Freeman, San Francisco, 1973).

[35] S. Chandrasekhar, The Mathematical Theory of Black Holes, (Clarendon Press, London, 1983).

[36] V. Frolov, S. Hendy and A.L. Larsen, "How to Create a 2-D Black Hole", to appear in Phys. Rev. D.

[37] A. Lawrence and E. Martinec, Phys. Rev. D 50 (1994) 2680.

[38] W. G. Unruh, Phys. Rev. Lett. 46 (1981) 1351; "Dumb Holes and the Effects of High Frequencies on Black Hole Evaporation". Preprint gr-qc/9409008;

[39] V.P. Frolov and I. Novikov, Phys. Rev. D 48 (1993) 1607.

[40] D. Gottesman, Phys. Rev. D 51 (1995) 4600.

Strings next to and inside black holes

H. J. de Vega[(a)] and I. L. Egusquiza[(b)]

*(a) Laboratoire de Physique Théorique et Hautes Énergies * (b) Fisika Teorikoaren saila*

Abstract

The classical string equations of motion and constraints are solved near the horizon and near the singularity of a Schwarzschild black hole.

In a conformal gauge such that $\tau = 0$ (τ = worldsheet time coordinate) corresponds to the horizon ($r = 1$) or to the black hole singularity ($r = 0$), the string coordinates express in power series in τ near the horizon and in power series in $\tau^{1/5}$ around $r = 0$.

We compute the string invariant size and the string energy-momentum tensor. Near the horizon both are finite and analytic. Near the black hole singularity, the string size, the string energy and the transverse pressures (in the angular directions) tend to infinity as r^{-1}. To leading order near $r = 0$, the string behaves as two dimensional radiation. This two spatial dimensions are describing the S^2 sphere in the Schwarzschild manifold.

---

*mail address: LPTHE, Université Paris VI, Tour 16, 1er étage, 4, Place Jussieu 75252 Paris, Cedex 05, FRANCE. Laboratoire Associé au CNRS UA 280.

†mail address: Fisika Teorikoaren saila Euskal Herriko Unibertsitatea
644 P.K. - 48080 BILBAO
SPAIN

I. INTRODUCTION AND MOTIVATIONS

The systematic investigation of strings in curved spacetimes started in [1,2] has uncovered a variety of new physical phenomena (see [3] for a review). These results are relevant both for fundamental (quantum) strings and for cosmic strings, which behave in an essentially classical way [4].

Black-hole spacetimes are probably the most relevant backgrounds to study strings propagating into a spacetime singularity (see [5] for plane wave null singularities). Strings in Schwarzschild black holes have been previously investigated in refs. [2,6–10]. Different kinds of symmetric Ansätze that **separate** the variables σ and τ have been proposed. In this way the string evolution equations (which are non-linear partial differential equations) become ordinary differential equations.

In ref. [6] stationary solutions are obtained with the spatial variables depending only on σ and the temporal variable set as $t = \tau$. In this way infinitely long string solutions outside the horizon were constructed. String solutions inside the horizon are obtained in ref. [9] by exchanging $\sigma \leftrightarrow \tau$ in the previous Ansatz. These string are also infinitely long. All solutions are expressed in terms of elliptic functions.

Ring solutions are obtained in ref. [8] through the Ansatz $\phi = n\sigma$ and assuming the other coordinates to depend only on τ. These solutions describe strings that can propagate from infinity. As they approach the black hole, they may fall or not into the singularity depending on the initial conditions. A special case describes strings oscillating inside the horizon [8,10]. This special solution expresses in terms of elementary functions.

We construct in this paper **generic** solutions of the classical string equations of motion and constraints in Schwarzschild black holes. We do that near the horizon and near the black hole singularity. We consider here closed strings. The generalization to open strings is straightforward.

Through appropriate conformal transformations on the world-sheet, we map in the generic case the intersection of the string world-sheet with the horizon into the line $\tau = 0$. A similar transformation can be performed for the black hole singularity $r = 0$. We then study the string equations of motion and constraints by expanding in τ around $\tau = 0$. Near the black hole singularity, it turns out that the string solutions possess a series expansion in integer powers of $\tau^{1/5}$. Around the horizon, strings solutions are analytic in τ and only integer powers of τ appear.

Calling upon the conformal freedom, our solutions depend on four arbitrary functions of σ. These functions can be expressed in terms of the initial data. We thus see that only the transverse coordinates are physical degrees of freedom.

In Kruskal-Szekeres coordinates u, v, θ, ϕ [see eq.(10)], a generic string solution behaves near the black hole singularity as follows:

$$
\begin{aligned}
u(\sigma,\tau) &= e^{a(\sigma)}\{1 - \frac{1}{4}\gamma(\sigma)^4\, \tau^{4/5}\left[1 + O(\tau^{2/5})\right] \\
&\quad - \gamma(\sigma)^6\, \frac{f'(\sigma)\nu(\sigma)\sin^2 g(\sigma) + \mu(\sigma)}{28\; a'(\sigma)}\, \tau^{7/5}\left[1 + O(\tau^{2/5})\right]\Big\}\,, \\
v(\sigma,\tau) &= e^{-a(\sigma)}\{1 - \frac{1}{4}\gamma(\sigma)^4\, \tau^{4/5}\left[1 + O(\tau^{2/5})\right]
\end{aligned}
$$

$$+ \gamma(\sigma)^6 \frac{f'(\sigma)\nu(\sigma)\sin^2 g(\sigma) + \mu(\sigma)}{28\, a'(\sigma)} \tau^{7/5} \left[1 + O(\tau^{2/5})\right]\Big\} ,$$

$$\theta(\sigma,\tau) = g(\sigma) + \tau^{1/5} \mu(\sigma) + O(\tau^{3/5}),$$

$$\phi(\sigma,\tau) = f(\sigma) + \tau^{1/5} \nu(\sigma) + O(\tau^{3/5}). \tag{1}$$

where

$$[2\,\gamma(\sigma)]^2 = \mu(\sigma)^2 + \nu(\sigma)^2 \sin^2 g(\sigma) . \tag{2}$$

In addition, the (time-like) coordinate r vanishes as,

$$r(\sigma,\tau) = \gamma(\sigma)^2 \tau^{2/5} + O(\tau^{4/5}) . \tag{3}$$

Notice that the angular coordinates θ, ϕ vary with τ near $\tau = 0, r = 0$ faster than the u, v or r coordinates.

The string behaviour near the horizon is analytic in τ. It is given in sec.III, eqs.(22-27).

We then compute the string size and energy-momentum near the black hole singularity. The invariant size S [3] grows as r^{-1} for $r \to 0$,

$$S = \frac{4\,a'(\sigma)^2}{r} + O(1) \tag{4}$$

The string infinitely stretches when it falls into the $r = 0$ singularity. This is due to the infinitely growing gravitational forces that act then on the string. The string stretching near $r = 0$ was observed in ref. [7] using perturbative methods.

The string energy diverges also as r^{-1} for $r \to 0$,

$$E = \frac{1}{2\pi\alpha'} \frac{2}{5\,r} \int_0^{2\pi} d\sigma\, |\gamma(\sigma)|^5 + O(1) \to +\infty , \tag{5}$$

where $2\pi\alpha'$ stands for the inverse string tension.

The pressure in the angular directions θ, ϕ tends to infinity with the same rate:

$$P_\phi = \frac{1}{2\pi\alpha'} \frac{1}{10\,r} \int_0^{2\pi} d\sigma\, \nu(\sigma)^2 \sin^2 g(\sigma)\, |\gamma(\sigma)|^3 \to +\infty ,$$

$$P_\theta = \frac{1}{2\pi\alpha'} \frac{1}{10\,r} \int_0^{2\pi} d\sigma\, \mu(\sigma)^2\, |\gamma(\sigma)|^3 \to +\infty. \tag{6}$$

Thus, to leading order,

$$E = P_\theta + P_\phi \quad \text{for} \quad r \to 0 .$$

which is the behaviour of a two-dimensional ultrarelativistic gas. This relation means that the string behaves to leading order as **two**-dimensional massless particles. This is the so-called dual to unstable behaviour [3,13] (here for two spatial dimension).

As is well known, the Schwarzschild manifold has a $M \otimes S^2$ structure, where M stands for the part described by the u, v coordinates. The present results indicate that test strings near the black hole singularity behave as massless particles propagating in the compact S^2 manifold.

Near the horizon both the string size and the energy-momentum tensor are finite and analytic in Kruskal-Szekeres coordinates. The trace of the energy-momentum tensor is positive at the horizon.

It must be noticed that the resolution method used here for strings in black hole spacetimes is analogous the expansions for $\tau \to 0$ used in ref. [16,17] for strings in cosmological spacetimes.

For strings at large distances from the black hole where the gravitational field is weak, the string propagation can be solved by perturbing the Minkowski solutions. Actually, one can take as zeroth order solution the center of mass motion, $q^A + 2p^A\alpha'\tau$ as it has been done in ref. [2].

In summary, the exact string behaviour near the black hole singularity and near the black hole horizon is presented in this paper.

II. EQUATIONS OF MOTION.

The Schwarzschild metric in Schwarzschild coordinates (t, r, θ, ϕ) takes the following form:

$$ds^2 = -\left(1 - \frac{1}{r}\right) dt^2 + \frac{dr^2}{1 - \frac{1}{r}} + r^2(d\theta^2 + \sin^2\theta\, d\phi^2)\,, \tag{7}$$

where we choose units where the Schwarzschild radius $R_s = 2m = 1$.

Since we are interested in the whole Schwarzschild manifold and not just in the external part $r > 1$ where the static Schwarzschild coordinates are appropriate, we consider the Kruskal-Szekeres coordinates (u, v, θ, ϕ) defined by

$$u = t_K - r_K \equiv \sqrt{1-r}\; e^{(r-t)/2} \quad , \quad v = t_K + r_K \equiv \sqrt{1-r}\; e^{(r+t)/2}\,. \tag{8}$$

for $v \geq 0, u \geq 0$ and by

$$u = t_K - r_K \equiv -\sqrt{r-1}\; e^{(r-t)/2} \quad , \quad v = t_K + r_K \equiv \sqrt{r-1}\; e^{(r+t)/2}\,. \tag{9}$$

for $v \geq 0, u \leq 0$. For $v \leq 0$ one just flips the sign of v in eq.(8) or (9) [11].

The coordinate t_K is a time-like coordinate, and r_K spacelike. In Kruskal-Szekeres coordinates the Schwarzschild metric takes the form,

$$ds^2 = -\frac{4}{r}\; e^{-r}\, du\, dv + r^2\, (d\theta^2 + \sin^2\theta\, d\phi^2)\,. \tag{10}$$

r is a function of the product uv defined by the inverse function of

$$uv = [1-r]\, e^r\,.$$

for $uv \geq 0$. The metric in such coordinates is regular everywhere except at its singularity, $r = 0$.

The string equations of motion in Schwarzschild coordinates and in the conformal gauge, are

$$
\begin{aligned}
r_\sigma t_\sigma - r_\tau t_\tau + r(r-1)(t_{\sigma\sigma} - t_{\tau\tau}) = 0\,, \\
\frac{2r}{1-r}(r_{\tau\tau} - r_{\sigma\sigma}) - \frac{1}{r^2}(t_\tau^2 - t_\sigma^2) + 2r(\theta_\sigma^2 - \theta_\tau^2) + \\
2\,r\sin^2\theta\,(\phi_\tau^2 - \phi_\sigma^2) + \frac{1}{(r-1)^2}(r_\tau^2 - r_\sigma^2) = 0\,.
\end{aligned}
\tag{11}
$$

$$
\begin{aligned}
r\sin\theta\,(\phi_{\tau\tau} - \phi_{\sigma\sigma}) + 2\,r\cos\theta\,(\phi_\tau\theta_\tau - \phi_\sigma\theta_\tau) + 2\,\sin\theta\,(r_\tau\phi_\tau - r_\sigma\phi_\sigma) = 0\,, \\
r\,(\theta_{\tau\tau} - \theta_{\sigma\sigma}) + 2(r_\tau\theta_\tau - r_\sigma\theta_\tau) - r\sin\theta\,\cos\theta\,(\phi_\tau^2 - \phi_\sigma^2) \quad = 0\,.
\end{aligned}
\tag{12}
$$

The constraints in Schwarzschild coordinates are

$$
\begin{aligned}
\frac{1-r}{r}(t_\sigma^2 + t_\tau^2) + \frac{r}{r-1}(r_\tau^2 + r_\sigma^2) + r^2\left[\theta_\tau^2 + \theta_\tau^2 + \sin^2\theta\,(\phi_\tau^2 + \phi_\sigma^2)\right] = 0\,, \\
\frac{1-r}{r}t_\tau t_\sigma + \frac{r}{r-1}r_\tau r_\sigma + r^2\left(\theta_\tau\theta_\tau + \sin^2\theta\,\phi_\tau\phi_\sigma\right) = 0\,.
\end{aligned}
\tag{13}
$$

The string equations of motion in Kruskal-Szekeres coordinates take the form (always in the conformal gauge),

$$
\begin{aligned}
u_{\tau\tau} - u_{\sigma\sigma} + \frac{1}{r}\left(1 + \frac{1}{r}\right)\,e^{-r}\,v\,\left[(u_\tau)^2 - (u_\sigma)^2\right] - \frac{r\,u}{2}\left[\theta_\sigma^2 - \theta_\tau^2 + \sin^2\theta\,(\phi_\tau^2 - \phi_\sigma^2)\right] = 0 \\
v_{\tau\tau} - v_{\sigma\sigma} + \frac{1}{r}\left(1 + \frac{1}{r}\right)\,e^{-r}\,u\,\left[(v_\tau)^2 - (v_\sigma)^2\right] - \frac{r\,v}{2}\left[\theta_\sigma^2 - \theta_\tau^2 + \sin^2\theta\,(\phi_\tau^2 - \phi_\sigma^2)\right] = 0\,,
\end{aligned}
\tag{14}
$$

plus eqs.(12) for the angular coordinates.

The constraints in Kruskal-Szekeres coordinates are

$$
\begin{aligned}
-\frac{4}{r}\,e^{-r}\,(u_\sigma\,v_\sigma + u_\tau\,v_\tau) + r^2\left[\theta_\tau^2 + \theta_\tau^2 + \sin^2\theta\,(\phi_\tau^2 + \phi_\sigma^2)\right] = 0\,, \\
-\frac{4}{r}\,e^{-r}\,(u_\tau\,v_\sigma + u_\sigma\,v_\tau) + r^2\left(\theta_\tau\theta_\tau + \sin^2\theta\,\phi_\tau\phi_\sigma\right) = 0\,.
\end{aligned}
\tag{15}
$$

Notice that both the equations of motion and constraints are invariant under the exchange $u \leftrightarrow v$.

Also notice that the equations of motion and constraints in Kruskal-Szekeres coordinates are regular everywhere except at the singularity $r = 0$.

We shall consider closed strings where the string coordinates must be periodic functions of σ:

$$
u(\sigma + 2\pi, \tau) = u(\sigma, \tau)\,,\ v(\sigma + 2\pi, \tau) = v(\sigma, \tau). \tag{16}
$$

Therefore, the angular coordinates θ, ϕ may be just quasiperiodic functions of σ:

$$
\theta(\sigma + 2\pi, \tau) = \theta(\sigma, \tau) + \mathrm{mod}\,2\pi\ ,\ \phi(\sigma + 2\pi, \tau) = \phi(\sigma, \tau) + 2n\pi, \tag{17}
$$

where n is an integer.

III. STRINGS ACROSS THE HORIZON

Let us study the string behaviour in the region near the horizon. The Kruskal-Szekeres coordinates are the appropriate ones here. The horizon is characterized by $uv = 0$. Let us consider a string crossing the right part of it, that is $u = 0, v > 0$.

We can write the curve describing the intersection of the horizon with the world-sheet as

$$x_+ = \chi(x_-), \tag{18}$$

whenever this intersection is nondegenerate. (Here $x_\pm \equiv \tau \pm \sigma$).

Upon a conformal transformation,

$$x_+ \to x'_+ = f(x_+) \quad , \quad x_- \to x'_- = g(x_-), \tag{19}$$

we can map the curve (18) into $\tau' = 0$ by an appropriate choice of f and g. For example, we can choose

$$f(x_+) = x_+ \quad , \quad g(x_-) = -\chi(x_-).$$

This defines our choice of gauge. From now on, we rename τ' and σ' by τ and σ, respectively. Notice that this choice does not completely fix the gauge. We can still perform transformations that leave the line $\tau = 0$ unchanged. This is the case for the following class of conformal mappings

$$x_+ \to x'_+ = \varphi(x_+) \quad , \quad x_- \to x'_- = -\varphi(-x_-), \tag{20}$$

where $\varphi(x)$ is an arbitrary function. Eq. (20) can be written as,

$$\begin{aligned}
\tau' &= \frac{1}{2}\left[\varphi(\tau+\sigma) - \varphi(\sigma-\tau)\right] = \tau\,\varphi'(\sigma) + \frac{1}{6}\tau^3\,\varphi'''(\sigma) + O(\tau^4) \\
\sigma' &= \frac{1}{2}\left[\varphi(\tau+\sigma) + \varphi(\sigma-\tau)\right] = \varphi(\sigma) + \frac{1}{2}\tau^2\,\varphi''(\sigma) + O(\tau^4)
\end{aligned} \tag{21}$$

The transformations (20) represent a diagonal subgroup of the set of left-right conformal transformations (19).

A. Equatorial Solutions

Let us first investigate strings on the plane $\theta = \pi/2$. This restriction is compatible with the equations of motion and constraints and it means that the string moves in an equatorial plane.

We can then propose the following expansion near $v = 0$,

$$\begin{aligned}
v(\sigma,\tau) &= e\left[\tau\ c_1(\sigma) + \tau^2\,c_2(\sigma) + O(\tau^3)\right] \\
u(\sigma,\tau) &= b_0(\sigma) + \tau\,b_1(\sigma) + \tau^2\,b_2(\sigma) + O(\tau^3) \\
\phi(\sigma,\tau) &= p_0(\sigma) + \tau\,p_1(\sigma) + \tau^2\,p_2(\sigma) + O(\tau^3)\ ,
\end{aligned} \tag{22}$$

where $e = 2.718281828459\ldots$. Inserting eqs.(22) in eqs.(12,14) and constraints (15) yields

$$\begin{aligned} c_2(\sigma) &= -b_0(\sigma)\, c_1(\sigma)^2 \,, \\ b_1(\sigma) &= \frac{c_1(\sigma)\,[b_0'(\sigma)]^2}{[p_0'(\sigma)]^2} + \frac{[p_0'(\sigma)]^2}{4\,c_1(\sigma)} \,, \\ p_1(\sigma) &= \frac{2c_1(\sigma)\, b_0'(\sigma)}{p_0'(\sigma)} \,. \end{aligned} \tag{23}$$

In addition $r(\sigma,\tau)$ is given by the expansion

$$r(\sigma,\tau) = 1 - b_0(\sigma)\, c_1(\sigma)\, \tau - \left\{ \frac{[c_1(\sigma)\, b_0'(\sigma)]^2}{[p_0'(\sigma)]^2} + \frac{1}{4}\,[p_0'(\sigma)]^2 \right\} \tau^2 + O(\tau^3) \,. \tag{24}$$

The expansion can be pushed to arbitrary high orders in τ yielding the series coefficients in terms of the arbitrary functions $b_0(\sigma), c_1(\sigma)$ and $p_0(\sigma)$. We have easily generated a number of terms with the help of Mathematica. The number of arbitrary functions can be reduced to two using the conformal freedom (21).

The boundary conditions (16-17) require here:

$$b_0(\sigma + 2\pi) = b_0(\sigma) \,,\; c_1(\sigma + 2\pi) = c_1(\sigma) \,,\; p_0(\sigma + 2\pi) = p_0(\sigma) + 2n\pi \,. \tag{25}$$

B. General String Solutions

Let us consider now a general string solution near the horizon $r = 1$. We keep working in a gauge such that $\tau = 0$ corresponds to $r = 1$.

Eqs. (22) are then supplemented by

$$\theta(\sigma,\tau) = t_0(\sigma) + t_1(\sigma)\,\tau + t_2(\sigma)\,\tau^2 + O(\tau^3) \,. \tag{26}$$

Inserting eqs.(22, 26) into eqs.(12,14) and constraints (15) yields

$$\begin{aligned} u(\sigma,\tau) &= b_0 + \\ &+ \frac{\tau}{4c_1}\left[(p_1 + p_0')^2 \sin^2 t_0 + (t_1 + t_0')^2 \right] + \\ &+ \frac{\tau^2}{4}\left[b_0\,({p_1}^2 - {p_0'}^2)\, \sin^2 t_0 + b_0\,({t_1}^2 - {t_0'}^2) + 2\, b_0'' \right] + O(\tau^3) \,, \\ v(\sigma,\tau) &= e\Big\{ c_1\tau - b_0\, {c_1}^2\, \tau^2 \\ &+ \frac{\tau^3}{12}\Big[6\, {b_0}^2\, {c_1}^3 + 4\, {c_1}^2\, b_0' + 2\, c_1'' - c_1\, p_0'\, (p_1 + p_0') \\ &- 2\, c_1\, t_0'(t_0' + t_1) + \cos(2\, t_0)\, c_1\, p_0'(p_1 + p_0') \Big] + O(\tau^4) \Big\} \,, \\ r(\sigma,\tau) &= 1 - b_0\, c_1\, \tau + \end{aligned}$$

$$-\frac{\tau^2}{4}\left[\sin^2 t_0\,(p_0'+p_1)^2+\ (t_0'+t_1)^2 c_1\, b_0'\right]+O(\tau^3)\ ,$$

$$\phi(\sigma,\tau)=p_0+p_1\tau+\left[b_0\,c_1\,p_1-(\,p_1\,t_1-\ p_0'\,t_0')\cot t_0+\frac{p_0''}{2}\right]\ \tau^2+O(\tau^3)\ ,$$

$$\theta(\sigma,\tau)=t_0+t_1\tau+\left[\frac{\sin(2\,t_0)}{4}({p_1}^2-{p_0'}^2)+b_0\,c_1\,t_1+\frac{t_0''}{2}\right]\ \tau^2+O(\tau^3)\ . \tag{27}$$

Here,

$$p_1(\sigma)=\frac{2\,c_1 b_0'-t_1 t_0'}{p_0'\ \sin^2 t_0}\ . \tag{28}$$

We have here five arbitrary functions $b_0(\sigma), c_1(\sigma), p_0(\sigma), t_0(\sigma)$ and $t_1(\sigma)$. This number can be reduced to four using the diagonal conformal transformations (21).

IV. STRINGS NEAR THE SINGULARITY $R=0$

Let us consider the solution of eqs.(12,14) and constraints (15), near $r=0$. That is to say, near $uv=1$.

For a generic world-sheet, we choose the gauge such that $\tau=0$ corresponds to the string at the singularity $uv=1$. This can be achieved as shown in sec. III for the horizon.

Notice that we use in sec. III and IV the same symbols (σ,τ) for different world-sheet coordinates. In sec. III, $\tau=0$ corresponds to $r=1$, whereas in sec. IV, $\tau=0$ corresponds to $r=0$.

A. Equatorial Solutions

Let us begin to investigate strings on the plane $\theta=\pi/2$. This restriction is compatible with the equations of motion and constraints and it means that the string moves in an equatorial plane.

Near the singularity $uv=1$, we propose for $\tau\to 0$ the Ansatz

$$u(\sigma,\tau)=e^{a(\sigma)}\left[1-\tau^{\alpha}\,\beta(\sigma)+\ldots\right]$$

$$v(\sigma,\tau)=e^{-a(\sigma)}\left[1-\tau^{\alpha'}\,\hat\beta(\sigma)+\ldots\right]$$

$$\phi(\sigma,\tau)=f(\sigma)+2\,\gamma(\sigma)\,\tau^{\lambda}+\ldots\ . \tag{29}$$

[The factor 2 is there just for future convenience].

Inserting eqs.(29) in eqs.(12,14) and constraints (15) yields

$$\alpha=\alpha'=4/5\quad,\quad\lambda=1/5,$$

$$\hat\beta(\sigma)=\beta(\sigma)\quad,\quad\beta(\sigma)=\frac{1}{4}\gamma(\sigma)^4. \tag{30}$$

The functions $a(\sigma), f(\sigma)$ and $\gamma(\sigma)$ are arbitrary.

The coordinate r then vanishes as

$$r(\sigma,\tau) = \gamma(\sigma)^2\, \tau^{2/5} + \dots\,. \tag{31}$$

We have pushed the resolution of eqs.(12,14) and constraints (15) to higher orders with the help of Mathematica. We find that the corrections to the leading behaviour appear as integer powers of $\tau^{2/5}$. That is, terms in $\tau^{6/5}, \tau^{8/5}, \tau^{10/5}, \dots$ in u and in v, and terms in $\tau^{3/5}, \tau^{5/5}, \tau^{7/5}, \dots$ in ϕ. The subdominant contributions start with the order $\tau^{7/5}$ in u and in v. The piece in ϕ of order τ^0 is directly connected with the subdominant contributions in u and in v.

We find,

$$\begin{aligned}
u(\sigma,\tau) &= e^{a(\sigma)}\,\{1 - \frac{1}{4}\gamma(\sigma)^4\,\tau^{4/5} - \frac{1}{21}\left(2\,\gamma(\sigma)^6 + \frac{75}{2}\frac{a'(\sigma)^2}{\gamma(\sigma)^4}\right)\tau^{6/5}\left[1 + O(\tau^{2/5})\right] \\
&\quad - \frac{\gamma(\sigma)^7 f'(\sigma)}{14\,a'(\sigma)}\,\tau^{7/5}\left[1 + O(\tau^{2/5})\right]\Big\}, \\
v(\sigma,\tau) &= e^{-a(\sigma)}\,\{1 - \frac{1}{4}\gamma(\sigma)^4\,\tau^{4/5} - \frac{1}{21}\left(2\,\gamma(\sigma)^6 + \frac{75}{2}\frac{a'(\sigma)^2}{\gamma(\sigma)^4}\right)\tau^{6/5}\left[1 + O(\tau^{2/5})\right] \\
&\quad + \frac{\gamma(\sigma)^7 f'(\sigma)}{14\,a'(\sigma)}\,\tau^{7/5}\left[1 + O(\tau^{2/5})\right]\Big\}, \\
\phi(\sigma,\tau) &= f(\sigma) + \frac{5}{21}\left(\frac{4\,f'(\sigma)\,a''(\sigma)}{7\,a'(\sigma)} + \frac{4\,f'(\sigma)\,\gamma'(\sigma)}{3\,\gamma(\sigma)} + \frac{53}{42}\,f''(\sigma)\right)\tau^2 + O(\tau^{12/5}) \\
&\quad + 2\,\gamma(\sigma)\,\tau^{1/5}\left\{1 + \frac{2}{21\,\gamma(\sigma)^8}\left[\gamma(\sigma)^{10} - 25\,a'(\sigma)^2\right]\tau^{2/5} + O(\tau^{4/5})\right\}\,,
\end{aligned} \tag{32}$$

and

$$\begin{aligned}
r(\sigma,\tau) &= \gamma(\sigma)^2\,\tau^{2/5}\left[1 + \frac{\tau^{2/5}}{7}\left(-\gamma(\sigma)^2 + \frac{25\,a'(\sigma)^2}{\gamma(\sigma)^6}\right) + O(\tau^{4/5})\right] \\
&\quad + \frac{25}{441}\,\tau^{11/5}\left[35\,\gamma'(\sigma)\,f'(\sigma) - \frac{6\,f'(\sigma)\,\gamma(\sigma)\,a''(\sigma)}{a'(\sigma)} - \gamma(\sigma)\,f''(\sigma) + O(\tau^{2/5})\right]\,.
\end{aligned} \tag{33}$$

The functions $f(\sigma), a(\sigma)$ and $\gamma(\sigma)$ are arbitrary and can be expressed in terms of the initial data. In the Appendix we give u, v, ϕ, r up to the order $\tau^{16/5}$.

Both the equations of motion and constraints are invariant under the exchange $u \leftrightarrow v$ but not the boundary conditions at $\tau = 0$. They differ by $a(\sigma) \leftrightarrow -a(\sigma)$ as we see from eqs.(29). Therefore one can obtain $u(\sigma,\tau)$ from $v(\sigma,\tau)$ and viceversa just by flipping the sign of $a(\sigma)$.

Notice that u/v is τ independent up to order $\tau^{7/5}$. Since $u/v = e^{-t}$, this imply that the spatial coordinate t is only σ-dependent up to $O(\tau^{7/5})$. More precisely,

$$t(\sigma,\tau) = \log\frac{v}{u} = -2\,a(\sigma) + \frac{\gamma(\sigma)^7 f'(\sigma)}{7\,a'(\sigma)}\,\tau^{7/5} + O(\tau^{9/5})\,. \tag{34}$$

In other words, $t(\sigma,\tau)$ varies slower than the other coordinates ϕ and r when the string approaches the black hole singularity ($\tau \to 0$).

The boundary conditions (16-17) require here:

$$a(\sigma+2\pi)=a(\sigma)\ ,\ \gamma(\sigma+2\pi)=\gamma(\sigma)\ ,\ f(\sigma+2\pi)=f(\sigma)+2n\pi\ . \tag{35}$$

Recall that we can still perform diagonal conformal transformations (20). It is easy to check that eqs.(32) keep their form under the transformations (21) when we expand in powers of τ.

We can use the arbitrary function $\varphi(\sigma)$ to fix one function among $f(\sigma), a(\sigma)$ and $\gamma(\sigma)$. For example, we can set to zero the $O(\tau^2)$ part in $\phi(\sigma,\tau)$ [see eq.(32)]. We are thus left with **two** arbitrary functions. They describe precisely the transverse degrees of freedom of the equatorial string.

B. General String Solutions

Let us consider now a general string solution near the singularity $r=0$. We keep working in a gauge such that $\tau=0$ corresponds to $r=0$.

Eqs.(32) generalize as follows for the dominant order

$$u(\sigma,\tau)=e^{a(\sigma)}\left[1\ -\frac{1}{4}\gamma(\sigma)^4\,\tau^{4/5}+\ldots\right],$$

$$v(\sigma,\tau)=e^{-a(\sigma)}\left[1\ -\frac{1}{4}\gamma(\sigma)^4\,\tau^{4/5}+\ldots\right],$$

$$\theta(\sigma,\tau)=g(\sigma)+\tau^{1/5}\,\mu(\sigma)+\ldots,$$

$$\phi(\sigma,\tau)=f(\sigma)+\tau^{1/5}\,\nu(\sigma)+\ldots. \tag{36}$$

Inserting eqs.(36) in eqs.(12,14) and constraints (15) yields

$$[2\,\gamma(\sigma)]^2=\mu(\sigma)^2+\nu(\sigma)^2\,\sin^2 g(\sigma)\ . \tag{37}$$

For $g(\sigma)\equiv\pi/2, \mu(\sigma)^2\equiv 0$ we get back the previous equatorial solution.

We can also find the ring solution of ref. [8] setting $f(\sigma)\equiv n\sigma, a(\sigma)\equiv 0, \mu(\sigma)=$ cte. $g(\sigma)=$ cte. and $\nu(\sigma)\equiv 0$.

The coordinate r vanishes here as for the equatorial solution

$$r(\sigma,\tau)=\gamma(\sigma)^2\,\tau^{2/5}+\ldots\ . \tag{38}$$

The string solution is completely fixed once the functions $f(\sigma), g(\sigma), a(\sigma), \mu(\sigma)$ and $\nu(\sigma)$ are given. These five functions are arbitrary and can be expressed in terms of the initial data.

Notice that ϕ and θ approach their limiting values with the same exponent $1/5$ in τ.

As in the equatorial case, the corrections to the leading behaviour appear as positive integer powers of $\tau^{2/5}$. The subdominant leading power in $u(\sigma,\tau)$ and $v(\sigma,\tau)$ is again $\tau^{7/5}$. We find with the help of Mathematica,

$$u(\sigma,\tau)=e^{a(\sigma)}\{1\ -\ \gamma(\sigma)^4\,\tau^{4/5}\left[1+O(\tau^{2/5})\right]$$

$$
\begin{aligned}
& - \ \gamma(\sigma)^6 \, \frac{f'(\sigma)\nu(\sigma)\sin^2 g(\sigma) + \mu(\sigma)}{28 \, a'(\sigma)} \, \tau^{7/5} \left[1 + O(\tau^{2/5})\right] \Big\} \, , \\
v(\sigma,\tau) &= e^{-a(\sigma)} \left\{1 - \ \gamma(\sigma)^4 \, \tau^{4/5} \left[1 + O(\tau^{2/5})\right] \right. \\
& + \ \gamma(\sigma)^6 \, \frac{f'(\sigma)\nu(\sigma)\sin^2 g(\sigma) + \mu(\sigma)}{28 \, a'(\sigma)} \, \tau^{7/5} \left[1 + O(\tau^{2/5})\right] \Big\} \, . \qquad (39)
\end{aligned}
$$

Using the diagonal conformal transformation (21), we can fix one of the arbitrary functions among $f(\sigma), g(\sigma), a(\sigma), \mu(\sigma)$ and $\nu(\sigma)$ keeping in mind the periodic boundary conditions:

$$
a(\sigma + 2\pi) = a(\sigma) \, , \ \nu(\sigma + 2\pi) = \nu(\sigma) \, , \ \mu(\sigma + 2\pi) = \mu(\sigma) \, ,
$$

$$
f(\sigma + 2\pi) = f(\sigma) + 2n\pi \, , \ g(\sigma + 2\pi) = g(\sigma) \bmod 2\pi \, . \qquad (40)
$$

We are left with **four** arbitrary functions of σ. This is precisely the number of transverse string degrees of freedom.

V. STRING ENERGY-MOMENTUM AND INVARIANT SIZE NEAR THE SINGULARITY AND THE NEAR THE HORIZON

The invariant string size ds^2 is defined using the metric induced on the string world-sheet [3]:

$$
ds^2 = G_{AB}(X) \, \dot{X}^A \, \dot{X}^B \left(d\tau^2 - d\sigma^2\right) . \qquad (41)
$$

We give the name *string size* to

$$
S = -G_{AB}(X) \, \dot{X}^A \, \dot{X}^B = \frac{4}{r} \, e^{-r} \, \dot{u}\dot{v} - r^2 \, \dot{\theta}^2 - r^2 \, \dot{\phi}^2 \, \sin^2\theta . \qquad (42)
$$

We find near the singularity at $r = 0$ using eqs.(32-33)

$$
\begin{aligned}
S &= \frac{4 \, u'(\upsilon)^2}{\gamma(\sigma)^2} \, \tau^{-2/5} - \frac{4}{7} \, a'(\sigma)^2 \left(6 + 25 \, \frac{a'(\sigma)^2}{\gamma(\sigma)^7}\right) + O(\tau^{2/5}) \\
&= \frac{4 \, a'(\sigma)^2}{r} - \frac{4}{7} \, a'(\sigma)^2 \left(6 + 25 \, \frac{a'(\sigma)^2}{\gamma(\sigma)^7}\right) + O(r) \, . \qquad (43)
\end{aligned}
$$

The invariant string size tends then to infinite when the string falls into the $r = 0$ singularity. This is due to the infinitely growing gravitational forces that act there on the string.

The string stretching near $r = 0$ was observed in ref. [7] using perturbative methods.

We find near the horizon using eqs.(22)-(24) and (42),

$$
S = [p_0'(\sigma)]^2 + O(\tau) = [p_0'(\sigma)]^2 + O(r - 1) \, . \qquad (44)
$$

As expected, we find a finite result since $r = 1$ is a regular point of the geometry.

The calculations in ref. [14] give stretching for the quantum string on the horizon. Such quantum stretching on the horizon is observed in the longitudinal coordinate for fiducial observers at fixed static position, using the light-cone gauge and the approximation that spacetime is almost flat even close to the horizon.

Our results are for the **invariant** string size and they are strictly classical. Hence they cannot be directly compared with the calculations in ref. [14].

The spacetime string energy-momentum tensor in four spacetime dimensions is given by

$$\sqrt{-G}\, T^{AB}(X) = \frac{1}{2\pi\alpha'}\int d\sigma d\tau \left(\dot{X}^A\dot{X}^B - X'^A X'^B\right)\delta^{(4)}(X - X(\sigma,\tau))\,. \tag{45}$$

Notice that X in eq.(45) is just a spacetime point whereas $X(\sigma,\tau)$ stands for the string dynamical variables. One sees from the Dirac delta in eq.(45) that $T^{AB}(X)$ vanishes unless X is exactly on the string world-sheet. We shall not be interested in the detailed structure of the classical strings. It is more useful to integrate the energy-momentum tensor (45) on a volume that completely encloses the string as in refs. [15], [3]. We obtain in this way a density Θ^{AB}.

Inside the horizon we can use t, θ, ϕ as spatial coordinates and r as a coordinate time. We find,

$$\sqrt{g_{rr}}\,\Theta^{AB}(r) = \frac{1}{2\pi\alpha'}\int d\sigma d\tau \left(\dot{X}^A\dot{X}^B - X'^A X'^B\right)\delta(r - r(\tau,\sigma)). \tag{46}$$

where $g_{rr} = r/(1-r) > 0$.

Let us first compute the trace of the energy-momentum tensor (45). We find

$$\sqrt{-G}\, T^A_A(X) = \frac{1}{\pi\alpha'}\int d\sigma d\tau\, G_{AB}(X)\,\dot{X}^A\dot{X}^B\,\delta^{(4)}(X - X(\sigma,\tau))\,. \tag{47}$$

where we used the string constraints. Notice that the integrand is just the (minus) string size [eq.(42)].

We have for the black hole case:

$$G_{AB}(X)\,\dot{X}^A\dot{X}^B = -\frac{r\dot{r}^2}{1-r} + r^2\,\dot{\theta}^2 + r^2\,\dot{\phi}^2\,\sin^2\theta + \frac{1-r}{r}\,\dot{t}^2\,. \tag{48}$$

Using eqs.(36) and (38) for $\tau \to 0$, we find that each of the first three terms grows as $\tau^{-4/5}$ whereas the last term vanishes as $\tau^{2/5}$. Moreover, the sum of the three terms $O(\tau^{-4/5})$ identically vanishes thanks to eq.(37). This cancellation in the trace tells us that near $r = 0$, the dominant (and divergent) components T^r_r, T^ϕ_ϕ and T^θ_θ yield a zero trace. This means that the string behaves to leading order as **two**-dimensional massless particles. This is the so-called dual to unstable behaviour [3] (here for two spatial dimension).

For $\tau \to 0$, $r \to 0$ we can use in eq.(46) the dominant behaviours:

$$r(\sigma,\tau) = \gamma(\sigma)^2\,\tau^{2/5} + O(\tau^{4/5}),$$

$$\theta(\sigma,\tau) = g(\sigma) + \mu(\sigma)\,\tau^{1/5} + O(\tau^{3/5}),$$

$$\phi(\sigma,\tau) = f(\sigma) + \nu(\sigma)\,\tau^{1/5} + O(\tau^{3/5}),$$

$$t(\sigma,\tau) = -2a(\sigma) + \gamma(\sigma)^6\,\frac{f'(\sigma)\nu(\sigma)\sin^2 g(\sigma) + \mu(\sigma)}{14\,a'(\sigma)}\,\tau^{7/5} + O\left(\tau^{9/5}\right)\,. \tag{49}$$

We thus find for $r \to 0$,

$$\begin{aligned}
2\pi\alpha'\,\Theta^{rr}(r) &= \frac{2}{5\,r^2}\int_0^{2\pi} d\sigma\,|\gamma(\sigma)|^5 + O(\frac{1}{r}) \to +\infty\,,\\
2\pi\alpha'\,\Theta^{\phi\phi}(r) &= \frac{1}{10\,r^3}\int_0^{2\pi} d\sigma\,\nu(\sigma)^2\,|\gamma(\sigma)|^3 + O(\frac{1}{r^2}) \to +\infty\,,\\
2\pi\alpha'\,\Theta^{\theta\theta}(r) &= \frac{1}{10\,r^3}\int_0^{2\pi} d\sigma\,\mu(\sigma)^2\,|\gamma(\sigma)|^3 + O(\frac{1}{r^2}) \to +\infty\,,\\
2\pi\alpha'\,\Theta^{tt}(r) &= -10\,r\int_0^{2\pi} d\sigma\,\frac{[a'(\sigma)]^2}{|\gamma(\sigma)|^5} + O(r^2) \to 0^-\,.
\end{aligned} \tag{50}$$

We can identify the string energy with the mixed component $-\Theta^r_r$. We define the mixed components $\Theta^B_A(r)$ by integrating $T^B_A(X)$ over the spatial volume.

This yields for $r \to 0$,

$$E \equiv -\Theta^r_r = \frac{1}{2\pi\alpha'}\,\frac{2}{5\,r}\int_0^{2\pi} d\sigma\,|\gamma(\sigma)|^5 + O(1) \to +\infty\,. \tag{51}$$

The transverse pressures are defined as the mixed components Θ^ϕ_ϕ and Θ^θ_θ. They diverge for $r \to 0$:

$$\begin{aligned}
P_\phi \equiv \Theta^\phi_\phi &= \frac{1}{2\pi\alpha'}\,\frac{2}{5\,r}\int_0^{2\pi} d\sigma\,\nu(\sigma)^2\,\sin^2 g(\sigma)\,|\gamma(\sigma)|^5 \to +\infty\,,\\
P_\theta \equiv \Theta^\theta_\theta &= \frac{1}{2\pi\alpha'}\,\frac{2}{5\,r}\int_0^{2\pi} d\sigma\,\mu(\sigma)^2\,|\gamma(\sigma)|^5 \to +\infty.
\end{aligned} \tag{52}$$

Thus, to leading order,

$$E = P_\theta + P_\phi \quad \text{for} \quad r \to 0\,.$$

exhibiting a two-dimensional ultrarelativistic gas behaviour. The tidal forces infinitely stretch the string near $r = 0$ in effectively only two directions: ϕ and θ.

We find for the off-diagonal components,

$$\begin{aligned}
2\pi\alpha'\,\Theta^{tr}(r) &= \frac{r^{1/2}}{10}\int_0^{2\pi}\frac{d\sigma}{a'(\sigma)}\gamma(\sigma)^4\,\left[f'(\sigma)\nu(\sigma)\sin^2 g(\sigma) + \mu(\sigma)\right] \to 0^+\,,\\
2\pi\alpha'\,\Theta^{t\theta}(r) &= \frac{2}{5}\int_0^{2\pi} d\sigma\,\frac{\mu(\sigma)}{a'(\sigma)}\,|\gamma(\sigma)|^3\,\left[f'(\sigma)\nu(\sigma)\sin^2 g(\sigma) + \mu(\sigma)\right] = O(1)\,,\\
2\pi\alpha'\,\Theta^{t\phi}(r) &= \frac{2}{5}\int_0^{2\pi} d\sigma\,\frac{\nu(\sigma)}{a'(\sigma)}\,|\gamma(\sigma)|^3\,\left[f'(\sigma)\nu(\sigma)\sin^2 g(\sigma) + \mu(\sigma)\right] = O(1)\,,\\
2\pi\alpha'\,\Theta^{r\phi}(r) &= \frac{1}{5\,r^{5/2}}\int_0^{2\pi} d\sigma\,\nu(\sigma)\,\gamma(\sigma)^4 \to \infty,\\
2\pi\alpha'\,\Theta^{r\theta}(r) &= \frac{1}{5\,r^{5/2}}\int_0^{2\pi} d\sigma\,\mu(\sigma)\,\gamma(\sigma)^4 \to \infty,
\end{aligned}$$

$$2\pi\alpha'\,\Theta^{\theta\phi}(r) = \frac{1}{10\,r^3}\int_0^{2\pi} d\sigma\,\mu(\sigma)\,\nu(\sigma)\,|\gamma(\sigma)|^3 \;\to \infty\,. \tag{53}$$

Notice that the invariant string size tends to infinity [see eq.(43)] with $4\,a'(\sigma)^2$ as proportionality factor. Since $-2a(\sigma)$ is the leading behaviour of $t(\sigma,\tau)$, this suggests us that the string stretches infinitely in the (spatial) t direction when $r \to 0$.

As a matter of fact, infinitely growing string sizes are not observed in cosmological spacetimes [3,13] for strings exhibiting radiation (dual to unstable) behaviour.

For particular string solutions the energy-momentum tensor and the string size can be less singular than in the generic case. For ring solutions [8], $\mu(\sigma) = \mu =$ constant, $g(\sigma) = g =$ constant, $a(\sigma) = \nu(\sigma) = 0$, there is no stretching and

$$S = r\sin g \to 0$$

$$E = P_\theta = \frac{1}{\alpha'}\,\frac{\mu^5}{80\,r} \to +\infty \quad , \quad P_\phi = 0\,.$$

There is no string stretching but the string keeps exhibiting dual to unstable behaviour. This is due to the balance of the tidal forces thanks to the special symmetry of the solution. It behaves in this special case as **one**-dimensional massless particles for $r \to 0$.

As is easy to see, setting $\mu(\sigma) = 0, g(\sigma) = \pi/2$ all equatorial string solutions behave as **one**-dimensional massless particles for $r \to 0$.

It must be noticed that the resolution method used here for strings in black hole spacetimes is analogous the expansions for $\tau \to 0$ used in ref. [16,17] for strings in cosmological spacetimes.

The string coordinates are regular near the horizon [see eqs. (22)]. Let us see that the energy-momentum tensor is also regular near the horizon. We use now Kruskal-Szekeres coordinates and define $\Theta^{AB}(v)$ by integration over u, θ and ϕ. [For simplicity, we consider the equatorial solution]. We find for the energy $\Theta^{t_K t_K}(v)$,

$$2\pi\alpha'\,\Theta^{t_K t_K}(v=0) = \frac{e^{-1/2}}{4}\int_0^{2\pi} d\sigma\,\frac{[b_1(\sigma) + e\,c_1(\sigma)]^2 - b_0'(\sigma)^2}{|c_1(\sigma)|} = O(1) \tag{54}$$

All other components of $\Theta^{AB}(t_K)$ can be analogously computed and take finite values. We find for $\Theta^{r_K r_K}(v)$,

$$2\pi\alpha'\,\Theta^{r_K r_K}(v=0) = \frac{e^{-1/2}}{4}\int_0^{2\pi} d\sigma\,\frac{[b_1(\sigma) - e\,c_1(\sigma)]^2 - b_0'(\sigma)^2}{|c_1(\sigma)|} = O(1) \tag{55}$$

For the energy-momentum trace we find

$$\pi\alpha'\,\Theta_A^A(t_K) = e^{1/2}\int_0^{2\pi} d\sigma\,[p_0(\sigma)']^2 = O(1) \tag{56}$$

We see that $\Theta_A^A(t_K) > 0$ near the horizon.

VI. STRINGS FAR AWAY FROM THE BLACK HOLE

At large distances from the black hole ($r >> 1$), the string equations of motion and constraints become those of Minkowski spacetime,

$$\partial_{-+} X^A(\sigma,\tau) = 0 \ , \quad 0 \leq A \leq 3,$$
$$\left[\partial_\pm X^0(\sigma,\tau)\right]^2 - \sum_{j=1}^{3} \left[\partial_\pm X^j(\sigma,\tau)\right]^2 = 0 \ . \tag{57}$$

The solution of eqs.(57) takes the customary form

$$X^A(\sigma,\tau) = q^A + 2p^A\alpha'\tau + i\sqrt{\alpha'}\sum_{n\neq 0}\frac{1}{n}\{\alpha_n^A \exp[in(\sigma-\tau)] + \tilde{\alpha}_n^A \exp[-in(\sigma+\tau)]\} \ , \tag{58}$$

where q^A and p^A stand for the string center of mass position and momentum and α_n^A and $\tilde{\alpha}_n^A$ describe the right and left oscillator modes of the string, respectively. In order to have $r >> 1$, we must assume $\tau >> 1$ so $|q^A + 2p^A\alpha'\tau| >> 1$. Therefore, we can get the large distance approximation here as a systematic expansion in inverse powers of τ. As zeroth order solution we can take:

$$q^A + 2p^A\alpha'\tau \ .$$

This is precisely the expansion constructed in ref. [2]. We refer to this article for strings propagating far away from the black hole. This regime is a weak field approximation. It can be considered as a newtonian approximation plus a post-newtonian one plus higher orders.

VII. AN EXACT STRING SOLUTION

An exact solution of the string equations of motion and constraints [eqs. (12,14-15)] is given by

$$u = f(x_-) \quad , \quad v = g(x_+) \ ,$$
$$\phi = 0 \quad , \quad \theta = \pi/2 \ . \tag{59}$$

where $f(x)$ and $g(x)$ are arbitrary periodic functions of x with period 2π. Even if the world-sheet reaches the $r = 0$ singularity, it is not necessarily true that we can perform a conformal transformation on this solution such that $\tau = 0$ corresponds to $r = 0$.

Let us see that this exact radial string solution has in fact zero energy-momentum. Inserting eqs.(59) in eq.(45), we see immediately that all components of $T^{AB}(X)$ vanish except possibly for $T^{UV}(X)$. For this one,

$$\begin{aligned} 2\pi\alpha'\sqrt{-G}\,T^{UV}(X) &= \int dx_+ dx_- \left(\partial_+ U \partial_- V + \partial_- U \partial_+ V\right) \delta^{(4)}(X - X(\sigma,\tau)) \\ &= \int dx_+ dx_- f'(x_-) g'(x_+) \delta^{(4)}(X - X(\sigma,\tau)) \\ &= \delta(\phi)\delta(\theta - \pi/2) \sum_i sign[f'(x_-^i(u))] \sum_j sign[g'(x_+^j(v))] \end{aligned} \tag{60}$$

where $x_-^i(u)$ and $x_+^j(v)$ are the real roots of

$$u = f(x_-) \quad , \quad v = g(x_+) \, , \tag{61}$$

for given u and v. Now, since $f(x_-)$ and $g(x_+)$ are periodic functions, eqs.(61) must have an even number of solutions. In addition, $f'(x_-^i(u))$ and $g'(x_+^j(v))$ will have positive and negative signs pairwise. Therefore, the sum over i, j will vanish in eq.(60):

$$T^{UV}(X) = 0 \, .$$

In summary, this exact radial solution does not carry energy-momentum. It is then a pure gauge object.

Solutions of this type have been previously discussed [18–20] and found to be either pure gauge or unphysical as they stand. In particular, the unphysical character of these solutions is suggested in ref. [19], insofar as they require (on top of the periodicity in σ) that the physical time coordinate be an increasing function of τ. This leads to folded string solutions, which are exact solutions parts of whose worldsheets are described by expressions (59), but with discontinuities in the first derivatives of the worldsheet to spacetime maps. In such a case, the stress-energy tensor is concentrated on the folds, as follows immediately from our computation above when applied to the specific situation at hand, thus leading to a system with closer resemblance to a grid than to a string worldsheet. That is to say, the lattice introduced for "mathematical convenience" only [19] has been mapped into a "grid" with physical significance in spacetime itself.

APPENDIX A

In this appendix we give the explicit expansion of the string coordinates $u(\sigma,\tau), v(\sigma,\tau), r(\sigma,\tau)$ and $\phi(\sigma,\tau)$ around the $r = 0$ singularity for the equatorial case. We choose a gauge on the world-sheet such that $r = 0$ corresponds to $\tau = 0$.

$$\begin{aligned}
e^a u = 1 &- \frac{\gamma^4}{4}\tau^{4/5} - \\
&-\left(\frac{2\,\gamma^6}{21} + \frac{25\,a'^2}{14\,\gamma^4}\right)\tau^{6/5} - \frac{\gamma^7\,f'}{14\,a'}\tau^{7/5} + \\
&+\left(-\frac{65\,\gamma^8}{4704} + \frac{250\,a'^2}{441\,\gamma^2} + \frac{3125\,a'^4}{588\,\gamma^{12}}\right)\tau^{8/5} - \\
&-\left(\frac{\gamma^9\,f'}{21\,a'} + \frac{25\,a'\,f'}{126\,\gamma}\right)\tau^{9/5} + \\
&+\left(\frac{71\,\gamma^{10}}{113190} + \frac{112975\,a'^2}{271656} - \frac{18750\,a'^4}{3773\,\gamma^{10}} - \frac{234375\,a'^6}{7546\,\gamma^{20}} - \frac{\gamma^{10}\,f'^2}{88\,a'^2} - \frac{5\,a'\,\gamma'}{3\,\gamma} + \frac{5\,a''}{6}\right)\tau^2 + \\
&+\left(-\frac{173\,\gamma^{11}\,f'}{12936\,a'} - \frac{650\,\gamma\,a'\,f'}{4851} + \frac{1250\,a'^3\,f'}{1617\,\gamma^9}\right)\tau^{11/5} + \\
&+\left(\frac{278263\,\gamma^{12}}{395531136} - \frac{117475\,\gamma^2\,a'^2}{18540522} + \frac{85896875\,a'^4}{40452048\,\gamma^8} + \frac{448750000\,a'^6}{9270261\,\gamma^{18}} + \frac{701171875\,a'^8}{3090087\,\gamma^{28}} - \right. \\
&\quad \left. - \frac{225\,\gamma^2\,f'^2}{1001} - \frac{12\,\gamma^{12}\,f'^2}{1001\,a'^2} - \frac{25\,\gamma\,a'\,\gamma'}{21} + \frac{625\,a'^3\,\gamma'}{63\,\gamma^9} + \frac{5\,\gamma^2\,a''}{21} - \frac{125\,a'^2\,a''}{63\,\gamma^8}\right)\tau^{12/5} +
\end{aligned}$$

$$
\begin{aligned}
&+\Big(-\frac{977\,\gamma^{13}\,f'}{588588\,a'}+\frac{285125\,\gamma^3\,a'\,f'}{3531528}-\frac{93125\,a'^3\,f'}{441441\,\gamma^7}-\frac{1390625\,a'^5\,f'}{294294\,\gamma^{17}}-\frac{\gamma^{13}\,f'^3}{1144\,a'^3}-\\
&\quad-\frac{125\,\gamma^2\,f'\,\gamma'}{126}+\frac{25\,\gamma^3\,f'\,a''}{147\,a'}+\frac{25\,\gamma^3\,f''}{882}\Big)\tau^{13/5}+\\
&+\Big(\frac{97897\,\gamma^{14}}{576816240}-\frac{191605625\,\gamma^4\,a'^2}{4153076928}-\frac{78452875\,a'^4}{194675481\,\gamma^6}-\frac{9069078125\,a'^6}{259567308\,\gamma^{16}}-\frac{10747656250\,a'^8}{21630609\,\gamma^{26}}-\\
&\quad-\frac{26869140625\,a'^{10}}{14420406\,\gamma^{36}}-\frac{3460\,\gamma^4\,f'^2}{63063}-\frac{3529\,\gamma^{14}\,f'^2}{672672\,a'^2}-\frac{21625\,a'^2\,f'^2}{42042\,\gamma^6}-\frac{95\,\gamma^3\,a'\,\gamma'}{252}+\\
&\quad+\frac{1250\,a'^3\,\gamma'}{1323\,\gamma^7}-\frac{31250\,a'^5\,\gamma'}{441\,\gamma^{17}}-\frac{2\,\gamma^2\,\gamma'^2}{3}-\frac{25\,\gamma^4\,a''}{504}-\frac{250\,a'^2\,a''}{1323\,\gamma^6}+\\
&\quad+\frac{6250\,a'^4\,a''}{441\,\gamma^{16}}-\frac{\gamma^3\,\gamma''}{6}\Big)\tau^{14/5}+\\
&+\Big(\frac{467\,\gamma^{15}\,f'}{5992896\,a'}+\frac{256225\,\gamma^5\,a'\,f'}{9270261}-\frac{50344625\,a'^3\,f'}{222486264\,\gamma^5}+\frac{42781250\,a'^5\,f'}{9270261\,\gamma^{15}}+\frac{213671875\,a'^7\,f'}{6180174\,\gamma^{25}}-\\
&\quad-\frac{5\,\gamma^{15}\,f'^3}{3003\,a'^3}-\frac{295\,\gamma^5\,f'^3}{24024\,a'}-\frac{28825\,\gamma^4\,f'\,\gamma'}{38808}-\frac{55625\,a'^2\,f'\,\gamma'}{6468\,\gamma^6}+\frac{170\,\gamma^5\,f'\,a''}{1029\,a'}+\\
&\quad+\frac{124625\,a'\,f'\,a''}{67914\,\gamma^5}-\frac{4525\,\gamma^5\,f''}{271656}-\frac{15625\,a'^2\,f''}{135828\,\gamma^5}\Big)\tau^3+O(\tau^{16/5})\,.
\end{aligned}
\tag{A1}
$$

The expression for the coordinate $v(\sigma,\tau)$ follows by changing $a\to -a, u\to v$ in both sides of eq.(A1).

We find for $r(\sigma,\tau)$ near $r=0$:

$$
\begin{aligned}
r=\;&\gamma^2\tau^{2/5}+\\
&+\Big(-\frac{\gamma^4}{7}+\frac{25\,a'^2}{7\,\gamma^6}\Big)\tau^{4/5}+\\
&-\Big(\frac{4\,\gamma^6}{147}+\frac{1850\,a'^2}{441\,\gamma^4}+\frac{2500\,a'^4}{147\,\gamma^{14}}\Big)\tau^{6/5}+\\
&+\Big(-\frac{89\,\gamma^8}{11319}+\frac{40850\,a'^2}{33957\,\gamma^2}+\frac{908125\,a'^4}{33957\,\gamma^{12}}+\frac{1390625\,a'^6}{11319\,\gamma^{22}}+\frac{\gamma^8\,f'^2}{44\,a'^2}\Big)\tau^{8/5}+\\
&+\Big(-\frac{2735\,\gamma^{10}}{1030029}+\frac{90400\,a'^2}{3090087}-\frac{142715000\,a'^4}{9270261\,\gamma^{10}}-\frac{811718750\,a'^6}{3090087\,\gamma^{20}}-\frac{1068359375\,a'^8}{1030029\,\gamma^{30}}+\\
&\quad+\frac{1475\,f'^2}{4004}+\frac{9\,\gamma^{10}\,f'^2}{2002\,a'^2}\Big)\tau^2+\\
&+\Big(\frac{125\,f'\,\gamma'}{63}-\frac{50\,\gamma\,f'\,a''}{147\,a'}-\frac{25\,\gamma\,f''}{441}\Big)\tau^{11/5}+\\
&+\Big(-\frac{105436\,\gamma^{12}}{108153045}-\frac{69500\,\gamma^2\,a'^2}{64891827}+\frac{739998250\,a'^4}{194675481\,\gamma^8}+\frac{45536000000\,a'^6}{194675481\,\gamma^{18}}+\frac{14182812500\,a'^8}{4991679\,\gamma^{28}}+
\end{aligned}
$$

$$+\frac{205929687500\,a'^{10}}{21630609\,\gamma^{38}}-\frac{18230\,\gamma^2\,f'^2}{63063}+\frac{38\,\gamma^{12}\,f'^2}{21021\,a'^2}+\frac{8375\,a'^2\,f'^2}{84084\,\gamma^8}+\frac{4\,\gamma'^2}{3}+\frac{\gamma\,\gamma''}{3}\Bigg)\tau^{12/5}+$$
$$+\Bigg(-\frac{4175\,\gamma^2\,f'\,\gamma'}{19404}+\frac{98125\,a'^2\,f'\,\gamma'}{9702\,\gamma^8}-\frac{40\,\gamma^3\,f'\,a''}{1029\,a'}-\frac{83375\,a'\,f'\,a''}{33957\,\gamma^7}+\frac{11125\,\gamma^3\,f''}{135828}+$$
$$+\frac{29375\,a'^2\,f''}{67914\,\gamma^7}\Bigg)\tau^{13/5}+O(\tau^{14/5})\,. \tag{A2}$$

We get for the angular coordinate $\phi(\sigma,\tau)$,

$$\phi = f+2\,\gamma\tau^{1/5}+$$
$$+\left(\frac{4\,\gamma^3}{21}-\frac{100\,a'^2}{21\,\gamma^7}\right)\tau^{3/5}+$$
$$+\left(\frac{34\,\gamma^5}{735}+\frac{940\,a'^2}{441\,\gamma^5}+\frac{4250\,a'^4}{147\,\gamma^{15}}\right)\tau+$$
$$+\left(\frac{164\,\gamma^7}{11319}-\frac{2600\,a'^2}{33957\,\gamma^3}-\frac{1037500\,a'^4}{33957\,\gamma^{13}}-\frac{2562500\,a'^6}{11319\,\gamma^{23}}-\frac{\gamma^7\,f'^2}{77\,a'^2}\right)\tau^{7/5}+$$
$$+\Bigg(\frac{2258\,\gamma^9}{441441}-\frac{25700\,a'^2}{1324323\,\gamma}+\frac{47665000\,a'^4}{3972969\,\gamma^{11}}+\frac{507625000\,a'^6}{1324323\,\gamma^{21}}+\frac{882031250\,a'^8}{441441\,\gamma^{31}}-\frac{500\,f'^2}{9009\,\gamma}-$$
$$-\frac{19\,\gamma^9\,f'^2}{3003\,a'^2}\Bigg)\tau^{9/5}+$$
$$+\left(\frac{20\,f'\,\gamma'}{63\,\gamma}+\frac{20\,f'\,a''}{147\,a'}+\frac{265\,f''}{882}\right)\tau^2+$$
$$+\Bigg(\frac{29912\,\gamma^{11}}{15450435}-\frac{48200\,\gamma\,a'^2}{9270261}-\frac{42554000\,a'^4}{27810783\,\gamma^9}-\frac{7690000000\,a'^6}{27810783\,\gamma^{19}}-\frac{303125000\,a'^8}{64827\,\gamma^{29}}-$$
$$-\frac{58421875000\,a'^{10}}{3090087\,\gamma^{39}}+\frac{745\,\gamma\,f'^2}{63063}-\frac{64\,\gamma^{11}\,f'^2}{21021\,a'^2}+\frac{7250\,a'^2\,f'^2}{21021\,\gamma^9}+\frac{4\,\gamma'^2}{3\,\gamma}+\frac{\gamma''}{3}\Bigg)\tau^{11/5}+$$
$$+\Bigg(-\frac{5275\,\gamma\,f'\,\gamma'}{19404}-\frac{26875\,a'^2\,f'\,\gamma'}{3234\,\gamma^9}+\frac{190\,\gamma^2\,f'\,a''}{3087\,a'}+\frac{235250\,a'\,f'\,a''}{101871\,\gamma^8}-\frac{325\,\gamma^2\,f''}{45276}-$$
$$-\frac{131875\,a'^2\,f''}{203742\,\gamma^8}\Bigg)\tau^{12/5}+O(\tau^{13/5})\,. \tag{A3}$$

We found analogous formulas for the general (non-equatorial) case with the help of Mathematica.

REFERENCES

[1] H. J. de Vega and N. Sánchez, Phys. Lett. **B 197**, 320 (1987).

[2] H. J. de Vega and N. Sánchez, Nucl. Phys. **B 309**, 552 and 577 (1988).

[3] H. J. de Vega and N. Sánchez in Proceedings of the Erice Schools: "String Quantum Gravity and Physics at the Planck Energy Scale", 21-28 June 1992, edited by N. Sánchez, World Scientific, 1993, and Third D. Chalonge School, 4-16 September 1994, edited by N. Sánchez, to appear.

[4] See for a review, T.W.B. Kibble, Erice Lectures at the Chalonge School on Astrofundamental Physics, N. Sánchez editor, World Scientific, 1992.

[5] H. J. de Vega and N. Sánchez, Phys. Rev. **D 45** , 2783 (1992).
H. J. de Vega, M. Ramón Medrano and N. Sánchez,
Class. and Quantum Grav. **10**, 2007 (1993).
G. T. Horowitz and A.R. Steif, Phys. Rev. Lett. **64**, 260 (1990) and Phys. Rev. **D 42** , 1950 (1990).

[6] V. P. Frolov, V. D. Skarzhinsky, A. I. Zelnikov and O. Heinrich,
Phys. Lett. **B 224**, 255 (1989).

[7] C. Loustó and N. Sánchez, Phys. Rev. **D 47**, 4498 (1993).

[8] H. J. de Vega and I.L. Egusquiza, Phys. Rev. **D49**, 763 (1994).

[9] A. L. Larsen and N. Sánchez, Phys. Rev. **D 51**, 6929 (1995).

[10] A. L. Larsen, Phys. Rev. **D50**, 2623 (1994).

[11] See for example,
C. W. Misner, K. S. Thorne and J. A. Wheeler, 'Gravitation', Freeman, 1973.

[12] I.L. Egusquiza, *Classical Strings in FRW spacetimes*, p. 86-110 in the Proceedings of the 2nd. Paris Cosmology Colloquium, Observatoire de Paris, 2-4 June 1994, eds. H. J. de Vega and N. Sánchez, World Scientific.

[13] H. J. de Vega and N. Sánchez, Phys. Rev. **D50**, 7202 (1994).

[14] L. Susskind, Phys. Rev. Lett. **71** (1993) 2367, Phys. Rev. **D49** (1994) 6606.

[15] H. J. de Vega and N. Sánchez, Int. J. Mod. Phys. **A 7**, 3043 (1992).

[16] N. Sánchez and G. Veneziano, Nucl. Phys. **B 333**, 253 (1990).

[17] M. Gasperini, N. Sánchez and G. Veneziano,
Int. J. Mod. Phys. **A 6**, 3853 (1991) and Nucl. Phys. **B364**, 365 (1991).

[18] H. J. de Vega, J. R. Mittelbrun, M. Ramón Medrano and N. Sánchez,
PAR-LPTHE-93/14, hep-th/9304002.

[19] I. Bars and J. Schulze, Phys. Rev. **D51**, 1854 (1995).

[20] I. Bars, p. 399-428 in the Proceedings of the 2nd. Paris Cosmology Colloquium, Observatoire de Paris, 2-4 June 1994, eds. H. J. de Vega and N. Sánchez, World Scientific.

STRING DYNAMICS IN COSMOLOGICAL AND BLACK HOLE BACKGROUNDS: THE NULL STRING EXPANSION

C. O. LOUSTO
Department of Physics, University of Utah
Salt Lake City, UT 84112, USA

N. SÁNCHEZ
Observatoire de Paris, DEMIRM, UA 336 Laboratoire associé au CNRS,
61, Avenue de l'Observatoire, F-75014 Paris, France

(May 13, 1996)

Abstract

We study the classical dynamics of a bosonic string in the D–dimensional flat Friedmann–Robertson–Walker and Schwarzschild backgrounds. We make a perturbative development in the string coordinates around a *null* string configuration; the background geometry is taken into account exactly. In the cosmological case we uncouple and solve the first order fluctuations; the string time evolution with the conformal gauge world-sheet τ–coordinate is given by $X^0(\sigma,\tau) = q(\sigma)\tau^{\frac{1}{1+2\beta}} + c^2 B^0(\sigma,\tau) + \cdots$, $B^0(\sigma,\tau) = \sum_k b_k(\sigma)\tau^k$ where $b_k(\sigma)$ are given by Eqs. (3.15), and β is the exponent of the conformal factor in the Friedmann–Robertson–Walker metric, i.e. $R \sim \eta^\beta$. The string proper size, at first order in the fluctuations, grows like the conformal factor $R(\eta)$ and the string energy–momentum tensor corresponds to that of a null fluid. For a string in the black hole background, we study the planar case, but keep the dimensionality of the spacetime D generic. In the null string expansion, the radial, azimuthal, and time coordinates (r,ϕ,t) are $r = \sum_n A^1_n(\sigma)(-\tau)^{2n/(D+1)}$, $\phi = \sum_n A^3_n(\sigma)(-\tau)^{(D-5+2n)/(D+1)}$, and $t = \sum_n A^0_n(\sigma)(-\tau)^{1+2n(D-3)/(D+1)}$. The first terms of the series represent a *generic* approach to the Schwarzschild singularity at $r = 0$. First and higher order string perturbations contribute with higher powers of τ. The integrated string energy-momentum tensor corresponds to that of a null fluid in $D-1$ dimensions. As the string approaches the $r = 0$ singularity its proper size grows indefinitely like $\sim (-\tau)^{-(D-3)/(D+1)}$. We end the paper giving three particular exact string solutions inside the black hole. They represent respectively straight strings across the origin, twisted, and rigidly rotating strings.

11.25.Db, 04.70.Bw, 98.80.Hw

I. INTRODUCTION

The investigation of strings in curved spacetime is currently the best framework to study the physics of gravitation in the context of string theory. The study of string propagation in curved spacetimes provides new insights with respect to string theory in flat, Minkowski spacetime (and with respect to quantum fields in curved spacetime) [See for example Refs. [1–4].] The results of this study apply to fundamental strings, as well as to cosmic strings, which behave essentially in a classical way.

The string equations of motion and constraints in curved spacetime are highly non linear and, in general, non exactly solvable. Different methods are available to solve this system: The string perturbation approach [5], the τ–expansion method [6] (which provides exact local solutions for any background), the null string approach [7], and the construction of global solutions (by solitonic and inverse scattering methods, for instance), which allowed to uncover the new feature of multistring solutions [8–11]. The expansion methods are described and classified by using the world sheet velocity of light, c, as an expansion parameter, in Sec. II below.

An approximate but general method is the expansion around the center of mass solution of the string (satisfying the geodesic equation of motion), the world sheet time τ being identified with the proper time associated to the center of mass trajectory. In this paper we are concerned with another general approximation method: The null string approach. In this approach, the string equations of motion and constraints are systematically expanded in powers of the parameter c (the speed of light in the world sheet). This corresponds to a small string tension expansion. To zeroth order the string is effectively equivalent to a continuous line–beam of massless particles labeled by the world sheet spatial parameter σ. The points on the "null" string do not interact among them, but they interact with the gravitational background. Note the extended character of the zeroth order solution in the null string approach, as opposed to the perturbation approach in which the zeroth order is given by the string's center of mass point–like approximation

We also study fluctuations around the null string configuration, which naturally appears as an expansion in powers of τ with precisely defined σ–dependent coefficients. The string coordinates are expressed as

$$X^A(\sigma,\tau) = A^A(\sigma,\tau) + c^2 B^A(\sigma,\tau) + c^4 C^A(\sigma,\tau) + ... \tag{1.1}$$

the zeroth order $A^A(\sigma,\tau)$, first and second order fluctuations $B^A(\sigma,\tau)$, $C^A(\sigma,\tau)$, satisfy Eqs. (2.3)–(2.4) as described in Sec. II. This null string approach is a powerful description for strings in the strong gravitational field regime, as is the case of black hole backgrounds near the $r = 0$ singularity and in inflationary cosmological spacetimes, at late times, where string instabilities develop. The string perturbation method around the string center of mass allows to detect the emergence of instabilities [12,13,11], but is unable to describe the highly unstable regime [6]. We apply this null string expansion to cosmological spacetimes and to strings in the interior of black holes, falling into the singularity at $r = 0$. In Friedmann–Robertson–Walker cosmological spacetimes with conformal expansion factor $R \propto \eta^\beta$, η being the conformal gauge world sheet time, the null string evolution can be expressed as

$$X^0(\sigma,\tau) = q(\sigma)\tau^{\frac{1}{1+2\beta}} + c^2 B^0(\sigma,\tau) + ... \tag{1.2}$$

where

$$B^0(\sigma,\tau) = \sum_k b_k(\sigma)\tau^k\ , \quad k = 2, 2 \pm \alpha/\beta, k_\pm\ . \tag{1.3}$$

The spatial coordinates $X^i(\sigma,\tau)$ are given by Eqs. (3.5) and (3.22) in Sec. III.

The proper size of the string grows proportional to the expansion factor $R(\eta)$ and the string energy momentum tensor corresponds to that of a null fluid.

In black hole spacetimes we find the null string expansion inside the Schwarzschild black hole. I is given by

$$t \doteq X^0(\sigma,\tau) = \sum_{n=1}^{\infty} A_n^0(\sigma)(-\tau)^{1+2n(D-3)/(D+1)} + c^2 B^0(\sigma,\tau) + \cdots \tag{1.4}$$

$$r \doteq X^1(\sigma,\tau) = \sum_{n=1}^{\infty} A_n^1(\sigma)(-\tau)^{2n/(D+1)} + c^2 B^1(\sigma,\tau) + \cdots \tag{1.5}$$

$$\phi \doteq X^3(\sigma,\tau) = \sum_{n=1}^{\infty} A_n^3(\sigma)(-\tau)^{(D-5+2n)/(D+1)} + c^2 B^3(\sigma,\tau) + \cdots \tag{1.6}$$

and we have taken, for simplicity, the string in the equatorial plane, i. e.

$$\theta^i \doteq A^i = \pi/2 \text{ for } \ i = 2, 4, 5, ..., D-1. \tag{1.7}$$

The τ–dependence exhibited above represents a generic behavior near the $r \to 0$ singularity ($\tau \to 0$), and generalizes to D–dimensions the case recently analyzed in Ref. [14]. Higher order fluctuations $c^2 B^\mu$, $c^4 C^\mu$, and so on, contribute with increasing orders of τ.

The proper string size grows indefinitely as the string approaches the $r \to 0$ singularity as

$$\left(\frac{dl}{d\sigma}\right) \to \left(\frac{R_S}{r}\right)^{(D-3)/2} X^{0\prime}(\sigma, \tau = 0) \to (-\tau)^{-(D-3)/(D+1)}\ . \tag{1.8}$$

The integrated string energy–momentum tensor for $r \to 0$ behaves as

$$\Theta_r^r \to -\frac{L}{2\pi\alpha' r} = -\Theta_\phi^\phi\ , \quad \Theta_t^t \to 0\ . \tag{1.9}$$

L being the string orbit angular momentum. In the infalling towards the $r \to 0$ singularity, the string behaves as a null fluid in $(D-1)$–dimensions; in particular, a planar string behaves as a two (spatially) dimensional null fluid. The emergence of string instabilities in Schwarzschild and Reissner–Nordström spacetimes, and the growing of the string proper size near the black hole singularity was first found in Ref. [13] using the perturbative expansion method around the string center of mass (although such expansion does not allow to describe the highly unstable regime and the full approach to the $r \to 0$ singularity).

This paper is organized as follows: In Sec. II we classify the different expansion methods to solve the string dynamics in curved spacetimes in terms of the parameter c (or equivalently, in terms of the ratio between $\dot{X}^A$ and $cX^{A'}$, the τ and σ derivatives of the string coordinates), with particular focusing on the null string expansion (See also Table II). In Sec. III we solve the null string expansion in cosmological spacetimes. In Sec. IV we find the null string expansion near the Schwarzschild $r \to 0$ singularity. In Sec. V we find particular exact solutions inside the black hole event horizon describing straight string across $r = 0$, twisted and rigidly rotating strings. Sec. VI summarizes our conclusions.

II. STRING DYNAMICS IN GRAVITATIONAL BACKGROUNDS

The action of a bosonic string in a D–dimensional curved manifold endowed with a metric G_{AB} $(0 \leq A, B \leq D-1)$ is given by [15,16]

$$S = -T_0 \int d\sigma d\tau \sqrt{-\det g_{\mu\nu}} \tag{2.1}$$

where $g_{\mu\nu} = G_{AB}(X)\partial_\mu X^A \partial_\nu X^B$ is the two dimensional world–sheet metric $(0 \leq \mu, \nu \leq 1)$ and $T_0 = 1/(2\pi\alpha')$ is the string tension.

By use of the reparametrization invariance of the world–sheet one can take the conformal gauge, i.e. $g_{\mu\nu} = \rho(\sigma,\tau)\eta_{\mu\nu}$, where $\eta_{\mu\nu}$ is the two-dimensional Minkowskian metric. In this gauge, the classical equations of motion derived from the action (2.1) read

$$\partial_\tau^2 X^A - c^2 \partial_\sigma X^A + \Gamma^A_{BC}\left[\partial_\tau X^B \partial_\tau X^C - c^2 \partial_\sigma X^B \partial_\sigma X^C\right] = 0 \ , \tag{2.2}$$

where Γ^A_{BC} are the Christoffel symbols associated to the metric G_{AB} and we have introduced the velocity of wave propagation along the string (velocity of light), c, as the second fundamental constant after the string tension.

Variation of action (2.1) with respect to the world–sheet variables yields the two non–trivial constraints

$$\partial_\tau X^A \partial_\sigma X^B G_{AB} = 0 \ , \tag{2.3}$$

$$[\partial_\tau X^A \partial_\tau X^B + c^2 \partial_\sigma X^A \partial_\sigma X^B] G_{AB} = 0 \ . \tag{2.4}$$

de Vega and Sánchez [5] have proposed a method to solve, both classically and quantum mechanically, the equations of motion and constraints of strings in curved spacetimes, i. e. Eqs. (2.2)-(2.4). This method treats the spacetime geometry exactly and considers the string excitations small as compared to the energy scales associated to the background gravitational metric. This method is particularly well suited to study strings in strong gravitational fields such us in black holes and in cosmological scenarios as opposed to the usual treatment in flat spacetime. The starting point is to consider a particular solution to the equations of motion and constraints, $X^A_{\text{part}}(\sigma,\tau) = A^A(\sigma,\tau)$ and then successively studying first, second and higher order fluctuations around this particular solution; to be denoted respectively as B^A, C^A, etc.

In Ref. [7] have been raised the interesting possibility of using c, the world–sheet light velocity, as the expansion parameter in the above development. This approach allow us to classify the different solutions proposed in the literature:

- i) $c \ll 1$ or equivalently, as can be seen from the field and constraint equations (2.2)-(2.4), $\partial_\tau A^A \gg c\partial_\sigma A^A$ (holding also this inequality for second derivatives).
- ii) $c = 1$ and thus all derivatives and other terms are, in principle, relevant in Eqs. (2.2)-(2.4).
- iii) $c \gg 1$ or equivalently, $c\partial_\sigma A^A \gg \partial_\tau A^A$ (holding also for second derivatives).

- iv) There can be, of course, several other possibilities such as hybrid starting solutions that for some components of A^A (and eventually B^A, C^A, etc) fulfill one of the cases i), ii) or iii) and for other components a different case.

More explicitly, we have
Case i) $c \ll 1$: The appropriate development reads

$$X^A(\sigma,\tau) = A^A(\sigma,\tau) + c^2 B^A(\sigma,\tau) + c^4 C^A(\sigma,\tau) + \ldots \tag{2.5}$$

Plugging this into Eqs. (2.2)-(2.4) one obtains [7], to zeroth order in c (since now on we adopt the notation $\dot{} = \partial_\tau$ and $' = \partial_\sigma$)

$$\ddot{A}^A + \Gamma^A_{BC}\dot{A}^B\dot{A}^C = 0\ , \tag{2.6}$$

$$\dot{A}^B\dot{A}^C G_{BC} = 0\ , \tag{2.7}$$

$$\dot{A}^B A'^C G_{BC} = 0\ . \tag{2.8}$$

Eq. (2.6) is the geodesic path followed by each point of the string. Eq. (2.7) means that this is a null geodesic and Eq. (2.8) equation states that the velocity is perpendicular to the string.

First order fluctuations around this particular solution $A(\sigma,\tau)$ can be obtained by retaining terms of order c^2 in Eqs (2.2)-(2.4)

$$\ddot{B}^A + 2\Gamma^A_{BC}\dot{A}^B\dot{B}^C + \Gamma^A_{BC,D}\dot{A}^B\dot{A}^C B^D = A''^A + \Gamma^A_{BC}A'^B A'^C\ , \tag{2.9}$$

$$(2\dot{A}^B\dot{B}^C + A'^B A'^C)G_{BC} + G_{BC,D}\dot{A}^B\dot{A}^C B^D = 0\ , \tag{2.10}$$

$$(\dot{B}^B A'^C + \dot{A}^B B'^C)G_{BC} + G_{BC,D}\dot{A}^B A'^C B^D = 0\ . \tag{2.11}$$

Higher order corrections can thus be systematically obtained. The interpretation of these equations is that they give the high energy (compared to the Planck energy) string behavior, corresponding to the limit of vanishing tension (and thus small c).
Case ii) $c = 1$, and thus all derivatives and other terms are, in principle, relevant in Eqs. (2.2)-(2.4). One can thus make developments around the center–of–mass motion as a physically appealing starting solution. In fact, de Vega and Sánchez [5] have proposed in his original approach

$$X^A = A^A(\sigma,\tau) + B^A(\sigma,\tau) + C^A(\sigma,\tau) + \ldots\ , \tag{2.12}$$

where, $A^A(\tau)$ follows the geodesic equation

$$\ddot{A}^A + \Gamma^A_{BC}\dot{A}^B\dot{A}^C = 0\ , \tag{2.13}$$

and the constraint

$$\dot{A}^B \dot{A}^C G_{BC} = 0 \ , \tag{2.14}$$

means that these geodesics are null.

First order fluctuation equations now read

$$\ddot{B}^A - B''^A + 2\Gamma^A_{BC} \dot{A}^B \dot{B}^C + \Gamma^A_{BC,D} \dot{A}^B \dot{A}^C B^D = 0 \ , \tag{2.15}$$

and

$$2G_{BC} \dot{A}^B \dot{B}^C + G_{BC,D} \dot{A}^B \dot{A}^C B^D = 0 \ , \tag{2.16}$$

$$G_{BC} \dot{A}^B B'^C = 0 \ . \tag{2.17}$$

Higher order corrections can be in this way systematically considered. Still, a Fourier series decomposition of the first order fluctuations can be made [5], i. e. $B^A(\sigma, \tau) = \Sigma_n \exp{(in\sigma)} \eta^A_n(\tau)$. Also, a decomposition in purely left $A^A(\sigma + \tau)$ or right movers $A^A(\sigma - \tau)$ can be used as a starting exact solution.

Case iii) $c \gg 1$ or equivalently, $c\partial_\sigma A^A \gg \partial_\tau A^A$ (holding also for second derivatives). The appropriate development reads in this case

$$X^A(\sigma, \tau) = A^A(\sigma, \tau) + \frac{1}{c^2} B^A(\sigma, \tau) + \frac{1}{c^4} C^A(\sigma, \tau) + \ldots \tag{2.18}$$

Plugging this expression into Eqs. (2.2)-(2.4) yields a set of equations that is equivalent to (2.6)-(2.11) upon the substitutions $\tau \leftrightarrow \sigma$ and $c \to c^{-1}$. The interpretation of this approximation is now exactly the opposite to case i), i. e. here the zeroth order dependence on σ overwhelms that on τ and thus represent very low energy strings, frozen at first approximation. Solutions to this case can be obtained from solutions to case i) by making the above mentioned substitutions. These represent, at zeroth order, stationary solutions as opposed to the dynamical ones. While solutions to case i) are appropriate to describe strings in strong gravitational fields, for instance, near black hole singularities, i.e. $R \to 0$, solutions to the case iii) are rather appropriate for strings far away from black holes, i.e. $R \to \infty$. We thus have an approximate duality here, represented by the transformation $\tau \leftrightarrow \sigma$ and $R \leftrightarrow R^{-1}$, which map the solutions to the cases i) and iii) into one another.

Case iv): This corresponds to an hybrid case. The circular string ansatz is an example of this situation:

$$X^0 = X^0(\tau) \ , \quad R = R(\tau) \ , \quad \psi = \psi(\sigma) \ , \quad \theta = \pi/2 \ . \tag{2.19}$$

Clearly X^0 and R coordinates follow criterion i) while ψ follows case iii) and θ could be assigned to case ii). The first order fluctuations around this starting solution have been studied in Ref. [17,11] with regards to the stability analysis as the string approaches the black hole singularity. The stationary string ansatz have been studied in Ref. [18].

Finally, one can consider exact or asymptotic particular string solutions that deserve study in its own. Asymptotic solutions in a Friedmann–Robertson–Walker background representing highly unstable strings have been found [6]. Exact string and multistring solutions in de Sitter spacetime have been found by solitonic methods (a single world-sheet describing multiple different and independent strings) [19,8,9,11]. Ring solutions in black holes [10] have also been studied. For the sake of completeness it is also worth to remark that the propagation of strings in shock wave and conical spacetimes has been exactly solved [20,21].

Two physical interesting quantities can be computed: The string energy–momentum tensor integrated over a spatial volume completely enclosing the string [22], at fixed time X^0

$$I^{AB}(X^0) = \int \sqrt{-G}T^{AB}(X)d^{D-1}\vec{X} \ , \tag{2.20}$$

where

$$\sqrt{-G}T^{AB}(X) = \frac{1}{2\pi\alpha'}\int d\sigma d\tau \left[\dot{X}^A\dot{X}^B - X'^A X'^B\right]\delta^{(D)}\left[X - X(\sigma,\tau)\right] \ , \tag{2.21}$$

and the invariant string size l [3]

$$dl^2 = -\dot{X}^A\dot{X}^B G_{AB}(X)d\sigma^2 \ . \tag{2.22}$$

[Note that the differential string size has the form of an effective mass $m_{\text{eff}}(\tau,\sigma)$ for the geodesic motion, and actually represents a projection from the target space onto the world sheet].

III. NULL STRING DYNAMICS IN COSMOLOGICAL BACKGROUNDS

Let us consider as a simple application of the formalism above the isotropic cosmological geometry given by the conformally flat Friedmann–Robertson–Walker metric

$$ds^2 = R^2(X^0)\left[-(dX^0)^2 + (dX^1)^2 + (dX^2)^2 + \cdots + (dX^{D-1})^2\right] \ , \tag{3.1}$$

where we parametrize the conformal factor as $R(X^0) = K(X^0/\beta)^\beta$.

We are interested in this paper in the null string dynamics and the fluctuations around it. *Zeroth order:* The equations of motion (2.6) have the following first integral of motion (conservation of the four-moment P_A)

$$R^2(A^0)\dot{A}^0(\sigma,\tau) = -P_0(\sigma) \ , \quad R^2(A^0)\dot{A}^i(\sigma,\tau) = P_i(\sigma) \ . \tag{3.2}$$

where from now on i,j run from 1 to $D-1$.

The constraints at $\tau = 0$ yield

$$(P_0)^2 = \sum_{i=1}^{D-1}(P_i)^2 \ , \tag{3.3}$$

$$-P^0(\bar{A}^0)' = \sum_{i=1}^{D-1} P^i(\bar{A}^i)' \ . \tag{3.4}$$

where we use the notation $A^A(\sigma,\tau=0) = \bar{A}^A$

From Eq. (3.2) we see that the only relevant τ−dependence is that of $A^0(\tau)$. In fact, we can write

$$A^i(\tau) - \bar{A}^i = \frac{P^i}{-P^0}\left[A^0(\tau) - \bar{A}^0\right] . \tag{3.5}$$

By direct integration of the zeroth component of Eq. (3.2) one obtains

$$A^0(\tau) - \bar{A}^0 = \left[\frac{(1+2\beta)\beta^{2\beta}P^0}{K^2}\tau\right]^{\frac{1}{1+2\beta}} \doteq q(\sigma)\tau^{\frac{1}{1+2\beta}} , \tag{3.6}$$

The particular time-dependence of the scale factor as usually referred to in the literature is recalled in Table I. The cases dealt with in Refs. [7] are in $D = 4$ and $\beta = -1/2$ and -1 respectively. Note that the cosmic time t is related to the conformal time $X^0 \approx A^0$ by

$$t - t_0 = \frac{K\left(A^0 - \bar{A}^0\right)^{1+\beta}}{\beta^\beta(1+\beta)} . \tag{3.7}$$

Thus, τ does *not* coincides with t.

We can now compute the zeroth order energy–momentum density tensor from Eq. (2.20)

$$I_A^B(\sigma) = \frac{P_A P_C \eta^{CB}}{-P_0\alpha'} . \tag{3.8}$$

Hence, at zeroth order, the string energy and momentum keep constant as the universe evolves. The trace of I_A^B vanishes to this zeroth order. Since, to this order, each point of the string follows a null geodesic, the proper size Eq. (2.22) also vanishes.

First order fluctuations: The 0 component of Eq. (2.9) yields

$$\ddot{B}^0 + \frac{2dR}{RdX^0}\left[\dot{A}^0\dot{B}^0 + \dot{A}^i\dot{B}^i\right] + \frac{d}{dX^0}\left(\frac{dR}{RdX^0}\right)\left[\left(\dot{A}^0\right)^2 + \left(\dot{A}^i\right)^2\right]B^0 =$$
$$A^{0''} + \frac{dR}{RdX^0}\left[\left(A^{0'}\right)^2 + \left(A^{i'}\right)^2\right] .$$

With the help of the constraint equation (2.10)

$$\left(2\dot{B}^A A^{B'} + \dot{A}^A B^{B'}\right)\eta_{AB} = 0 , \tag{3.9}$$

we obtain an uncoupled equation for B^0

$$\ddot{B}^0 + \frac{4\alpha}{\tau}\dot{B}^0 - \frac{2\alpha^2}{\beta\tau^2}B^0 = A^{0''} + \frac{2\left(A^{i'}\right)^2}{A^0} \doteq F^0(\sigma,\tau) , \tag{3.10}$$

where $\alpha = \beta/(\beta+1)$, and we have used that

$$\frac{dR}{RdX^0} = \frac{\beta}{X^0} , \quad \frac{d}{dX^0}\left(\frac{dR}{RdX^0}\right) = \frac{-\beta}{(X^0)^2} . \tag{3.11}$$

We can write the general solution to Eq. (3.10) as

$$B^0(\sigma,\tau) = \sum_k b_k(\sigma)\tau^k \tag{3.12}$$

where b_{k_+} and b_{k_-} are arbitrary constants corresponding to the general solution to the homogeneous equation associated to Eq. (3.10); $k_\pm = -(4\alpha-1)/2 \pm \sqrt{(2\alpha-1/2)^2 + 2\alpha^2/\beta}$. The other powers of k correspond to the series development of the function $F^0(\sigma,\tau)$ and give the particular solution to the inhomogeneous equation.

To provide a simpler solvable example we take $\bar{A}^0 = 0$. In this case F^0 takes the following form:

$$F^0(\sigma,\tau) = \left[q'' + \frac{2\beta}{q}\left(\frac{P_i q}{P_0}\right)'^{\,2}\right]\tau^{\alpha/\beta} - \frac{4\beta\bar{A}^{i'}}{q}\left(\frac{P_i q}{P_0}\right)' + \frac{\beta\left(\bar{A}^{i'}\right)^2}{q}\tau^{-\alpha/\beta} . \tag{3.13}$$

Plugging this into the right hand side of Eq. (3.10) and using the form of B^0 given by Eq. (3.12) we obtain

$$\sum_k b_k[k(k-1) + 4k\alpha - 2\alpha^2/\beta]\tau^{k-2} = F^0(\sigma,\tau) \tag{3.14}$$

Thus, there are only three values of k given by $k-2 = 0, \pm\alpha/\beta$ with the corresponding b_k as defined by Eqs. (3.13) and (3.14)

$$\begin{aligned} b_2 &= -\frac{4\beta\bar{A}^{i'}}{q[2+8\alpha-2\alpha^2/\beta]} \\ b_{2+\alpha/\beta} &= \frac{q'' + \frac{2\beta}{q}\left(\frac{P_i q}{P_0}\right)'^{\,2}}{(2+\alpha/\beta)(1+4\alpha+\alpha/\beta) - 2\alpha^2/\beta} \\ b_{2-\alpha/\beta} &= \frac{\beta\left(\bar{A}^{i'}\right)^2}{q\left[(2-\alpha/\beta)(1+4\alpha-\alpha/\beta) - 2\alpha^2/\beta\right]} \end{aligned} \tag{3.15}$$

Let us now consider the i components of Eq. (2.9) for the first order fluctuations:

$$\ddot{B}^i + \frac{2}{R}\frac{dR}{dX^0}\left[\dot{A}^0\dot{B}^i + \dot{A}^i\dot{B}^0\right] + 2\frac{d}{dX^0}\left(\frac{dR}{RdX^0}\right)\dot{A}^0\dot{A}^i B^0 = A^{i''} + 2\frac{dR}{RdX^0}A^{0'}A^{i'} . \tag{3.16}$$

With the help of Eq. (3.11)

$$\ddot{B}^i + \frac{2\alpha}{\tau}\dot{B}^i = A^{i''} + \frac{2\beta A^{i'}A^{0'}}{A^0} + \frac{2\alpha P_i}{P_0\tau}\left(\dot{B}^0 - \frac{\alpha B^0}{\beta\tau}\right) \doteq F^i(\sigma,\tau) , \tag{3.17}$$

Known $B^0(\sigma,\tau)$ the general solution to this equation can be written in terms of quadratures

$$B^i(\sigma,\tau) = \bar{B}^i + \dot{\bar{B}}^i\frac{\tau^{1-2\alpha}}{1-2\alpha} + \int^\tau d\tilde{\tilde{\tau}}\tilde{\tilde{\tau}}^{-2\alpha}\int^{\tilde{\tilde{\tau}}} d\tilde{\tau}\tilde{\tau}^{2\alpha}F^i(\sigma,\tilde{\tau}) , \tag{3.18}$$

where $\bar{B}^i = B^i(\sigma,0)$. Note that $\bar{B}^i$ and $\dot{\bar{B}}^i$ can be absorbed into $\bar{A}^i$ and P_i respectively. We are thus left with the two constants $b_{k_\pm}$ which are fixed by the two first order constraints equations (2.10) and (2.11)

$$\left(\dot{A}^A\dot{B}^B + A^{A'}A^{B'}\right)\eta_{AB} = 0 \ ,$$
$$\left(2\dot{B}^A A^{B'} + \dot{A}^A B^{B'}\right)\eta_{AB} = 0 \ . \tag{3.19}$$

We are thus sure we are not adding any spurious degree of freedom at first order.

We can again show the explicit dependence for the case $\bar{A}^0 = 0$. In this case F^i takes the following form

$$F^i(\sigma,\tau) = \bar{A}^{i''} + 2b_2\alpha\tfrac{P_i}{P_0}(2-\alpha/\beta) + 2\beta\tfrac{q'}{q}\bar{A}^{i'} -$$
$$\left[\left(\tfrac{P_i q}{P_0}\right)'' + 2\beta\tfrac{q'}{q}\left(\tfrac{P_i q}{P_0}\right)' - 4\alpha b_{2+\alpha/\beta}\tfrac{P_i}{P_0}\right]\tau^{\alpha/\beta} +$$
$$4\alpha b_{2-\alpha/\beta}\tfrac{P_i}{P_0}(1-\alpha/\beta)\tau^{-\alpha/\beta} + 2\alpha b_{k_\pm}\tfrac{P_i}{P_0}\left((k_\pm)-\alpha/\beta\right)\tau^{k_\pm - 2} \ , \tag{3.20}$$

or in a more compact way

$$F^i(\sigma,\tau) = \sum_n F^i_n(\sigma)\tau^n \ , \quad n = 0, \pm\alpha/\beta, (k_\pm) - 2 \ , \tag{3.21}$$

where the notation F^i_n clearly refers to the coefficient of the n-th power of τ in Eq. (3.20).

Plugging this into the right hand side of Eq. (3.18) and using the form of B^0 given by Eqs. (3.12) and (3.15), we obtain

$$B^i(\sigma,\tau) = \bar{B}^i + \bar{\bar{B}}^i\frac{\tau^{1-2\alpha}}{1-2\alpha} + \sum_n \frac{F^i_n \tau^{n+2}}{(n+2)(2\alpha+n+1)} \ , \quad n = 0, \pm\alpha/\beta, (k_\pm) - 2 \ . \tag{3.22}$$

The proper string size that vanished to zeroth order now takes the following form

$$dl^2 = R^2\eta_{AB}A^{A'}A^{B'}d\sigma^2 \ , \tag{3.23}$$

where we have used the constraint (3.19). We see that the string size grows like the conformal factor of the Friedmann–Robertson–Walker metric. This unstable behavior was already observed in Ref. [12] in the fluctuations around the center of mass zeroth order solution.

The coordinate fluctuations B^0 and B^i evolve with several powers of τ and for a given value of the conformal factor exponent α, we will have several regimes. We can see from the above expressions (Eq. (3.22)) that in the regime of R large, that is $\tau \to \infty$, for $\alpha/\beta > 0$, i.e. $\beta > -1$, the overwhelming power will be $(2+\alpha/\beta)$ or $(2+k_+)$. When $\alpha/\beta < 0$, i.e. $\beta < -1$, the regime of R large is reached when $\tau \to 0$, then the predominant power will be $2+\alpha/\beta$ or $2+k_-$. Note also that these results do not depend explicitly on the dimension, D, since the metric coefficients neither do.

IV. BLACK HOLE SPACETIMES

We are particularly interested in studying the dynamics of strings in black hole spacetimes and its approach to the singularity at radial coordinate $r = 0$.

Let us consider a D–dimensional black hole background [23],

$$ds^2 = -g(r)(dX^0)^2 + g(r)^{-1}dr^2 + r^2 d\Omega_{D-2} \ , \tag{4.1}$$

where for the Reissner-Nordström metric

$$g(r) = 1 - \left(\frac{R_S}{r}\right)^{D-3} + \left(\frac{\tilde{Q}}{r}\right)^{2(D-3)} \quad , \quad \tilde{Q}^{2(D-3)} = \frac{8\pi G Q^2}{(D-2)(D-3)} \, , \tag{4.2}$$

with M and Q being respectively the mass and charge of the black hole.

We can now study the string dynamics by using the null string approach described in Sec. II.

We then propose the following development for the string coordinates

$$X^A(\sigma,\tau) = A^A(\sigma,\tau) + c^2 B^A(\sigma,\tau) + \ldots \tag{4.3}$$

Replacing this into the field equations (2.2), with (4.1) acting as the curved background, we find to zeroth order the following first integrals of motion

$$\dot{\theta} \simeq \dot{A}^2 = 0 \; , \quad \dot{\phi} \simeq A^3 = \frac{L(\sigma)}{r^2} \, , \tag{4.4}$$

where for simplicity we consider a string lying on the equatorial plane. $L(\sigma)$ is the angular momentum of each point forming the string. The other first integrals of motion are

$$\dot{X}^0 \simeq \dot{A}^0 = E(\sigma) g^{-1}(r) \; , \quad \dot{r}^2 \simeq (\dot{A}^1)^2 = E^2(\sigma) - \frac{L^2(\sigma)}{r^2} g(r) \; . \tag{4.5}$$

From this last equation one can see that for the Reissner-Nordström black hole, if $L(\sigma) \neq 0$, the string will only skirt the singularity. Since we are interested in the string approach to the singularity, we consider from now on Schwarzschild black holes (in D dimensions).

The remaining constraint equation (2.8) at $\tau = 0$ takes the following simple form

$$\bar{A}^{0\prime}(\sigma) = \frac{L(\sigma)}{E(\sigma)} \bar{A}^{3\prime}(\sigma) \; . \tag{4.6}$$

We can further integrate expression (4.5) near the black hole singularity to obtain

$$A^1(\sigma,\tau) \simeq p(\sigma)(-\tau)^{\frac{2}{D+1}} \; , \quad p(\sigma) = [\frac{1}{2}(D+1) R_S^{\frac{D-3}{2}} L(\sigma)]^{\frac{2}{D+1}} \tag{4.7}$$

where we have chosen $\tau \leq 0$ such that $A^1(\tau = 0) = 0$. τ measures the proper time of infall to the singularity at $r = 0$.

By use of the above expression we find

$$A^0(\sigma,\tau) \simeq \bar{A}^0 + \frac{(D+1)}{(3D-5)} E \left(\frac{p(\sigma)}{R_S}\right)^{D-3} (-\tau)^{\frac{3D-5}{D+1}} \, , \tag{4.8}$$

and

$$A^3(\sigma,\tau) \simeq \bar{A}^3 - \frac{(D+1)}{(D-3)} \frac{L(\sigma)}{p^2(\sigma)} (-\tau)^{\frac{D-3}{D+1}} \, , \tag{4.9}$$

$$A^i = \pi/2 \text{ for } \; i = 2, 4, 5, \ldots, D-1 \tag{4.10}$$

Comparison of these expressions with the corresponding ones found in Ref [13], (Eq. (88)), shows that they have the same τ dependence. The difference lies in the prefactors: In Eqs. (4.7)–(4.8) they are σ–dependent while in those of Ref [13] they are constant. This is due to the different zeroth order solution we have chosen. In Ref. [13] we have taken the center of mass motion which is a point–like solution, while now we have a null string, an extended zeroth order solution.

We can consider the approach to the singularity with higher powers of A^1 (our radial coordinate). This yields the following power dependence in τ

$$A^1 = \sum_{n=1}^{\infty} A_n^1(\sigma)(-\tau)^{2n/(D+1)} , \tag{4.11}$$

and then

$$A^3 = \sum_{n=1}^{\infty} A_n^3(\sigma)(-\tau)^{(D-5+2n)/(D+1)} , \tag{4.12}$$

$$A^0 = \sum_{n=1}^{\infty} A_n^0(\sigma)(-\tau)^{1+2n(D-3)/(D+1)} . \tag{4.13}$$

This τ dependence seems to be a quite general behavior near singularities and generalize to D–dimensions the case analyzed recently in Ref ([14]). There will still be higher order corrections coming form the first order fluctuations B^μ, second order fluctuations C^μ, and so on. We have already seen that in the cosmological case, five different powers of τ contribute to the first order.

The proper string size that at zeroth order vanished for null string, does not vanish at higher order. In order to compute the first order correction to the string size we do not need to explicitly compute the B^A terms. In fact, by use of the constraint (2.4), Eq. (2.22) can be rewritten as

$$\left(\frac{dl}{d\sigma}\right)^2 = X^{A'}X^{B'}G_{AB}(X) . \tag{4.14}$$

Plugging expressions (4.5)–(4.4) into this equation and making use of the constraint (2.8) to eliminate $A^{1'}$, we obtain

$$\left(\frac{dl}{d\sigma}\right)^2 = -g(A^{0'})^2 + (A^1)^2(A^{3'})^2 + g\frac{(EA^{0'} - LA^{3'})^2}{(E^2 - L^2 g/(A^1)^2)} . \tag{4.15}$$

As the string approachs the singularity it diverges in the following way

$$\left(\frac{dl}{d\sigma}\right)^2 \to \left(\frac{R_S}{r}\right)^{D-3} (\bar{X^0}')^2 \to (-\tau)^{-2(D-3)/(D+1)} . \tag{4.16}$$

This unstable behavior have already been discovered in Ref [13], where we studied string propagation in D–dimensional Reissner–Nordström black holes under the approximation labeled by $c = 1$. We solved there the first order string fluctuations around the center of mass motion at spatial infinity, near the event horizon and at the spacetime singularity. In Reissner–Nordström black holes the radial components and angular string components develop instabilities as the string approachs the singularity. The string motion is like that of a particle in a strongly attractive potential proportional to $-(\Delta\tau)^{-2}$. As $\Delta\tau \to 0$ the string ends trapped by the black hole singularity.

We can also compute the integrated energy–momentum tensor from expressions (2.20) and (2.21). Using Eqs. (4.5)–(4.4) we find

$$\Theta^{AB} = \frac{dI^{AB}}{d\sigma} \approx \frac{1}{2\pi\alpha'} \frac{\sqrt{-g}\dot{A}^A \dot{A}^B}{\sqrt{E^2 - L^2 g/(A^1)^2}} , \tag{4.17}$$

where we have taken into account that inside the black hole A^0 is spacelike and A^1 is timelike.

In the limit $r \to 0$, we have

$$\Theta_1^1 \to -\frac{L}{2\pi\alpha' r} = -\Theta_3^3 , \quad \Theta_0^0 \to 0 . \tag{4.18}$$

We first observe that the results are independent of the spacetime dimensionality, D. The facts that the trace vanishes (since we are dealing with a null system), and that $\Theta_0^0 \to 0$, reduce the system to a $D-1$–dimensional null system (in our example of a planar string, this reduces the system to a two dimensional null string).

V. EXACT STRING SOLUTIONS INSIDE A BLACK HOLE

In four dimensions the equations of motion of a string in the curved background (4.1) read [18]

$$\begin{aligned}
&\ddot{t} - t'' + \frac{g_{,r}}{g}(\dot{t}\dot{r} - t'r') = 0, \\
&\ddot{r} - r'' - \frac{g_{,r}}{2g}(\dot{r}^2 - r'^2) + \frac{gg_{,r}}{2}(\dot{t}^2 - t'^2) - gr(\dot{\theta}^2 - \theta'^2) - gr\sin^2\theta(\dot{\phi}^2 - \phi'^2) = 0, \\
&\ddot{\theta} - \theta'' + \frac{2}{r}(\dot{\theta}\dot{r} - \theta'r') - \sin\theta\cos\theta(\dot{\phi}^2 - \phi'^2) = 0, \\
&\ddot{\phi} - \phi'' + \frac{2}{r}(\dot{\phi}\dot{r} - \phi'r') + 2\cot\theta(\dot{\theta}\dot{\phi} - \theta'\phi') = 0,
\end{aligned} \tag{5.1}$$

and the constraints

$$\begin{aligned}
&- g\dot{t}t' + \frac{1}{g}\dot{r}r' + r^2\dot{\theta}\theta' + r^2\sin^2\theta\dot{\phi}\phi' = 0, \\
&- g(\dot{t}^2 + t'^2) + \frac{1}{g}(\dot{r}^2 + r'^2) + r^2(\dot{\theta}^2 + \theta'^2) + r^2\sin^2\theta(\dot{\phi}^2 + \phi'^2) = 0.
\end{aligned} \tag{5.2}$$

We are interested in finding solutions inside the black hole, where we know that the r coordinate becomes timelike while the t coordinate is spacelike. We thus look for solutions of the following particular form

$$r = r(\tau) \doteq \mathcal{T}(\tau) \quad \text{and} \quad t = t(\sigma) \doteq \mathcal{R}(\sigma) . \tag{5.3}$$

Using this ansatz the first equation of motion in (5.1) give us

$$\mathcal{R}(\sigma) = E\sigma + t_0 \ , \tag{5.4}$$

where E and t_0 are arbitrary constants.

From the first constraint Eq. (5.2) we have

$$\mathcal{T}(\tau)^2 \sin^2\theta \dot{\phi}\phi' = 0 \ . \tag{5.5}$$

In terms of Kruskal coordinates we have

$$u(\sigma,\tau) = \left(1 - \frac{\mathcal{T}(\tau)}{2M}\right)^{1/2} e^{\frac{\mathcal{T}(\tau)}{4M}} \sinh\left(\frac{\mathcal{R}(\sigma)}{4M}\right) \tag{5.6}$$

$$v(\sigma,\tau) = \left(1 - \frac{\mathcal{T}(\tau)}{2M}\right)^{1/2} e^{\frac{\mathcal{T}(\tau)}{4M}} \cosh\left(\frac{\mathcal{R}(\sigma)}{4M}\right) \tag{5.7}$$

Eq. (5.5) allows us to analyze three cases:

a) $\theta =$ constant $\neq \pi/2$. Then ϕ is also constant and

$$\dot{\mathcal{T}}(\tau) = \pm E g(\mathcal{T}) \tag{5.8}$$

represents a straight string in the two dimensional $\mathcal{R}$–ϕ space.

b) For $\theta = \pi/2$, i.e. the planar case, we have:

b1) $\dot{\phi} = 0$. Then, from the equation of motion of ϕ we deduce $\phi = n\sigma + \phi_0$, where n and ϕ_0 are arbitrary constants. While for the timelike coordinate $\mathcal{T}(\tau)$ we have

$$\dot{\mathcal{T}}(\tau)^2 = E^2 g(\mathcal{T})^2 - n^2 \mathcal{T}(\tau)^2 \ . \tag{5.9}$$

This solution represents a straight twisted string in the $\mathcal{R}$–ϕ space.

b2) $\phi' = 0$. Then, from the equation of motion of ϕ we deduce $\dot{\phi} = L/\mathcal{T}(\tau)^2$. Here L is a constant and now

$$\dot{\mathcal{T}}(\tau)^2 = E^2 g(\mathcal{T})^2 - g(\mathcal{T}) \frac{L^2}{\mathcal{T}(\tau)^2} \ . \tag{5.10}$$

This solution represents a rigid rotating straight string in the $\mathcal{R}$–ϕ space.

VI. CONCLUSION

We have performed a systematic study of the null string expansion method to solve the string equations of motion and constraints in curved spacetimes. This corresponds to a small string tension expansion (inverse powers of α'). The perturbative expansion series is conveniently described in terms of the parameter c, the wave propagation velocity along the string. The zeroth order describes a null string configuration; first and higher order fluctuations around it are systematically constructed. The null string expansion series in D-dimensional Friedmann–Robertson–Walker and Schwarzschild spacetimes have been computed. In the latter case, the first terms of the series represent a generic approach to the string falling towards the $r = 0$ singularity. The string integrated energy-momentum tensor and string proper size have been computed and analyzed.

The different methods to solve the string dynamics in curved apacetime have been characterized in terms of the parameter c (or equivalently, in terms of the ratio between $\dot{X}^A$ and $cX^{A'}$), as summarized in Sec. II and Table II:

(i) "Impulsive", $c \ll 1$ (null string expansion);

(ii) "Perturbative", $c = 1$ (center of mass expansion);

(iii) "Adiabatic", $c \gg 1$ (dual to (i), initial configuration expansion);

(iv) "Composite", ($\tau \to 0$ expansion, ring string expansion).

The null string expansion is well appropriate to describe the string propagation in the strong gravitational field regime and well encompasses the $\tau \to 0$ expansion in cosmologiacal and black hole spacetimes, in the string unstable regime where the proper size of the string grows like the scalar factor, and like $(R_S/r)^{(D-3)/2} \sim (-\tau)^{-(D-3)/(D+1)}$ inside the Schwarzschild black hole (near the $r = 0$ singularity).

ACKNOWLEDGMENTS

C.O.L was supported by the NSF grant PHY-95-07719 and by research founds of the University of Utah. He also thanks the DEMIRM–Observatoire de Paris, where part of this work was done, for kind hospitality and use of its working facilities.

TABLES

TABLE I. Here we recall the explicit conformal time (η) and cosmic time (t) –dependence of the flat Friedmann–Robertson–Walker metric as usually referred to in the literature [6].

$R(\eta) = (\eta/\beta)^\beta$	$R(t) \sim t^{\alpha=\beta/(\beta+1)}$	Spacetime
constant, $\beta \to 0$	constant	Flat space
η^β, $0 < \beta < \infty$	t^α, $0 < \alpha < 1$	Standard cosmology
e^η, $\beta \to \infty$	t	Linear expansion
η^β, $-1 < \beta < 0$	t^α, $-\infty < \alpha < 0$	Super inflation
η^{-1}	e^{Ht}	de Sitter
η^β, $-\infty < \beta < -1$	t^α, $1 < \alpha < \infty$	Power-law inflation

TABLE II. In this table we show the different approaches to study the string dynamics in curved spacetimes. $c \ll 1$, $c = 1$ and $c \gg 1$ correspond to $\partial_\tau X^A \gg c\partial_\sigma X^A$, $\partial_\tau X^A \sim c\partial_\sigma X^A$ and $\partial_\tau X^A \ll c\partial_\sigma X^A$ respectively. Hybrid cases are still possible when some components of the string coordinate X^A satisfy eventually one of the inequalities and other coordinate components do not. The regimes of stability of the first order fluctuations are also shown as well as the references in which the approaches (in Cosmology and Black Holes) have been originally developped.

	Black Holes		Cosmology	
Approach	Zeroth Order	Fluctuations $R \to 0$	Zeroth Order	Fluctuations $R \to \infty$
(i) Impulsive, $c \ll 1$ $\dot{X}^A \gg cX^{A\prime}$	null string[a]	unstable	null string[b]	unstable
(ii) Perturbative, $c = 1$ $\dot{X}^A \sim cX^{A\prime}$	center of mass[c]	unstable	center of mass[d]	stable/unstable
(iii) Adiabatic, $c \gg 1$ $\dot{X}^A \ll cX^{A\prime}$	initial configuration[a]	unstable	initial configuration[a]	unstable
(iv) Hybrid	ring[e]	unstable	asymptotic[f]	stable/unstable

[a]This paper
[b]Ref. [7].
[c]Ref. [13].
[d]Ref. [5].
[e]Ref. [17].
[f]Ref. [6].

REFERENCES

[1] M. B. Green, J. H. Schwarz and E. Witten, *Superstring Theory*, (Cambridge Univ. Press, Cambridge, 1987).

[2] N.D.Birrell and P.C.W.Davies, Quantum Field Theory in Curved Space (Cambridge U.P., Cambridge, 1982).

[3] H. J. de Vega and N. Sánchez in Proceedings of the Erice Course: "String Quantum Gravity and Physics at the Planck Energy Scale", 21-28 June 1992, edited by N. Sánchez, World Scientific, 1993.

[4] Proceedings of the Erice Course: "String Quantum Gravity and Physics at the Planck Energy Scale", 4-15 September 1995, edited by N. Sánchez and A. Zichichi, NATO ASI Series, C476, 1996.

[5] H.J.de Vega and N.Sánchez, Phys.Lett., B197, 320 (1987).

[6] M.Gasperini, N.Sánchez and G.Veneziano, Int.J.Mod.Phys., A6, 3853 (1991); Nucl. Phys. B364, 365 (1991).

[7] H.J.de Vega and A. Nicolaidis, Phys.Lett., B295, 214 (1992); H.J.de Vega, I.Giannakis and A. Nicolaidis, Modern Phys. Lett, A10, 2479 (1995).

[8] F. Combes, H. J. de Vega, A. V. Mikhailov and N. Sánchez, Phys. Rev. D50, 2754 (1994).

[9] H. J. de Vega, A. L. Larsen and N. Sánchez, Nucl. Phys., B427, 643 (1994).

[10] H. J. de Vega and I. L. Egusquiza, Phys. Rev. D49, 763 (1994).

[11] A. L. Larsen and N. Sánchez, Phys. Rev. D50, 7493 (1994).

[12] N. Sánchez and G. Veneziano, Nucl. Phys., B333, 253 (1990).

[13] C. O. Lousto and N. Sánchez, Phys. Rev. D47, 4498 (1993).

[14] H. J. de Vega and I. L. Egusquiza, preprint PAR-LPTHE/95-33, EHU-FT/9509, hep-th/9506214.

[15] C. Lovelace, Phys.Lett., B135, 75 (1984).

[16] E. S. Fradkin and A. A. Tseylin, Nucl. Phys., B261, 1 (1985).

[17] A. L. Larsen, Phys. Rev. D50, 2623 (1994).

[18] A. L. Larsen and N. Sánchez, Phys. Rev. D51, 6929 (1995).

[19] H. J. de Vega, A. V. Mikhailov and N. Sánchez, Theor. Math. Phys., 94, 166 (1993).

[20] H. J. de Vega and N. Sánchez, Phys.Lett., B244, 215 (1990).

[21] H. J. de Vega and N. Sánchez, Phys. Rev. D42, 3969 (1990).

[22] H. J. de Vega and N. Sánchez, Int.J.Mod.Phys., A7, 3043 (1992).

[23] R. C. Myers and M. J. Perry, Ann. of Physics, Vol. 172, 304 (1986).

THE BLACK HOLE: SCATTERER, ABSORBER AND EMITTER OF PARTICLES

N. SANCHEZ

Observatoire de Paris-DEMIRM,
61 Avenue de l'Observatoire, 75014 PARIS, FRANCE

Abstract

Accurate and powerful computational methods developped by the author, based on the analytic resolution of the wave equation in the black hole background, allow to obtain the highly non trivial **total absorption spectrum** of the Black Hole. As well as phase shifts and cross sections (elastic and inelastic) for a wide range of energy and angular momentum, the angular distribution of absorbed and scattered waves, and the Hawking emission rates. The total absorption spectrum of waves by the Black Hole is known exactly. It presents as a function of frequency a remarkable **oscillatory** behaviour characteristic of a diffraction pattern. It oscillates around its optical geometric limit ($\frac{27}{4}\pi {r_s}^2$) with decreasing amplitude and almost constant period. This is an **unique** distinctive feature of the black hole absorption, and due to its $r = 0$ singularity. Ordinary absorptive bodies and optical models do not present these features.

The Hamiltonian describing the wave-black hole interaction is non hermitian (despite being real) due to its singularity at the origin ($r = 0$). The unitarity optical theorem of scattering theory is generalized to the black hole case explicitely showing that absorption takes place only at the origin ($r = 0$).

All these results allow to **understand** and **reproduce** the Black Hole absorption spectrum in terms of Fresnel-Kirchoff diffraction theory: interference takes place between the absorbed rays arriving at the origin by different optical paths.

These fundamental features of the Black Hole Absorption will be present for generic higher dimensional Black Hole backgrounds, and whatever the low energy effective theory they arise from.

In recent and increasing litterature devoted to compute absorption cross sections ("grey body factors") of black holes (whatever ordinary, stringy, D-braned), the fundamental remarkable features of the Black Hole Absorption spectrum are overlooked.

Contents

THE BLACK HOLE: SCATTERER, ABSORBER AND EMITTER OF PARTICLES

N. SANCHEZ

Observatoire de Paris-DEMIRM
61, Avenue de l'Observatoire, 75014 Paris - FRANCE

I INTRODUCTION AND RESULTS

I shall report here about some of my results on the physics of black holes and the dynamics of fields in the vicinity of such objects, describing at the same time, the Black Hole under its triple aspect of Scatterer, Absorber and Emitter of particles.

I shall first report about the Absorption, it appears in the concept of black hole itself, the gravitational field being so intense that even light can not escape of it. Absorption is one of the properties that characterizes the black hole description in classical physics: black holes absorb waves but they can not emit them. If a quantum description of perturbation fields is considered, black holes also emit particles. For a static black hole, the quantum particle emission rate $H(k)$, and the classical wave absorption cross section $\sigma_A(k)$ are related by the Hawking's formula (1975, [1])

$$H(k) = \frac{\sigma_A(k)}{e^{8\pi kM} - 1} , \tag{1}$$

the factor relating them being planckien. Here k and M stands for the frequency of waves and for the mass of the hole respectively.

We see the role played by the absorption in the emission. In spite of the extensive litterature discussing about the interaction of waves with black holes, the absorption spectrum $\sigma_A(k)$ (and other scattering parameters), as well as their theoretical foundations, was largely unsolved. In fact, the complexity of analytic solutions of the perturbation fields equations made this problem very difficult.

We have studied in detail the absorption spectrum of the black hole and we have found the total absorption cross section $\sigma_A(k)$ in the Hawking formula, obtaining a very simple expression (N. Sánchez, 1978 [2]), which is valid to very high accuracy over the entire range of k, namely

$$\sigma_A(k) = 27\pi M^2 - 2\sqrt{2}M\frac{\sin\left(2\sqrt{27}\pi kM\right)}{k} \quad , \qquad kM \geq 0.07 \tag{2}$$

$$\sigma_A(0) = 16\pi M^2$$

The absorption spectrum presents, as a function of the frequency, a remarkable oscillatory behaviour characteristic of a diffraction pattern (Fig.2). It oscillates around its constant geometrical optics value $\sigma(\infty) = 27\pi M^2$ with decreasing amplitude (as $\frac{1}{(\sqrt{2}kM)}$) and constant period ($\frac{2}{3}\sqrt{3}$). The value of $\sigma_A(0)$ is exactly given by $16\pi M^2$. See below.

We have also calculated the Hawking radiation. This is only important in the interval $0 \leq k \leq \frac{1}{M}$. The emission spectrum (Fig.1) does not show any of the interference oscillations characteristic of the absorption cross section, because the contribution of the S-wave dominates the Hawking radiation. The rapidly decrease of the Planck factor for $kM \geq 1$ supresses the contribution of higher partial waves.

Thus, for a black hole the emission follows a planckian spectrum, given by eq. (1), (Fig.1), and the absorption follows an oscillatory spectrum, given by eq.(2), (Fig.2).

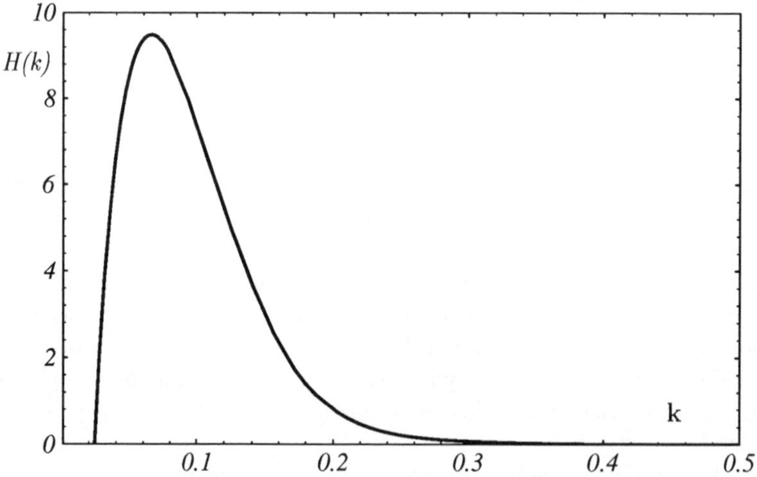

Figure 1: EMISSION BY A BLACK HOLE

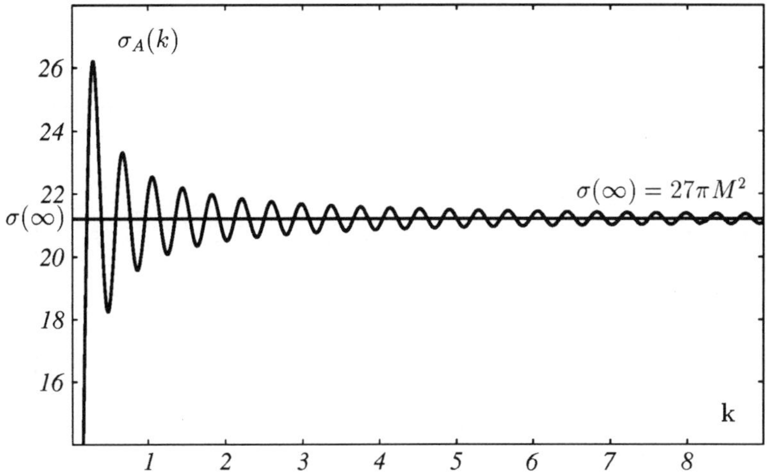

Figure 2: ABSORPTION BY A BLACK HOLE

It is interesting to compare the absorption by a black hole with that of other physical systems. Fig.3 shows the total absorption cross section for an ordinary material sphere with a complex refraction index. It is a monotonically increasing function of the frequency. It attains its geometrical optics limit without any oscillation.

Comparison of Fig.2 with Fig.3 shows the differences between the absorptive properties of a black hole and those corresponding to ordinary absorptive bodies or described by complex potentials (optical models). For a black hole, the presence of absorption processes is due to the non hermitian character of the effective potential that describes the wave - black hole interaction [3]. The effective Hamiltonian is non hermitian, despite of being real, due to its singularity at the origin ($r = 0$), as we have shown in ref.[3], so that the absorption takes place only at the origin.

We have generalized to the black hole case the well known unitarity theorem (optical theorem) of the elastic potential scattering theory, explicitly relating the presence of a non zero absorption cross section to the existence of a singularity in the space-time [3]. All these results allowed me to give a simple physical interpretation of the total absorption cross section $\sigma_A(k)$ in the context of the Fresnel-Kirchoff diffraction theory. The oscillatory behaviour of $\sigma_A(k)$ is explained in a good approximation by the interference of the absorbed rays arriving at the origin through different optical paths. [2]

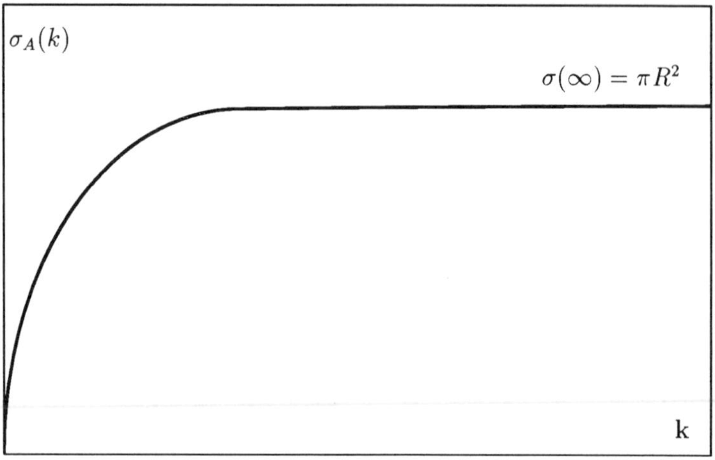

Figure 3: ABSORPTION BY A MATERIAL SPHERE WITH A COMPLEX REFRACTION INDEX

Usually, in scattering theory, absorption processes are related to complex (and non-singular) potentials. On the contrary, in the black hole case, the potential is real and singular at the origin. All partial absorption amplitudes have absolute maxima at the frequence $k = \frac{3}{4}(\frac{\sqrt{3}}{M})(l + \frac{1}{2})$. By summing up for all angular momenta, each absolute maximum of partial absorption cross section produces a relative maximum in the total absorption spectrum giving rise to the presence of oscillations.

It can be pointed out that associated to the planckian spectrum, the black hole has a temperature equal to $\frac{1}{(8\pi M)}$. In what concerns the absorption spectrum it is not possible to associate a refraction index to the black hole. For optical materials, the absorption takes place in the whole volume, whereas for the black hole, it takes place only at the origin.

It is also interesting to calculate the angular distribution of absorbed waves. For it one must study too the black hole as elastic scatterer.

The distribution of scattered waves, as a function of the scattering angle θ, has been computed in a wide range of the frequency [3],[4]. It presents a strong peak ($\sim \theta^{-4}$) characteristic of long range interactions in the forward direction, and a "glory" in the backward, characteristic of the presence of strongly attractive interactions for short distances. For intermediate θ, it shows a complicated behaviour with peaks and drops that disappear only at the geometrical-optics limit.

The angular distribution of absorbed waves is shown in [4]. It is isotropic for low frequencies and gradually shows features of a diffraction pattern, as the frequency increases. It presents an absolute maximum in the forward direction which grows and narrows as the frequency increases. In the geometrical-optics limit, this results in a Dirac Delta distribution. The analytic behaviour expresses in terms of the Bessel function $\mathcal{J}_1$, as given by eq. (8) below.

In the course of this research, we have developed accurate and useful computational methods based on the analytical resolution of the wave equation, which in addition, have allowed us to determine the range of validity of different approximations for low and high frequencies made by other authors (Starobinsky, Sov. Phys. JETP **37**, 1, 1973; Unruh, Phys. Rev. **D14**, 3251, 1976), respectively, and by ourselves [5]. It follows that the analytical computation of elastic scattering parameters for low frequencies is a rather open problem.

We have also obtained several properties concerning the scattering, absorption and emission parameters in a partial wave analysis. They are repported in references [2] and [4]. Some of them are also reported in references [6] and [7].

The work presented here has also a direct interest for the field and string quantization in curved space-times, related issues and other current problems. See the Conclusions Section at the end of this paper.

II PARTIAL WAVE ANALYSIS

The partial scattering matrix is given by

$$S_l = e^{2i\delta_l}$$

$$\delta_l = \eta_l + i\beta_l$$

We have found ([2], [4]) that the real and imaginary parts of the Black Hole phase shifts δ_l are odd and even functions of the frequency respectively:

$$\eta_l(x_s) = -\eta_l(-x_s) \tag{3}$$

$$\beta_l(x_s) = \beta_l(-x_s) \quad , \quad x_s \equiv kr_s = 2kM$$

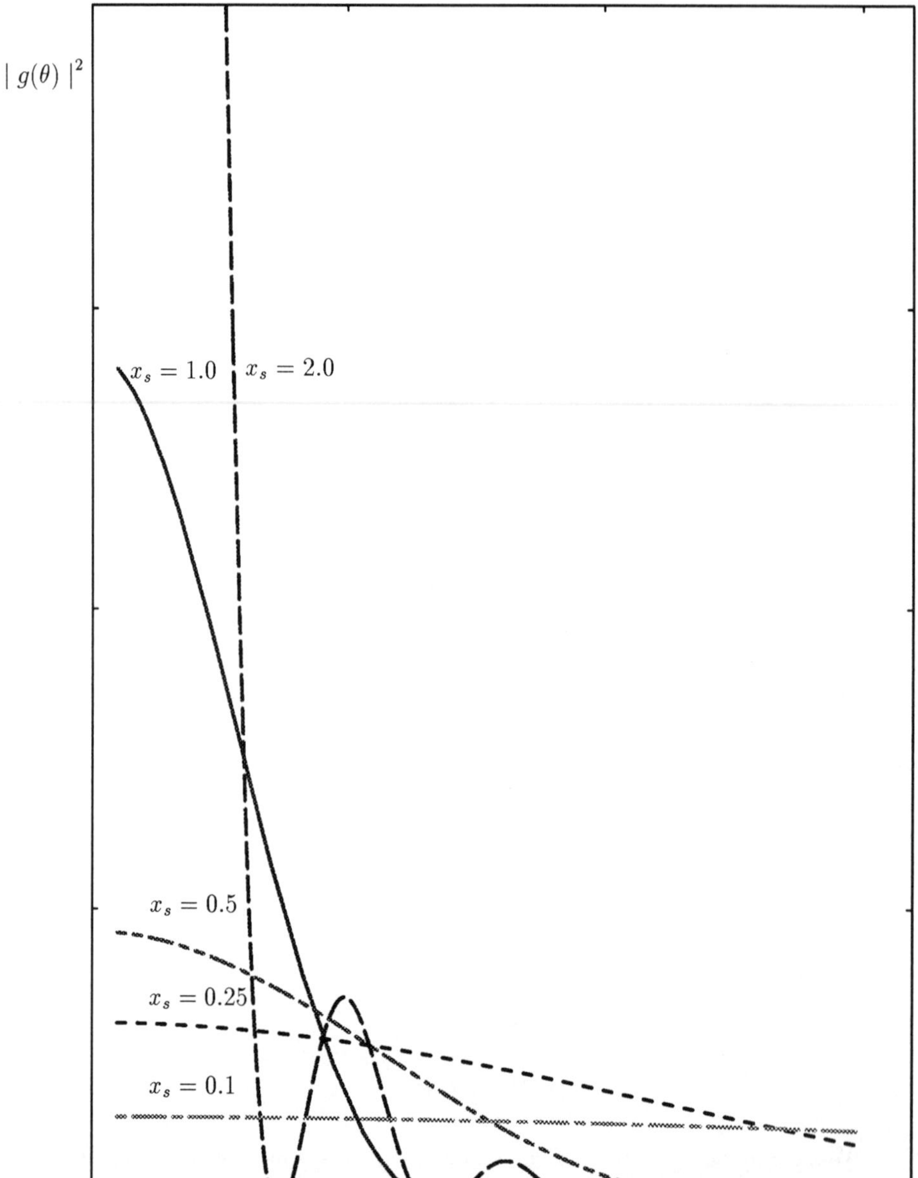

Figure 4: ANGULAR DISTRIBUTION $| g(\theta) |^2$ OF ABSORBED WAVES

In terms of the phase shifts, the partial elastic and absorption cross sections are respectively given by:

$$\epsilon_l = \frac{\pi}{{x_s}^2}(2l+1)(1 - e^{-2\beta_l}\cos 2\eta_l + e^{-4\beta_l}) \qquad \text{partial elastic cross section}$$

$$\sigma_l = \frac{\pi}{{x_s}^2}(2l+1)(1 - e^{-4\beta_l}) \qquad \text{partial absorption cross section}$$

II.1 Absorption cross sections

For all value of the angular momentum l, the imaginary part of the phase shifts, $\beta_l(x_s)$, is a monotonically increasing function of x_s.

All $\beta_l(x_s)$ are zero at $x_s = 0$ and tend to infinity linearly with x_s as x_s increases to infinity.

Low frequencies: For low frequencies, $(x_s \ll 1)$, $\beta_l(x_s)$ behaves as:

$$\beta_l^{exact}(x_s \ll 1) = C_l {x_s}^{2l+2} \qquad x_s \ll 1\ ,\ x_s \equiv k\, r_s$$

$$C_l = \frac{2^{2l}(l!)^6}{[(2l)!]^2[(2l+1)!]^2} \tag{4}$$

We have found for C_l values in agreement with Starobinsky's formulae (Starobinsky, Sov. Phys. JETP **37** (1973) 1), for $x_s = 0$ and $l = 0$. However, the Starobinsky's approximation is accurate only in a small neighborhood of $x_s = 0$. For example, the ratio

$$\frac{\beta_0^{exact}(x_s) - C_0 {x_s}^2}{\beta_0^{exact}(x_s)} \equiv X_0$$

varies as

$$0.15 \leq X_0 \leq 0.5 \qquad \text{for} \qquad 0.05 \leq x_s \leq 0.1$$

For $l = 1$:

$$0.18 \leq X_1 \leq 0.6 \qquad \text{for} \qquad 0.05 \leq x_s \leq 0.1$$

For small x_s, the inaccuracy of Starobinsky's approximation increases with l.

At $x_s = 0$, all absorption cross sections $\sigma_l(x_s)$ are zero, except for $l = 0$. For the S-wave:

$$\sigma_0(0) = 4\pi \tag{5}$$

The presence of a pole at $x_s = 0$ for $l \geq 1$ in the Jost function of the Black Hole ([2], [4]) means that waves with very small frequency and $l \neq 0$ are repelled out of the vecinity of the black hole.

High frequencies: For high frequencies, ($x_s \gg 1$), the imaginary part of the phase shifts are given by

$$\beta_l^{exact}(x_s) = \beta_l^{as}(x_s) + O(\frac{l+1/2}{x_s^{3/2}}) \qquad x_s \gg 1 \tag{6}$$

β_l^{as} is the asymptotic expression derived with the DWBA (Distorted Wave Born Approximation):

$$\beta_l^{as}(x_s) = \pi x_s - 1/4 \ln 2 - 1/16(\pi/x_s)^{1/2} - \frac{\pi}{2\sqrt{2}} \frac{(l+1/2)^2}{x_s} \tag{7}$$

There is very good agreement between β_l^{exact} and β_l^{as}. For example:

$$\beta_0^{exact}(2) = 5.79 \quad , \quad \beta_0^{as}(2) = 5.89$$

$$\beta_1^{exact}(2) = 4.98 \quad , \quad \beta_1^{as}(2) = 4.78$$

For all x_s, $\beta_l(x_s)$ are described in ref. [2].

The differential absorption cross section per unit solid angle $d\Omega$ for the Black Hole

$$\frac{d\sigma_A(\theta)}{d\Omega} = |\sum_{l=0}^{\infty}(2l+1)g_l(x_s)P_l(\cos\theta)|^2$$

is expressed in terms of the Bessel function J_1 as [4]:

$$\frac{d\sigma_A(\theta)}{d\Omega} \overset{x_s \gg 1\ \theta\to 0}{\simeq} \frac{27}{4}\frac{x_s^2}{\theta^2}[J_1(\frac{\sqrt{27}}{2}x_s\theta)]^2 \tag{8}$$

II.2 Elastic Scattering

For all angular momenta $l \neq 0$, the real part of the phase shifts, $\eta_l(x_s)$, as a function of x_s has three zeros in the range $0 \leq x_s \leq \infty$ [4].

For the S-wave, $\eta_0(x_s)$ has two zeros.

We denote as $x_s^i(l)$, the frequencies at which $\eta_l(x_s)$ vanishes ($i = 0,1,2$ stands for first, second or third zero, respectively).

$$\eta_l(x_s^{l(i)}) = 0 \quad , \quad i = 1,2,3.$$

The first zero of η_l is at $x_s = 0$.

Low frequencies: For low x_s, $x_s \ll 1$:

$$\eta_l(x_s) \overset{x_s \ll 1}{\simeq} a_l\, x_s \quad , \qquad \text{for all l} \tag{9}$$

The detailed behavior is discussed in ref.[4]. Usually, the presence of long-range interactions produces a divergence in the low-frequency behaviour of the phase shifts. This is not the case here since the Coulomb interaction vanishes when the wave energy tends to zero.

In the interval $0 \leq x_s \leq {x_s}^{(1)}(l)$:

$$\eta_l < 0 \qquad \text{for low } l \neq 0$$

$$\eta_0 > 0 \qquad \text{for } l = 0$$

and very nearly equal to zero.

For increasing l, η_l becomes more and more negative according to the general variation $\Delta\eta_l(x_s)$, due to variations ΔV_{eff} of the effective potential:

$$\Delta\eta_l = -\frac{1}{k}\int_0^\infty \frac{dr}{1-\frac{r_s}{r}}\left(\frac{R_l}{r}\right)^2 \Delta V_{eff}$$

where R_l is the solution to radial wave equation and

$$V_{eff}(x^*) = (1-\frac{x_s}{x})\left[\frac{x_s}{x^3}+\frac{l(l+1)}{x^2}\right] \quad , \quad x^* = x + x_s \ln(1-\frac{x_s}{x}) \ .$$

High frequencies: For large x_s, $x_s \gg 1$:

$$\eta_l^{exact}(x_s) = \eta_l^{as}(x_s) + O((\frac{1}{x_s})^{3/2}) \quad , \quad l \ll x_s \tag{10}$$

where

$$\eta_l^{as}(x_s) = \frac{\delta_c}{2} - x_s + \frac{\pi}{2}(l+\frac{3}{2}) + \frac{1}{16}(\frac{\pi}{x_s})^{1/2} + O(\frac{1}{{x_s}^{3/2}}) \tag{11}$$

$$\begin{aligned}\delta_c &= Im \ln\Gamma(\frac{1}{2} - 2i\, x_s) \\ &= -2x_s \ln(\frac{2x_s}{e}) + O(\frac{1}{x_s}).\end{aligned}$$

η_l^{as} is the asymptotic formula derived by us in the DWBA (Distorted Wave Born Approximation) scheme [5].

η_l^{as} and η_l^{exact} are in good agreement. For example,

$$\eta_0^{exact}(2.3) = -1.2 \quad , \quad \eta_0^{as}(2.3) = -1.18$$

$$\eta_0^{exact}(2.5) = -1.6 \quad , \quad \eta_0^{as}(2.5) = -1.60$$

We have

$$\eta_l(x_s \to \infty) \to -x_s \ln(2x_s) \to -\infty \quad , \quad \text{for fixed } l \text{ and } x_s \to \infty$$

$$\eta_l(x_s) \to -x_s \ln(l + \frac{1}{2}) \quad , \quad \text{for fixed } x_s \text{ and } l \to \infty$$

High Angular Momenta: High-order partial waves give an important contribution to the elastic scattering amplitude. This fact lead us to study $\eta_l(x_s)$ for $l \geq 2$ and fixed x_s. We found:

$$\eta_l(x_s) = \eta_l^{coul}(x_s) + \alpha_0(x_s) + \Delta_l(x_s) \quad , \quad l \gg x_s$$

where

$$\eta_l^{coul}(x_s) = \arg\Gamma(l + 1 - ix_s) \tag{12}$$

$$\Delta_l(x_s) = \frac{\alpha_1(x_s)}{(l+\frac{1}{2})} + \frac{\alpha_2(x_s)}{(l+\frac{1}{2})^2} + \frac{\alpha_3(x_s)}{(l+\frac{1}{2})^3} + O(\frac{1}{(l+\frac{1}{2})^4})$$

$$\alpha_l(x_s) \sim (x_s)^{l+1}$$

$$\alpha_0(x_s) \simeq \frac{1}{5}x_s \ , \quad \alpha_1(x_s) \simeq \frac{3}{2}x_s^2$$

As could be expected, the leading behaviour of $\eta_l(x_s)$ for large l is the same as that of the Coulomb phase shift η_l^{coul}.

The difference

$$[\eta_l^{exact}(x_s) - \eta_l^{coul}(x_s)] = x_s f(b)$$

can be written as x_s times a function $f(b)$ of the impact parameter $b = \frac{(l+\frac{1}{2})}{x_s}$, which is analytic at $b = 0$.

The fact that η_l does not vanishes in the infinite energy limit is a consequence of the strong attractive character of the black hole interaction at short distances (the term of the type $-\frac{1}{(r-r_s)^2}(k_{r_s}{}^2 + \frac{r_s{}^2}{4r^2})$ in the radial wave equation).

III DIFFERENTIAL ELASTIC CROSS SECTION

The scattering amplitude, whose squared modulus gives the differential elastic cross section, is given by

$$f(\theta) = \sum_{l=0}^{\infty} \frac{(2l+1)}{2ix_s} (\exp(-2\beta_l)\exp(2i\eta_l) - 1) P_l(\cos\theta)$$

For all values of x_s and small θ, the differential elastic cross section behaves as

$$\mid f(\theta) \mid^2 = \frac{4}{\theta^4} - \frac{C_1(x_s)}{\theta^3} - \frac{4/3}{\theta^2} + \frac{C_2(x_s)}{\theta} + C_3(x_s) + O(\theta) \tag{13}$$

The expressions for $C_i(x_s)$, $i = 1, 2, 3$ are given by [4]:

$$\begin{aligned}
C_1(x_s) &= \frac{8\alpha_1}{x_s} \cos(2\gamma_{(-)}) \\
C_2(x_s) &= \left(1 + \frac{4}{x_s} - \frac{15}{4{x_s}^2}\right) 2\alpha_1 \sin(2\gamma_{(-)}) + \\
&+ \left(1 + \frac{1}{4{x_s}^2}\right) \alpha_1 \sin(2\gamma_{(+)}) + \\
&- 4{\alpha_1}^3 \cos(2\gamma_{(-)}) \\
C_3(x_s) &= \frac{{x_s}^2}{18} + \frac{4}{3}\frac{\alpha_1}{{x_s}^2} + \frac{363}{72} + \frac{97}{288}\frac{1}{{x_s}^2} + \\
&- \frac{7}{6}\frac{\alpha_2}{{x_s}^3} + \frac{{\alpha_1}^4 - {\alpha_2}^2}{{x_s}^4}
\end{aligned}$$

where

$$\gamma_{(\pm)} = \arg\left(\Gamma(\frac{1}{2} + i\,x_s) \pm \Gamma(i\,x_s)\right)$$

$$\alpha_0 \sim \frac{1}{5} x_s \quad , \qquad \alpha_1 \sim \tfrac{3}{2} {x_s}^2 \quad , \qquad \alpha_2 \sim \frac{7}{4} {x_s}^3$$

For intermediate angles, $\mid f(\theta) \mid^2$ has a complex behaviour with peaks and drops which disappear at the geometrical-optics limit.

IV HAWKING EMISSION RATES

The energy emitted by the black hole in each mode of frequency k and angular momentum l is given by the Hawking's formula [1]

$$dH_l(k) = \frac{\mathcal{P}_l(k)}{(\exp(4\pi k r_s) - 1)} \frac{(2l+1)}{\pi} k\, dk$$

Let us recall our first analytic expression for the partial absorption rate $\mathcal{P}_l(k)$ (Sánchez, 1976) [5]. This is a formula for high frequencies ($kr_s \gg 1$) obtained within the DWBA scheme:

$$\mathcal{P}_l(k) \overset{kr_s \gg 1}{=} \frac{1 - \exp(-4\pi k r_s)}{1 + \exp(-4\pi k r_s)} \qquad \text{for } l \ll kr_s$$

$$\mathcal{P}_l(k) \overset{kr_s \gg 1}{=} \frac{1}{1 + \exp\{(2l+1)\pi[1 - \frac{27k^2 {r_s}^2}{(2l+1)^2}]\}} \qquad \text{for } l \gg 1 \tag{14}$$

In terms of this formula, $dH_l(k)$ can be expressed very simply by

$$dH_l(k) \overset{kr_s \gg 1}{=} \frac{1}{(\exp(4\pi k r_s) + 1)} \frac{2l+1}{2\pi} k dk \qquad l \ll kr_s$$

$$dH_l(k) \overset{kr_s \gg 1}{=} \frac{(2l+1)kdk}{(\exp(4\pi k r_s) - 1)\{1 + \exp[(2l+1)\pi(1 - \frac{27k^2 {r_s}^2}{(2l+1)^2})]\}} \qquad l \gg 1 \tag{15}$$

In order to compute the total emission $H(k)$, the total absorption $\sigma_A(k)$ eq.2, and the set of properties discussed in the preceding section are needed.

By using the absorption cross section eq.2 and the partial wave results reported in the preceding sections, we have calculated the Hawking emission for a wide range of frequency and angular momentum (Sánchez, 1978) [2].

The total emission spectrum as a function of x_s is plotted in Fig.1. It does not show any of the interference oscillations characteristic of the total absorption cross section eq.2, Fig.2. This is related to the fact that the S-wave contribution predominates in Hawking radiation. For example, the maxima of $H_l(k)$ for $l = 0, 1, 2$ are in the ratio

$$1 \ : \ \frac{1}{11} \ : \ \frac{1}{453}.$$

For angular momenta higher than two, $H_l(k)$ is extremely small.

The spectrum of total emission has only one peak following closely the S-wave absorption cross section behaviour. Its maximum lies at the same point the maximum of σ_0 (${x_s}^{max} = 0.23$).

The peaks of σ_1 and σ_2 turn out to have no influence on $H(k)$.

In conclusion, Hawking emission is only important in the frequency range

$$0 \ \leq \ k \ \leq \frac{1}{r_s} \tag{16}$$

V REMARKS ON APPROXIMATIONS

The analytic computation of the real part of the phase shift for low frequencies is a rather non-trivial problem.

Although the power-series expansion around $x_s = 0$ for the regular solution is known [8], the phase shifts cannot be obtained directly from it, because the asymptotic limit $r \to \infty$ cannot be taken term by term. Thus, the phase shifts are usually calculated by a standard procedure in which the radial equation is solved approximately in two or more regions of the positive real axis. By matching these solutions in the overlapping regions, approximated expressions for the phase shifts are found. By this procedure, several authors (Starobinsky [9] and Unruh[10]) have obtained approximate expressions for the imaginary part of the phase shifts.

In Starobinsky's approximation, effects connected to the Coulomb tail of the interaction have not been taken into account. With his approximation, it is possible to find for the real part of the phase shift a linear behaviour in x_s ($\eta_l \sim a_l x_s$), but inaccurate values for the coefficient a_l are obtained.

In the context of a massive field, Unruh included the Coulomb interaction. However, his approximation is not sufficiently accurate to give the real part of the phase shift at least for $l = 0$ in the zero-mass case.

The discrepancy between Unruh's approximation with the exact calculation can be explained as follows: in Unruh's approach, for $r \gg r_s$, all terms of order higher than $(\frac{r_s}{r})^2$ are neglected in the exact radial wave equation written as

$$\frac{d^2}{dr^2}(\xi R_l)+\left[k^2+\frac{2k^2 r_s}{r-r_s}+\frac{k^2 {r_s}^2}{(r-r_s)^2}+\frac{1}{4}\frac{{r_s}^2}{r^2(r-r_s)^2}-\frac{l(l+1)}{r(r-r_s)}\right](\xi R_l)=0 \quad (17)$$

where

$$\xi=\frac{1}{k[r(r-r_s)]^{\frac{1}{2}}}.$$

In Unruh's approximation, the exact solutions (Coulomb functions) of the approximate wave equation for $r \gg r_s$ are also used in the region $kr \ll 1$. However, the term $\frac{\frac{1}{4}{r_s}^2}{r^2(r-r_s)^2}$ is much smaller than the Coulomb term, only for $k^2r^2 \gg \frac{1}{8}\frac{r_s}{r}$. Then the following double inequality must hold:

$$1 \;\gg\; (kr) \;\gg\; \frac{(kr_s)^{\frac{1}{3}}}{2} \quad (18)$$

Thus, in his approximation, one cannot expect to obtain good results for x_s which are not extremely small. [If one consideres that the symbol $\gg$ indicates a difference of one order of magnitude, inequality (18) implies $x_s \leq 10^{-6}$].

Finally, concerning the $(\frac{r_s}{r})^2$ order term $\frac{k^2\,{r_s}^2}{(r-r_s)^2}$ in Eq.(17), it cannot be neglected obviously for $l=0$.

In a different approach (Persides, [8]), it has been shown that the Wronskian of the exact radial wave equation can be expanded in a double power series of x_s and $x_s \ln x_s$, although explicit expressions for the coefficients are not known. In this way, if we calculate the phase shifts the leading behaviour obtained for the real part η_l at low frequencies shows a linear and a cubic term in x_s plus $O({x_s}^3 \ln x_s)$.

For large values of x_s, ($x_s > {x_s}^{(2)}(l)$, where ${x_s}^{(2)}(l)$ is the frequency at which the second zero of η_l occurs), it follows from our results [4] that η_l is negative and a monotonically decreasing function os x_s. Here $({x_s}^{(2)})^2 \gg V_{eff}(max)$ and η_l tends to $-\infty$ as x_s increases to ∞.

For large x_s, we find a good agreement between our exact values eq.(10) and the asymptotic formula eq.(11) derived by us [5] in the Approximation DWB (Distorted Wave Born Approximation).

VI CONCLUSIONS

Accurate and powerful computational methods, based on the analytic resolution of the wave equation in the black hole background, developed by the present author allow to obtain the total absorption spectrum of the Black Hole. As well as phase shifts and cross sections (elastic and inelastic) for a wide range of energy and angular momentum, the angular distribution of absorbed and scattered waves, and the Hawking emission rates.

The total absorption spectrum of the Black Hole is known exactly. The absorption spectrum as a function of the frequency shows a **remarkable oscillatory** behaviour characteristic of a diffraction pattern. The absorption cross section oscillates around its optical geometric limit with decreasing amplitude and almost constant period. Such oscillatory absorption pattern is an unique distinctive feature of the Black Hole. Absorption by ordinary bodies, complex refraction index or optical models do not present these features.

For ordinary absorptive bodies, the absorption takes place in the whole medium while for the Black Hole it takes place only at the origin ($r = 0$).

For the Black Hole, the effective Hamiltonian describing the wave-black hole interaction is non-hermitian, despite of being real, due to its singularity at the origin ($r = 0$). The well known unitarity (optical) theorem of the potential scattering theory can be generalized to the Black Hole case, explicitly relating the presence of a non zero absorption cross section to the existence of a singularity ($r = 0$) in the space time.

All partial absorption amplitudes have absolute maxima at the frequence $k = \frac{3\sqrt{3}}{4M}(l + \frac{1}{2})$. By summing up all angular momenta, each absolute partial wave maximum, produces a relative maximum in the total cross section giving rise to the presence of oscillations.

All these results allow to **understand** and **reproduce** the exact absorption spectrum in terms of the Fresnel-Kirchoff diffraction theory. The oscillatory behaviour of $\sigma_A(k)$ is due to the interference of the absorbed rays arriving at the origin ($r = 0$) through different optical paths.

Semiclassical WKB Approximation for the Scattering by Black Holes only gives information about the high frqeuency ($kr_s \gg 1$), (and high angular momenta), of the elastic (real part) of the phase shifts, but fail to describe well the absorption properties (and low partial wave angular momenta).

DWBA (Distorted Wave Born Approximation) for the Black Hole as it was implemented by the present author more than twenty years ago [5] is an accurate better approximation for high frequencies ($kr_s \gg 1$) to compute the absorption (imaginary part) phase shifts and rates, both for high ($l \gg kr_s$) and low ($l \ll kr_s$) angular momenta.

Approximative expressions (whatever they be), for very high frequencies, or for low frequencies **do not** allow to find the remarkable **oscillatory** behaviour of the total absorption cross section as a function of frequency, of the Black Hole. The knowledge of the highly non trivial total absorption spectrum of the Black Hole needed the development of computational methods [2] more powerful and accurate than the commonly used approximations.

The angular distribution of absorbed and elastically scattered waves have been also computed with these methods.

The conceptual general features of the Black Hole Absorption spectrum will survive for higher dimensional ($D > 4$) generic Black Holes, and including charge and angular momentum. They will be also present for Black Hole backgrounds solutions of the low energy effective field equations of string theories and D branes.

The Absorption Cross Section is a classical concept. It is exactly known and understood in terms of classical physics (classical perturbations around fixed backgrounds). (Although, of course, it is possible to rederive and compute magnitudes from several different ways and techniques).

An increasing amount of paper [11] has been devoted to the computation of absorption cross sections ("grey body factors") of Black Holes, whatever D-dimensional, ordinary, D-braneous, stringy, extremal or non extremal. All these papers [11] deal with approximative expressions for the partial wave cross sections. In all these papers [11] the fundamental remarkable features of the Total Absorption Spectrum of the Black Hole are overlooked.

References

[1] S.Hawking, *"Particle Creation by Black Holes"*, Comm. Math. Phys. **43** (1975) 199.

[2] N.Sánchez, *"Absorption and Emission Spectra for a Schwarzschild Black Hole"*, Phys.Rev.**D18** (1978) 1030.

[3] N.Sánchez, *"Wave Scattering Theory and the Absorption Problem for a Black Hole"*, Phys.Rev.**D16** (1977) 937.

[4] N.Sánchez, *"Elastic Scattering of Waves by a Black Hole"*, Phys.Rev.**D18** (1978) 1798.

[5] N.Sánchez, *"Scattering of Scalar Waves from a Schwarzschild Black Hole"*, J.Math.Phys. **17** (1976) 688.

[6] N.Sánchez, *"Sur la Physique des Champs et la Géométrie de l'Espace-Temps"*, Thèse d'Etat, Paris (1979).

[7] J.A.H.Futterman, F.A.Handler and R.A.Matzner, *"Scattering from Black Holes"*, Cambridge University Press, Cambridge, U.K. (1988), and references therein.

[8] S.Persides, Int.Jour.Math.Phys.**48** (1976) 165; **50** (1976) 229.

[9] A.A.Starobinsky, Zh.Eks.Teor.Fiz. **64** (1973) 48. (Sov.Phys.-JETP **37**(1973)1).

[10] W.G.Unruh, Phys.Rev.**D14** (1976) 3251.

[11] For example:

S.S.Gubser and I.R.Klebanov, Phys.Rev.Lett.**77** (1996) 4491;

J.Maldacena and A.Strominger, Phys.Rev.**D55** (1997) 861 and Phys.Rev.**D56** (1997) 4975;

S.S.Gubser, Phys.Rev.**D56** (1997) 4984 and **hep-th/9706100**;

M.Civetič and F.Larsen, Phys.Rev.**D56** (1997) 4994 and **hep-th/9705192**;

I.R.Klebanov and S.D.Mathur, **hep-th/9701187**;

S.Das, A.Dasgupta and T.Sarkar, Phys.Rev.**D55** (1997) 12;

S.P. de Alwis and K.Sato, Phys.Rev.**D55** (1997) 6181;

R.Emparan, **hep-th/9704204**;

H.W.Lee, Y.S.Myung and J.Y.Kim, **hep-th/9708099**.

Une approche du temps et des fréquences. Vers le mètre et la seconde

SUZANNE DÉBARBAT ET MICHEL GRANVEAUD

Observatoire de Paris- DANOF/URA 1125 du CNRS
61 Avenue de l'Observatoire de Paris
F-75014 PARIS

A l'occasion du "Second Paris Cosmology Colloquium" tenu à l'Observatoire de Paris en 1994 (2-4 juin), une visite de l'Observatoire de Paris a fait connaître aux participants les richesses de l'Etablissement dont la première pierre fut posée le 21 juin 1667, et où une activité de recherche n'a pas cessé d'y être menée. Le descriptif de la visite conduite par l'un d'entre nous (S. Débarbat) est paru aux pages 528 à 531 des Actes du Colloque (World Scientific, 1995). Plusieurs des photographies reproduites aux pages 533 à 543 permettent d'apprécier "l'état des lieux", tandis que la page IV de couverture représente la pendule "astronomique" de 1710 qui orne l'une des salles du Bâtiment Perrault.

Le "Third Paris Cosmology Colloquium" à l'Observatoire de Paris (7-9 juin), se tenait l'année même (1995) du bicentenaire de la définition de la nomenclature du Système métrique fixée en 1795, en même temps que la longueur du "mètre provisoire" était déterminée. Il a été proposé aux participants une visite plus spécifiquement centrée sur la présentation des documents et objets mis en place (N. Daliès, J. Alexandre, S. Débarbat), dans la Grande galerie et sa salle annexe au premier étage du Bâtiment Perrault. La visite conduite par l'un d'entre nous (S. Débarbat), se terminait par une présentation (M. Granveaud) des activités actuellement développées à l'Observatoire de Paris au sein du Laboratoire Primaire du Temps et des Fréquences, laboratoire lié par convention au Bureau National de Métrologie. Ces travaux relèvent de plusieurs domaines (temps, définition de la seconde, temps légal français, fréquences, réalisation du mètre....) qui assurent la continuité des recherches menées depuis plus de trois siècles dans l'Etablissement. Le compte rendu de cette visite est paru dans les pages 387 à 400, des Actes du Colloque (World Scientific, 1996).

En 1996, à l'occasion du "Network Meeting 'String Gravity' " (juin 6-7) tenu à l'Observatoire de Paris, et dans la continuité de la présentation des activités de cet Etablissement sur son site parisien, une visite du Laboratoire Primaire du Temps et des Fréquences (LPTF) du Département d'Astronomie fondamentale/URA 1125 du Centre National de la Recherche Scientifique "Systèmes de références spatio-temporels", a été organisée. Les participants ont été accueillis en trois sites du laboratoire, par des collaborateurs de l'un d'entre nous (M. Granveaud).

En parallèle, S. Débarbat présentait les médaillons "Arago" jalonnant, du boulevard périphérique nord au boulevard périphérique sud, le méridien de Paris qui coïncide avec l'axe du Bâtiment Perrault. Il existe 121 médaillons identiques répartis principalement près de grandes institutions parisiennes comme la Bibliothèque municipale de Montmartre, le Louvre ou, encore, l'Institut de France sans oublier les sept de l'Observatoire de Paris. Ces symboles, représentatifs de l'oeuvre d'Arago qui en 1806 a assuré, avec Biot, la prolongation des travaux de Delambre et de Méchain le long de la Méridienne de France, avaient été proposés par Jan Dibbets (né en Hollande en 1941) dans le cadre d'un concours ayant pour objet de remplacer la statue d'Arago. Cette dernière érigée en 1893, sur le méridien de Paris, au sud de l'Observatoire, disparut en 1942 pour être fondue. Les médaillons constituent selon J. Dibbets un "monument imaginaire réalisé sur le tracé d'une ligne imaginaire, le Méridien de Paris". Cette ligne, parcourue par Delambre et Méchain puis prolongée par Biot et Arago, a vu sa longueur servir de base au Système métrique décimal et à son "mètre" de 1799, base du "Système International" (SI) d'unités de 1960.

L' Observatoire de Paris est, depuis sa création, engagé dans des recherches fondamentales dans différents domaines. En 1996, les participants de la rencontre des 6 et 7 juin ont donc visité les laboratoires où s'élaborent le hertz, la seconde, les références de fréquence (infrarouges et visibles) qui servent à matérialiser la définition du mètre, les références primaires de temps et de fréquence.

Jusqu'en 1967, la métrologie du temps a été rattachée directement à l'astronomie, aussi bien pour les concepts fondamentaux des références de temps, pour les définitions des unités de temps et de fréquence, que pour les applications scientifiques. L'accès permanent aux références de temps et de fréquence passait par le fonctionnement continu de garde-temps - les horloges - qui assuraient la continuité entre deux observations astronomiques.

Dans la première moitié du XXème siècle, les progrès techniques des garde-temps ont été considérables; en particulier, le développement d'oscillateurs (c'est-à-dire de générateurs de fréquence) à quartz a marqué une étape essentielle:c'est, en effet, grâce aux qualités métrologiques-spécialement la stabilité de fréquence, c'est-à-dire la capacité à générer la même fréquence moyenne- des oscillateurs à quartz utilisés comme garde-temps que certaines irrégularités de la rotation de la Terre ont pu être observées et mesurées dès 1937. Ces résultats montraient les limites des réalisations des références de temps et de fréquence par des processus astronomiques, c'est-à-dire grâce à des événements liés aux mouvements de corps célestes. Les avancées de la physique quantique associées aux développements des oscillateurs à quartz ont précipité l'abandon de la référence de temps astronomique comme meilleure réalisation d'un temps physique idéal. Le temps atomique a pris la relève avec une première phase en 1967, année de la nouvelle définition de la seconde et du hertz: l'étalon à césium réalisant la seconde devient l'étalon de référence. La deuxième phase se situe en 1972 avec la définition du temps atomique qui est reconnu sur le plan international comme la référence officielle de temps.

De 1967 à 1996 les performances des étalons réalisant la seconde - qui sont pratiquement des étalons de fréquence - se sont améliorées d'un facteur 10 000 passant d'une exactitude de 10^{-11} à environ 10^{-15} de nos jours. Les améliorations ont été possibles grâce aux apports de la technologie (notamment les diodes laser) et de la physique atomique, notamment le pompage optique d'A. Kastler et, plus récemment, le refroidissement des atomes de césium en mélasses optiques. Les possibilités des atomes froids sont particulièrement exploitées dans les fontaines atomiques dont le premier prototype, à atomes de césium fonctionne au LPTF. Il réalise la seconde à 3 x 10^{-15} près.

A côté de ces étalons primaires le physicien a besoin de références secondaires de fréquence pour les travaux de spectroscopie et pour la réalisation de la définition du mètre de 1983,....Ces étalons secondaires sont des lasers, par exemple laser H_eN_e/CH_4, $H_eN_e/I2$ ou des diodes lasers asservies sur des transitions atomiques. Le LPTF a notamment développé un système diode laser/rubidium à 385 TH_z qui affiche une exactitude de $5x10^{-12}$. Il est utilisé dans le laboratoire de physique de l'Ecole Normale Supérieure comme référence de fréquence pour la détermination de la constante de Rydberg.

Les applications des étalons de fréquence et des horloges atomiques sont multiples en dehors de la métrologie. Dans les télécommunications ils servent de références pour l'écoulement des paquets de données qui transitent sur les lignes. En géodésie, en navigation, les mesures de distance sont fondamentalement des mesures d'intervalles de temps qui s'appuient sur des horloges atomiques. En astronomie, la datation des signaux pulsars, la corrélation de signaux de quasars demandent des étalons atomiques. Les applications militaires font aussi appel aux systèmes temps/fréquence de haute précision: références locales de temps dans les sous-marins, aides à la poursuite de missiles, identification de mobiles amis/ennemis...

A l'occasion des futurs colloques de cosmologie, les participants pourront découvrir les autres recherches menées sur le site parisien de l'Observatoire de Paris.